ATOMIC COLLISIONS AND SPECTRA

U. FANO

Department of Physics and James Franck Institute
University of Chicago
Chicago, Illinois

A. R. P. RAU

Department of Physics and Astronomy
Louisiana State University
Baton Rouge, Louisiana

1986

ACADEMIC PRESS, INC.
Harcourt Brace Jovanovich, Publishers
Orlando San Diego New York Austin
London Montreal Sydney Tokyo Toronto

ACADEMIC PRESS, INC.
Orlando, Florida 32887

United Kingdom Edition published by
ACADEMIC PRESS INC. (LONDON) LTD.
24–28 Oval Road, London NW1 7DX

Library of Congress Cataloging in Publication Data

Fano, Ugo.
Atomic collisions and spectra.

Includes index.
1. Collisions (Nuclear physics) 2. Atomic spectra.
I. Rau, A. Ravi P. II. Title.
QC794.6.C6F36 1986 539.7'54 85-9135
ISBN 0–12–248460–6 (alk. paper)
ISBN 0–12–248461–4 (paperback)

PRINTED IN THE UNITED STATES OF AMERICA

86 87 88 89 9 8 7 6 5 4 3 2 1

CONTENTS

3 CHARGED-PARTICLE COLLISIONS IN THE BORN APPROXIMATION

Part B SINGLE-CHANNEL PHENOMENA

4 ELASTIC SCATTERING BY A SHORT-RANGE CENTRAL POTENTIAL

5 INFLUENCE OF LONG-RANGE POTENTIALS

6 CALCULATION OF THE PHASE SHIFTS

Part C MULTICHANNEL PHENOMENA

7 NONRESONANT CONTINUA AND DISCRETE SPECTRA

8 RESONANCES AND THE ELIMINATION OF CHANNELS

9 CALCULATION OF INTERACTION PARAMETERS

Part D MULTIELECTRON EXCITATIONS

10 DOUBLY EXCITED He AND H^-

11 EXCITATIONS WITH A STRUCTURED CORE

12 CONCLUSIONS AND PREVISIONS

PREFACE

The introduction of new lines of experimentation on atoms and molecules since the early 1960s has revealed a wealth of novel phenomena. Novel conceptual schemes arose from the interpretation of these phenomena; some of the new concepts fitted within the mainstream of the traditional independent particle model of atoms, while many others complemented this model by striking in new directions. An objective of this book is to stress and clarify the new directions so as to attract to them a larger subset of atomic theorists.

The present book grew out of lecture notes for a succession of courses at the University of Chicago in 1967 – 1979, which reported the new material as it was taking a definite form. It has been enriched since 1980, as the subject matured and continued to expand. Nevertheless it covers but a portion of the material encompassed by its broader framework.

We have endeavored to explain our criteria for the selection and organization of the material in the initial and final sections of Chapter 1. A more detailed view should be obtained by a perusal of the table of contents. Briefly, the book is organized into four parts. Part A deals briefly with rather elementary items of general information. Part B then takes up in considerable detail those aspects of single-electron scattering whose mastery is essential for treating multielectron processes. Part C, the core of the book, deals with multielectron processes with a residual—if often realistic—restriction, namely, that the multielectron interactions remain confined within a core region from which only a single electron escapes into alternative channels of a long-range field. Part D surveys the more recent and not fully developed study of double (or multiple) escape of electrons from a core; strong correlations persist then in the long-range field of the ionic core.

The book is intended for multiple use as a graduate school text, a tool for independent study, or a reference for particular topics. The original courses at Chicago extended over one quarter (33 hours of class), but the material in this book is substantially longer. A limited selection of topics from the book could be covered in one semester, a larger one in two semesters. Small type has been used to set apart material dealing with details, specialized subjects, or anything not immediately relevant to the main development of the text. Some of this material consists of a few paragraphs in a section, while some consists of entire sections appended to various chapters. Problems are included in many chapters, both for classroom use and for testing the reader's understanding of specific topics.

References have been kept to a minimum, generally only as pointers to the literature. Where we draw on a source for a specific point (including such items as figures and tabulated data) or feel that more detailed consultation of the original is desirable, we specify a reference. Most of the remaining references are review articles from which the development and status of the particular subject can be traced. We have not viewed references as a device for assigning credit with the result that the original contributors have often not been cited.

Much of the material in this book stems from research supported by the U.S. Department of Energy, Office of Basic Energy Sciences, and by its predecessor agencies, at Chicago and other locations, as well as by the U.S. National Science Foundation. Indeed the very preparation of the text spawned further research at various stages. The production of countless editions of lecture notes and of the newer text hinged on the intelligent and dedicated labor of Helen Barileau and of G. Zaransky in its final stages. Much help has been received from several colleagues, notably in the final version by Chris H. Greene.

1 INTRODUCTION

1.1 INTRODUCTION

Atomic collision physics is a portion of the general physics of atoms, molecules, and condensed matter. It deals with collision processes of electrons, photons, atoms, and molecules and with related spectroscopic phenomena such as photoabsorption, autoionization, resonances, and discrete spectra. It aims at a description and analysis of phenomena that reveal the paths along which multiparticle systems evolve from one state to another. This emphasis on understanding the dynamics of atomic processes distinguishes our subject from the broader field of atomic physics, which includes, e.g., quantum optics, quantum electronics, and the "Rabi physics." The last of these regards atoms and molecules as well-understood structures that serve as laboratories for the study of particles, nuclei, the electromagnetic field, and their interactions. The physics of atomic collisions and spectra may be studied for its intrinsic interest as well as for its applications to basic chemistry, astrophysical and thermonuclear plasmas, and atomic engineering which develops new materials with properties hinged on the use of specific atoms.

Atomic theory has proceeded traditionally along the microscopic approach of formulating the Schrödinger equation for a set of electrons and nuclei and trying to solve it as accurately as possible. Macroscopic physics, by contrast, would proceed inductively by identifying the variables appropriate to describe any particular phenomenon and determining the equations that govern them, empirically or analytically. The emphasis in this book will follow the macroscopic approach because the transformations of atomic and molecular states turn out to be described effectively by small sets of collective variables or parameters—typically quantum defects—that depend on the positions and interactions of several particles. Even when the Schrödinger equation has been solved adequately, it is only a few collective, average features of the solution that are observable. Furthermore, it is these features that can be more readily extrapolated from one atom to another. The microscopic and macroscopic approaches are in fact compatible; they will be pursued concurrently by stressing the description of phenomena by simple equations in terms of collective parameters.

Our general approach, and most of the results that we develop, are applicable equally to atoms, molecules, and larger aggregates. Following the inductive approach, most of the analytical treatment will be developed in the context of specific phenomena. These phenomena will generally concern single atoms for the sake of simplicity. We hope, however, that the reader may utilize the treatment in broader contexts.

Two main unifying themes in our current understanding of the subject and in our current emphasis are that

(1) the joint state of two colliding systems is viewed as belonging to the continuous spectrum of the transient complex formed by their merger, a "complex" that may also possess discrete bound state levels, and

(2) the discrete spectra of atomic and molecular levels, the subject of spectroscopy, are then viewed as appendages of the continuous spectra, differing only in that their energy is not sufficient to become ionized or dissociated.

This book covers only a fraction of the subject to which the title could reasonably apply. The selection has been guided by the primary goal of providing the underpinnings of excitation and chemical transformation processes, even though we hardly deal with molecules or collisions between atoms explicitly. Accordingly, we deal mainly with the range extending from thermal energies to approximately the ionization threshold. Higher-energy processes that have been particularly fruitful for understanding are nevertheless included, as well as an outline of the perturbative treatment that applies to the collision of fast charged particles. With this high-energy ground thus under control, we have concentrated on the low-energy processes, feeling that the intermediate energy range—an energy decade above the ionization threshold—cannot be fruitfully attacked until both ends of the spectrum are adequately mastered. Inner-shell phenomena have not been dealt with in general because of their lesser relevance to our main goal.

Thus, the primary treatment in this book is of electron–atom or-ion interactions involving the outer shells. In terms of energy, this amounts to a range of energies from a few electron volts (eV) to a few tens of electron volts. Bound-state spectral levels, photoionization, and electron scattering constitute our subject matter, first in the context of potential models and later taking into account the full complexity of the many-body problem that an atomic system represents.

1.2 EXPERIMENTAL ASPECTS

New experimental techniques and the results obtained with them in the past 20 years have provided the basis of our present picture of atomic dynamics. As a bow to this critical role experiments have played, it is appropriate, therefore, to begin with a brief survey of the experimental side, highlighting the particular developments in apparatus and techniques since 1960 that have contributed significantly to our present understanding of atomic dynamics. A major aspect both in experiment and in theory is the contrast between collisions with fast charged particles (which can be treated perturbatively) and those with slow particles in which the effects are strongly nonperturbative. On the experimental side, this contrast between low- and high-energy physics is that any target material absorbs heavily at low energies.

To derive information about an elementary collision, a major requirement is to ensure that we are looking at single collisions. This is achieved by studying its cross section as a function of pressure. To reduce multiple collisions, low-energy experiments necessarily call for high-vacuum environments (some experiments have been done at 10^{-9} torr). Furthermore, the feature that low-energy charged projectiles are very easily disturbed calls for extreme care in screening against stray fields. All metal surfaces the electrons are likely to encounter or pass by in their paths must be free of magnetic impurities, suitably coated, and periodically baked to avoid the buildup of surface charge and surface potentials. These requirements are, of course, more stringent in the measurement of angular distributions and differential cross sections.

Another basic requirement is a high energy resolution in the preparation and detection of electron beams. An essential step was the development since the 1960s of instruments such as the spherical analyzer (Fig. 1.1) and the 127° cylindrical electron spectrometer, which enabled us to achieve resolutions of a few tens of millielectron volts in electron beams of a few electron volts. Many of the phenomena, such as the resonances due to multiply excited states, would have escaped detection otherwise.

Another key development was the use of crossed and merged beams (Fig. 1.2). Here, in an interesting way, the development of molecular-beam techniques in "Rabi physics" has come back to enrich the field of atomic dynamics. In contrast to experiments in which the sample vapor under study is held in a cell, the use of an atomic or molecular beam allows much better

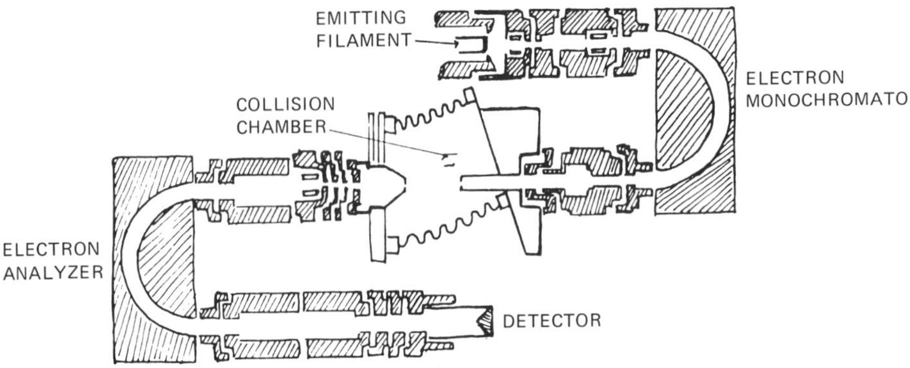

Fig. 1.1. Sketch of a spherical electron spectrometer. (From Fano and Fano [1]. Copyright © 1972 The University of Chicago.)

control of the geometry and precise definition of the scattering volume. This is particularly valuable in the measurement of absolute cross sections. Use of a beam also permits the introduction of powerful modulation techniques (e.g., through mechanical chopping) so that we can discriminate against spurious effects due to any background gas. There is also the problem that a closed-volume cell with slits gives rise to unwanted scattered electrons and

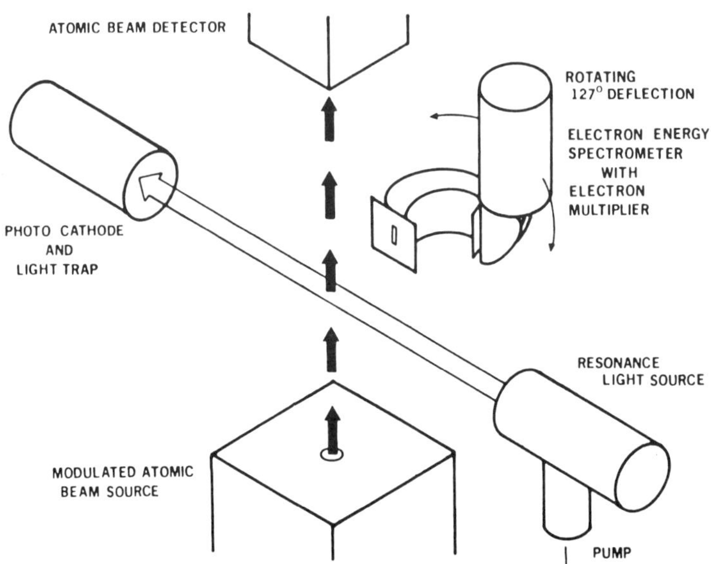

Fig. 1.2. Schematic arrangement for the study of the angular distribution of photoelectrons using atomic or molecular beams. (From Cairns *et al.* [2].)

unwanted potentials that distort angular distribution studies, and these too are avoided by modulated crossed-beam techniques. Beams also permit other sophistications such as the study of low-energy collisions with high-velocity beams: merging of two beams at a very small angle reduces the reactant velocity in their center-of-mass frame.

Very recently, the use of spin-polarized beams (electron, atomic/molecular, or both) is bringing in another dimension in which collisions can be studied in specific spin states. Coincident detection [e.g., of a scattered electron and of a secondary electron or photon (Fig. 1.3)] increases further the number of observable parameters of a collision. Even more exotic techniques have entered the field, such as the use of extremely high-energy relativistic beams which can be crossed with other particle or photon (laser) beams. This permits an extremely precise tuning of the collision energy across a wide range through the Doppler effect simply by changing the angle between the colliding beams.

Turning to collisions with photons, a major advance came in the 1960s with the opening up of the spectral region between hard X rays and the extreme ultraviolet, roughly the range between 1 and 100 nm (Fig. 1.4). This region had been essentially unexplored in previous decades because of the absence of convenient sources in this wavelength range where all substances absorb strongly. This absorption reflects the fact of atomic structure that ionization potentials of atoms lie around 10 eV, beyond which a strong continuum absorption generally prevails. A corollary is that most of the oscillator strengths of optical electrons are in this range and had remained essentially unstudied. Furthermore, the range is also the seat of multiple excitations so that many of the degrees of freedom of the complex and attendant dynamics that are of concern in this book escaped attention before the opening up of this "vacuum ultraviolet" region. The two crucial experimental developments that brought these phenomena into view were the advent of synchrotron light sources as convenient continuum radiation sources in this range and the ability to do windowless absorption studies with a gas target through the use of fast differential pumping.

As detectors, the photographic plate, photomultipliers, and counters continue to be used, although technical improvements have made it possible to achieve higher resolutions and sensitivities. Once again, for instance, windowless "free-flow" counters using differential pumping must be used at the lower wavelengths. In the vacuum ultraviolet region, the gratings are used at grazing incidence to minimize low reflectances. The attendant problems of order overlap, aberrations, and light losses impose strict mechanical tolerances. At one time good gratings were only available in-house at a couple of laboratories, but now carefully designed, ruled, and blazed gratings for precision work are more easily available.

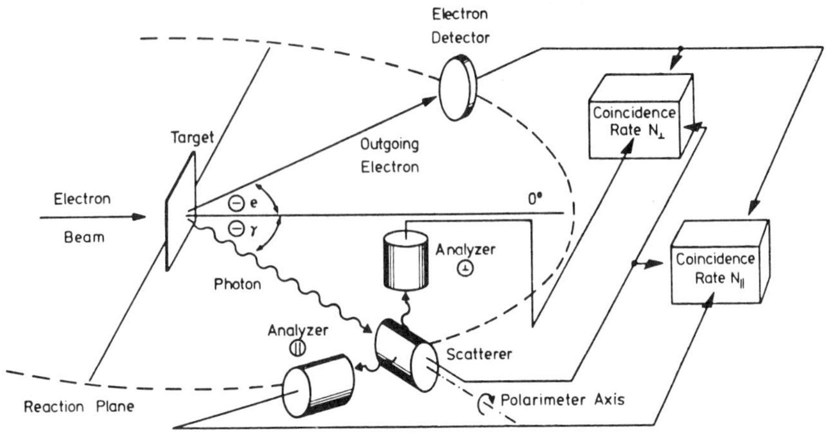

(a)

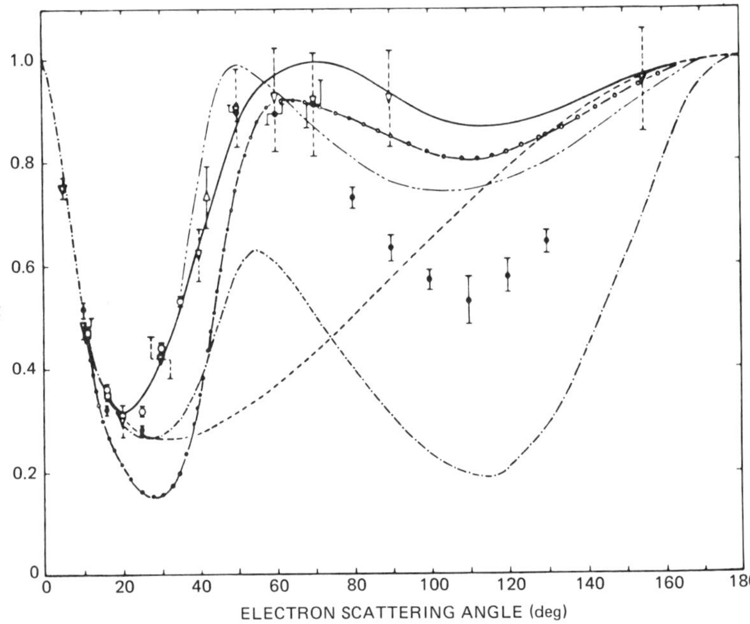

(b)

Fig. 1.3. (a) Schematic view of an experimental arrangement for coincident detection of scattered electrons and fluorescent light. (From Nakel [3].) (b) Fraction of fluorescence λ attributable to a Hertzian dipole oriented along the direction of incidence for e − He scattering at 81.2 eV. Note the variation in experimental results and the departure from the curves that represent the predictions of alternative, approximate theories. (From Hollywood *et al.* [4].)

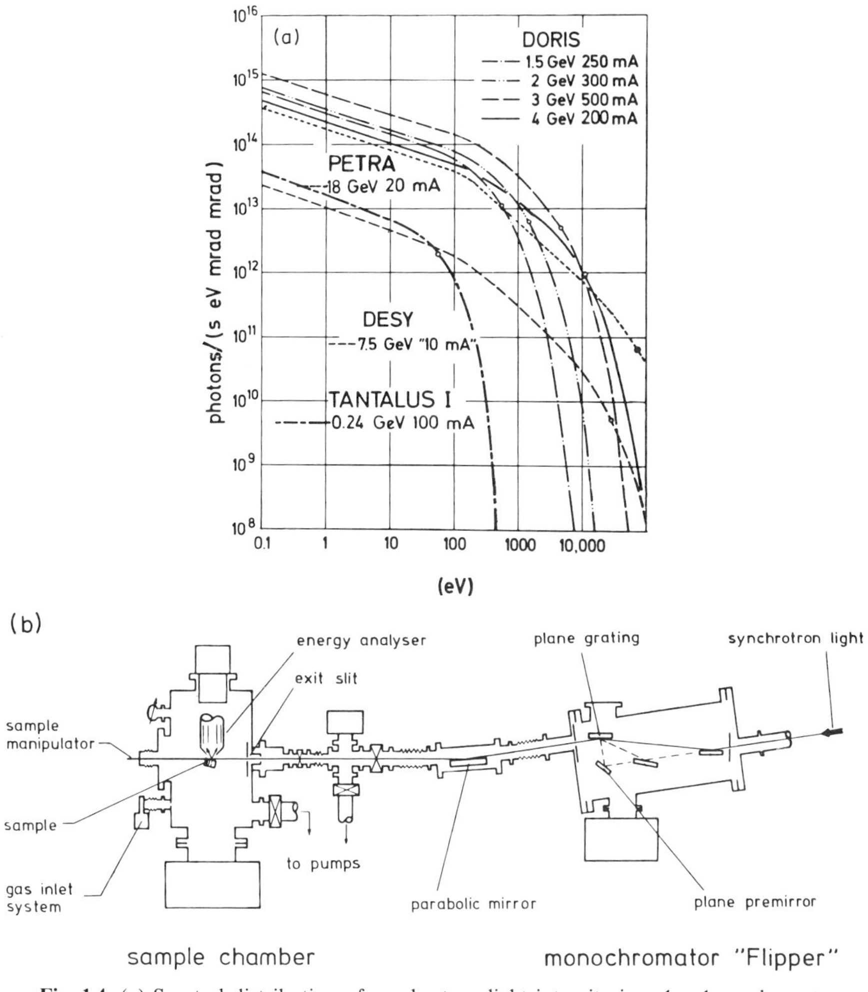

Fig. 1.4. (a) Spectral distribution of synchrotron light intensity in a 1 × 1-mrad aperture showing the dependence of usable photon energies on electron current and energy. (b) Schematic arrangement for vacuum ultraviolet spectroscopy with a synchrotron source. (From Kunz [5].)

On the other side of experiments with photon impact, the continued development of lasers has, of course, revolutionized the field in the range of the visible spectrum, particularly with regard to high-resolution work. With the continued push to shorter wavelengths in the blue and violet region and with laser intensity levels that remain usable even after frequency doublings, it is now possible to reach even high-lying states of the one-electron excitation spectrum of most atoms. High resolution and tunability permit here selective excitation to very specific levels. In recent years, special detection techniques like the use of electric field ionization of high Rydberg states as a highly selective way of labeling and detecting these states down to every individual quantum number have permitted extremely detailed studies of atomic spectra and atomic collisions.

The combination of photon sources or detectors with those of charged particles have added new levels of sophistication to the study of collisions and spectroscopy. Thus, laser excitation within the collision chamber permits the study of specifically selected states of atoms in their interactions either with electron or other atomic beams. High-lying states prepared by the absorption of one or more photons, which subsequently decay by emitting electrons, can be studied by combining aspects of laser and electron spectroscopy. Such combinations of spectroscopic and collisional techniques permit more detailed investigation of the pathways for the flow of energy in atomic systems.

1.3 ATOMIC COMPLEX: INPUT AND EXIT CHANNELS

Atomic theory developed initially through independent particle models, which account to a large extent for spectra and elastic collisions. Combination of these models with treatment of particle interactions by lowest-order perturbation theory has extended the range of successful applications, particularly to include inelastic collisions by fast particles and the Auger effect.[†] On the other hand, the experimental progress of the past 20 years has presented us with a rapidly expanding body of detailed evidence on phenomena in which interactions play a more elaborate role, thus exceeding the scope of ordinary perturbation treatments. These interactions govern the transfer of energy, momentum, or angular momentum from one particle to another. They are seen—if only dimly as yet—to channel such transfers through rather specific pathways. Outlining our present knowledge of these more complex phenomena (which are often characterized as "correlation effects") is a main objective of this book.

Foremost among these phenomena is the broad field of inelastic collisions at low energy, because it includes the processes of chemical transformation. Unraveling the mechanisms of these processes and describing the basic chemical process in which two reactants are transformed into other products may be regarded as the central challenge of low-energy physics. Accordingly, we look at the processes to be discussed as sources of insight on the largely unknown pathways through which many-electron systems evolve from one

[†] In the Auger effect, a vacancy in an inner shell is filled by an outer electron, with the energy difference transferred to another outer electron, thus leading to its ejection.

stable (or metastable) state to another one. This view complements Bohr's emphasis on how the discrete nature of atomic spectra underlies the stability of matter. We are concerned here with the pathways by which this stability is circumvented in moving from one stable state to another.

This point of view lends remarkable unity to our subject, which might otherwise seem utterly diverse and disconnected. Indeed it involves radiative and collision processes, ionic and neutral systems, valence- and inner-shell and even nuclear processes, and energy exchanges ranging from fractions of an electron volt to many kiloelectron volts. As an aid to perceiving both the diversity and the unity, we consider the many-branched reaction diagrams

$$\begin{array}{ccc} & h\nu + \mathrm{A}^* & \\ & \upharpoonleft\downharpoonright & \\ h\nu + \mathrm{A} \rightleftharpoons & \mathrm{A}^{**} & \rightleftharpoons e + \mathrm{A}^{+*}, \\ & \upharpoonleft\downharpoonright & \\ & e + \mathrm{A}^+ & \end{array} \qquad \begin{array}{ccc} h\nu + \mathrm{A}^- \rightleftharpoons & \mathrm{A}^{-**} & \rightleftharpoons e + \mathrm{A}^* \\ & \upharpoonleft\downharpoonright & \\ & e + \mathrm{A} & \end{array} \tag{1.1}$$

and their isoelectronic analogs with positive charge. In (1.1), each * means excitation of one electron and we have restricted ourselves to monoatomic systems indicated by A. Diversity of the subject results from the multiplicity of alternative entrance and exit channels and alternative excitation levels; its unity results from the common role of the central complex, which is generally multiply excited as indicated by the double asterisk. All the physically distinct processes in (1.1) can be viewed as transitions between different states of the complex, whether discrete or continuum or multiply excited.

The presence of two (or more) excited electrons in the central complex of (1.1) is essential for the mechanisms of electron correlations. Whereas the electrons of inner or valence shells are strongly constrained in their motion by the exclusion principle and by nuclear attraction, excited electrons move over a large volume and are influenced by their mutual repulsion no less than by the screened field of the nucleus. Early insights into electron correlations emerged in fact from the study of doubly excited states of atoms in the 1960s. Multiple excitations bring in a considerable increase in the richness and variety of phenomena. If we view the ground state of an atom as a system with many degrees of freedom locked in, multiple excitations "unfreeze" many more—and qualitatively new—degrees of freedom than do single-electron excitations.

Mathematically, the joint motion of N_e electrons and N_n nuclei is described by solutions of a nonseparable Schrödinger equation in $3(N_e + N_n - 1)$ variables that span a multidimensional configuration space. These solutions are not represented by analytical formulas nor can they be profitably constructed by purely numerical methods, except in a few instances. The theory of this motion is instead developing in a piecewise fashion, by identifying the factors that prevail in different regions of the configuration space and developing analytical and numerical methods appropriate to their description. Thus, we treat separately the regions in which one variable becomes asymptotically much larger than the others, or instead where two or more particles approach closely (whereby their Coulomb and centrifugal potentials diverge), as well as intermediate ranges in which approximate separation of variables and/or WKB treatments apply. Physically, the study of the Schrödinger equation in each of these regions amounts to identifying a separate phenomenon and solving the relevant theoretical problem. In other words, we follow the development of a process from the point of view of the particles participating in it. Thus, as a particle moves from one range to the other, we switch between the variables and descriptions appropriate to each region. What may have looked at the outset as a dry and intractable task of applied mathematics has thus resolved itself into a varied and still expanding set of problems. From these problems have also emerged suggestive implications for entirely different branches of physics.

1.4 SYMMETRIES

The physicomathematical analysis of our problems—specifically their fractionation into separate problems—is facilitated by the invariance of interactions under various operations. Invariance under translation in time permits us to develop the theory in terms of stationary states. Among these states invariance under interchange of in- and outgoing wave components selects those represented by standing waves, and invariance under space rotations selects the eigenstates of the total squared angular momentum. Sorting out such special states has also been found to reduce the risk of self-amplifying numerical errors.

On the other hand, taking advantage of these opportunities, which we shall do systematically, tends to obscure the connection of theory with the physical development of the phenomena of interest. Collision processes develop in the course of time toward a final state that generally differs substantially from the initial state. Such unsymmetric processes are described by quantum-mechanical superposition of eigenstates of different symmetries. We can in fact trace a correspondence between the observable features of collisions and relevant features of symmetry eigenstates, but it takes some skill to perceive this correspondence through the veil of the superposition process.†

† This skill may be developed by study of illustrative examples such as are developed in Fano and Fano [1].

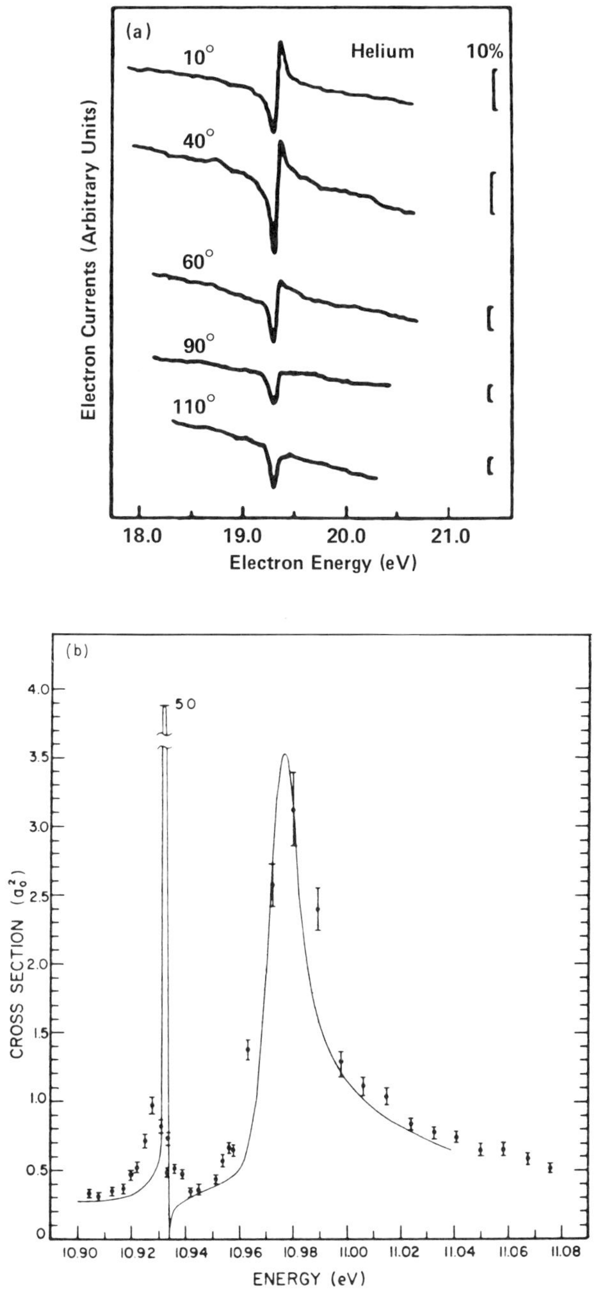

Fig. 1.5. Resonances in $e - \text{He}$ and $e - \text{H}$ scattering. (a) 19.3 resonance He^{-**} ^2S at different angles of observation. (From Andrick and Ehrhardt [6].) (b) Two resonances of H^{-**} in the vicinity of the $\text{H}(n = 2)$ threshold. (From Bryant *et al.* [7].)

For purposes of illustration, consider one of the systems in (1.1). The central complex alone A** is inherently unstable and accordingly represented by a superposition of stationary states that covers a broad spectral range if the complex is short-lived and a narrow range if it is long-lived. The complex may well be described as an eigenstate of angular momentum, as in the example of the He^{-}**($1s2s^2$) complex formed in collisions He($1s^2$) + e(19.37 eV) with zero angular momentum. Superposition of angular momenta is more apparent, on the other hand, in the analogous process H(1s) + e(10.15 eV) = H^{-}**, where s and p orbitals are strongly mixed in the final state of each electron (Fig. 1.5).

The wave function that represents a complex as an unstable, nonstationary state vanishes at large radial distances as a result of destructive interference between components of different energies. Alternatively, and more conveniently, we may instead represent a complex as a particular portion—the "core"—of a stationary state that extends outside the volume of A** into the asymptotic region. Here, for example, an electron lies far out of the rest of the system and thus represents an $A^+ + e$ channel; this portion has a standing waveform if the state has even parity under time reversal. This same wave function must also consist of separate components corresponding to the absence or presence of a photon in the electromagnetic field. The former represents A** and $A^+ + e$, while the latter represents the channels $A + h\nu$ and $A^* + h\nu$.

As another class of examples, in this case concerning changes in symmetries and their associated quantum numbers from one region of configuration space to another, consider the channels in (1.1) that correspond to the fragmentation of the complex such that there is finally one electron at infinite separation from the rest. The nature of the angular momentum coupling may, in general, be quite different when the electron is close to all the other electrons than when it recedes to infinity; a typical example, with a transition from LS to jj coupling will be considered in detail in Chapter 7. A more dramatic example, also considered in Chapter 7, arises in molecular contexts in which it is appropriate to switch from a rotating "molecular frame" description to a "laboratory frame" when the electron is at infinity. Considerable economy in the number of states that must be retained at any stage can result by switching in this manner between alternative descriptions and relevant symmetries.

1.5 A ONE-DIMENSIONAL ILLUSTRATION

Developing procedures that serve to implement alternative descriptions of a wave function for a realistic system is a major goal of this book. Essential aspects of such procedures are already displayed by the following one-

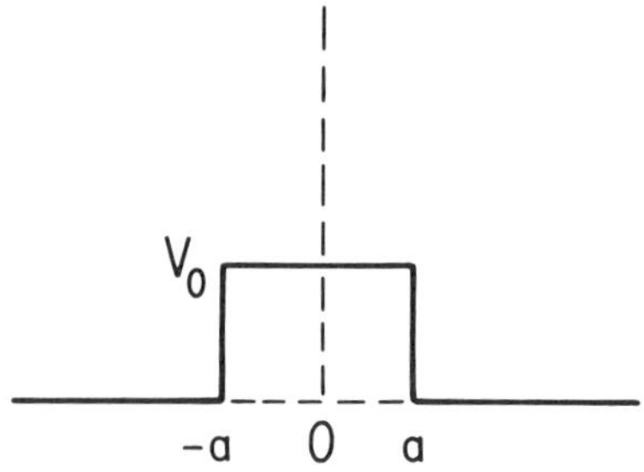

Fig. 1.6. One-dimensional square potential.

dimensional model (Fig. 1.6). Consider a particle of mass m and energy $E > 0$ that moves freely along the x axis except where it traverses a potential $V(x) = V_0$ for $|x| \leq a$, with V_0 of either sign. The region $|x| \leq a$ corresponds to the central complex of (1.1); the sections to the right ($x > a$) and left ($x < -a$) correspond to entrance or exit channels. We can consider various alternative pairs of orthogonal states that are characterized by different symmetries or asymptotic conditions. Thus, defining as usual

$$E = \hbar^2 k^2/2m, \qquad E - V_0 = \hbar^2 k_0^2/2m$$

(with the proviso that, for $V_0 > E$, we set $k_0 = i\kappa_0$ and let trigonometric functions in k_0 become hyperbolic ones in κ_0), we can enumerate the following:

(1) Two eigenstates of reflection at $x = 0$ (parity):

$$\begin{aligned} u_+(x) &= N_E \cos(kx - \delta_+), & u_-(x) &= -N_E \cos(kx - \delta_-), & x &\leq -a, \\ &= A \cos k_0 x, & &= B \sin k_0 x, & |x| &< a, \\ &= N_E \cos(kx + \delta_+), & &= N_E \cos(kx + \delta_-), & x &\geq a. \end{aligned} \tag{1.2}$$

Here $u_\pm$ have eigenvalues ± 1 with respect to parity, N_E is a normalization constant to be discussed in detail in Chapter 4, while the constants A, B, and $\delta_\pm$ (which represent the so-called asymptotic phase shifts) are determined by matching the logarithmic derivatives of the wave function at $x = \pm a$:

$$k_0 \tan k_0 a = k \tan(ka + \delta_+), \qquad k_0 \cot k_0 a = -k \tan(ka + \delta_-), \tag{1.3a}$$

$$A \cos k_0 a = N_E \cos(ka + \delta_+), \qquad B \sin k_0 a = N_E \cos(ka + \delta_-). \tag{1.3b}$$

The functions (u_+, u_-) represent a particular base pair of standing-wave states. They can be used to form any other pair of orthogonal solutions because a general, normalized solution would be given by

$$u(x) = c_+ u_+(x) + c_- u_-(x), \qquad |c_+|^2 + |c_-|^2 = 1, \tag{1.4}$$

with $c_\pm$ as complex constants. For $|x| > a$, (1.4) can be written as

$$\begin{aligned} u(x) &= \tfrac{1}{2}N_E[(c_+e^{i\delta_+} + c_-e^{i\delta_-})e^{ikx} + (c_+e^{-i\delta_+} + c_-e^{-i\delta_-})e^{-ikx}], \qquad x > a \\ &= \tfrac{1}{2}N_E[(c_+e^{-i\delta_+} - c_-e^{-i\delta_-})e^{ikx} + (c_+e^{i\delta_+} - c_-e^{i\delta_-})e^{-ikx}], \qquad x < -a. \end{aligned} \tag{1.5}$$

(2) Other pairs of standing-wave states, which are now no longer parity eigenstates, are given by (1.4) with real values of c_+/c_- so that $u(x)$ remains a real function apart from an overall phase factor. There is, therefore, a one-parameter family of such standing-wave solutions.

(3) Traveling-wave solutions in which the particle approaches the well only from $x = -\infty$ or only from $x = +\infty$. These can be read off from (1.5). The former are specified by $c_+e^{-i\delta_+} + c_-e^{-i\delta_-} = 0$ and the latter by $c_+e^{-i\delta_+} - c_-e^{-i\delta_-} = 0$. With c_+/c_- thus fixed, the transmission and reflection coefficients can be read off from (1.5). Of course, as we would expect, these coefficients are the same in both cases—the reflection coefficient is $\sin^2(\delta_+ - \delta_-)$.

(4) A pair of states corresponding to particle escape from the well only in the direction $x = -\infty$ or $x = +\infty$. These are complementary to the ones in (3) and again follow from inspection of (1.5). They are given by $c_+e^{i\delta_+} + c_-e^{i\delta_-} = 0$ and $c_+e^{i\delta_+} - c_-e^{i\delta_-} = 0$, respectively. The ratios c_+/c_- are complex conjugates of the corresponding values in (3), as expected from the feature that (4) and (3) are time-reversal conjugates.

Physically, the traveling-wave solutions in (3) arise in scattering problems where a particle is incident on the potential well and scatters from it. In (1.1) this corresponds to the channels in which the complex is formed starting with an electron in the incident channel. This situation is commonly referred to as the "outgoing wave" boundary condition. On the other hand, example (4) with the "ingoing wave" boundary condition arises in processes such as photoionization, where the entrance channel does not involve any free electron; such an electron is instead released from the complex by the action of some other agent (photon, proton, etc.) and is to be detected in a specific direction.

1.6 OUTLINE OF LATER CHAPTERS

The preceding one-dimensional example displays many of the features that will run as a common thread throughout our discussion of atomic collision physics. The enumeration of alternative sets of orthogonal states of interacting particles and their use in describing the development of the physical processes will be recurring themes. Passage between alternative basis sets becomes particularly necessary and important when we are dealing with effects that are far from being perturbative. By a judicious choice of an appropriate basis, we may only need to deal with a few states, whereas a very large number of states would be required to describe the process in another basis.

Having said this, we now sketch the plan of this book. Even though perturbation approximations are generally inadequate to treat low-energy collisions, it is necessary to consider how certain processes can in fact be treated from an independent particle point of view supplemented by standard perturbation theory. Part A (Chapters 2 and 3) deals with two such major classes of processes, namely, photoabsorption and the collisions of fast charged particles. In both instances, what makes perturbation theory applicable is that the incident particles (whether photons or fast charged particles) interact only weakly with atomic systems. In the electromagnetic case this arises because the typical fields of an electromagnetic wave are very much smaller than the internal fields in an atom, while in the case of fast-charged-particle impact the perturbative nature derives from the short duration of interaction. In fact, the two instances are well understood to be intimately related because the dominant effect of a fast charged particle going by an atom is the electromagnetic pulse caused by it at the site of the atom.

A primary consequence of weak interaction is that the entrance channel of the complex is not affected by the complexity of collision physics (at least when the atom is initially in its ground state). The complicated—and more interesting—aspects manifest themselves in the exit channels, that is, within the complex formed in (1.1) *after* it has absorbed the incident radiation and in its subsequent breakup into the exit channels. Experiments on photoabsorption and fast particle collisions, thus, have a particularly instructive role, which has been termed the study of "half-scattering" processes.

Actually, this aspect of unity, that it is often irrelevant by which entrance channel the complex was initially formed, also arises at lower energies, particularly just above inelastic thresholds. A case to be considered in detail in Chapter 7 is the photodetachment of the F^- ion. Just above the detachment threshold, the escaping electron leaves in an s state of long wavelength. In that sense, this exit channel of the fragmentation of the complex (e + F) formed on absorption of the photon looks just like half of e + F elastic scattering. The breakup of the negative ion, therefore, may be described by the same few parameters that describe s-wave elastic electron scattering at thermal energies. The earlier discussion of the intimate connection between cases (4) and (3) of the one-dimensional potential also illustrates the link between these "half" and "full" scattering cases, respectively.

As the next step in complexity, Part B (Chapters 4–6) deals with elastic or single-channel collisions in which an incident particle may interact strongly with the atom but still does not exchange with it any energy, angular momentum, or parity. The study of this limited class leads us to single out collision parameters, mainly the scattering phase shifts, which appear later in more complicated settings. It also introduces the connection between scattering parameters and spectral levels. Such a connection can be expected on the grounds that the bound states and the continuum are part of the spectrum of the same Hamiltonian operator. In the context of the Coulomb potential, this connection underlies the development dubbed "quantum defect theory." The powerful handle this connection provides on pulling together diverse aspects of collisions and spectra is the main focus of later chapters together with generalizations to potentials other than Coulomb.

The study of elastic scattering in Chapters 4–6 also brings in the aspect of the energy dependence of phase shifts, particularly at low incident energies ("threshold behavior"). Finally, these chapters outline the connection, indeed the basic equivalence, between the calculation of the phase shifts for purely elastic scattering and the calculation of the spectral levels of corresponding atomic systems (mainly the alkalis) by standard methods such as Hartree–Fock, configuration mixing, and variational procedures.

Part C (Chapters 7–9) represents the third, and main, part of the book in which the previous restriction to no exchanges of energy, angular momentum, and parity is removed, thereby allowing the full dynamics of the interaction between the colliding systems to come into play. Indeed, as in a basic chemical process, the reactants emerging from the collision may even be quite different from those in the incident channel. The connections between the asymptotic states in the entrance and exit channels and different stable (or metastable) states of the central complex and the passage between these states are the objects of study. Gradually, by tying together the different pieces of information contained in different channels, we build up knowledge of the complete spectrum (discrete as well as continuum) of states of the complex and the transitions between them. This, in turn, helps connect the results of physically different experimental arrangements. The observable effects of a collision, embodied in partial scattering cross sections for transitions from any one channel to any other one, are cast in more refined form as scattering amplitudes whose aggregate, a "scattering matrix," represents the transformations between the channels branching from the complex.

The construction of such a scattering matrix (or rather of its equivalent factorized form in terms of "Jost matrices") is, therefore, the goal of these chapters. The main task is to factor the scattering matrix into parameters that represent separate dynamical effects. Some of these parameters represent purely geometrical effects of angular momentum coupling between the reactants and are common to different physical systems that share the same quantum numbers of angular momentum and spin. Filtering out these parameters helps in reducing the number of relevant parameters that are essential to a description of the dynamics of the particular system under study. Other parameters, akin to phase shifts, characterize the forces operating within the individual channels or in the central complex.

As remarked earlier about relevant symmetries often being different in different regions or in different channels, quantum numbers appropriate to any one entrance or exit channel differ generally from those relevant to the complex at small r. Thus, in Section 7.2, which deals with elastic electron–hydrogen atom collisions, the separate reactants are characterized by their respective spin orientations and the H^- complex by the invariant labels of singlet and triplet. A similar example occurs next for the photodetachment of a negative ion where the exit channel is described in jj coupling whereas the complex is labeled by quantum numbers in the LS scheme. For electron–H_2^+ scattering, the initial and final channels are described in terms of the orbital angular momentum of the electron and the rotational quantum number of the molecule, whereas the central complex has as its invariant the projection of the angular momentum of the electron on the internuclear axis.

In all these cases, the channels are connected to the eigenstates of the central complex by a quantum-mechanical "frame transformation." The unitary matrix that describes this transformation is one factor (a relatively trivial one) of the scattering matrix, other factors being made up of the phase shift contributions of the channels. Such an analysis of collision processes forms the phenomenological core of multichannel quantum defect theory (MQDT). This theory serves to represent by a limited set of parameters a large body of collision and spectral data (energy levels, oscillator strengths, resonance positions, their widths and profiles, branching ratios for particle ejection into alternative continua, etc.). A further key feature is the weak energy dependence of these parameters on the total energy of the complex. Chapters 7–9 review examples of phenomenological MQDT analysis of experimental data as well as the diverse *ab initio* approaches to calculating MQDT parameters or other observable properties. Illustrative examples of comparison of experimental and calculated data are included.

Part D (Chapters 10–11) introduces an aspect of collision processes that has been glossed over up to here: (1.1) implies that the excitation energy permits only a single electron to emerge in the exit channel. In fact, the energy may suffice for two or more particles to move to appreciable distances from the rest, even escaping to infinity. The transition between complex and channel functions must then be described in much greater detail; the current state of this task will be described. In particular, we find it convenient for this purpose to describe configuration space in terms of the so-called hyperspherical variables for two (or more) electrons. Chapter 10 deals with doubly excited states of He and H^-, in which two electrons move in the field of a bare nucleus. New quantum numbers serve to classify these states. The extreme case of two-electron escape will be touched on briefly. Chapter 11 introduces the extension to other atoms with a structured ionic core.

This introductory chapter has been illustrated with references to (1.1) representing an atomic central complex that can eject an electron or emit a photon. Molecular complexes also occur that can dissociate into two atoms (or into smaller molecular species) and can be formed by collisions of two atoms or molecules. These polyatomic processes are in fact more important than monoatomic ones from the chemical point of view. The considerations presented in this introduction are generally appropriate to their treatment, too, but the mechanics are naturally more complicated for molecular systems than for single atoms. As noted at the outset, the explicit scope of this book will be generally limited to the treatment of monoatomic systems.

Problems

1.1 In the model one-dimensional problem of Section 1.5 four pairs of states were defined in (1)–(4). For each of these four pairs verify the orthogonality by evaluating the overlap integral $\int_{-\infty}^{\infty} u_1(x)u_2(x)\,dx$, where $u_1(x)$ and $u_2(x)$ are the two wave functions of the pair.

1.2 *Mapping of states of a two-level system on the orientation states of a spin-$\frac{1}{2}$ particle* Set the eight states considered in (1)–(4) of Section 1.5 in one-to-one correspondence with states of orientation of a spin-$\frac{1}{2}$ particle identified by the polar coordinates (θ, ϕ) of the spin direction. Define this correspondence by the equation between spinors

$$\begin{vmatrix} C_+ \\ C_- \end{vmatrix} \equiv \begin{vmatrix} \cos\frac{1}{2}\theta \exp(i\frac{1}{2}\phi) \\ \sin\frac{1}{2}\theta \exp(-i\frac{1}{2}\phi) \end{vmatrix}.$$

(a) This mapping of each of the degenerate states to one direction $\hat{u} = (\theta, \phi)$ of physical space begins by assigning the two states of one orthogonal pair to specific values of (θ, ϕ). Begin with the states (u_+, u_-) in (1.2) and find their coordinates (θ, ϕ).

(b) For the pair of standing waves described in (2) in Section 1.5, find the coordinates (θ, ϕ).

(c) Find the coordinates of each of the two traveling waves in (3) and (4).

(d) Verify that the squared overlap of any two states u_1 and u_2, namely, $|(\hat{u}_1|\hat{u}_2)|^2$ equals $\frac{1}{2}(1 + \hat{u}_1 \cdot \hat{u}_2)$.

(e) Time-reversal conjugates are described in (4) as pairs of states with equal currents in opposite directions. Identify the operation that interchanges the states of such a pair geometrically with reference to the spin orientation map and also analytically.

1.3 *Nonsymmetric potentials* Consider the effect of a potential well that does not fulfill the symmetry requirement $V(-x) = V(x)$ under reflection at $x = 0$. Such a well is obtained, for example, from the one in Section 1.5 by a translation of coordinates that replaces x by $x' = x + \Delta$. Show that:

(a) The wave functions in (1) remain valid throughout the range of $|x| > a + \Delta$, and remain symmetric or antisymmetric with respect to the new origin of coordinates, albeit with different phase shifts.

(b) The new values of these phase shifts are determined by enforcing the condition

$$d \ln u/dx|_{x=b} = -d \ln u/dx|_{x=-b} = d \ln u/d(-x)|_{x=-b}$$

with $b \geq a + \Delta$. Note that the first and last expressions, which are to be set equal, can be written in a common form $d \ln u/dn$ in terms of a variable $n = \pm x$ for $x \gtrless 0$, defined so as to increase with increasing distance from the well regardless of direction. This equality of "normal derivatives" of the logarithm of the wave function extends to problems in a greater number of dimensions.

1.4 *Bound-state eigenvalues* When $V_0 < 0$ ("potential well") in the model considered in Section 1.5, the solutions with $E > 0$ can be extended to negative energies, $E < 0$, so long as $E - V_0$ remains positive. Thus, the values of $u_\pm$ in (1.2) remain valid as solutions for $V_0 < E < 0$, with the definition $k \equiv i\kappa$. For $|x| > a$, the trigonometric functions become hyperbolic ones. In order that the solutions be finite, the coefficient of a piece of these hyperbolic functions that becomes exponentially large at $|x| \to \infty$ must be set equal to zero. This condition on the phase shifts $\delta_\pm$ translates through (1.3a) into an eigenvalue condition for the allowed values of $E < 0$.

Derive the transcendental equations for the eigenvalues of even and odd parity in the one-dimensional potential well.

1.5 *The 127° electrostatic analyzer* This instrument for high resolution work with low-energy electron beams is built on the following principle. The electric field between two concentric cylinders of radii a and b with a potential difference V is everywhere radial and given by $-AV/r$, where A is a geometrical factor, depending on a and b.

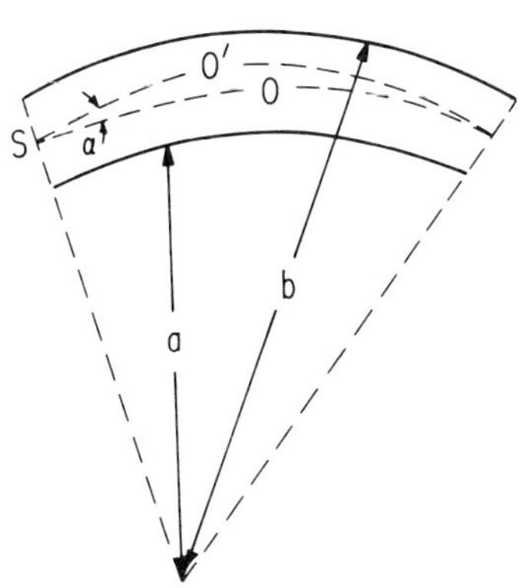

Verify that the electrical and centrifugal forces balance for electrons entering at any radius r tangential to the equipotentials and with a velocity v such that $mv^2 = AeV$. One such trajectory is indicated by O.

A working instrument for detecting electrons of equal v must, however, be capable of accepting particles entering through a slit S of finite aperture, all of which have the same speed v but velocity vectors distributed over a narrow range of acceptance angles α. Consider electrons entering at $r_0 = \frac{1}{2}(a + b)$ with mutual angles as shown for the trajectories O and O'.

Show that the trajectories with different α come together again after an angle $\pi/\sqrt{2} \simeq 127°$.

1.6 *Spectrum of synchrotron radiation* An electron circulating in a magnetic field at frequency $\omega = eB/mc$ emits radiation of this frequency as a Hertzian dipole in its own frame. For relativistic electrons, this radiation appears in the laboratory frame as a narrow beam of high-frequency synchrotron light. As illustrated in Fig. 1.4a, the range of frequencies over which useful intensities are available increases rapidly with the energy E of the electrons. By following the line of argument given below, show that this frequency range increases as γ^3, where $\gamma \equiv E/m_0c^2$ is the relativistic Lorentz factor.

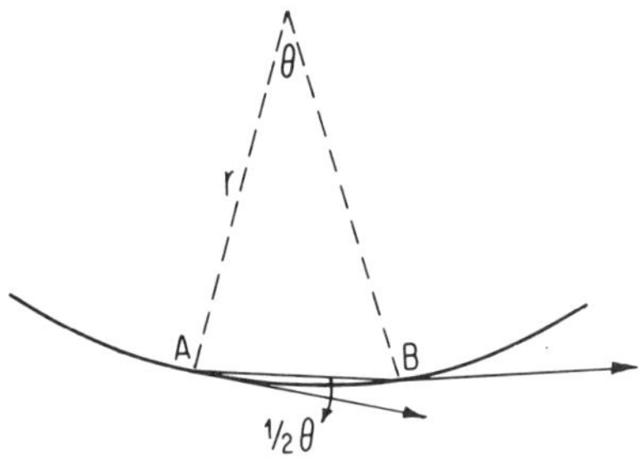

Show that the beam is confined to a narrow cone of aperture $\theta \simeq 1/\gamma$ by using the relativistic Doppler formula relating angles in the frame of the electron to those in the laboratory frame. The observer receives radiation as a pulse emitted while the electron travels between A and B of the orbit. Determine the width of this pulse as seen by the observer, considering the time taken by the particle to go from A to B. Fourier resolution translates the width into a band of frequencies of the emitted radiation.

References

1. U. Fano and L. Fano, "Physics of Atoms and Molecules," Appendix A. Univ. of Chicago Press, Chicago, Illinois, 1972.
2. R. B. Cairns, H. Harrison, and R. L. Schoen, *Adv. At. Mol. Phys.* **8**, 131 (1972).
3. W. Nakel *in* "Coherence and Correlation in Atomic Collisions," (H. Kleinpoppen and J. F. Williams, eds.), p. 187. Plenum, New York, 1980.

4. M. T. Hollywood, A. Crowe, and J. F. Williams *in* "Coherence and Correlation in Atomic Collisions," (H. Kleinpoppen and J. F. Williams, eds.), p. 97. Plenum, New York, 1980.
5. C. Kunz, "Synchrotron Radiation." Springer-Verlag, Berlin and New York, 1979.
6. D. Andrick and H. Ehrhardt, *Z. Phys.* **192**, 99 (1966).
7. H. C. Bryant, B. D. Dieterle, J. Donahue, H. Sharifian, H. Tootoonchi, D. M. Wolfe, P. A. M. Gram, and M. A. Yates-Williams, *Phys. Rev. Lett.* **38**, 228 (1977).

Part A

PERTURBATION TREATMENTS

2 PHOTOABSORPTION

2.1 INTRODUCTION

The phenomenon of photoabsorption has been mentioned in Chapter 1 in two separate but related contexts. It constitutes a first example of atomic processes amenable to simple quantum perturbation theory because the electric fields of ordinary light sources are much weaker than those prevailing inside an atom. (The action of intense laser sources requires a special treatment.) More important, photoabsorption can be regarded as a direct method of preparing a complex in a rather well-defined state of excitation, whose fragmentation can then be studied experimentally. In this respect, photoabsorption complements electronic or atomic collisions, which usually prepare the complex in an ill-defined superposition of stationary states of different angular momenta; for this reason, photoabsorption has been dubbed a "half-scattering" process. In other words, photoabsorption tests the linear response of matter to a well-defined disturbance.

2.2 BASIC THEORY

Let us consider how light produces transitions of an atom (or molecule) to states of its continuous spectrum. (Transitions to discrete states depend on a line profile and will be treated as transitions to a continuum with a peaked profile.) The cross section for such a transition is defined by

$$\sigma = P/I, \tag{2.1}$$

where P is the transition probability per unit time and

$$I = c\langle|\mathbf{E} \times \mathbf{H}|\rangle/4\pi\hbar\omega = c\langle|\mathbf{E}|^2 + |\mathbf{H}|^2\rangle/8\pi\hbar\omega \tag{2.2}$$

is the quantum-mechanical mean (or expectation) value of the incident light intensity expressed in photons per unit area per unit time. According to perturbation theory

$$P = 2\pi\hbar^{-1}|(f|U|i)|^2, \tag{2.3}$$

where U is the light–atom interaction and where the final state of the atom, included in $(f|$, is normalized per unit energy. [This normalization incorporates the density-of-states factor, which should otherwise be added to (2.3).]

We treat only dipole interactions, that is, those in which the radiation field is practically uniform over the target atom, setting

$$U = e\mathbf{E}\cdot\mathbf{r} = e\mathbf{E}\cdot\Sigma_j\mathbf{r}_j, \tag{2.4}$$

where $\mathbf{r}_j$ is the position of the jth atomic electron. The initial state i includes an initial (ground) state of the atom, $|0)$ (whose orientation quantum numbers may have to be averaged over) and N radiation photons; thus we set $|i) \equiv |0)|N)$. The final state f consists of $N - 1$ photons and of the atom in a state of energy ϵ above the ground state, i.e., $(f| \equiv (\epsilon|(N - 1|$; summation over degenerate $(\epsilon|$ is understood unless a particular final state is to be considered.

We sketch here the quantum-mechanical evaluation of the radiation-dependent factors of σ. Consider a plane wave of light with electric field

$$\mathbf{E} \equiv (V^{-1/2}E_0 \exp(ikz), 0, 0), \tag{2.5}$$

where V is a large quantization volume, and the amplitude E_0 is treated as an operator. Due to Maxwell's equations the corresponding magnetic field is

$$\mathbf{H} \equiv \left(0, V^{-1/2}\frac{i}{kc}\frac{dE_0}{dt}\exp(ikz), 0\right). \tag{2.6}$$

The field Hamiltonian integrated over the volume V,

$$V\frac{|\mathbf{E}|^2 + |\mathbf{H}|^2}{8\pi} = \frac{1}{2}\frac{1}{4\pi(kc)^2}\left(\frac{dE_0}{dt}\right)^2 + \frac{1}{2}\frac{1}{4\pi}E_0^2, \tag{2.7}$$

is equivalent to that of a harmonic oscillator with amplitude E_0, mass $1/[4\pi(kc)^2]$, force constant $1/(4\pi)$, and frequency $\omega = kc$ [1].

In the initial state, the eigenvalue of (2.7) is $N\hbar\omega$ above the state with no photon and the value (2.2) of I is cN/V. The matrix element in (2.3) factors into atomic and radiation parts. In the dipole approximation we set $\exp(ikz) = 1$ and have

$$\begin{aligned}|(f|U|i)|^2 &= e^2|(\epsilon|x|0)|^2\,|(N-1|E_0|N)|^2/V \\ &= e^2|(\epsilon|x|0)|^2\,2\pi\hbar\omega N/V,\end{aligned} \tag{2.8}$$

where $(N - 1|E_0|N)$ has been evaluated as the dipole matrix element for an oscillator of force constant $1/(4\pi)$. Substitution in (2.1) yields

$$\sigma = \frac{2\pi}{\hbar} e^2 |(\epsilon|x|0)|^2 \frac{2\pi\hbar\omega N/V}{cN/V}$$
$$= 4\pi^2 e^2 (\omega/c) |(\epsilon|x|0)|^2. \tag{2.9}$$

The cancellation of the photon density between field expectation values in the numerator and denominator makes it possible to obtain correct results by treating the field as a macroscopic variable.

For purposes of quantitative estimation of σ, consider that $|(\epsilon|x|0)|^2$, with $(\epsilon|$ normalized per unit energy, has the dimensions of area per unit energy and the product $\epsilon|(\epsilon|x|0)|^2 = \hbar\omega|(\epsilon|x|0)|^2$ may be no larger than the geometrical cross section of an atom ($\sim 10^{-16}$ cm^2 = 100 Mb). Hence, σ may be expected to be of order $\lesssim (4\pi^2/137)$ 100 Mb.

2.3 MACROSCOPIC ANALYSIS

2.3.1 Oscillator Strength

The photoabsorption formula (2.9) may be illustrated by relating it to the energy absorbed by a classical oscillator (with mass m, charge e, and displacement x) when it is driven by an electric field $A \cos \omega t$. The amplitude *and* the phase of the oscillator's response to the field are obtained together by solving its equation of motion in the form

$$m \frac{d^2x}{dt^2} = -m\omega_s^2 x - m\gamma_s \frac{dx}{dt} + eAe^{-i\omega t}. \tag{2.10}$$

Here the first term on the right represents the restoring force on the oscillator which sets its resonant frequency at ω_s; the second one represents a friction, and the last one represents an exponential component of the driving force. [Use of the complex exponential drive enables us to obtain the phase lag of the oscillation as a phase of the complex displacement $x(t)$.] The friction force represents energy dissipation to unspecified degrees of freedom; its quantum-mechanical interpretation will be discussed in Section 2.3.2.

The solution of (2.10) that vanishes for $A = 0$ is

$$x(t) = \frac{e}{m} \frac{1}{\omega_s^2 - \omega^2 - i\gamma_s\omega} Ae^{-i\omega t}. \tag{2.11}$$

The polarizability of the oscillator, defined as the ratio of the induced dipole moment $ex(t)$ to the driving field $Ae^{-i\omega t}$, is then

$$\alpha(\omega) = \frac{ex(t)}{Ae^{-i\omega t}} = \frac{e^2}{m} \frac{1}{\omega_s^2 - \omega^2 - i\gamma_s\omega}$$
$$\xrightarrow[\gamma_s \ll \omega_s]{} \frac{1}{2} \frac{e^2}{m\omega_s} \frac{1}{\omega_s - \omega - \frac{1}{2}i\gamma_s}. \tag{2.12}$$

Our goal here is to compare the power dissipated by the field, through its action on the oscillator with polarizability $\alpha(\omega)$, with the energy spent in photoabsorption per unit time, namely, $P\hbar\omega = \sigma I\hbar\omega$ according to (2.1), (2.2), and (2.9). The average power dissipation is

$$T^{-1}\int_0^T dt\, e\left(\frac{dx}{dt}\right)A\cos\omega t = T^{-1}\int_0^T dt\,\frac{1}{2}(i\omega\alpha^* Ae^{i\omega t} - i\omega\alpha Ae^{-i\omega t})A\cos\omega t$$
$$\xrightarrow[T\to\infty]{} \frac{1}{2}\omega \operatorname{Im}\alpha(\omega)A^2$$
$$\xrightarrow[\gamma_s \ll \omega_s]{} \frac{1}{4}\frac{e^2}{m}\frac{\frac{1}{2}\gamma_s}{(\omega_s - \omega)^2 + \frac{1}{4}\gamma_s^2}A^2. \tag{2.13}$$

Expression (2.13) represents the familiar result that the absorption spectrum of the oscillator spreads over a line with approximately Lorentzian profile and width equal to the friction coefficient γ_s (on a frequency scale). Since the integral over this profile $\int_0^\infty d\omega \frac{1}{2}\gamma_s/[(\omega_s - \omega)^2 + \frac{1}{4}\gamma_s^2]$ equals π, we introduce the parameter

$$\frac{df}{d\omega} = \frac{1}{\pi}\frac{\frac{1}{2}\gamma_s}{(\omega_s - \omega)^2 + \frac{1}{4}\gamma_s^2}, \tag{2.14}$$

with unit integral, to characterize the oscillator strength, that is, the spectral distribution of the driven oscillator. The power dissipation (2.13) is expressed in terms of this parameter by

$$P = \frac{\pi}{4}\frac{e^2}{m}A^2\frac{df}{d\omega}. \tag{2.15}$$

The photoabsorption by any atomic system will be characterized by an equivalent oscillator strength parameter. The definition of the oscillator strength in (2.14) is also supported by considering the power absorbed by a free electron considered as an oscillator; its expression coincides with the limit of (2.13) at high ω.

To establish the equivalence of (2.15) and (2.9) we note that if the driving field $A \cos \omega t$ belongs to an electromagnetic wave represented by $A \cos \omega(t - z/c)$, the energy flux of this wave (equivalent to $I\hbar\omega$) amounts to $cA^2/8\pi$. The power dissipation in photoabsorption is accordingly represented by multiplying (2.9) by this energy flux to get

$$P\hbar\omega = \sigma I\hbar\omega = 4\pi^2 e^2 \frac{\omega}{c} |(\epsilon|x|0)|^2 \frac{c}{8\pi} A^2 = \frac{\pi}{4} \frac{e^2\hbar}{m} A^2 \frac{df}{d\epsilon}, \tag{2.16}$$

which is in agreement with (2.15), save that the oscillator strength distribution is expressed per unit energy ($\epsilon = \hbar\omega$) rather than per unit ω, and is defined as

$$\frac{df}{d\epsilon} = \frac{2m\epsilon}{\hbar^2} |(\epsilon|x|0)|^2. \tag{2.17}$$

The cross section (2.9) is then expressed in terms of the oscillator strength by

$$\sigma = 2\pi^2 e^2 \frac{\hbar}{mc} \frac{df}{d\epsilon} = 1.098 \times 10^{-16}\ \text{cm}^2\ \text{eV} \frac{df}{d\epsilon}. \tag{2.18}$$

We may express σ in atomic units, $(0.53)^2 \times 10^{-16}$ cm^2, and the spectral density of oscillator strength per atomic energy unit, that is, per 27.2 eV; Eq. (2.18) then reduces to $\sigma = (2\pi^2/137)\, df/d\epsilon$.

2.3.2 Quantum-Mechanical Calculation

It is instructive to develop the quantum-mechanical counterpart of the classical derivation of the polarizability and oscillator strength by perturbation theory. Consider the perturbation of a state ψ_0 with energy E_0 by the driving field $A(e^{i\omega t} + e^{-i\omega t})$. The perturbed state is represented as a superposition of the complete set of energy eigenstates of the atom as

$$\psi_0 e^{-iE_0 t/\hbar} + \sum_s \psi_s c_s(t) e^{-iE_s t/\hbar}. \tag{2.19}$$

The time-dependent Schrödinger equation with the full Hamiltonian of atom plus electromagnetic field, $H = H_0 - exA(e^{i\omega t} + e^{-i\omega t})$, translates into a first-order differential equation for the coefficients $c_s(t)$ with solutions

$$c_s(t) = (e/\hbar)(s|x|0)A\left[\frac{\exp[i(\omega + \omega_s - \frac{1}{2}i\gamma_s)t]}{\omega_s + \omega - \frac{1}{2}i\gamma_s} + \frac{\exp[-i(\omega - \omega_s + \frac{1}{2}i\gamma_s)t]}{\omega_s - \omega - \frac{1}{2}i\gamma_s}\right], \tag{2.20}$$

where we have written $E_s - E_0 = \hbar(\omega_s - \frac{1}{2}i\gamma_s)$.†

The induced dipole moment due to the mixing into ψ_0 of the states ψ_s is given by

$$\begin{aligned} e\langle x\rangle &= e\sum_s[(0|x|s)c_s\exp[-i(\omega_s - \tfrac{1}{2}i\gamma_s)t] + c_s^\dagger\exp[i(\omega_s + \tfrac{1}{2}i\gamma_s)t](s|x|0)] \\ &= \frac{e^2}{\hbar}\sum_s(0|x|s)\left[\frac{Ae^{i\omega t}}{\omega_s^2 - \omega^2 + i\gamma_s\omega} + \frac{Ae^{-i\omega t}}{\omega_s^2 - \omega^2 - i\gamma_s\omega}\right](s|x|0). \end{aligned} \tag{2.21}$$

[Note how pairs of the resonance denominators of (2.20) have combined into the quadratic expressions of (2.21).] A quantum-mechanical definition of polarizability emerges by identifying the quantum dipole moment (2.21) as $A[\alpha(\omega)e^{-i\omega t} + \alpha^* e^{i\omega t}]$ and comparing the result with the classical expression (2.11) and (2.12). The quantum expression of the polarizability with the same structure as (2.12) is

$$\alpha(\omega) = \frac{e^2}{m}\sum_s \frac{f_s}{\omega_s^2 - \omega^2 - i\gamma_s\omega}. \tag{2.22}$$

However, the unit coefficient in the numerator of (2.12), pertinent to the treatment of *one* oscillator, is now replaced by the numerical parameter

$$f_s = (2m\omega_s/\hbar)|(s|x|0)|^2, \tag{2.23}$$

which represents the *fraction* of the atomic electrons that participate in the mixing of ψ_0 and ψ_s and is accordingly called the *oscillator strength* of the $\psi_0 \leftrightarrow \psi_s$ transition. This parceling of an integer classical quantity into a

† The introduction of an imaginary part in the energy raises even more questions in the quantum description than the similar *ad hoc* introduction of a damping constant in the classical equation [Eq. (2.10)]. How are we to fit in a frictional force at a microscopic level, especially with a Hermitian Hamiltonian? Its origin actually lies in the coupling of the system with its environment whose enormous number of degrees of freedom may be ignored as of no detailed relevance. As will be seen later, such complex eigenvalues and non-Hermitian behavior result from the elimination of ignorable degrees of freedom that, nevertheless, act as sinks to drain away some of the excitation; information concerning the excitation of the environment flows out of the part of the system that is of interest *without ever returning to it*. This loss of information is represented by a boundary condition that leads to the introduction of the imaginary term. Such a description of friction and of information sinks with a continuous and degenerate spectrum is a fundamental point of physics. Its appearance in the context of collision physics for systems with continuous spectra is dealt with in Section 8.6.

spectral distribution is typical of the passage from classical mechanics to quantum mechanics. The distribution of classical polarizability over the width of the resonance has been represented in Section 2.3.1 by the distribution (2.14) of oscillator strength over differential intervals $\omega \to \omega + d\omega$ with unit integral. The corresponding quantum distribution consists of the product of the expressions (2.14) and (2.23)

$$\left(\frac{df}{d\omega}\right)_{\omega \sim \omega_s} = \frac{1}{\pi} \frac{\frac{1}{2}\gamma_s f_s}{(\omega_s - \omega)^2 + \frac{1}{4}\gamma_s^2}. \tag{2.23'}$$

2.4 GENERAL PROPERTIES OF THE SPECTRUM

The study of the spectral density of oscillator strength of a system, $df/d\epsilon$, is equivalent to the study of its photoabsorption cross section σ since these two quantities are proportional to one another according to (2.18) with a ratio consisting of standard parameters. We shall consider $df/d\epsilon$ because its properties have more general significance and more direct interpretation. For instance, as we shall see, $df/d\epsilon$ puts the discrete and continuous spectrum on a common footing. This study, aimed at providing information on atomic mechanisms, forms the subject of a review [2], of which we abstract important aspects. This review will be referred to as FC throughout this chapter.

The remainder of this chapter describes general aspects of the spectrum of $df/d\epsilon$ for purposes of orientation and qualitative information. Aspects that have a more direct bearing on the mechanics of collisions and spectra will be taken up in the relevant chapters.

Recall here that (1) $df/d\epsilon$ is defined by (2.17) in terms of matrix elements of a dipole operator for an atomic system, while (2) the concept of oscillator strength was introduced through (2.13) and (2.14) to characterize the spectrum of Im $\alpha(\omega)$, the absorptive part of the polarizability response of a classical harmonic oscillator driven by an external field. The third approach, (3), calculates the polarizability $\alpha(\omega)$ of an atomic system subjected to an external field $A \cos \omega t$ by quantum-mechanical perturbation theory; the result [(2.23)] leads to the same relation Im $\alpha(\omega) = \pi(e^2/m\omega)\, df/d\omega$ that holds for a classical oscillator, but $\hbar^{-1}\, df/d\omega$ is defined here by (2.17).

Knowledge of the oscillator strength distribution of a system over the entire spectrum amounts to knowledge of the system's response to a uniform weak perturbation of its electrons with any time dependence. For example, the passage of a fast charged particle outside an atom, to be discussed in Chapter 3, exerts a force of short duration on all atomic electrons; this disturbance of atomic equilibrium gives rise to intra-atomic currents whose variations in the course of time represent the Fourier transform of the oscillator strength distribution (see Section 2.3 of FC).

The Fourier transform relationship between the spectrum of oscillator strength of a system and its response to short disturbances manifests itself through numerous integral properties of the spectrum called sum rules (see Section 2.5 of FC). The most famous of these, the Thomas–Reiche–Kuhn rule, relates to our having so defined the oscillator strength distribution (2.14) of a classical oscillator that its integral equals unity; quite generally the integral over the whole spectrum of the oscillator strength (2.17) of an atomic system equals the number of electrons in the system†

$$\int_0^\infty d\epsilon \frac{df}{d\epsilon} = N. \tag{2.24}$$

The weighted integrals over the oscillator strength distribution,

$$S_r = \int_0^\infty d\epsilon\, \epsilon^r \frac{df}{d\epsilon}, \tag{2.25}$$

with various integer values of r of either sign, are related to geometrical, mechanical, or optical properties of an atomic system (for example, its size, its electron kinetic energy, and its rotatory power in the Faraday effect).

These relations imply that mechanical properties of an atom manifest themselves in particular features of the absorption spectrum. For example, it turns out that the integral in (2.25) diverges for $r \geq 3$ while for $r = 2$ it equals the electron density at the atomic nucleus multiplied by a standard coefficient. This implies that $df/d\epsilon$ decreases as ϵ^{-p} when $\epsilon \to \infty$ with $3 < p \leq 4$; a determination of $\epsilon^p(df/d\epsilon)$ would thus measure the electron density at the nucleus. Classical mechanics interprets this result by remarking than an electron absorbs radiation of high frequency ($\omega = \epsilon/\hbar$) only insofar as it is subject to a force of strength sufficient to make it resonate with that frequency. Strong forces occur only near the nucleus; hence the larger ω is, the smaller will be the portion of an atom that contributes to photoabsorption. Wave mechanics interprets the same result by remarking that the wave function of the final state $(\epsilon|$ in (2.17) has shorter and shorter wavelength as ϵ increases. This rapid oscillation tends to cancel the contributions of most of the atomic volume to the integral $(\epsilon|x|0)$ in (2.17), except for the portion very close to the nucleus where the wavelength of the ground state $|0)$ is itself very short.

Relativistic effects actually reduce the exponent of ϵ^{-p} to $p = \frac{1}{2}$ for very

† The formal proof of this formula (see Section 62 of Bethe and Salpeter [3]) rests only on the Heisenberg commutation law $[p, q] = \hbar/i$; in fact, however, Heisenberg arrived at this law in reverse, by postulating that atomic mechanics must satisfy (2.24).

large ϵ and even cause S_0 to diverge, but they are not to be interpreted within our oscillator point of view, which rests on the nonrelativistic equation (2.10); we disregard these effects.

2.5 DISCRETE SPECTRUM

The spectral distribution of oscillator strength concentrates into discrete lines in the range of energies ϵ below the ionization threshold of the atom or molecule under consideration. The profile of each line coincides with the profile (2.14) of an oscillator (in the absence of Doppler broadening). On the other hand, the integral of the oscillator strength over the line width does not amount to unity but to a number that is indicated by f_s and is given, in analogy to (2.17), by (2.23). The aggregate oscillator strength distribution over the discrete spectrum is thus represented by

$$\frac{df}{d\epsilon} = \sum_s f_s \frac{\frac{1}{2}\gamma_s \hbar}{(\epsilon_s - \epsilon)^2 + \frac{1}{4}\gamma_s^2 \hbar^2} \frac{1}{\pi}. \tag{2.26}$$

The study of the variations of f_s from line to line and of their connection with the oscillator strength $df/d\epsilon$ in the continuum will be developed stepwise in this book. Here we only anticipate the application of a procedure developed in Chapter 5 and applicable to simple Rydberg series of lines such as occur in the alkali absorption spectra. Figure 2.1 shows oscillator strength distributions for H and Li. The abscissa represents the excess of the excitation energy over the ionization threshold, $E = \epsilon - I$. The ordinate represents the oscillator strength density for $\epsilon \geq 0$. For $\epsilon < 0$ the values of f_s are represented by histogram blocks that make them comparable to $df/d\epsilon$ through the following device. The energy $\epsilon_s - I$ of each level is first plotted on the lower part of the figure against the index s; the slope of the plot (dE/ds) is then measured for each value of s. The base of each histogram block is taken equal to dE/ds and its height equal to $f_s(ds/dE)$. Successive intersections of the tangents drawn to the lower curve at different s determine the precise location of each histogram block. An algebraic equivalent to this geometrical prescription for the histogram plot is that the left edge of the first block must be at $-\sum_{s=2}^{\infty} (dE/ds)$. In H, this sum is $1 - \zeta(3)$ in atomic units or -5.5 eV. The df/dE curve for $E > 0$ extrapolates through the histogram smoothly for H; also for Li except for a major irregularity at $s = 2$ that is discussed further on.

Photoabsorption by an electron may leave the rest of an atom or molecule in an excited state. (This occurs, for example, when the electron was initially in an inner shell.) The ionization threshold is raised accordingly by the excitation energy of the residue. Discrete lines then occur in the spectrum

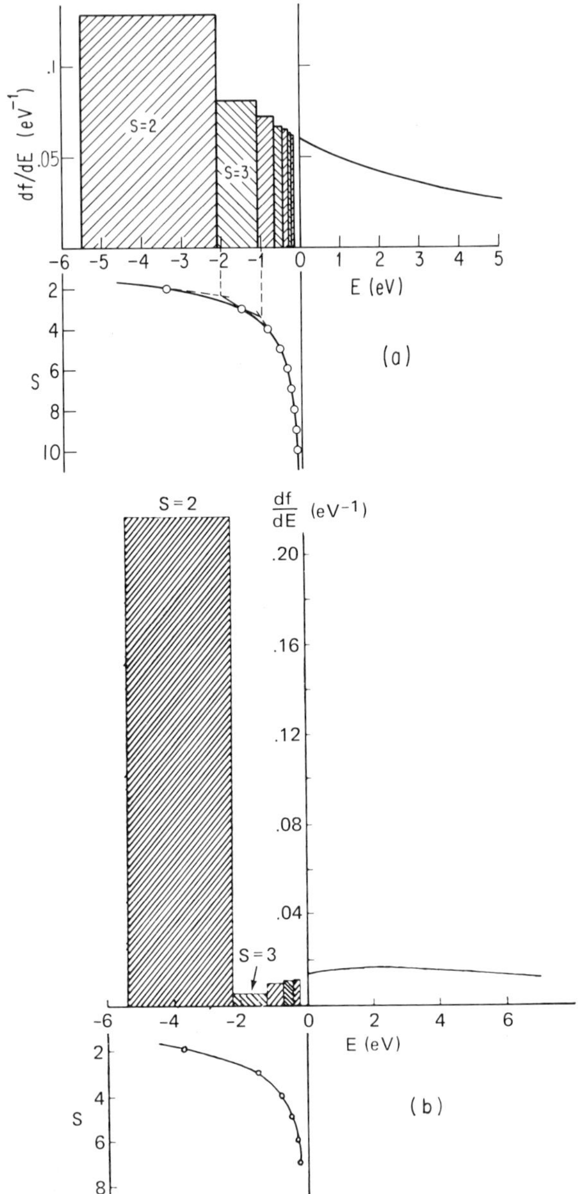

Fig. 2.1. Oscillator strength distribution in the discrete and the continuous spectra of (a) H (theory) and (b) Li (experiment). Oscillator strengths for discrete transitions are plotted as histogram blocks of height $f_s/\Delta E_s$, where the base of each block is determined by the geometrical construction of tangents to the E_s versus s curve as shown in the lower part of (a). Note the continuity of the plot across the ionization threshold. (From Fano and Cooper [2].)

below such higher ionization thresholds, as shown in Fig. 2.2. The excited energy levels ϵ_s corresponding to these lines would be stable in an independent particle approximation; in actual fact, the excitation generally goes over to another channel with lower ionization threshold, thus resulting in a process of *autoionization* with electron ejection. This decay of the excitation increases the width $\hbar\gamma_s$ of the absorption line, generally by many orders of magnitude, that is, it manifests itself as additional "friction" of the driven oscillation. The accessibility of these observations and their direct bearing on interchannel processes have contributed much evidence to the study of atomic phenomena.

2.6 PARTIAL OSCILLATOR STRENGTHS

We have seen how absorption of a photon of sufficient energy $\epsilon = \hbar\omega$ can excite an atom or molecule into different channels with different allotment of the energy between an electron and the rest of the system. The oscillator strength corresponding to the total absorption cross section can accordingly be regarded as the sum of components pertaining to alternative channel states with equal energy, $(\epsilon 1|$, $(\epsilon 2|, \ldots$. However, the measurement of such partial oscillator strengths requires separate measurements of the probability of transition to each channel.

A straightforward method of achieving this measurement consists of

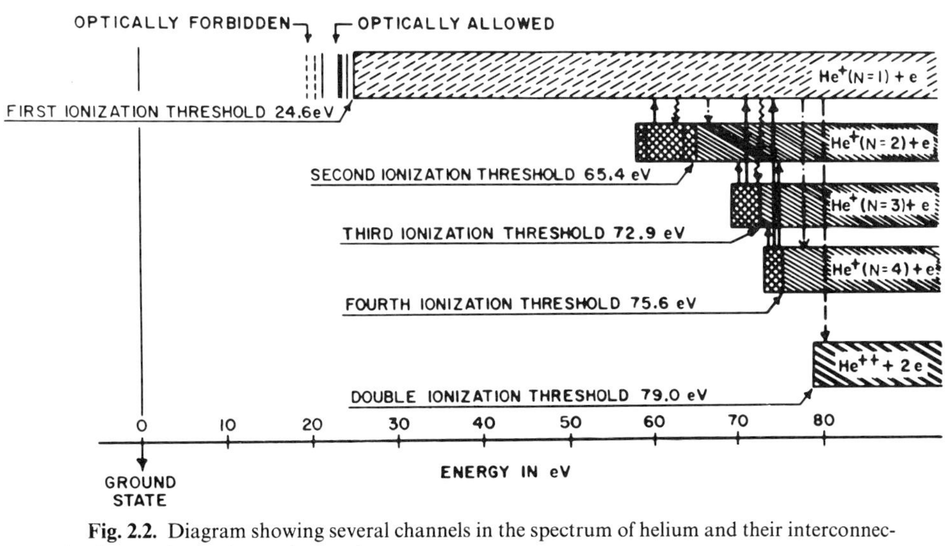

Fig. 2.2. Diagram showing several channels in the spectrum of helium and their interconnection by autoionization and related processes.

observing the energy spectrum of photoelectrons. Detection of photoelectrons with kinetic energies E_1, $E_2, \ldots$ implies that their ejection required threshold energies $I_1 = \epsilon - E_1, I_2 = \epsilon - E_2, \ldots$. The intensity of the electron groups with these respective energies measures the partial strengths $df_1/d\epsilon$, $df_2/d\epsilon, \ldots$.

The threshold energies I_1, $I_2, \ldots$ may be interpreted as the binding energies of different atomic electrons before photoabsorption, but only within the independent electron approximation and the validity of Koopman's theorem. Actually, it is the differences of these energies, $I_2 - I_1, I_3 - I_1, \ldots$ that have a precise meaning. They represent excitation energy levels of the residue left behind by the photoelectron. This method of spectral analysis, called electron spectroscopy for chemical applications (ESCA) has been utilized widely, especially to measure higher levels of molecular spectra that are less accessible to other methods; the freedom in choosing the photoabsorption energy ϵ is also helpful.

2.7 SPECIFIC FEATURES

Contributions to the spectrum from excitations that remove electrons from different shells or subshells of an atom (or molecule) should be considered separately for purposes of analysis, especially because they generally center in different regions of the spectrum. The sum rules on the oscillator strength distribution, discussed earlier and in Section 2.6 of FC, do *not* apply correctly to each separate component of this distribution (see Section 4.9 of FC), but their separate application does nevertheless provide order-of-magnitude estimates. Thus, the integrated strength of the component corresponding to excitation of a given shell will be comparable to the number of electrons in that shell, and the mean value of the excitation energy ϵ for a shell will be comparable to the mean kinetic energy of one of its bound electrons. These aspects are illustrated, for example by comparing the oscillator strength distributions of the two 10-electron gases CH_4 and Ne in Fig. 2.3. Both the valence and the K shells of Ne are more compact than for CH_4, and their electrons are faster; accordingly, the spectral contributions of Ne center at higher energy than for CH_4.

The spectrum of each component reflects two distinguishable sets of dynamical factors arising from the photoabsorption proper, which takes place within the shell where an electron was initially confined, and the escape of the excited electron away from that shell's volume. The dynamics of the photoabsorption proper is represented by the overlap in space of the initial and final state wave functions, ψ_0 and ψ_ϵ, in the matrix element $(\epsilon|x|0)$. The excited electron's escape is represented instead by the wave function ψ_ϵ of the final state $(\epsilon|$ at larger radial distances, at which many factors affect the

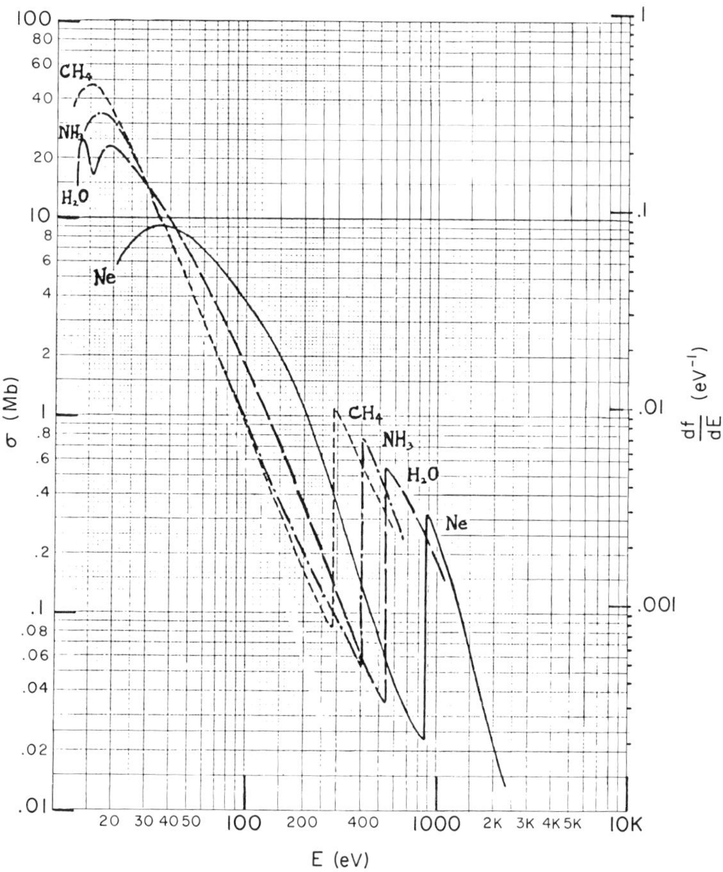

Fig. 2.3. Comparison of photoabsorption cross sections for different ten-electron gases; the scale on the right gives the oscillator strengths. (Courtesy D. A. Harmin; data from Berkowitz [4].)

function ψ_ϵ while the initial state's ψ_0 is vanishingly small. The behavior of ψ_ϵ at larger r affects the value of $(\epsilon|x|0)$ through a single parameter, namely, the normalization of ψ_ϵ in the region of its overlap with ψ_0. Since this normalization is established for $r \to \infty$, the effects of escape on the oscillator strength are actually represented by the single parameter

$$|\psi_\epsilon(r \sim 0)/\psi_\epsilon(r \sim \infty)|^2. \tag{2.27}$$

The study of this ratio is a major subject of later chapters.

The photoabsorption proper can be treated approximately in the independent electron approximation, as detailed in Section 4.2ff. of FC. In the evaluation of $(\epsilon|x|0)$, the angular part of the integration proceeds analytically for single atoms and yields the selection rule on the orbital quantum number of the excited electron, $l_\epsilon = l_0 \pm 1$. A special sum rule [3] also shows that the oscillator strength for the channel with $l_\epsilon = l_0 + 1$ is usually considerably larger than the one for $l_\epsilon = l_0 - 1$. This can be understood in dynamical terms as saying that an increase in the electron's energy tends to be accompanied by an increase of orbital momentum contributed by the radiation field that drives the electrons. The radial integral

$$\int_0^\infty dr\, P_\epsilon(r) r P_0(r) \tag{2.28}$$

must instead generally be evaluated numerically. Its dependence on l_ϵ and especially on the energy ϵ is not readily anticipated, since both P_0 and P_ϵ are oscillatory functions of r and the wavelength of P_ϵ varies with ϵ.

Yet an important regularity, whose physical interpretation remains elusive, emerges from the analytical evaluation of (2.28) for the H atom and persists in modified form for other atoms. The modulus of (2.28) for H neither vanishes nor goes through a minimum as ϵ increases. For other atoms this rule holds only when $P_0(r)$ is nodeless; otherwise (2.28) generally starts with a large absolute value at low ϵ but then vanishes reversing its sign at a single value of ϵ, called a "Cooper zero," not much larger than the ionization threshold. (Only for Li and Be does this "zero" lie at $\epsilon < I$, as evidenced by the minimum strength shown at $s = 3$ in Fig. 2.1b.) This phenomenon is described and discussed extensively in Sections 4.5 and 4.6 of FC and is illustrated by the occurrence of successive maxima in the M-shell component of the Ar spectrum in Fig. 2.4. (Only a "Cooper minimum" appears here, in

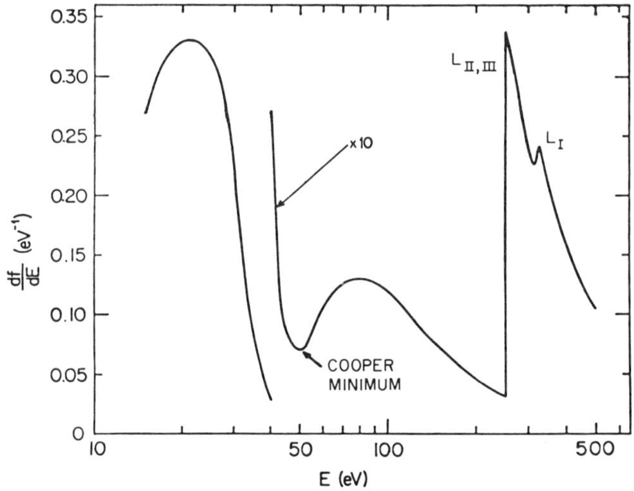

Fig. 2.4. Oscillator strength distribution of argon. Note the change of scale, the Cooper minimum, and L edges.

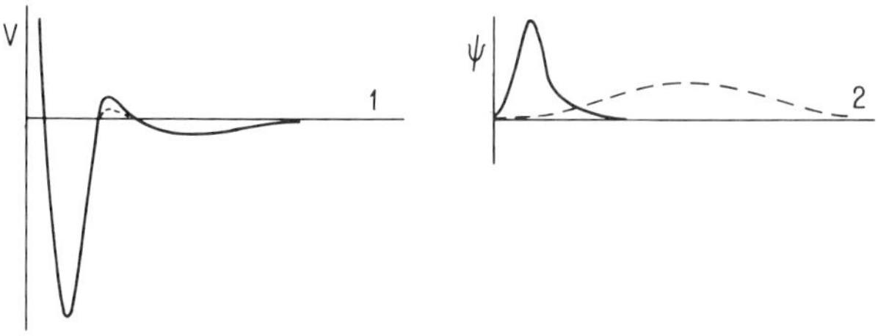

Fig. 2.5. Two-valley potential with an intermediate barrier for $l = 3$ electrons generated by the combination of atomic and centrifugal potentials and its influence on wave functions near zero energy. Note how a small change in the potentials shown as full and dashed lines is magnified in the extreme in the resulting wave functions. (Adapted from Connerade [5].)

place of a zero, because the spectrum consists of unresolved components with zeros at different ϵ.) A few subshells yield spectra with two Cooper zeros.

The passage of electrons across the valence shell of atoms is hindered greatly by centrifugal potential barriers (Fig. 2.5) for low values of the escape energy $\epsilon - I$ and for $l_\epsilon \geq 2$, as detailed in Sections 4.4, 4.7, and 4.8 of FC. This effect results in low values of the ratio (2.27) and manifests itself in depressed values of the oscillator strength spectrum near the relevant thresholds, as shown in Fig. 2.3 for the L-shell threshold of Ne and in Fig. 2.4 for the M threshold of Ar [2]. (This effect is obscured by the logarithmic scale of energies for the L shell of Ar, but is suppressed in the CH_4 curve of Fig. 2.3 because the centrifugal potential is greatly reduced at the large radius of this molecule.) Electron escape from a molecule may be hindered further by an outer layer of electronegative atoms, to the extent of forming discrete levels above the ionization threshold (Fig. 2.6). Beyond these effects that can be

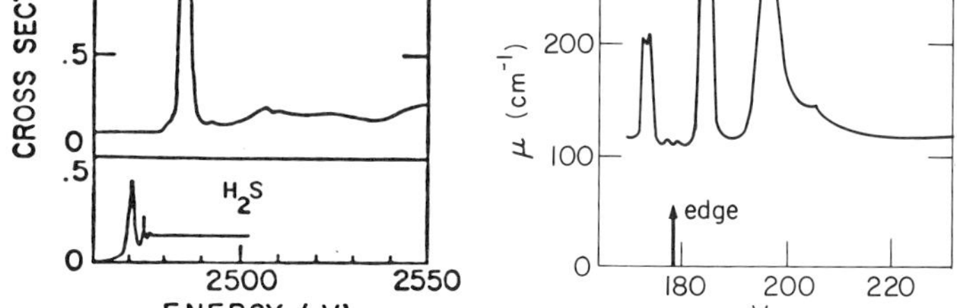

Fig. 2.6. (a) Comparison of H_2S and SF_6 absorption spectra near the K edge of sulfur. (b) Absorption spectrum of SF_6 near the $L_{II,III}$ edge of sulfur. (From LaVilla and Deslattes [6].)

represented in terms of a real potential affecting the excited electron, the spectra often exhibit extensive evidence of energy exchanges between the excited electron and other particles, which is considered in later chapters.

Problems

2.1 Consider radiation from a distant vertical antenna emitting 1 μW of radiation at 100 MHz $\pm$ 1 kHz at a distance of 300 km from the observer. Calculate the photon flux and compare it with the flux of blackbody photons in that frequency range due to the cosmic 3 K microwave background.
2.2 Starting with the definition of moments of the oscillator strength distribution (2.25), prove that $S_{-1} = 2a_0/e^2$ for hydrogen.
2.3 Express S_{-2} in terms of matrix elements and an energy denominator. Observing that this denominator is smallest for the $n = 2$ level, obtain an upper bound on the value of S_{-2} for hydrogen.
2.4 (a) Use the Thomas–Reiche–Kuhn sum rule (2.24) to derive a lower bound on S_1 for hydrogen.

(b) Evaluate S_1 by relating it to the binding energy of the hydrogen atom, using the virial theorem. Compare this result with the one obtained in (a). [Hint: Transform the expression for the oscillator strength in (2.23) by means of the commutation relation between p and x into one in terms of matrix elements of p.]

(c) Under what assumptions will the preceding relation between S_1 and the binding energy remain valid for a many-electron system?
2.5 *Dipole matrix element of a hydrogenic transition* Calculate the matrix element $(210|z|100)$ of the z coordinate of an electron between the ground state ($n = 1, l = m = 0$) and the excited state ($n = 2, l = 1, m = 0$) of a hydrogen atom. The wave functions of the two states are

$$\psi_{100} = \pi^{-1/2} a_0^{-3/2} \exp(-r/a_0),$$
$$\psi_{210} = \pi^{-1/2} (2a_0)^{-5/2} \exp(-r/2a_0) r \cos\theta,$$

where a_0 is the Bohr radius. Note that $\int_0^\infty x^n \exp(-x)\,dx = n!$
2.6 *Connection of discrete and continuous spectra* The dipole matrix elements for H atom transitions 1s $\to$ np and 1s $\to$ Ep of the discrete and continuum are given by Eqs. (63.4) and (71.4) of Bethe and Salpeter [3] as

$$(n\text{p}|z|1\text{s})^2 = \frac{2^8}{3} \frac{n^7 (n-1)^{2n-5}}{(n+1)^{2n+5}} a_0^2, \qquad E_n = -\frac{e^2}{2a_0 n^2}$$

$$(E\text{p}|z|1\text{s})^2 = \frac{2^8}{3} \left(\frac{n'^2}{1+n'^2} \right)^5 \frac{\exp(-4n' \operatorname{arc\,cot} n')}{1 - \exp(-2\pi n') St(E)} \frac{a_0^3}{e^2}, \qquad E = \frac{e^2}{2a_0 n'^2}$$

where $St(E) = 1$ for $E \geq 0$, $St(E) = 0$ for $E < 0$.

Using these formulas verify the law

$$(n\text{p}|z|1\text{s})^2 = (E\text{p}|z|1\text{s})^2_{E=E_n} \frac{dE_n}{dn}$$

with $(Ep|z|1s)^2$ extrapolated to negative values of E. (Recall the definition of the exponential function as the limit of an algebraic expression for the large integers n.)
2.7 Using the algebraic procedure mentioned in Section 2.5 for laying out the histogram blocks of Fig. 2.1a, verify that their locations are indeed as shown.

References

1. U. Fano and L. Fano, "Physics of Atoms and Molecules," Section 12.5. Univ. of Chicago Press, Chicago, Illinois, 1972.
2. U. Fano and J. W. Cooper, *Rev. Mod. Phys.* **40**, 441 (1968). This work will be referred to as FC. A more recent review with somewhat the same coverage is A. F. Starace, "Handbuch der Physik," Vol. 31 (S. Flügge and W. Mehlhorn, eds.). Springer-Verlag, Berlin and New York, 1982.
3. H. A. Bethe and E. E. Salpeter, "Quantum Mechanics of One- and Two-Electron Atoms," Eqs. 61.4 and 61.5 and Section 62.8. Springer-Verlag, Berlin and New York, 1957.
4. J. Berkowitz, "Photoabsorption, Photoionization, and Photoelectron Spectroscopy." Academic Press, New York, 1979.
5. J. P. Connerade, *J. Phys. B.* **15**, L367 (1982).
6. R. E. LaVilla and R. D. Deslattes, *J. Chem. Phys.* **44**, 4399 (1966).

3 CHARGED-PARTICLE COLLISIONS IN THE BORN APPROXIMATION

3.1 INTRODUCTION

The interaction between an incident particle and an atomic target is treated as a weak perturbation in the Born approximation. The validity of this approximation for an incident charged particle does not hinge on the strength of the interaction being weak but on the combination of this strength with the high speed of the particle, which limits the momentum transferred in the collision. (The special and infrequent case of collisions with high momentum transfer will be discussed separately.)

The criterion used for the applicability of the Born approximation is that the particle velocity be fast with respect to the internal motion of the atomic electron. Collisions with momentum transfers much smaller than the momenta of intra-atomic motions correspond in a classical picture to incident particle trajectories passing far outside the target, with "impact parameters" ranging up to ~ 100 nm when the particle speed approaches the light velocity. The treatment of fast collisions in the Born approximation was first developed by Bethe in 1930, but the key physical considerations and their quantitative effects had been spelled out by Bohr as early as 1913. The results are important in radiological physics, for collision cross sections and the slowing down of fast particles. They are also of interest in atmospheric physics, astrophysics, plasma physics, and electron microscopy.

3.2 THE BORN APPROXIMATION

In the Born approximation we do not regard the incident particle as forming a temporary "complex" with the target. On the contrary, the crucial simplification is that the particle remains an external agent separate from the target electrons throughout the course of the interaction. As a result, we represent the state of the system with reference to independent stationary states of a free particle and of the target system. Exchange of an incident electron with target electrons will accordingly be disregarded here as a very unlikely event, with an exception to be mentioned in Section 3.5.2.

Thus, we designate the initial state of particle + target as $|i) \equiv |\mathbf{p}_i)|n_i)$ meaning that (a) its wave function is simply the product of the incident particle's $\exp(i\mathbf{p}_i \cdot \mathbf{r}/\hbar)$ and an initial state wave function ψ_i of the target and (b) its energy is the sum of the separate energies, $E_i = p_i^2/2M + \epsilon_i$, where M is the reduced mass of particle and target. The complete state to be constructed will be a superposition of the initial state and other similar states $|j) \equiv |\mathbf{p}_j)|n_j)$, where n_j indicates another state of the target that may be excited or may also coincide with $|n_i)$. The complete state is represented by

$$\sum_j |j)a_{ji}. \tag{3.1}$$

The corresponding representation of the Hamiltonian for the combined system particle + target is

$$H = T + H_0 + V, \tag{3.2}$$

where T is the kinetic energy of the particle, H_0 the Hamiltonian of the target, and V the particle–target interaction. States $|j) \equiv |\mathbf{p}_j)|n_j)$ are eigenstates of $T + H_0$ with eigenvalue E_j. The Schrödinger equation for an eigenstate of (3.2) with the form (3.1) and with eigenvalue E_i equal to the unperturbed energy of the initial state is

$$H \sum_j |j)a_{ji} = \sum_j (E_j + V)|j)a_{ji} = \sum_j |j)a_{ji} E_i. \tag{3.3}$$

We reduce this equation to an algebraic system by projecting it on a final state $(f|$. Since $(f|j) = \delta_{fj}$, this operation yields

$$E_f a_{fi} + \sum_j (f|V|j)a_{ji} = a_{fi} E_i, \tag{3.3$'$}$$

where the symbol $(f|V|i)$ represents a multiple integral over the particle's position and all the atomic variables τ, namely,

$$(f|V|i) = \int d\mathbf{r} \int d\tau \exp(-i\mathbf{p}_f \cdot \mathbf{r}/\hbar)\psi_f^*(\tau)V(\mathbf{r}, \tau)\psi_i(\tau) \exp(i\mathbf{p}_i \cdot \mathbf{r}/\hbar). \tag{3.3$''$}$$

Equation (3.3) involves no approximation yet. The approximation is introduced by seeking a solution of the form

$$a_{fi} = \delta_{fi} + a_{fi}^{(1)} + \cdots, \tag{3.4}$$

where $a_{fi}^{(1)}$ is a small number. The product of the two small numbers $a_{fi}^{(1)}$ and

$(\mathrm{f}|V|j)$ is to be disregarded. Equation (3.3) then reduces to

$$(E_{\mathrm{i}} - E_{\mathrm{f}})a_{\mathrm{fi}}^{(1)} = (\mathrm{f}|V|\mathrm{i}). \tag{3.5}$$

Here it is essential to consider that $E_{\mathrm{i}} - E_{\mathrm{f}}$ may vanish; in fact, this difference does vanish just when energy is conserved, i.e., where $E_{\mathrm{f}} = E_{\mathrm{i}}$. Dividing (3.5) by $E_{\mathrm{i}} - E_{\mathrm{f}}$ is accordingly nontrivial. The general solution of (3.5) is nevertheless expressed by a mathematical notation due to Dirac

$$a_{\mathrm{fi}}^{(1)} = \left[P\frac{1}{E_{\mathrm{i}} - E_{\mathrm{f}}} + b\delta(E_{\mathrm{f}} - E_{\mathrm{i}})\right](\mathrm{f}|V|\mathrm{i}), \tag{3.6}$$

where b is an integration constant, $\delta(E_{\mathrm{f}} - E_{\mathrm{i}})$ is the Dirac delta function, and the symbol P specifies that principal part integration should be taken over the singularity of $1/(E_{\mathrm{i}} - E_{\mathrm{f}})$ when summing over alternative final states $|\mathrm{f})$. The value of b will be set in the next section by enforcing the physically appropriate boundary conditions on (3.1).

3.3 CONSTRUCTION OF THE WAVE FUNCTION

The connection between the value of b and the characteristics of an energy eigenstate emerges from the construction of the state's wave function. This construction is carried out by writing the wave functions of the final states $|f)$ as $(2\pi\hbar)^{-3/2}\exp(i\mathbf{p}_{\mathrm{f}}\cdot\mathbf{r}/\hbar)\psi_{\mathrm{f}}$. The normalization factor $(2\pi\hbar)^{-3/2}$ ensures that a $\sum_{\mathrm{f}}$ will include an integration over $\mathbf{p}_{\mathrm{f}}$ without further coefficients. Here we emphasize the essential aspects of the treatment; further details are found in textbooks [1, 2].

In writing out the complete final state wave function (3.1), the integration over the term δ_{fi} of (3.4) gives simply $(2\pi\hbar)^{-3/2}\exp(i\mathbf{p}_{\mathrm{i}}\cdot\mathbf{r}/\hbar)\psi_{\mathrm{i}}$. In the integration over $a_{\mathrm{fi}}^{(1)}$ we set $d\mathbf{p}_{\mathrm{f}} = p_{\mathrm{f}}^2 dp_{\mathrm{f}} d\phi\, d\cos\theta$ in polar coordinates with axis parallel to the position vector $\mathbf{r}$. The wave function (3.1) has the explicit form

$$\begin{aligned}
&(2\pi\hbar)^{-3/2}\left[\exp(i\mathbf{p}_{\mathrm{i}}\cdot\mathbf{r}/\hbar)\psi_{\mathrm{i}} + \sum_{\mathrm{f}}\exp(i\mathbf{p}_{\mathrm{f}}\cdot\mathbf{r}/\hbar)\psi_{\mathrm{f}}a_{\mathrm{fi}}^{(1)}\right] \\
&\quad = (2\pi\hbar)^{-3/2}\Bigg[\exp(i\mathbf{p}_{\mathrm{i}}\cdot\mathbf{r}/\hbar)\psi_{\mathrm{i}} \\
&\qquad + \sum_{n_{\mathrm{f}}}\int_0^\infty p_{\mathrm{f}}^2\, dp_{\mathrm{f}}\left[P\frac{1}{E_{\mathrm{i}} - E_{\mathrm{f}}} + b\delta(E_{\mathrm{i}} - E_{\mathrm{f}})\right] \\
&\qquad \times \int_0^{2\pi} d\phi \int_{-1}^{1} d\cos\theta\, \exp(ip_{\mathrm{f}}r\cos\theta/\hbar)\psi(\mathbf{p}_{\mathrm{f}}n_{\mathrm{f}}|V|\mathrm{i})\Bigg].
\end{aligned} \tag{3.7}$$

The first important step consists of carrying out the integration over $\cos\theta$ by parts by utilizing the elementary integral $\int d\cos\theta \exp(ip_f r\cos\theta/\hbar) = (\hbar/ip_f r)\exp(ip_f r\cos\theta/\hbar)$, which is proportional to $1/r$. Repeated steps of this procedure yield an asymptotic form of (3.7) for large r, expanded into powers of $\hbar/p_f r$. Since all we need of the scattered wave function is its asymptotic part, we retain only the first term of this expansion, namely,

$$\sum_{n_f}(2\pi\hbar)^{-3/2}\psi_f\int_0^\infty p_f^2\,dp_f\left[P\frac{1}{E_i-E_f}+b\delta(E_i-E_f)\right]$$
$$\times\int_0^{2\pi}d\phi(\hbar/ip_f r)[e^{ip_f r/\hbar}(p_f\hat{r},n_f|V|i)-e^{-ip_f r/\hbar}(-p_f\hat{r},n_f|V|i)]$$
$$+\,O[(\hbar/p_f r)^2]. \tag{3.8}$$

Notice that this wave function consists of two terms: The first one represents the outgoing spherical wave emerging from the collision process, but the second one represents an ingoing wave foreign to the specification of our problem according to which the wave function (3.7) should consist only of the incident plane wave and of a scattered outgoing wave. To fulfill this specification the integration constant b will be set below so as to *cancel the amplitude* of the ingoing wave. The integrand of (3.8) is now independent of ϕ; hence we have $\int_0^{2\pi} d\phi = 2\pi$. The remaining $\int_0^\infty dp_f$ is transformed into an integral over $E_f = p_f^2/2M + \epsilon_f$ by writing $p_f^2\,dp_f = Mp_f\,dE_f$.

Here we consider the two terms in the brackets of (3.8) separately. The term with the structure $b\int_0^\infty dE_f\delta(E_i-E_f)f(p_f)$ yields $bf(\bar{p}_f)$ with $\bar{p}_f$ defined by $\bar{p}_f^2/2M+\epsilon_f=E_i$. In the term with the factor $P[1/(E_i-E_f)]$, which is odd in the difference E_i-E_f, we set $\exp(\pm ip_f r/\hbar) = \exp(\pm i\bar{p}_f r/\hbar)\{\cos[(p_f-\bar{p}_f)r/\hbar]\pm i\sin[(p_f-\bar{p}_f)r/\hbar]\}$ and note that the cosine term is even in $(p_f-\bar{p}_f)$ and oscillates very fast as p_f varies at large r, whereby it yields no net contribution. The sine term is, instead, of the form

$$P\int_0^\infty dE_f\frac{\pm i\sin[(p_f-\bar{p}_f)r/\hbar]}{E_i-E_f}g(p_f)\xrightarrow[r\to\infty]{}\mp i\pi g(\bar{p}_f). \tag{3.9}$$

The full expression (3.8) thus reduces to

$$(2\pi\hbar)^{-1/2}\sum_{n_f}\psi_f M(ir)^{-1}[e^{i\bar{p}_f r/\hbar}(-i\pi+b)(\bar{p}_f\hat{r},n_f|V|i)$$
$$+\,e^{-i\bar{p}_f r/\hbar}(i\pi+b)(-\bar{p}_f\hat{r},n_f|V|i)]+O[(\hbar/\bar{p}_f r)^2]. \tag{3.10}$$

Note how the two terms in the braces of (3.8) being odd and even, respectively, with respect to E_i-E_f contribute components 90° out of phase.

As noted earlier, the state resulting from scattering of an incident plane wave should display only a wave traveling away from the scatterer in directions other than $\mathbf{p}_i$; accordingly cancellation of the ingoing wave requires us to set $b = -i\pi$. This determination of b is analogous to the determination of the coefficients for the waves of type 3 in the one-dimensional model problem of Section 1.5. With this value of b, the final expression of the outgoing scattered wave is

$$-(2\pi/\hbar)^{1/2} M \sum_{n_f} (\bar{p}_f \hat{r}, n_f | V | i) \psi_f e^{i\bar{p}_f r/\hbar}/r + O[(\hbar/\bar{p}_f r)^2]. \tag{3.11}$$

3.4 DIFFERENTIAL CROSS SECTION

We determine now the differential cross section for a collision that leaves the atom in its state $(n_f|$ and the particle traveling in the direction $\hat{r}$ within a solid angle element $d\Omega_f$. This cross section equals the ratio of the number of particles scattered per second in the direction $\hat{r}$ to the incident particle flux in the direction $\mathbf{p}_i$. The scattered flux equals the squared coefficient of $\psi_f \exp[i\bar{p}_f r/\hbar]/r$ in (3.11) multiplied by the scattered particle velocity $\bar{p}_f/M$. The incident flux, for the initial state normalized per unit $\mathbf{p}$, is $(2\pi\hbar)^{-3} p_i/M$. The corresponding differential cross section takes the form of Eq. 17.65 of Fano and Fano [1],

$$\frac{d\sigma}{d\Omega_f} = (4\pi^2 M\hbar)^2 |(\bar{f}| V | i)|^2 \frac{\bar{p}_f}{p_i}, \tag{3.12}$$

where $(\bar{f}| \equiv (\bar{p}_f \hat{r}, n_f|$, $d\Omega_f$ is an element of solid angle in the direction of $\mathbf{p}_f$, $\bar{p}_f/p_i$ is the velocity ratio after and before scattering, and the final state energy is set at $\bar{E}_f = \bar{p}_f^2/2M + \epsilon_f = E_i$.

3.5 MOMENTUM TRANSFER AND FORM FACTORS

The motion of the colliding particle affects the Born approximation scattering coefficient $(f| V |i)$ defined in (3.3″) only through the product of the plane waves representing the particle's initial and final states, $\exp[i(\mathbf{p}_i - \bar{\mathbf{p}}_f) \cdot \mathbf{r}/\hbar]$. This expression depends in turn only on the momentum $\mathbf{p}_i - \bar{\mathbf{p}}_f$ imparted by the particle to the target rather than on the separate momenta $\mathbf{p}_i$ and $\bar{\mathbf{p}}_f$ (Fig. 3.1). Indeed, a practical test of the Born approximation consists of verifying to what extent a collision cross section, apart from kinematic terms such as $\bar{p}_f/p_i$ in (3.12), depends only on the momentum transfer.

The probability of the target being excited to one or another final state $(n_f|$ hinges on the response of its constituents to the momentum transfer. To

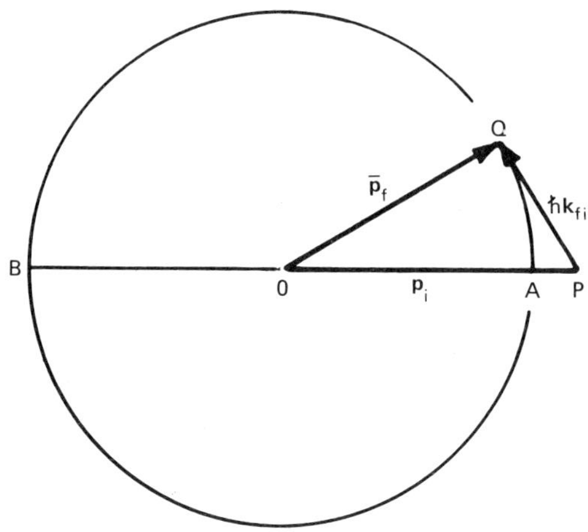

Fig. 3.1. Kinematics of a collision. Here $\mathbf{p}_i$ and $\bar{\mathbf{p}}_f$ are the initial and final momenta of a fast particle in an inelastic collision and $\hbar\mathbf{k}_{fi}$ the momentum transfer. Energy conservation restricts the tip Q of $\bar{\mathbf{p}}_f$ to a circle with center O. The minimum and maximum values of $\hbar k_{fi}$ are then PA and PB, but Coulomb interaction weights heavily towards small values of k_{fi} corresponding to a small arc around A, that is, to forward-angle scattering; it also weights the cross section heavily to small energy losses corresponding to values of PA much smaller than as shown.

develop this aspect we resolve the matrix element $(\mathrm{f}|V|\mathrm{i})$ into contributions from the interactions of the incident particle with separate target constituents $1, 2, \ldots s, \ldots$ and analyze these interactions into their separate Fourier components, that is, by setting

$$V = \sum_s V_s(\mathbf{r} - \mathbf{r}_s) = \sum_s \int d\mathbf{k}\; v_s(\mathbf{k}) e^{i\mathbf{k}\cdot(\mathbf{r}_s - \mathbf{r})}. \tag{3.13}$$

In the expression (3.3″) of $(\mathrm{f}|V|\mathrm{i})$—with an additional factor $(2\pi\hbar)^{-3}$ from the normalization in Section 3.3—the integration over the incident particle's position $\mathbf{r}$ now yields a nonzero contribution only for $\mathbf{k} = \mathbf{k}_{fi} \equiv (\mathbf{p}_i - \bar{\mathbf{p}}_f)/\hbar$, namely,

$$(\mathrm{f}|V|\mathrm{i}) = \sum_s v_s(k_{fi})\hbar^{-3}(n_f|\exp(i\mathbf{k}_{fi}\cdot\mathbf{r}_s)|n_i). \tag{3.14}$$

The Coulomb interaction between an incident particle with charge ze and each target charge $z_s e$ yields $v_s(k) = zz_s e^2/2\pi^2 k^2$ and hence

$$\begin{aligned}(\mathrm{f}|V|\mathrm{i}) &= \frac{ze^2}{2\pi^2\hbar^3 k_{fi}^2} \sum_s z_s(n_f|\exp(i\mathbf{k}_{fi}\cdot\mathbf{r}_s)|n_i) \\ &= \frac{ze^2}{2\pi^2\hbar^3 k_{fi}^2} F_{n_f n_i}(\mathbf{k}_{fi}).\end{aligned} \tag{3.15}$$

This expression exhibits explicitly a characteristic feature of the Born approximation, namely that the scattering cross section in (3.12) is of the form of a product of two distinct factors, one dealing with the incident particle only and the other, $F_{n_f n_i}$, dealing with the target alone. The expression $F_{n_f n_i} = \sum_s z_s (n_f| \exp(i\mathbf{k}_{fi} \cdot \mathbf{r}_s)|n_i)$ represents an important property of the target, namely, a Fourier component—with wave vector $\mathbf{k}_{fi} = (\mathbf{p}_i - \bar{\mathbf{p}}_f)/\hbar$—of the electric charge distribution associated with transition of the target from the state $|n_i)$ to $(n_f|$. It is called the *form factor* of this transition. When $(n_f|$ coincides with $(n_i|$, $F_{n_i n_i}(\mathbf{k}_{fi})$ is a Fourier component of the charge distribution within the target's ground state.

Scattering experiments conducted with selected momentum transfers $\hbar\mathbf{k}_{fi}$ probe the charge structure within a target and its variations that accompany excitation processes. Notice that most of our information on the microstructure of matter and its dynamics stems from actual experimental determination of form factors. Note also that for the same momentum transfer, different excitations of the atom are possible. This lack of any unique correspondence between momentum and energy transfer is, of course, a reflection of the motions of the electrons inside the target (Fano and Fano [1], Sections 4.2, 4.4, and 4.5).

Substitution of (3.15) into the cross section (3.12) yields

$$\frac{d\sigma}{d\Omega_f} = \left[\frac{2Mze^2}{|\mathbf{p}_i - \bar{\mathbf{p}}_f|^2}\right]^2 |F_{n_f n_i}(\mathbf{k}_{fi})|^2 \frac{\bar{p}_f}{p_i}. \tag{3.16}$$

Once again, a remarkable feature of this result is that its first factor coincides with the Rutherford cross section for scattering of the incident particle by a target of charge e with momentum transfer $\mathbf{p}_i - \bar{\mathbf{p}}_f$. (Indeed the Rutherford formula holds for all values of its numerical parameter $z_1 z_2 e^2/\hbar v$, which is very large in the classical limit and very small in the Born approximation limit.) The two main characteristics of the cross-section formula (3.16) are the sharp peaking of its first factor for small momentum transfer and its dependence on a single property of the target, represented by the form factor. All the projectile parameters (z, p_i, and $\bar{p}_f$) are incorporated into the first, relatively trivial, Rutherford factor and the main object of study for fast collisions becomes the form factor of the target. Thus, the study of collisions with fast charged particles becomes a study of target properties, a kind of spectroscopy. The applications and implications of (3.16) have been reviewed extensively [3].

3.5.1 Small Deflections

Small momentum transfers correspond to classical trajectories that pass far outside the target. For elastic collisions, with $n_f \equiv n_i$ and $\bar{p}_f = p_i$, the Rutherford factor $|\mathbf{p}_i - \bar{\mathbf{p}}_f|^{-4}$ actually diverges sharply at zero deflection.

However, the form factor vanishes in this limit for a neutral target since it reduces to the sum of all charges, $\sum_s z_s$ (see Problem 3.1). This factor also vanishes to order $|\mathbf{p}_i - \bar{\mathbf{p}}_f|^2$ for targets without any net dipole moment. The cross section is thus seen to tend to a finite limit at zero deflection for neutral nonpolar targets.

For inelastic collisions with small deflections we have $\bar{p}_f < p_i$ and

$$|\mathbf{p}_i - \bar{\mathbf{p}}_f|^2 = p_i^2\left[\left(\frac{\epsilon_f - \epsilon_i}{p_i v}\right)^2 + \theta^2\right] + O(\theta^4), \tag{3.17}$$

where $\epsilon_f - \epsilon_i$ is the target excitation energy and v the impact velocity. The zero-angle parameter $\theta_0 = (\epsilon_f - \epsilon_i)/p_i v$ is often very small; for example, it amounts to 10^{-4} for $\epsilon_f - \epsilon_i = 10\,\text{eV}$ and an incident electron of 50 keV. (A corollary implication of this statement for the experimental study of the form factors as a function of the momentum transfer is the need for high angular resolution. For a given projectile velocity, θ_0 is even smaller for a heavier projectile. Experimental data on the form factor are derived, therefore, primarily with electrons as the projectile.) The squared form factor also becomes small for small deflections but only in proportion to k_{fi}^2 while the Rutherford factor rises as k_{fi}^{-4}. The cross section has a sharp peak at small θ with width θ_0. This width θ_0 is roughly the ratio of the wavelength of the incident particle to the size of the scatterer.

The form factor $F_{n_f n_i}(\mathbf{k}_{fi})$ of an inelastic transition is seen from (3.15) to become proportional to the dipole matrix element for the same transition in the limit of small k_{fi}. Such a connection is intuitively plausible because a small momentum transfer corresponds to a large impact parameter. The force exerted on the atom by a distant charged particle is uniform over the spatial extent of the atom in the same manner as the force exerted by light with long wavelength ($\lambda \gg a_0$). Furthermore, the force is sharply peaked in time because of the high velocity of the projectile, which means that its Fourier components are widely and smoothly distributed in frequency. Thus, charged-particle collisions with small k_{fi} are equivalent to irradiation with a broad spectrum photon flux of long wavelength. Accordingly, we consider the parameter [compare with (2.17)]

$$f_{n_f n_i}(\mathbf{k}_{fi}) \equiv \frac{2M(\epsilon_f - \epsilon_i)}{\hbar^2}\left|\frac{F_{n_f n_i}(\mathbf{k}_{fi})}{k_{fi}}\right|^2 \xrightarrow[k_{fi}\to 0]{} \left.\frac{df}{d\epsilon}\right|_{\epsilon = \epsilon_f - \epsilon_i} \tag{3.18}$$

as a generalized oscillator strength. Its study serves purposes analogous to those discussed in Chapter 2; it is, in fact, more general since it extends beyond the range of small k_{fi}.

For collisions at relativistic velocities and with small deflection the particle target interaction includes an additional magnetic component, which couples the current generated by the particle motion to internal currents of the target.

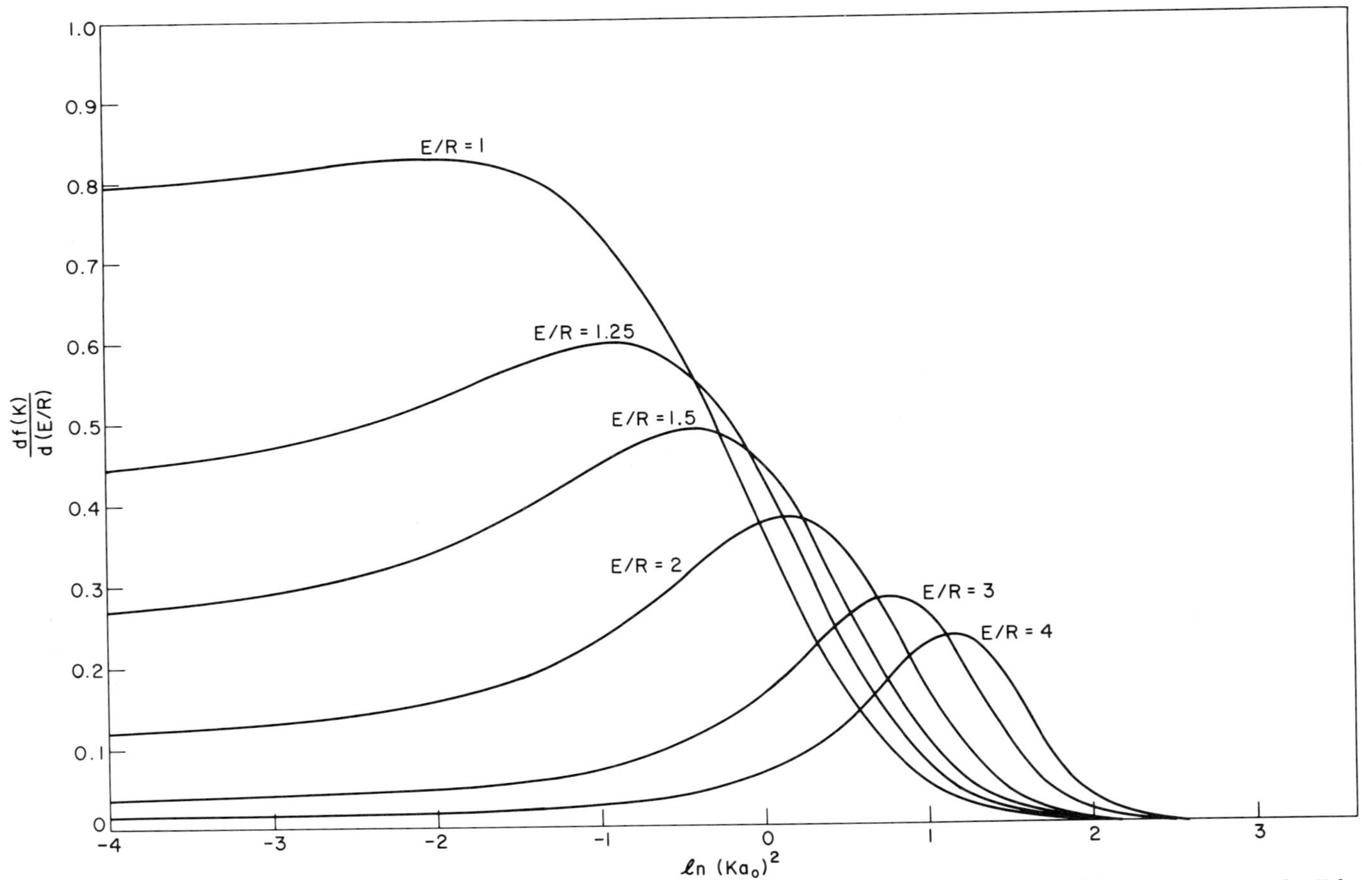

Fig. 3.2. Distribution of the generalized oscillator strength per unit range of excitation energy E as a function of the momentum transfer K for transitions from the ground state into the continuum of atomic hydrogen. Note the sharp falloff at large momentum transfer and the increasing peaking of the distribution with increasing E as given by (3.21). (From Inokuti [3].)

This interaction contributes to the cross section an additional term, proportional to v^4 for $v \ll c$ and proportional to a squared magnetic form factor that is a Fourier component of the current distribution within the target.

3.5.2 Large Deflections

We have introduced the Born approximation as relevant primarily to collisions with small deflections, in which each incident particle acts at long range with a small probability of affecting the target. The basic result (3.16), nevertheless, often also remains relevant in the opposite limit of large deflections, albeit with a different foundation. This is the limit in which the momentum transfer $\hbar\mathbf{k}_{fi}$ greatly exceeds the initial root mean square momentum of the target constituent that receives the momentum. This constituent then recoils with high probability clear out of the target, almost freely of its initial bond.

The formal basis of (3.16) is now the following. Its first (Rutherford) factor correctly represents the cross section for momentum transfer from the incident particle to an electron (or other constituent) of the target, as though this electron were free, for any value of $\mathbf{k}_{fi}$. The form factor, instead, nearly vanishes owing to the rapid oscillations of $\exp i\mathbf{k}_{fi}\cdot\mathbf{r}$ unless the target state $(n_f|$ has an electron recoiling with momentum close to $\hbar\mathbf{k}_{fi}$. Thus, the form factor and, thereby, the generalized oscillator strength will be sharply peaked around the value of ϵ_f such that $\epsilon_f - \epsilon_i = \hbar^2 k_{fi}^2/2M$. The distribution of values of $f_{n_f n_i}(\mathbf{k}_{fi})$ among different final states around this n_f represents the residual bonding effect upon the recoiling particle. Figure 3.2 illustrates the generalized oscillator strength for ionization of hydrogen. Note the evolution of the peak in the distribution with increasing energy.

The effect of exchange between the incident and recoiling particles is no longer negligible when these particles are identical and have come into close interaction attaining comparable velocities. The Rutherford factor in (3.16) must accordingly be replaced by the corresponding Mott formula (see Section 4.4 of Inokuti [3]).

3.6 TOTAL CROSS SECTION

To obtain the total cross section for scattering of the incident particle from $\mathbf{p}_i$ with excitation of the target atom from state n_i to n_f, the differential cross section in (3.16) must be integrated over the scattering angle. Azimuthal symmetry is generally obtained, either because of spherical symmetry of the atoms themselves in the initial state or because of their random orientations. The form factor, or generalized oscillator strength in (3.18), is then a function only of the *magnitude* k_{fi} of the momentum transfer. Integration over the azimuthal angle contained in $d\Omega_f$ in (3.16) is immediate, thus yielding a factor of 2π. The remaining integration over the scattering angle θ can be cast instead in terms of the scalar variable k_{fi} (Fig. 3.1).

Thus, on writing

$$d\Omega_f = \pi\hbar^2\, d(k_{fi}^2)/p_i\bar{p}_f \tag{3.19}$$

and expressing the form factor F in terms of the generalized oscillator strength (3.18), Eq. (3.16) can be recast as

$$d\sigma_{n_f n_i} = \frac{4\pi a_0^2 z^2}{(\hbar v/e^2)^2}\left[\frac{f_{n_f n_i}(k_{fi})}{(\epsilon_f - \epsilon_i)/Ry}\right]^2 d[\ln(k_{fi}a_0)^2], \tag{3.20}$$

where $Ry \equiv 13.6\,\text{eV}$. Once again, the factors outside the term in brackets, depending only on the particle and not on the target, constitute the Rutherford cross section for transferring a momentum k_{fi} to a target with charge e. The key variable here is the velocity of the incident particle in the center-of-mass frame, expressed in (3.20) as a dimensionless speed through the use of $e^2/\hbar = c/137$.

The term in the brackets in (3.20), on the other hand, reflects target properties alone, thus expressing the probabilities for various internal excitations following receipt of any particular momentum k_{fi}. Its distribution reflects the internal dynamics of the electrons in the target. As stated in Section 3.5.2, the term in brackets peaks sharply at the energy–momentum-conserving value.

$$\epsilon_f - \epsilon_i = \hbar^2 k_{fi}^2/2M, \tag{3.21}$$

only in the limit of large k_{fi}, when one of the electrons inside the target recoils freely, its binding becoming negligible in this limit. Note in (3.20) that $4\pi a_0^2 Ry \simeq 4.788 \times 10^{-15}\ \text{cm}^2\ \text{eV}$.

The expression in (3.20) must now be integrated over the kinematically allowed limits of k_{fi}. These follow simply from the conservation of energy,

$$(p_i^2/2M) + \epsilon_i = (\bar{p}_f^2/2M) + \epsilon_f, \tag{3.22a}$$

and from the geometry shown in Fig. 3.1,

$$\hbar^2 k_{fi}^2 = p_i^2 + \bar{p}_f^2 - 2p_i\bar{p}_f\cos\theta. \tag{3.22b}$$

Together, these equations give (recall that $p_i = Mv$)

$$(k_{fi}a_0)^2 = \frac{Mv^2}{Ry}\left(\frac{M}{m}\right)\left[1 - \frac{\epsilon_f - \epsilon_i}{Mv^2} - \cos\theta\left(1 - \frac{\epsilon_f - \epsilon_i}{\frac{1}{2}Mv^2}\right)^{1/2}\right], \tag{3.23}$$

with limiting values at $\theta = 0$ and π, shown also geometrically in Fig. 3.1. In the limit $Mv^2 \gg \epsilon_f - \epsilon_i$, these limiting values of k_{fi} take a simple form

$$(k_{fi}a_0)^2_{min} = \tfrac{1}{2}(\epsilon_f - \epsilon_i)^2/Ry(mv^2), \tag{3.24a}$$

$$(k_{fi}a_0)^2_{max} = 2(M/m)^2(mv^2/Ry). \tag{3.24b}$$

For plots of the kinematically allowed regions when the incident particle is an electron or proton and for the effect of relativistic kinematics on (3.23) and (3.24), see Inokuti [3].

As illustrated in Fig. 3.2, the generalized oscillator strength drops off very sharply at high values of k_{fi}. Therefore, the contribution to the integral of (3.20) over k_{fi} becomes negligible long before the maximum allowed kinematical value (3.24b), particularly for impact by a heavy particle like the proton. Stated in terms of the scattering angle, most of the contribution comes from a small range of forward angles. The leading contribution is from the first term in the expansion of $f_{n_f n_i}(k_{fi})$ into powers of k_{fi}. The first term in this expansion [see (3.18)] is the optical oscillator strength $f_{n_f n_i}$ and succeeding terms involve powers of $(k_{fi}a_0)^2$.

For optically allowed transitions, therefore, we have

$$\sigma_{n_f n_i} = \frac{4\pi a_0^2 z^2}{(\hbar v/e^2)^2}\frac{Ry}{\epsilon_f - \epsilon_i} f_{n_f n_i} \int d[\ln(k_{fi}a_0)^2] + O\left(\frac{\epsilon_f - \epsilon_i}{Mv^2}\right). \tag{3.25}$$

The limits on the integral in (3.25) are $(k_{fi}a_0)^2_{min}$ and a suitable cutoff value of $(k_{fi}a_0)^2$. The ratio of these two values is customarily expressed as $4C_{n_f n_i}(\hbar v/e^2)^2$ to display explicitly the dependence on v contained in (3.24a); $C_{n_f n_i}$ is a constant characteristic of the transition. The total cross section is, therefore,

$$\sigma_{n_f n_i} = \frac{4\pi a_0^2 z^2}{(\hbar v/e^2)^2}\frac{Ry}{\epsilon_f - \epsilon_i} f_{n_f n_i} \ln\left[4C_{n_f n_i}\left(\frac{\hbar v}{e^2}\right)^2\right] + O\left(\frac{\epsilon_f - \epsilon_i}{Mv^2}\right). \tag{3.26}$$

A plot of $(\hbar v/e^2)^2 \sigma_{n_f n_i}$ versus $\ln(\hbar v/e^2)^2$ gives a straight line, the slope being a measure of the optical oscillator strength (Fig. 3.3).

For optically forbidden transitions, the expansion of $f_{n_f n_i}(k_{fi})$ at small k_{fi} begins with the power $(k_{fi}a_0)^2$. Its integral, therefore, over $d[\ln(k_{fi}a_0)^2]$ as in (3.25) no longer depends on the excitation energy or the incident velocity in lowest order, and the total cross section becomes

$$\sigma_{n_f n_i} = \frac{4\pi a_0^2 z^2}{(\hbar v/e^2)^2} b_{n_f n_i} + O\left(\frac{\epsilon_f - \epsilon_i}{Mv^2}\right). \tag{3.27}$$

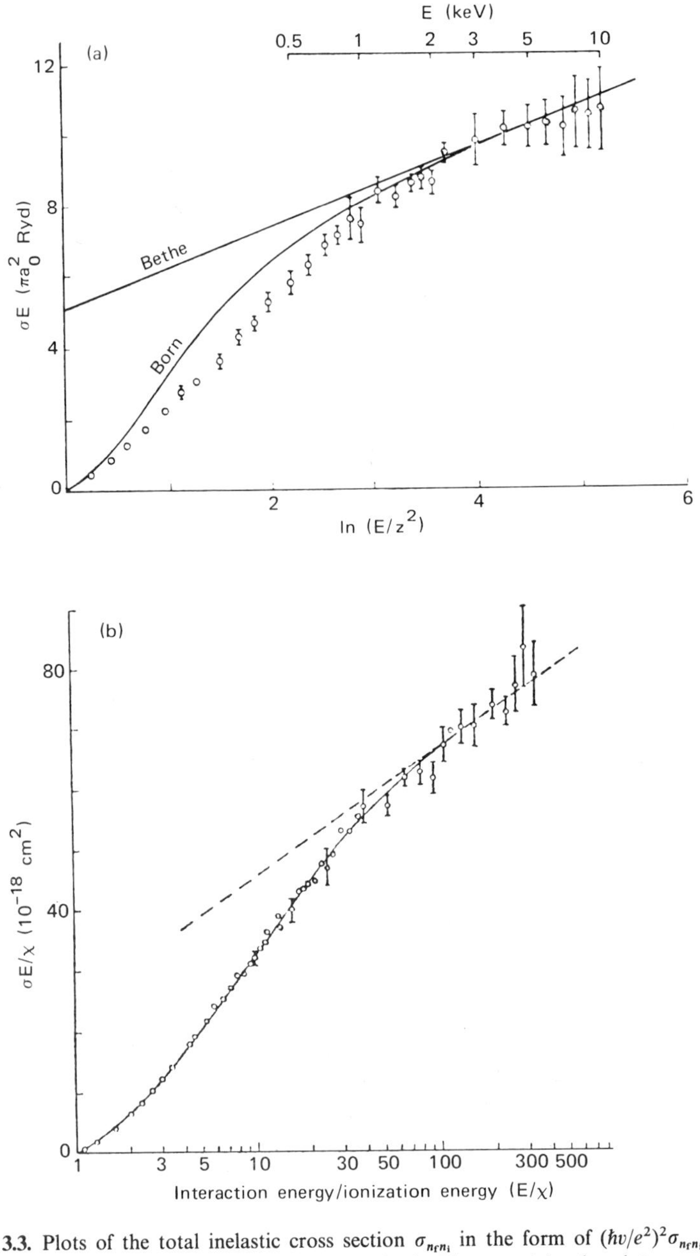

Fig. 3.3. Plots of the total inelastic cross section $\sigma_{n_f n_i}$ in the form of $(\hbar v/e^2)^2\sigma_{n_f n_i}$ versus $\ln(\hbar v/e^2)^2$. At asymptotically high velocities v of the incident particle, the plots approach a straight line whose slope equals the optical oscillator strength for the transition according to (3.26) and (3.27). (a) $e - He^+$ ionization and (b) $e - Li^+$ ionization. Here E/χ is the interaction energy in units of the ionization energy. (From Dolder and Peart [4].)

3.6.1 Mean Energy Loss

The energy lost by the colliding particle is another quantity of interest that can be obtained from the differential cross section in (3.20) and is important in problems of radiation effects and measurements. We start with atoms in initial state n_i but focus on the energy lost by the incident charged particle rather than on any specific state of excitation of the atom. The loss of energy per target atom is indexed by the so-called effective retardation

$$d\kappa = \sum_{n_f} (\epsilon_f - \epsilon_i)\, d\sigma_{n_f n_i}, \tag{3.28}$$

where $d\kappa$ relates to the element of solid angle into which the particle is scattered. If N is the number density of atoms, $N\, d\kappa$ is the energy lost by the incident particle per unit path length as it scatters into the given element of solid angle. On substituting (3.20) into (3.28), the summation over n_f can be carried out through the sum rule analogous to (2.24) and treated in Problem 3.3, namely,

$$\sum_{n_f} f_{n_f n_i}(k_{fi}) = Z, \tag{3.29}$$

where Z is the number of electrons in the atom (note that the summation includes integration over continuum states n_f). This sum rule expresses an important point of physics, first recognized and emphasized by Bohr, that regardless of the distribution of oscillator strength over various excitations of the atom for given k_{fi} (as in Fig. 3.2), the average of the energy transfer to the atom over all modes of excitation must be the same as the energy transfer to Z free electrons. This fact holds irrespective of the dynamics (Section 2.2 of Inokuti [3]).

As a consequence of (3.28) and (3.29), we have

$$N\, d\kappa = \frac{16\pi a_0^2 z^2 (Ry)^2}{mv^2} NZ \frac{dk_{fi}}{k_{fi}}. \tag{3.30}$$

The stopping power of the gas, defined as the loss of energy of the incident particle per unit path length, follows on integrating over all possible k_{fi}. This integration is readily carried out using (3.24). Because final states n_f have been summed over, we introduce a mean excitation energy $\bar{I}$ (empirically well described by $10Z$ eV) in place of $\epsilon_f - \epsilon_i$ in (3.24a) and write

$$\int N\, d\kappa = 16\pi a_0^2 z^2 (Ry)^2 (NZ/mv^2) \ln (mv^2/\bar{I}). \tag{3.31}$$

A closely related concept is that of the "range" of a charged particle in a medium. An order of magnitude estimate is obtained by dividing the initial energy $\frac{1}{2}Mv^2$ by the energy lost per unit path length as given in (3.31). These are key parameters in radiation physics.

3.7 REMARKS ON THE BORN APPROXIMATION

The Born approximation, introduced here in the special context of collisions of charged particles, has an important role throughout the subject of this book. That role will first emerge explicitly in Section 6.3.2 in which the Born approximation appears as the first term of the expansion of an exact collision parameter (the reaction matrix) into powers of the interaction V between a particle and an atom. Successive terms of the same expansion are called second- or higher-order Born approximations.

The accuracy of the Born approximation is often assessed conveniently in the framework of the expansion in partial waves that underlies the following chapters. Interaction of a charged particle at large impact parameters, indicated at the onset of this chapter as appropriate for application of the Born approximation, implies that the interaction bears on numerous partial waves with increasing values of their orbital quantum number l. Accordingly, each partial wave may contribute a small amount (represented accurately by a matrix element $|-\pi(\mathrm{f}|V^{(l)}|\mathrm{i})| \ll 1$ in the language of Section 6.3.2) to a collision, even though the aggregate of all these contributions is far from negligible. In the example of elastic collision, $\arctan[-\pi(\mathrm{i}|V^{(l)}|\mathrm{i})]$ represents the Born approximation value of the phase shift δ_l to be described in Section 4.7ff. The cumulative action of many small phase shifts generally yields large cross sections for small-angle scattering.

Conversely, resorting to second-order (or higher) Born approximations is often required in the absence of such cumulative action. Small deflections contribute to inelastic collisions through high-l waves by dipole interaction only, as noted in Section 3.5.1, their effect being thus restricted by the optical selection rule. Quadrupole excitation of a target is thus achieved by low-l partial waves only, in the ordinary Born approximation, but the small cross section of this direct process is often overshadowed by the second-order action of two successive dipole excitations contributed by high-l partial waves. Similarly, inelastic scattering of a charged particle with large deflection generally results from a two-step process of elastic scattering by a large angle due to action of the strong nuclear field on a low-l wave combined with inelastic contributions of many high-l waves.

Problems

3.1 (a) By solving Poisson's equation for the electrostatic potential due to the charge distribution in the ground state of a hydrogen atom, show that the potential seen by an incident electron is given by

$$V(r) = -e^2[(1/r) + (1/a_0)] \exp(-2r/a_0),$$

ignoring the incident electron's effect on the atom.

(b) The scattering amplitude in the Born approximation for potential scattering is given by

$$f(\mathbf{p}_i, \mathbf{p}_f) = -\frac{2m}{4\pi\hbar^2}\int e^{-i\mathbf{p}_f\cdot\mathbf{r}/\hbar}V(\mathbf{r})e^{i\mathbf{p}_i\cdot\mathbf{r}/\hbar}\,d\mathbf{r}.$$

Using the result of part (a), verify explicitly that the elastic form factor or scattering amplitude for electron–hydrogen scattering is a function only of the momentum transfer $\mathbf{k}_{fi} \equiv (\mathbf{p}_f - \mathbf{p}_i)/\hbar$ and that it vanishes in the forward direction (as $\mathbf{k}_{fi} \to 0$). This is illustrated in the results shown in Fig. 7.1.

Show also that the form factor depends only on the magnitude of the momentum transfer and not on the vector direction. This is generally true when either the charge distribution of the atom is spherically symmetric as is the case here or when the atoms are randomly oriented and an averaging over atomic orientations is assumed.

3.2 (a) Consider the inelastic form factor defined in (3.15) for the H atom. If $|n_i)$ is the ground state, evaluate this form factor for excitation to an arbitrary state in the discrete spectrum. The evaluation is best carried out in parabolic coordinates and involves integrals over confluent hypergeometric functions as given in Landau and Lifshitz [5].

(b) Several parabolic states are degenerate within any given manifold of principal quantum number n. As a result, what is often of interest is a sum over these degenerate states of the result obtained in (a). Evaluate this sum and compare with Eq. (3.9) of Inokuti [3].

3.3 Prove the Thomas–Reiche–Kuhn sum rule in its extended form given in (3.29) for the generalized oscillator strength. [Hint: Starting with the definition in (3.18), recast $(\epsilon_f - \epsilon_i)F_{n_f n_i}$ as $\sum_s z_s(n_f|[H_0, \exp(i\mathbf{k}_{fi}\cdot\mathbf{r}_s)]|n_i)$. Simplification of the commutator and application of the closure relation $\sum_{n_f}|n_f)(n_f| = 1$ leads to the result.]

References

1. U. Fano and L. Fano, "Physics of Atoms and Molecules," Section 17.42. Univ. of Chicago Press, Chicago, Illinois, 1972.
2. H. A. Bethe and E. E. Salpeter, "Quantum Mechanics of One- and Two-Electron Atoms," Section 9. Springer-Verlag, Berlin and New York, 1957.
3. M. Inokuti, *Rev. Mod. Phys.* **43**, 297 (1971); *Rev. Mod. Phys.* **50**, 23 (1978).
4. K. T. Dolder and B. Peart, *Rep. Prog. Phys.* **39**, 693 (1976).
5. L. D. Landau and L. M. Lifshitz, "Quantum Mechanics. Non-Relativistic Theory," Appendix f. Pergamon, Oxford, 1977.

Part B

SINGLE-CHANNEL PHENOMENA

4 ELASTIC SCATTERING BY A SHORT-RANGE CENTRAL POTENTIAL

4.1 INTRODUCTION

This chapter reviews aspects of the elementary theory of scattering of a particle by a central potential well, such as partial wave expansions and the normalization of continuum wave functions, with emphasis on the close connections between the bound and continuum states adjoining a common threshold. These aspects remain relevant to more complex collisional and spectral phenomena. In fact, key aspects that are crucial to our view of collisions and bound states of many-electron atoms are exemplified in a relatively straightforward and familiar context by the partial wave analysis of potential scattering.

These include the feature of working with eigenstates of an angular momentum operator, reducing the integrations to run over a single radial variable with range from 0 to ∞, and dealing with a single equation that describes the scattering at energies above the detachment threshold and the bound states below the threshold. These features will be extended to deal with multielectron systems in Part D by introducing a generalized angular momentum $\mathbf{\Lambda}^2$ and a hyperspherical radius R.

4.2 PARTIAL WAVE EXPANSION

Let us consider wave functions of a free particle, i.e., solutions of the wave equation $(\Delta + k^2)\psi = 0$ that represent energy eigenfunctions for a particle with mass m and energy $E = \hbar^2k^2/2m$. The plane wave solutions have the familiar expansion into spherical waves

$$
\begin{aligned}
e^{ikz} &= \sum_{l=0}^{\infty} i^l(2l+1)j_l(kr)P_l(\cos\theta) \\
&= 4\pi \sum_l \sum_{m=-l}^{l} i^l Y^*_{lm}(\theta_k, \phi_k) Y_{lm}(\theta, \phi) j_l(kr), \qquad (4.1)
\end{aligned}
$$

where $z = r\cos\theta$, $P_l(\cos\theta)$ and Y_{lm} are standard spherical functions, and $j_l(kr)$ is a spherical Bessel function.† In the last expression, (θ_k, ϕ_k) are polar coordinates of the direction of propagation $\mathbf{k}$ of the plane wave; having $\mathbf{k}$ directed along the z axis means that $\theta_k = 0$ and $Y^*_{lm}(0, \phi_k) = [(2l+1)/4\pi]^{1/2}\delta_{m0}$. The last expression in (4.1) thus represents the expansion of the more general plane wave $\exp(i\mathbf{k}\cdot\mathbf{r})$ and is designed to show how the spherical function $i^l Y^*_{lm}(\theta_k, \phi_k)$ plays the role of a coefficient in the expansion of a momentum eigenfunction into the complete set of angular momentum eigenfunctions $j_l(kr)Y_{lm}(\theta, \phi)$.

An essential point governing the usefulness of partial wave analysis is that the squared orbital momentum $|\mathbf{l}|^2$ and its component l_z remain constant under interaction with a central potential. Therefore, we can study the influence of a central potential *separately* for each eigenstate of $|\mathbf{l}|^2$ and l_z. In atomic collisions we generally do not deal with a central potential, but the squared *total* angular momentum $|\mathbf{J}|^2$ of the colliding systems is conserved, as is J_z; to this extent the study of a central potential is relevant to all collisions. Furthermore, in most situations in which the potential is noncentral, physical circumstances are usually obtained such that any partial wave couples appreciably only to a few others and the partial wave expansion still retains its usefulness. For example, the diamagnetic energy $(e^2B^2/8mc^2)r^2 \sin^2\theta$ of an electron in a magnetic field $\mathbf{B}$ along the z axis only couples states with $\Delta l = 0, \pm 2$.

Whereas the collision phenomena take place within a limited volume, the connection between eigenstates of momentum and angular momentum can be established for large values of the radial coordinate r. Accordingly, it is valuable to consider once again the expansion (4.1) in its large-r form, which is obtained from the asymptotic form of j_l [NBS (10.1.8)] in terms of ingoing and outgoing spherical waves characterized by exponentials $\exp(\mp ikr)$, respectively,

$$
\begin{aligned}
e^{ikz} &\xrightarrow[kr\to\infty]{} \frac{1}{2ikr}\sum_l (2l+1)P_l(\cos\theta)[-(-1)^l e^{-ikr} + e^{ikr}] \\
&= \frac{1}{2ikr}\sum_l (2l+1)[-P_l(-\cos\theta)e^{-ikr} + P_l(\cos\theta)e^{ikr}] \\
&= \frac{2}{ikr}[-\delta(1+\cos\theta)e^{-ikr} + \delta(1-\cos\theta)e^{ikr}]. \qquad (4.2)
\end{aligned}
$$

† We follow here the conventions of Eq. (10.1.1) of Abramowitz and Stegun [1], to be quoted as NBS.

Here δ indicates the usual Dirac function; note how the delta functions identify the directions of the particle source ($\theta = \pi$) and that of the unscattered particle flight ($\theta = 0$).

4.3 SCATTERING WAVE FUNCTIONS

As we start considering the effect of a short-range central potential, two key points stand out: (1) spherical symmetry permits us to treat each orbital momentum eigenfunction separately, and (2) this eigenfunction must remain a solution of the same radial equation as the Bessel $j_l(kr)$ throughout the range of radial distances exceeding the radius r_0 of the potential well. A general solution of the equation for $j_l(kr)$ is representable as a superposition of any base set of two independent solutions; standard practice is to choose the pair $\{j_l(kr), y_l(kr)\}$, where y_l is the function of the second kind [NBS (10.1.1)]. The radial function *outside the well* is accordingly represented by the superposition

$$F_l(kr) = j_l(kr)\cos\delta_l - y_l(kr)\sin\delta_l \xrightarrow[r\to\infty]{} j_l(kr + \delta_l), \tag{4.3}$$

whose coefficients have been so chosen that the asymptotic form departs from a Bessel function only by the *phase shift* δ_l. This single parameter δ_l thus represents the *entire effect* of scattering upon the eigenfunction $F_l(kr)$ outside the well; it depends, of course, on the potential at $r < r_0$ and on the particle energy, which is represented here by its wavenumber k.

4.3.1 Alternative Boundary Conditions

The standing waves $F_l(kr)Y_{lm}(\theta, \phi)$, with $F_l(kr)$ defined by (4.3) and with fixed k and different values of l, form a complete set of degenerate eigenfunctions of the wave equation $[\Delta + k^2 - 2V(r)]\psi = 0$ in the range $r > r_0$ where $V(r)$ vanishes. With respect to the manifold of all solutions of this equation, the functions $\{j_l, y_l\}$ in (4.3) perform the same role as the pair of solutions of the one-dimensional model considered in case (1) in Section 1.5. That is, any superposition of these states also represents an energy eigenstate.

In collision problems, as in all wave phenomena in infinite space, the states of interest are identified by boundary conditions at $r \to \infty$. Consider then the superposition analogous to (4.1) and (4.2) for $r > r_0$, i.e.,

$$\sum_l c_l i^l (2l + 1) P_l(\cos\theta)[j_l(kr)\cos\delta_l - y_l(kr)\sin\delta_l]$$

$$\xrightarrow[r\to\infty]{} \frac{1}{2ikr}\sum_l c_l(2l + 1)[-P_l(-\cos\theta)e^{-i(kr+\delta_l)} + P_l(\cos\theta)e^{i(kr+\delta_l)}]. \tag{4.4}$$

The expression on the right differs from (4.2) by the existence of the still undetermined coefficients c_l and the phase shift coefficients $\exp(\pm i\delta_l)$. The most familiar expression of this type is that representing the scattering of a particle as an incident plane wave accompanied by scattered outgoing waves. Hence, it consists of the following two terms: an expression identical to (4.2) and a superposition of *outgoing* waves only. This is analogous to case (3) in the one-dimensional problem discussed in Section 1.5. Its expression is obtained by equating the coefficients of the ingoing wave terms in (4.2) and (4.4), that is, by setting $c_l \exp(-i\delta_l) = 1$, or

$$c_l = e^{i\delta_l}. \tag{4.5}$$

With this choice of coefficients, (4.4) becomes

$$\sum_l e^{i\delta_l} i^l (2l+1) P_l(\cos\theta)[j_l(kr)\cos\delta_l - y_l(kr)\sin\delta_l]$$

$$\xrightarrow[r\to\infty]{} \frac{1}{2ikr}\sum_l (2l+1)[-P_l(-\cos\theta)e^{-ikr} + e^{2i\delta_l}P_l(\cos\theta)e^{ikr}]$$

$$= e^{ikz} + \frac{e^{ikr}}{kr}\sum_l \frac{e^{2i\delta_l}-1}{2i}(2l+1)P_l(\cos\theta). \tag{4.6+}$$

The energy eigenstate thus identified is said to be normalized according to *outgoing wave boundary conditions* and is often identified by a plus sign, for example by replacing the symbol c_l in (4.5) by $c_l^{(+)}$.

A different type of state is relevant to problems in which the particle does not actually arrive on the scatterer from infinite distance by traveling in a given direction (the $\hat{z}$ axis) but is released from within the scatterer by a different process (e.g., photoionization). We may then wish to specify that the particle travels *away* from the scatterer in a given direction, for example, again along the $\hat{z}$ axis. (This specification corresponds to escape to $+\infty$ only in case (4) of the Section 1.5 problem.) In this case it is the outgoing-wave part of (4.4) that must coincide with the corresponding part of (4.2); this condition requires that

$$c_l \to c_l^{(-)} = e^{-i\delta_l}. \tag{4.5−}$$

Thus, we obtain the wave function normalized according to *ingoing wave boundary conditions,*

$$\sum_l e^{-i\delta_l} i^l (2l+1) P_l(\cos\theta)[j_l(kr)\cos\delta_l - y_l(kr)\sin\delta_l]$$

$$\xrightarrow[r\to\infty]{} e^{ikz} - \frac{e^{-ikr}}{kr}\sum_l \frac{e^{-2i\delta_l}-1}{2i}(2l+1)P_l(-\cos\theta). \tag{4.6−}$$

Note that wave functions (4.6+) obtained from one another by arbitrary rotations of the coordinate axes, that is, by replacing $\exp(ikz)$ by $\exp(i\mathbf{k}\cdot\mathbf{r})$

with variable direction (θ_k, ϕ_k), are orthogonal and form a complete set of degenerate wave functions. The same holds for wave functions of the set similarly obtained from (4.6−). On the other hand, any wave function of one set is *not orthogonal* to any function of the other set. The sets (4.6+) and (4.6−) are analogous to the sets (3) and (4) in the model one-dimensional problem of Section 1.5. They are, as expected, time-reversal conjugates, the passage from one to the other being accomplished by complex conjugation together with a reversal in the vector direction of $\mathbf{k}$.

In concluding these considerations, which are elementary but of central importance to collision theory, let us stress again that the nature of the potential $V(r)$ at $r < r_0$ influences them only through the values of the phase shifts δ_l and especially through their dependence on the particle energy. For many purposes we may think, even within the scope of this chapter, that $V(r)$ represents the net interaction of an electron with a neutral spherical atom. At the simplest level, neglecting all exchange and dynamical effects, the potential is just the electrostatic potential due to the charge distribution in the atom as in Problem 3.1. The structure of this atom, as an aggregate of electrons, would manifest itself at higher energies with the onset of inelastic collisions, which fall within the scope of later parts of the book.

4.3.2 Normalization

The coefficients $i^l Y^*_{lm}(\theta_k, \phi_k)$ on the right-hand side of (4.1) may be regarded as elements of a unitary transformation matrix $(lm|\hat{k})$, which connects a set of eigenfunctions of momentum $\hbar\mathbf{k}$, that is, of plane waves, to the set of eigenfunctions of angular momentum $j_l(kr)Y_{lm}(\theta, \phi)$ or, equivalently, $(\mathbf{r}|lm)$. (The unit vector symbol $\hat{k}$ is used here to indicate that the transformation affects only the direction of the momentum $\hbar\mathbf{k}$ but preserves its magnitude.) In the presence of a potential well the same role is played by the coefficients

$$i^l e^{i\delta_l} Y^*_{lm}(\theta_k, \phi_k) = (lm|\hat{k}), \tag{4.7}$$

except that $\hbar\mathbf{k}$ now represents only the momentum of the unscattered part of the wave, since momentum conservation is spoiled by the presence of a potential well at $r < r_0$. Since unitary transformations preserve the normalization of a set of functions, (4.1) and (4.6) should display the connection between the normalizations of plane and spherical waves.

The set of plane waves $\exp(i\mathbf{k}\cdot\mathbf{r})$ is not quite normalized because

$$\int d\mathbf{r} \exp[i(\mathbf{k} - \mathbf{k}')\cdot\mathbf{r}] = (2\pi)^3 \delta(\mathbf{k} - \mathbf{k}'). \tag{4.8}$$

Accordingly, (4.1) and (4.6) will represent a transformation between normalized sets when divided by $(2\pi)^{3/2} = 4\pi(\pi/2)^{1/2}$. This division cancels the 4π

factor on the right-hand side of (4.1) and replaces it with a factor $(2/\pi)^{1/2}$; the radial wave function of (4.1) is thus renormalized to $(2/\pi)^{1/2}j_l(kr)$.

The normalization integral for this function is obtained by transforming the Dirac function $\delta(\mathbf{k} - \mathbf{k}')$ with the unitary matrix (4.7), i.e.,

$$\int d\hat{k} \int d\hat{k}'(lm|\hat{k})\delta(\mathbf{k} - \mathbf{k}')(\hat{k}'|l'm') = \delta_{ll'}\delta_{mm'}k^{-2}\delta(k - k'). \tag{4.9}$$

The normalization integral resulting from this transformation of (4.8) depends only on the asymptotic form of the radial function

$$(2/\pi)^{1/2}j_l(kr) \to (2/\pi)^{1/2}\sin(kr - l\tfrac{1}{2}\pi)/kr, \tag{4.10}$$

because it extends over the infinite range of r. For the same reason it is unaffected by the presence of a scattering potential at $r < r_0$ which only introduces a phase shift according to (4.3). Notice further that the square of the factor k in the denominator of (4.10) matches the factor k^{-2} on the right of (4.9) and that the square of the factor r cancels upon integration with the volume differential $r^2\,dr$. We conclude that the normalization integral of any radial function with the asymptotic form $(2/\pi)^{1/2}\sin(kr + \delta)$ with the *linear* differential dr yields $\delta(k - k')$, that is, that such functions are normalized "per unit k" [2]. Verification of the normalization of continuum wave functions by direct integration over the radial variable involves a double limiting process over the infinite range of this variable and the continuous variation of the spectral variable; we defer this study to Chapter 5.

We show here, on the other hand, how to adjust the normalization of these wave functions when the wave number k is replaced by the particle energy $E = \hbar^2k^2/2m$, or by some other spectral parameter $f(k)$. The adjustment utilizes the general property of the Dirac function $\delta[f(k) - f(k')] = (df/dk)^{-1}\delta(k - k')$, specifically $\delta(E - E') = (m/\hbar^2k)\delta(k - k')$. Accordingly, a wave function $(2/\pi)^{1/2}\sin kr$, normalized per unit k with the linear differential dr, is renormalized per unit energy on multiplication by $(m/\hbar^2k)^{1/2}$.

4.4 TIME DELAY

We have treated thus far only stationary states of a particle scattered by a potential. The time development of the scattering process is described by a superposition of stationary states, as noted in Section 1.4. In this superposition the wave function of each stationary component displays explicitly its time-dependent factor $\exp(-iEt/\hbar)$, which is otherwise irrelevant. The combination of this factor with the energy dependence of other terms of the wave function brings out the physical significance of this dependence. This fundamental procedure will be illustrated here by showing how any delay experienced by a particle in rebounding from a potential field is implied by the energy dependence of phase shifts $\delta_l(E)$ in the last term of (4.6+).

Consider a wave packet represented by the superposition of a set of waves (4.6+) with wave numbers $k \sim \bar{k}$ and energies

$$E = \frac{\hbar^2 k^2}{2m} = \frac{\hbar^2 \bar{k}^2}{2m} + \frac{\hbar^2}{2m}(k^2 - \bar{k}^2) \sim \bar{E} + \bar{v}\hbar(k - \bar{k}), \tag{4.11}$$

where $\bar{v} = d\bar{E}/\hbar d\bar{k} = \hbar\bar{k}/m$ is the group velocity, i.e., the particle velocity. Call $a(k)$ the coefficients of the superposition chosen so that $a(k)$ vanishes unless $k \sim \bar{k}$ and set

$$\int_0^\infty a(k)e^{ikz}\,dk = e^{i\bar{k}z}\int_0^\infty a(k)e^{i(k-\bar{k})z}\,dk = f(z)e^{i\bar{k}z}, \tag{4.12}$$

where $f(z)$ represents the profile of the wave packet and vanishes except for $z \sim 0$. The broader the distribution of $a(k)$ around $k \sim \bar{k}$, the narrower will be the profile $f(z)$. To represent the motion of the wave packet we must multiply $a(k)$ by $\exp(-iEt/\hbar)$ with E represented by (4.11); the wave packet (4.12) then takes the traveling-wave form

$$\int_0^\infty a(k)e^{i(kz - Et/\hbar)}\,dk = e^{i(\bar{k}z - \bar{E}t/\hbar)}\int_0^\infty a(k)e^{i(k-\bar{k})(z - \bar{v}t)}\,dk$$
$$= f(z - \bar{v}t)e^{i(\bar{k}z - \bar{E}t/\hbar)}. \tag{4.13}$$

On applying this construction to the wave function (4.6+), which includes both the incident wave $\exp(ikz)$ and the scattered waves, we must take into account the energy dependence of the phase shifts represented by

$$\delta_l(E) \sim \delta_l(\bar{E}) + \left(\frac{d\delta_l}{dE}\right)_{\bar{E}} \hbar\bar{v}(k - \bar{k}). \tag{4.14}$$

The completed wave is represented by replacing $\exp(ikz)$ in (4.13) by the asymptotic expression in the middle part of (4.6+) using the linear approximation (4.14) for the dependence of the phase shift on energy. The superposed packets of ingoing and outgoing spherical waves then become

$$\frac{1}{2ikr}\sum_l (2l+1)\{-P_l(-\cos\theta)f(r + \bar{v}t)e^{-i\bar{k}(r+\bar{v}t)}$$
$$+ e^{2i\delta_l(\bar{E})}P_l(\cos\theta)f[r - \bar{v}(t - 2\hbar\, d\delta_l/dE)]$$
$$\times e^{i\bar{k}[r - \bar{v}(t - 2\hbar d\delta_l/dE)]}\}e^{-i\bar{E}t/\hbar}, \qquad r \to \infty. \tag{4.15}$$

The first term in the braces represents a wave packet that propagates inward (toward $r = 0$) with velocity $\bar{v}$. The second term represents a packet of equal profile that propagates outward at a later time but with its time variable shifted by an amount that depends on l and is proportional to $(d\delta_l/dE)_{\bar{E}}$.

This shift represents a delay when δ_l rises with increasing E, thus implying that the particle is delayed in the potential well region $r < r_0$ before reemerging from it. Examples of this behavior will be shown in the following; they are particularly conspicuous at "resonances" when one of the δ_l increases rapidly by nearly 180°. (See Section 8.3.) By contrast, a negative value of $d\delta_l/dE$ implies that the particle reemerges from the well sooner than we would expect if it penetrated all the way to $r = 0$ with the incident velocity $\bar{v}$. This may occur because it does not in fact penetrate that deeply owing to repulsion and/or because it is accelerated during the scattering process. Indeed, the values of $d\delta_l/dE$ are generally negative for most values of E.

In the case of a potential with finite range r_0 the particle is not expected to rebound before having reached values of $r \leq r_0$. Wigner has pointed out that the value of $2\hbar\, d\delta_l/dE$ must accordingly remain $\geq -2r_0/\bar{v}$. The equality results for a well that has an infinitely repulsive barrier at r_0. This represents the extreme time advance that can be achieved. No opposite limit, namely, an upper limit on the time delay, can be expected in general because an appropriately chosen attractive well can retard the scattered component by an arbitrarily long time.

4.5 COLLISION PARAMETERS AND MATRICES

We have stressed that the set of phase shifts δ_l contains all the information about the effect of the potential well that can be gathered by experimental observations at distant points. This information is manifested more directly through the ratios of coefficients of different terms in the expansion (4.4) and (4.6) (recall the appearance of c_+/c_- in reflection and transmission coefficients in the model problem of Section 1.5), namely:

(1) the *scattering amplitude* $e^{2i\delta_l}$, i.e., the amplitude ratio of the outgoing and ingoing wave components for each value of l;

(2) *the transition amplitude* $(e^{2i\delta_l} - 1)/2i$, i.e., the ratio of the amplitude of each outgoing wave component to the incident plane wave amplitude; and

(3) the *reaction amplitude* $\tan \delta_l$, i.e., the ratio of the amplitudes of the standing wave component $y_l(kr)$ [which lags in phase by 90° with respect to the unperturbed $j_l(kr)$] and $j_l(kr)$ itself.

Each of these collision parameters, which represent alternative but equivalent forms of the effect of the potential, takes a definite value for each eigenstate of the squared orbital momentum $\mathbf{l}^2$, as do δ_l itself and the time delay $2\hbar\, d\delta_l/dE$. A basic difference between one- and three-dimensional problems now enters in that the angular momentum eigenstates in our three-dimensional analysis do not generally coincide with the physical states in which the incident particle is prepared or the scattered ones detected.

Therefore, the representation of a collision process in terms of these eigenstates is rather indirect; the experimental conditions are usually such that the parameters come under observation in other representations, mainly for eigenstates of the momentum $\hbar\mathbf{k}$ of the incident or scattered particle. To perform the transformation of a collision parameter we regard its set of values for different l as the eigenvalues of an operator, arranged along the diagonal of a matrix with rows and columns labeled by $(lm, l'm')$. Alternative matrix representations of the operator are obtained thereafter through the usual procedure of quantum-mechanical transformation theory, namely, by unitary transformations.

These operators are most familiar in their wave vector matrix representation, namely,

(1) the *scattering matrix*,

$$S \equiv \sum_{lm} (\hat{k}'|lm) e^{2i\delta_l} (lm|\hat{k}), \tag{4.16a}$$

(2) the *transition matrix*,

$$T \equiv \sum_{lm} (\hat{k}'|lm) \frac{e^{2i\delta_l} - 1}{2i} (lm|\hat{k}) = \frac{S - 1}{2i}, \tag{4.16b}$$

(3) the *reaction matrix*,

$$K \equiv \sum_{lm} (\hat{k}'|lm) \tan \delta_l (lm|\hat{k}) = -i \frac{S - 1}{S + 1}, \tag{4.16c}$$

(4) the *time delay matrix*,

$$\sum_{lm} (\hat{k}'|lm) 2\hbar \frac{d\delta_l}{dE} (lm|\hat{k}) = -i\hbar S^{-1} \frac{dS}{dE}. \tag{4.16d}$$

The phase shift itself is seldom used as an operator. In the following we shall refer to the K, S, and T matrices in a great many different representations. However, the representation (4.16) in terms of matrix elements between wave vectors $\hat{k}'$ and $\hat{k}$ is often implied unless otherwise stated.

4.6 COLLISION CROSS SECTION

The cross section for elastic scattering in the direction (θ, ϕ) per unit solid angle equals the squared modulus of the amplitude of the outgoing spherical wave in the last term in the expression (4.6+), i.e.,

$$\frac{d\sigma}{d\Omega} = \frac{1}{k^2} \left| \sum_l \frac{e^{2i\delta_l} - 1}{2i} (2l + 1) P_l(\cos\theta) \right|^2. \tag{4.17}$$

Note that here, as in (4.6+), each partial wave contributes to the scattered amplitude in proportion to its transition amplitude. The angular distribution of the scattering results from the interference of all partial waves. [This cross section is actually defined as the ratio of the particle flux scattered toward (θ, ϕ) to the incident flux. Hence, it is generally proportional to the speed of the scattered particles and inversely proportional to the speed of the incident particles, as in (3.7). These two factors cancel out in the present case of elastic scattering, but Section 7.9 will deal with the more general context.]

The total cross section for elastic scattering is obtained by integrating (4.17) over all directions. Cross terms in the expansion of (4.17), which represent the interference of different partial waves, cancel out on integration, owing to the orthogonality of spherical harmonics

$$\int_0^{2\pi} d\phi \int_0^{\pi} \sin\theta \, d\theta \, P_l(\cos\theta) P_{l'}(\cos\theta) = \frac{\delta_{ll'} 4\pi}{(2l+1)}. \tag{4.18}$$

Thus, we find the simple result

$$\sigma = \frac{4\pi}{k^2} \sum_l (2l+1) \sin^2 \delta_l, \tag{4.19}$$

which shows that each partial wave contributes to σ separately in proportion to $\sin^2 \delta_l$, the squared modulus of its transition amplitude, and to its statistical weight $2l + 1$.† The coefficient 4π represents the integral over the solid angle of scattering, while $1/k^2 = (\lambda/2\pi)^2$ provides the dimensional element, namely, the squared wavelength of the incident particles.

4.7 CALCULATION OF PHASE SHIFTS: EXAMPLE OF A SPHERICAL WELL

The calculation of phase shifts, or equivalent parameters, for collision and spectral problems generally presents a major problem. However, within the limited scope of scattering by a central potential field $V(r)$ confined to $r < r_0$, this problem reduces to solving the radial equation

$$[\tfrac{1}{2}(d^2/dr^2) + \tfrac{1}{2}k^2 - (m/\hbar^2)V(r) - \tfrac{1}{2}l(l+1)/r^2] rF_l(r) = 0, \tag{4.20}$$

with the boundary condition $rF_l(r) \to 0$ as $r \to 0$. Accordingly, the problem is readily solved, at least numerically. Once $F_l(r)$ and dF_l/dr are obtained at $r = r_0$, δ_l is found by matching of the radial function (4.4) to the logarithmic

† Since $(2l + 1) = (l + 1)^2 - l^2$, $(\pi/k^2)(2l + 1)$ represents the area of an annular ring bounded by impact parameters $(l + 1)/k$ and l/k.

derivative $d \ln F_l(r)/dr|_{r=r_0}$. The next more general problem, calculating δ_l for electron scattering by a spherically symmetric atomic core, will be considered in Chapter 6.

Here we only illustrate the more elementary problem by the example of a spherical-well potential

$$V(r) = -V_0 \quad \text{for} \quad r < r_0, \qquad V(r) = 0 \quad \text{for} \quad r > r_0. \tag{4.21}$$

The wave function for $r < r_0$ is then $rj_l(k_0 r)$, where k_0 is the wave number inside the well given by the energy equation

$$\frac{\hbar^2}{2m} k_0^2 - V_0 = \frac{\hbar^2}{2m} k^2. \tag{4.22}$$

The phase shift is determined by matching at $r = r_0$ the logarithmic derivatives of the wave functions $rj_l(k_0 r)$ within the well and $r[j_l(kr) \cos \delta_l - y_l(kr) \sin \delta_l]$ outside the well, that is, by solving the equation

$$\left\{\frac{d}{dr} \ln[rj_l(k_0 r)]\right\}_{r=r_0} = \left\{\frac{d}{dr} \ln[rj_l(kr) \cos \delta_l - ry_l(kr) \sin \delta_l]\right\}_{r=r_0} \tag{4.23}$$

for δ_l. The solution δ_l can be obtained numerically by using formulas and tables from Ref. [1]. Figure 4.1a shows the resulting dependence of δ_l on $E = \hbar^2 k^2/2m$ for given values of V_0 and r_0.

4.8 PHENOMENA AT NEAR-ZERO ENERGY

When the total energy expression (4.22) is small as compared with the magnitude V_0 of a negative (i.e., attractive) potential within the well, the wave number k_0 within the well will depend weakly on k^2 over an appreciable range of the total energy. Therefore, the boundary parameter (4.23) varies slowly with the energy, owing to its dependence on the slowly varying k_0 on the left-hand side; the rapid variations on k on the right-hand side must be compensated accordingly by corresponding variations of δ_l. These changes are reflected in conspicuous phenomena that depend primarily on the properties of the wave functions outside the well but only weakly on the characteristics of the inner well, that is, only through the value of (4.23) or of analogous parameters. The details of the particle dynamics in the inner well are then irrelevant for the phenomena to be discussed here.

A corollary of the weak dependence on k^2 of phenomena within the well is that there exists a close relationship between continuum ($k^2 \gtrsim 0$) and bound ($k^2 \lesssim 0$) states around the threshold. In this chapter we are considering the highly schematized spherical well model only for definiteness. However, this

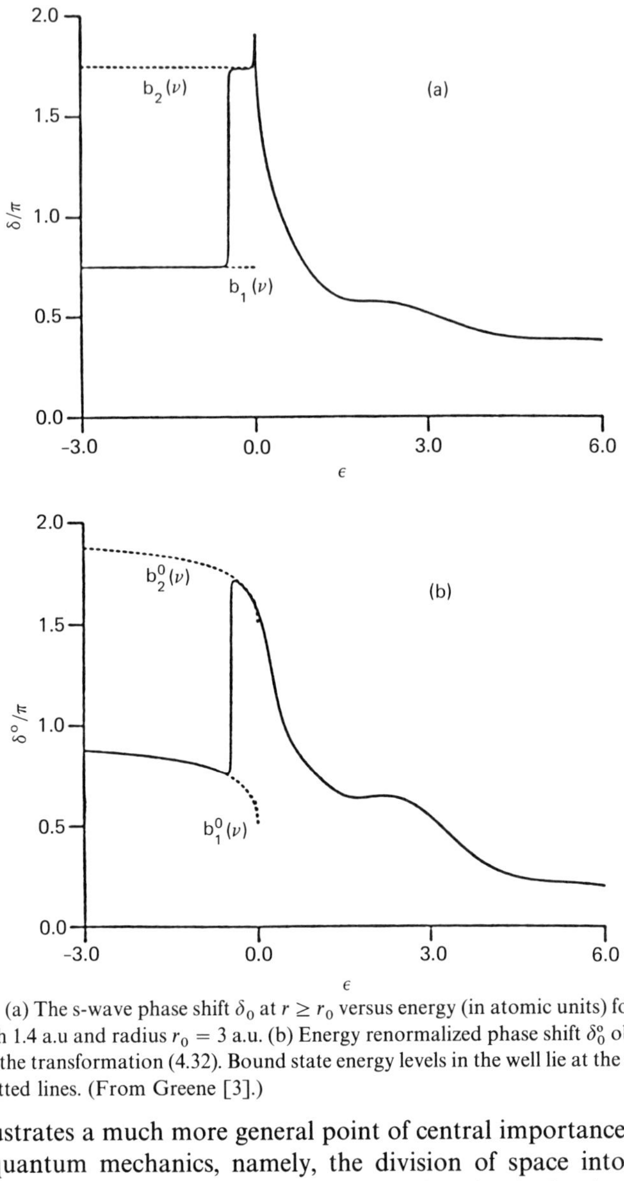

Fig. 4.1. (a) The s-wave phase shift δ_0 at $r \geq r_0$ versus energy (in atomic units) for a spherical well of depth 1.4 a.u and radius $r_0 = 3$ a.u. (b) Energy renormalized phase shift δ_0° obtained from (a) through the transformation (4.32). Bound state energy levels in the well lie at the intersections with the dotted lines. (From Greene [3].)

model illustrates a much more general point of central importance to many-particle quantum mechanics, namely, the division of space into an inner region of complicated dynamics and an outer region that is simpler, involving only one (or a few) particles separated from the rest.

This approach to problems, which has been most clearly emphasized by Wigner, serves two main purposes. On the one hand, certain questions, such as the near-threshold phenomena we study in this chapter, rest primarily on the motion in the outer region for which a detailed handling of the complicated inner region is unnecessary . For those questions that rest on a full treatment of the many-particle dynamics, on the other hand, consideration of an inner region of finite (and small) volume reduces the most complicated parts of the dynamics to the study of a discrete spectrum. Numerical handling will also be facilitated by the restrictions to a small range of radial integrations. The problem of the inner region will be taken up in Chapter 6.

4.8.1 Small-k Behavior of Spherical Bessel Functions

The radial wave functions in the brackets on the right-hand side of (4.23) were initially normalized per unit k. It is preferable henceforth to use wave functions normalized per unit energy according to Section 4.3.2, i.e.,

$$(2mk/\pi\hbar^2)^{1/2} r[j_l(kr)\cos\delta_l - y_l(kr)\sin\delta_l]. \tag{4.24}$$

We are interested here in a range of small energies for which $kr_0 \ll 1$. In this range, (4.24) has the power expansion

$$(2m/\pi\hbar^2)^{1/2}[k^{l+1/2}r^{l+1}(2l+1)!!^{-1}\cos\delta_l + \cdots \\ + k^{-(l+1/2)}r^{-l}(2l-1)!!\sin\delta_l + \cdots]. \tag{4.25}$$

This expression confirms that the determination of δ_l from (4.23) will depend strongly on k. To distinguish the phenomena connected with the phase shift from those connected with normalization, we factor a coefficient out of (4.25), thus rewriting it in the form

$$\{(2l+1)!!^{-1}r^{l+1} + \cdots + (2l-1)!!r^{-l}[k^{-(l+1/2)}\tan\delta_l k^{-(l+1/2)}] + \cdots\} \\ \times (2m/\pi\hbar^2)^{1/2}k^{l+1/2}\cos\delta_l. \tag{4.25$'$}$$

The terms of (4.25′) indicated by dots are smaller than the others by powers of kr and will be disregarded for the time being.

4.8.2 Threshold Behavior of the Phase Shift

The logarithmic derivative to be entered on the right-hand side of (4.23) depends only on the factor in the braces of (4.25′). We see now that the solution of (4.23) at low energy will actually determine a slowly varying value of the expression $[k^{-(l+1/2)}\tan\delta_l k^{-(l+1/2)}]$ rather than of δ_l itself. [We have kept the two factors $k^{-(l+1/2)}$ separate here because they will take different values in multichannel applications.] It follows that

$$\tan\delta_l \propto k^{2l+1} \qquad \text{for small} \quad k, \tag{4.26}$$

whereby δ_l departs from a multiple of π only by an amount proportional to

k^{2l+1}. This phenomenon is quite evident in the plots of many experimental results on phase shifts (Fig. 4.2; see also Rau [5]). In the important case of $l = 0$, in which the departures from $n\pi$ are largest, we obtain from (4.23) and (4.25′)

$$\lim_{k=0}(\delta_0/k) = -r_0 + k_0^{-1} \tan k_0 r_0 = -a. \tag{4.27}$$

The parameter a, which is called the "scattering length," owing to its role in a schematic plot of the wave function (Fig. 4.3), determines the zero energy value of the elastic scattering cross section (4.19)

$$\lim_{k=0} \sigma = \lim[(4\pi/k^2) \sum_l (2l+1) \sin^2 \delta_l] = 4\pi a^2. \tag{4.28}$$

[Note that a is sometimes defined in the literature with a sign opposite to that given in (4.27).]

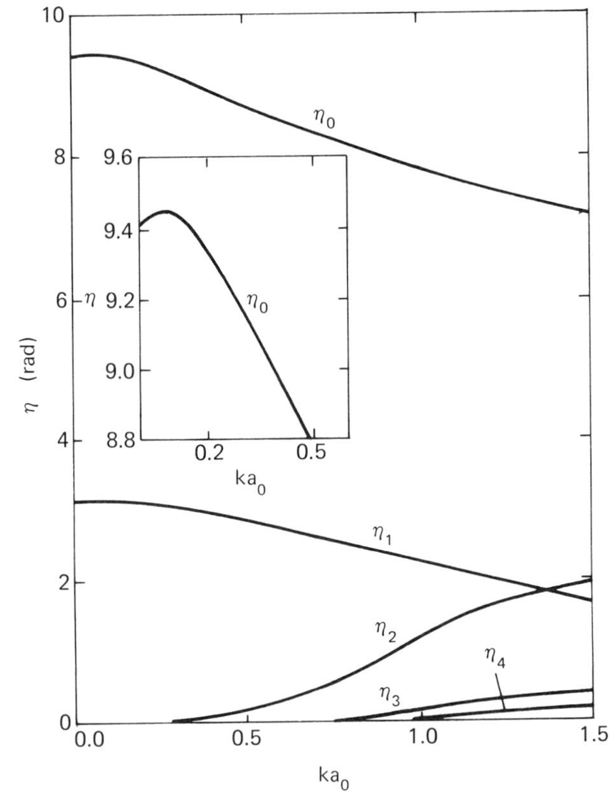

Fig. 4.2. Phase shifts in argon for different partial waves. Note the values $n\pi$ at zero energy ($k = 0$). (From Massey and Burhop [4].)

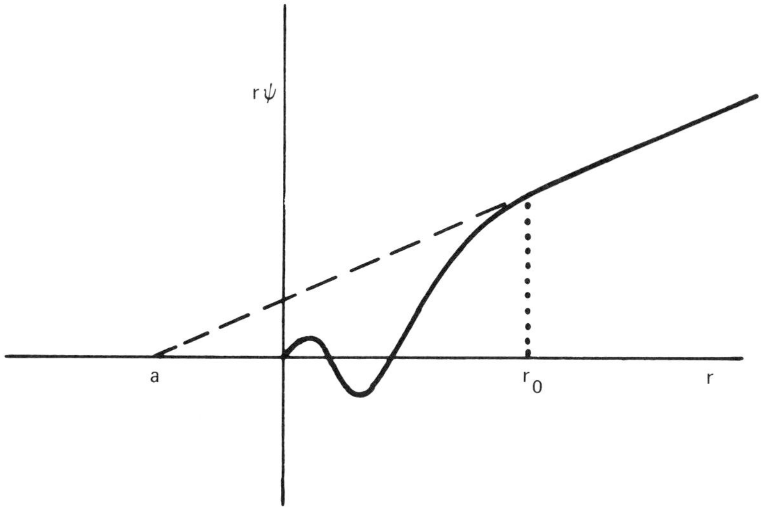

Fig. 4.3. Scattering length a for an electron with low kinetic energy in an attractive, spherically symmetric potential of radius r_0. Since the value of a is defined as the abscissa of the intercept, the scattering length shown is negative. (Adapted from Fermi [6].)

The scattering length provides ready information on the nature and strength of the potential. A negative value, as in Fig. 4.3, signifies an attractive potential that fails to support a bound state of near-zero energy; the wave function fails to turn downward before the edge of the potential, so that it cannot be matched to a falling exponential that characterizes a bound state function at $r > r_0$. If the potential is made more attractive, the wave length of $r\psi$ is reduced, thus making a more negative. Finally, a very large negative a signifies that the well is on the verge of supporting a bound state with $r\psi$ almost flat at r_0; in this event, we may speak of a virtual state. An even stronger potential will result in the slope of the wave function at r_0 turning over so that a bound state is in fact supported and a is positive. The scattering length is also positive (but small) for repulsive potentials as a result of the wave function being pushed out of the well and rising at r_0 with positive slope.

At higher values of energy, corrective terms to Eqs. (4.26–4.28) can be expressed as a series in powers of k^2 provided that the scattering potential $V(r)$ vanishes at least as $\exp(-\alpha r)$ at $r > r_0$ for some α. The first-order correction to (4.27) takes the form

$$k \cot \delta_0 \sim -(1/a) + \tfrac{1}{2} r_e k^2, \tag{4.29}$$

where r_e is called the "effective range" of the well and is independent of details of its shape. The effect of potentials $V(r)$ that vanish more slowly than $\exp(-\alpha r)$ will be treated in Section 5.5.

4.8.3 Alternative Normalization at Small Radial Distances

The vanishing of phase shifts at zero energy is seen from (4.26) to result from factors $k^{\pm(l+1/2)}$ in the wave function pair $\{j_l, y_l\}$ rather than from effects of particle interaction within the potential well. According to (4.25) both $j_l(kr)$ and $y_l(kr)$ vary rapidly as functions of the energy $E = \frac{1}{2}k^2$ at values of r sufficiently low to yield $kr \ll 1$. Indeed, this variation is nonanalytic, owing to the factors $k^{\pm 1/2} \propto E^{\pm 1/4}$. These rapid variations stem from the normalization of $\{j_l, y_l\}$, which is due, in turn, to circumstances that prevail at large r but are not immediately relevant to the phenomena within or near the potential well.

The study of the collisional process at low energies is accordingly developed more transparently by replacing the pair of functions $\{j_l, y_l\}$ with an alternative pair $\{j_l^{\mathrm{o}}, y_l^{\mathrm{o}}\}$ that is independent of energy in the limit of $r \to 0$ and is defined by

$$j_l^{\mathrm{o}}(k, r) = (kr_0)^{-[l+(1/2)]} j_l(kr), \; y_l^{\mathrm{o}}(k, r) = (kr_0)^{l+(1/2)} y_l(kr). \tag{4.30}$$

The eigenfunction $F_l(r)$ then has the expression, alternative to (4.3),

$$F_l(r) = [j_l^{\mathrm{o}}(k, r) - y_l^{\mathrm{o}}(k, r) \tan \delta_l](kr_0)^{l+(1/2)} \cos \delta_l, \qquad r > r_0. \tag{4.31}$$

Here

$$\tan \delta_l^{\mathrm{o}} = (kr_0)^{-[l+(1/2)]} \tan \delta_l (kr_0)^{-[l+(1/2)]} \tag{4.32}$$

is a new reaction parameter that remains finite at threshold and reduces to (4.27) for $l = 0$ and $k = 0$. This parameter and the corresponding renormalized reaction matrix K^{o} will be used extensively in later chapters. Figure 4.1b exhibits the effect of this renormalization for the spherical well of Section 4.7.

The wave function (4.31) is a variant of (4.25′). It is designed to show explicitly how the determination of $\tan \delta_l^{\mathrm{o}}$ by (4.23) yields a result that depends on k_0 and r_0, rather than on k, and hence depends smoothly on the energy E near its zero value. On the other hand, the coefficient $k^{l+(1/2)}$ of the eigenfunction F_l still causes its magnitude to vanish at threshold, as long as F_l remains normalized per unit energy. The amplitude of F_l within the well accordingly vanishes too, since it is matched to that of (4.31) at r_0.

This phenomenon manifests itself strikingly in phenomena in which a particle is ejected from the interior of a system but escapes with near-zero energy, typically in the photodetachment of electrons from negative ions near threshold. The cross section of this process is proportional to the squared amplitude of F_l for the ejected electron within the field of a residual neutral atom and, therefore, behaves at $E \sim 0$ as

$$\sigma \propto (k^{l+(1/2)})^2 \propto E^{l+(1/2)} \qquad \text{(Wigner threshold law).} \tag{4.33}$$

A factor $E^{1/2}$ (essentially dE/dk) in this formula stems from the normalization of F_l per unit energy and reflects the reduction of the phase space available to escaping slow particles. The remaining factor E^l stems from the hindrance to escape of slow particles in the presence of a centrifugal barrier [5].

The threshold factor $E^{l+(1/2)}$ also affects the cross section of inverse processes, in which a slow incident particle penetrates into the core of a neutral atom, where it is captured in a further reaction. The effect is partly canceled in this case by the additional factor $k^2 \propto E$ that occurs in the denominator of (4.17) and boosts the cross section for processes initiated by an incident slow particle. Finally, the factor $E^{l+(1/2)}$ occurs twice in elastic scattering, once due to the hindrance to the penetration of the centrifugal barrier by the ingoing particle on its way to the potential well and again during the escape of the particle after scattering. The factor of $k^2 \propto E$ in the denominator of the cross section (4.19) yields E^{2l} as the threshold dependence for elastic scattering of a particle with angular momentum l by a neutral system.

All these Wigner threshold laws are only valid for potentials vanishing like $\exp(-\alpha r)$—"short-range potentials." Modifications due to longer-range potentials will be considered in Section 5.5. In particular, the suppression caused by penetration through the centrifugal barrier suggests that the elastic threshold law is drastically modified for higher partial waves by a long-range potential $1/r^n$, $n > 2$. All the Wigner threshold laws are completely modified when the potential is of even longer range ($n < 2$) than the centrifugal potential [5].

The depression of cross sections near threshold is also compensated in part by the occasional occurrence of very large values of the reaction parameter $\tan \delta_l^\circ$ in (4.31) of $F_l(r)$, corresponding to large values of the expression (4.23) and (for $l = 0$) the scattering length a in (4.27). In this case the process under consideration is said to be "enhanced by resonance" of the particle within the potential well. An example of this physicomathematical effect is developed in Problem 4.1.

4.9 BOUND-STATE LEVELS

Extrapolation of a scattering wave function to represent a bound-state wave function in the range of negative energies becomes possible after casting it in the form (4.31). The essential reaction parameter $\tan \delta_l^\circ$ is now free from any singularity at $E = 0$ and may be extrapolated linearly over an apprecia-ble range, as shown by (4.29) for the important case of $l = 0$. The wave number k in the functions j_l^o and y_l^o becomes imaginary at negative energies as indicated by

$$E = \hbar^2k^2/2m \to -\hbar^2\kappa^2/2m, \qquad k \to i\kappa. \tag{4.34}$$

The trigonometric functions in the large-r expressions of $\{j_l, y_l\}$ now become hyperbolic functions of κr that generally diverge as $r \to \infty$. Bound states exist only at energy eigenvalues in which the coefficients of the rising exponential $\exp \kappa r$ of the two functions j_l^o and y_l^o of (4.31) happen to cancel. This condition determines the energy eigenvalues through the requirement that the wave number $i\kappa$ and the reaction parameter $\tan \delta_l^o$ satisfy a certain linear relationship. Standard formulas (e.g., NBS 10.2.9–10.2.10) readily show this relationship to be

$$1 - i(kr_0)^{2l+1} \tan \delta_l^o = 0, \tag{4.35}$$

and to be expressed in terms of the scattering length, in the most interesting case of $l = 0$, by

$$1 - \kappa a + O(\kappa^3) = 0. \tag{4.36}$$

An alternative form of (4.35) is obtained on combination with (4.32), namely,

$$1 - i \tan \delta_l = 0. \tag{4.35$'$}$$

Note the similarity to the result in Problem 1.4. Equation (4.36) sets the value of κ for a bound state at $1/a$; its energy level is then, according to (4.34),

$$E_b = -\hbar^2/2ma^2, \tag{4.37}$$

and the corresponding wave function tends to $\sim\exp(-\kappa r)$. A classic application of this result occurs in the study of the interaction of slow neutrons with protons. The value of the scattering length for collisions between these particles in the triplet state, $a = 5.4 \times 10^{-13}$ cm, and the energy of their bound state, the deuteron, $E = -2.2 \times 10^6$ eV are related by (4.37). On the other hand, the much larger negative scattering length for scattering in the singlet state, -24×10^{-13} cm, implies the absence of a bound state in the interaction potential between a neutron and a proton in the singlet configuration. This extrapolation of collision data to negative energies yields far more extensive results in related problems, as we shall see in the following chapters.

4.10 LEVINSON'S THEOREM†

We develop here an additional implication of the threshold relations between scattering states and bound states. The scattering length (4.27), regarded as a function

† Newton [7].

of the width r_0 of the potential well, diverges whenever k_0r_0 is an odd multiple of $\frac{1}{2}\pi$, i.e., when

$$k_0r_0 = (2mV_0)^{1/2}r_0/\hbar = (n + \tfrac{1}{2})\pi. \tag{4.38}$$

The threshold cross section also diverges in this case according to (4.28).

The half-integer $n + \frac{1}{2}$ in (4.38) represents the number of half-wavelengths of the wave function in the well, i.e., $rj_0(k_0r)$ for $l = 0$, in the range $0 \leq r \leq r_0$. The bound state condition (4.35) shows that a bound state occurs at a small negative energy, that is, for a small positive value of κ, whenever $\delta_0^\circ = k_0r_0$ is somewhat larger than $(n + \frac{1}{2})\pi$, which permits the wave function to extrapolate as $e^{-\kappa r}$ at $r > r_0$. We can see intuitively, and verify through the generalized form of (4.35), that the bound-state condition will be satisfied again at lower energies, that is, at lower values of k_0. This will occur once, and only once, as the value of k_0r_0 decreases through each successive range $(n + 1)\pi > k_0r_0 > (n + \frac{1}{2})\pi$, that is, when the number of half-wavelengths within the well is somewhat less than an integer.

In this sense the value of $\delta_0^\circ = k_0r_0$ at threshold serves as an index of the number of bound states. This index can be sharpened by interpreting (4.32) in the sense that δ_l remains in the same quadrant as δ_l° while $k \to 0$ and $\delta_l \to n\pi$; this integer n equals the number of bound states. The divergence of the cross section $4\pi a^2$ at threshold, as a function of r_0, indicates that the well is very nearly wide enough to accommodate an additional bound state. Similar considerations apply for $l > 0$, but the centrifugal potential reduces the number of half-wavelengths by $\sim l$. These results reflect the fact that the entire spectrum of bound and scattering states of a potential well forms a single orthogonal family.

4.11 GENERAL ASPECT OF THRESHOLD BEHAVIOR

The threshold laws have been derived in Section 4.8 from prior knowledge of the pair of functions $\{j_l(kr), y_l(kr)\}$, but it is instructive to rederive them here directly so as to stress their general character [5].

In the region $r > r_0$, where the short-range potential vanishes, the radial equation (4.20) has the general solution at zero energy

$$rF_l(r) = Ar^{l+1} + Br^{-l}. \tag{4.39}$$

Proceeding to nonzero but small energies, the normalization of rF_l may well show a strong energy dependence, but the ratio B/A must be weakly dependent on k^2 Dimensional inspection of (4.20) also shows that rF_l must depend on k and r in their combination kr, wherever $V(r)$ vanishes. On the other hand, the general expression (4.3) of the radial function at $r > r_0$ in terms of a regular and irregular function amounts to

$$rF_l(r) = (\text{regular}) - \tan\delta_l(\text{irregular}) \xrightarrow[kr \ll 1]{} (kr)^{l+1} - \tan\delta_l(kr)^{-l}. \tag{4.40}$$

In order that B/A be essentially independent of k, we must have $\tan\delta_l \propto k^{2l+1}$ in accordance with (4.26).

Problems

4.1 *Enhancement factor of an s wave function for a spherical well* Consider a particle of positive energy E subjected to an attractive spherical potential $V(\mathbf{r}) = -V_0$ for $r \leq r_0$. Normalize its wave function per unit energy and calculate its probability density at the center of the well ($r = 0$). Show under what condition this density has a sharp peak as a function of energy at a small value of E and relate this result to the value of the scattering length a.

4.2 Evaluate the time delay for scattering by the potential well in the previous problem. How does it relate to the enhancement factor considered there?

4.3 Consider a particle of small positive energy $E = \hbar^2 k^2/2m$ that starts at $r = r_0$ and moves out to infinity. If the particle has angular momentum l, this propagation entails the tunneling through the angular momentum barrier. Show that the tunneling probability has the threshold dependence k^{2l+1} in accordance with the remarks in Section 4.8.3, using the WKB barrier penetration formula and recalling that the WKB approximation involves the replacement $l(l+1) \to (l+\frac{1}{2})^2$.

4.4 For the short-range potential considered in Sections 4.7 and 4.8, and for $l = 0$, establish the effective range result quoted in (4.29) by retaining terms of higher order in k^2 in (4.23) and (4.25).

References

1. M. Abramowitz and I. A. Stegun, eds. "NBS Handbook of Mathematical Functions," Applied Mathematics Series 55. U.S. Govt. Printing Office, 1964.
2. H. A. Bethe and E. E. Salpeter, "Quantum Mechanics of One- and Two-Electron Atoms," p. 23. Springer-Verlag, Berlin and New York, 1957.
3. C. H. Greene, *Phys. Rev. A* **20**, 656 (1979).
4. H. S. W. Massey and E. H. S. Burhop, "Electronic and Ionic Impact Phenomena," p. 403. Clarendon, Oxford, 1969.
5. A. R. P. Rau, *Comments. At. Mol. Phys.* **44**, 285 (1984).
6. E. Fermi, *Nuovo Cimento* **11**, 157 (1934).
7. R. G. Newton, "Scattering Theory of Waves and Particles." Springer-Verlag, Berlin and New York, 1982.

5 INFLUENCE OF LONG-RANGE POTENTIALS

5.1 INTRODUCTION

The scattering of slow particles by a spherically symmetric neutral atom in its ground state is equivalent to their scattering by a short-range potential well, as we have mentioned, to the extent that either process can be described in terms of a set of energy-dependent phase shifts δ_l. The calculation of these phase shifts is, of course, far more complicated in the case of scattering by an atom, due to its multielectron structure, but we defer this problem to Chapter 6.

Here we deal instead with the additional effects that occur when the atom is surrounded by a potential field that extends far beyond the atomic radius r_0. This long-range field modifies the scattering and may also bind a particle in a finite, or even infinite, set of stationary states. The most familiar example is provided by an electron moving in the Coulomb field of a positive spherical ion in its ground state. Another example is seen in the relative motion of two slow atoms in the field of their mutual van der Waals potential.

The principal effect of the long-range field, at $r > r_0$, will be represented by a separate phase shift $\delta_l^{(l)}$ which combines additively with the contribution of short-range forces, to be called $\delta_l^{(s)}$ henceforth. The principal reason for keeping $\delta_l^{(s)}$ and $\delta_l^{(l)}$ separate is that they reflect influences operating in different regions of space and with quite *different sensitivity* to the particle energy E and to the orbital number l. However, the exact subdivision of the combined effect into $\delta_l^{(s)}$ and $\delta_l^{(l)}$ is arbitrary to the extent that it depends on the separation of the interaction at $r < r_0$ into short-range and long-range components; it also depends on the value of r_0 itself.

We shall concern ourselves in this chapter with the energy dependence of

$\delta_l^{(1)}$ for the Coulomb and other fields, with several matters incidental to extending the procedures of Chapter 4, and with the spectrum of bound state levels, which has only been touched on in Chapter 4 but is very rich, for example, for the Coulomb field. The first two sections deal with the effects of the long-range potential alone. The combination of these effects with those of short-range interactions are treated in the rest of the chapter.

The initial concepts of our approach date from Rydberg's and Bohr's work early in the century. Extensive analytical development for the Coulomb field was done by Ham and especially by Seaton and his co-workers in the 1950s under the name of "quantum defect theory" [1]. Formal extension to arbitrary fields [2] and to relativistic problems [3] is, however, quite recent.

5.2 ALTERNATIVE PAIRS OF BASE FUNCTIONS

We consider solutions of the radial Schrödinger equation in atomic units,

$$\{\tfrac{1}{2}(d^2/dr^2) - \tfrac{1}{2}[l(l+1)/r^2] - v(r) + \epsilon\} f(\epsilon, l, r) = 0, \tag{5.1}$$

where $v(r)$ indicates the "long-range" potential energy function (extrapolated to $r < r_0$) and $\epsilon = \frac{1}{2}k^2$ is the total energy of a particle. Note that the extrapolation to $r < r_0$ is largely arbitrary and that the replacement of symbols $E \to \epsilon$ and $V \to v$ implies the use of atomic units, $e = \hbar = m = 1$.

Various base pairs of solutions of this second-order differential equation play a role in describing the parameters of bound states and of scattering in the potential $v(r)$. One such pair to be considered later, normalized per unit energy, corresponds to the pair of Bessel functions indicated in Chapter 4 by $\{rj_l, ry_l\}$ for the case $v = 0$. [The extra factor r accompanies the removal of a d/dr term in constructing (5.1) and causes $rj_l(r)$ to represent the probability amplitude of the particle per unit r and per unit solid angle rather than per unit volume of space.] Such a pair consists of eigenfunctions of ϵ, $\mathbf{l}^2$, and time reversal.

Collision processes are *not* symmetric under time reversal as noted in Chapter 1; the initial state of a particle is instead usually an eigenstate of momentum. The connection between the two types of states was established in Chapter 4 through an analysis in which the real standing wave of the base pair $\{rj_l, ry_l\}$ is split into its complex ingoing and outgoing wave components $\{e^{-ikr}, e^{ikr}\}$. Likewise, we introduce here a general base pair of traveling wave solutions of (5.1) identified by their analytic form at large r,

$$\{f^-, f^+\} \underset{r\to\infty}{\longrightarrow} \{e^{-ikr}r^{-i\zeta}, e^{ikr}r^{i\zeta}\}. \tag{5.2}$$

The index ζ is introduced to handle situations in which $v(r)$ has pieces that fall off asymptotically like $1/r$ or slower. It is the root of an indicial-type equation and equals $-z_1 z_2/k$ for a Coulomb field, where z_1 and z_2 are the charges of the interacting particles.

As in Chapter 4, the base pair $\{f^-, f^+\}$ in (5.2), satisfying boundary conditions at large radial distances, serves to define the phase shift. Comparison of the relative amount of f^- and f^+ contained in a solution of (5.1) with the similar amounts in a plane wave in the absence of interactions defines the phase shift due to $v(r)$. To make this comparison, we need a solution of (5.1) that satisfies boundary conditions at the origin and is then expressed at large r as a superposition of $f^{\pm}$. The different nature of the boundary conditions at the origin and at infinity thus requires different pairs of functions to carry out the analysis. Our procedures will deal, therefore, with alternative sets of standard base pairs to handle the radial propagation of the electron from small to large r. The two ranges have different aspects and are treated separately. Thus, for example, solutions either are regular or irregular at the origin or, as in (5.2), represent ingoing or outgoing fluxes at infinity. We consider next the solutions that satisfy boundary conditions at the origin.

In accordance with the introductory remarks to this chapter, we aim to keep separate the contributions to the phase shift from long-range and short-range interactions. The immediate purpose of this and the subsequent section is to derive scattering parameters due to the *long-range* $v(r)$ *alone*. Therefore, we need initially only the solution of (5.1) that is regular at the origin. A companion irregular solution will become necessary later on to formulate the phase shift due to additional short-range interactions confined to a range $r < r_0$.

Since the phase shift $\delta_l^{(1)}$ contributed by the long-range potential $v(r)$ is unaffected by an overall multiplicative factor in the wave function, our immediate purpose is served by any solution of (5.1) that is regular at the origin and is independent of any specific choice of normalization. With an eye to later extensions when short-range interactions are present, it is reasonable to choose a solution with energy-independent normalization.

At the origin, the term in ϵ in (5.1) is totally overwhelmed by the potential terms so that any energy dependence introduced into the solutions through normalization factors necessarily reflects influences from larger distances. Our choice, therefore, for the regular function f° is the counterpart not of rj_l of Chapter 4 but rather of rj_l°, called the "renormalized function" in Section 4.8.3. This normalization of f° removes influences from larger radial distances that are extraneous to the small-r region and that would otherwise introduce dependences on phenomena extraneous to the effects of a short-range scatterer.

As discussed in Chapter 4, there is merit in separating out the singularities introduced by long-range influences on the scattering parameters defined by

(4.26) and (4.33), the Wigner threshold laws. These threshold singularities (at $\epsilon = 0$) are real but not specific to the collision proper, as evidenced by the fact that the same threshold laws obtain regardless of the nature of the short-range interaction. Being dominant at $\epsilon \sim 0$, they often obscure more regular and smooth features of the collision. For these reasons renormalizations such as the $k^{\pm[l+(1/2)]}$ in (4.30) serve the useful purpose of filtering out the dynamical effects of zero velocity and of forces at large r [centrifugal in the case of (4.30)] from the dynamics specific to the short-range collision proper. More general renormalizations when $v(r)$ contains asymptotically Coulomb or attractive $1/r^2$ fields have been given in Greene *et al.* [2] and analytical f°'s have been defined for general potentials. Generally, f° is defined through its boundary condition

$$f^{\circ} \xrightarrow[r\to 0]{} r^{l+1}, \tag{5.3}$$

that is, with unit normalization.

5.3 JOST FUNCTIONS

The next step is to match the behavior at large and small r, that is, to express f° in terms of the base pair $\{f^-, f^+\}$ defined with respect to boundary conditions at infinity [4].

$$f^{\circ}(\epsilon, l, r) = J^+(k, l) f^+(\epsilon, l, r) + J^-(k, l) f^-(\epsilon, l, r). \tag{5.4}$$

The coefficients $J^{\pm}$ are called Jost functions, because Jost emphasized their important properties:

(1) $J^{\pm}(k, l) k^{l+1}$ is an analytic function of k and l for short-range potentials;
(2) $J^{\pm}(k, l) \propto \lim_{r\to 0} r^l f^{\mp}(\epsilon, l, r)$ ensures that f° vanishes at $r = 0$;
(3) $J^-(k^*) = [J^+(k)]^*$ at $\mathrm{Re}\, \epsilon > 0$.

The Jost functions can be expressed as Wronskians, recalling that $W(f^{\pm}, f^{\pm}) = 0$, and $W(f^{\pm}, f^{\mp}) = \mp 2ik$ from expression (5.2) of these functions at large r. Taking the Wronskian of both sides of (5.4) with f^+ and f^- yields

$$J^{\pm}(k, l) = \pm W(f^{\mp}, f^{\circ})/2ik. \tag{5.5}$$

Comparison of (5.4) with the standard asymptotic form for scattering [see Eq. (4.4)] shows that $J^{\pm}$ is proportional to $\exp(\pm i\delta_l^{(1)})$ for positive energies where $\delta_l^{(1)}$ is the phase shift due to the long-range field. Explicit expressions for

$J^{\pm}$ when $v(r)$ is some combination of Coulomb and dipole fields have been developed in Greene *et al.* [2] and are given in Table 5.1. The Jost functions are more convenient objects of study than, for instance, the scattering matrix $S_l(k)$, which is proportional to J^+/J^-. They have more desirable analytical properties: Whereas S_l is singular both at the singularities of J^+ and at zeroes of J^-, physical significance is attached only to the latter set. As we will see in Section 5.3.2, the zeroes of J^- are associated with bound states and resonances.

A major aspect of Jost functions is that they describe the effects of a "one-way traversal" of the field from small to large r. This is illustrated by (5.4), wherein f° at small r is related to the large-r solutions $f^{\mp}$. As a result, it is the Jost functions that are directly relevant to physical phenomena such as photoionization that involve only a one-way motion: a particle released by a photon near the origin moves out from this region to infinity. The description of bound states also has a one-way character in that for $\epsilon < 0$, a wave function regular at small r must connect at large r to an exponentially decaying function at infinity. Once again, therefore, bound states find their natural description in terms of Jost functions. The scattering matrix, on the other hand, depending as it does on the ratio of the coefficients of f^- and f^+ in (5.4), is relevant to the "two-way traversals" of a scattering process in which a particle starts at large r, goes into the collisional region, and is again monitored at large r.

5.3.1 The Quantum Defect Approach

The essence of quantum defect procedures lies in expressing scattering and bound-state properties in terms of Jost functions. Different combinations of J^+ and J^- play a role. We have already noted that the scattering matrix is given by the ratio J^+/J^-. For the description of the wave function itself, however, we need full knowledge of J^+ and J^- as in (5.4). Quantum defect procedures handle the effects of a short-range scatterer economically by restricting the radial integration of the Schrödinger equation to the small range of r covered by this collisional region (usually the most complicated region of the problem). Radial integration in the outer potential is handled separately and often analytically. All spectral properties are expressed in terms of the standard J^+ and J^-, which connect f° to a base pair for the long-range field, along with the superposition parameters involved in writing the small-r solutions in terms of such a base pair.

It is customary to choose for the basic parameters of quantum defect theory (QDT) the various combinations of J^+ and J^- that can be ascribed physical significance. The phase shift δ, which involves the ratio J^+/J^- is one such parameter. The product J^+J^- is another and is related to the parameter

Table 5.1

QDT Parameters for Coulomb and "Dipole" Fields and in the WKB Approximation[a]

Parameter[b]	Coulomb[c]	Dipole[d]	WKB expression[e]
$B(k, l)$	$\dfrac{\Gamma(l+1+iZ/k)\Gamma(l+1-iZ/k)}{[\Gamma(2l+2)]^2}\dfrac{e^{\pi Z/k}}{\pi}(2k)^{2l+1}$	$\dfrac{2\sinh\pi\alpha}{\pi\alpha[\cosh\pi\alpha+\cos\{2\alpha\ln(2/k)-2\chi_\alpha\}]}$	$[\pi(2l+1)]^{-1}\exp\left[-2\lim_{r\to 0}\left\{\int_r^{r_1}\lvert k(r')\rvert dr'+(l+\frac{1}{2})\ln r\right\}\right]$
$A(\kappa, l)$	$2\,\mathrm{Re}[B(i\kappa, l)]$		$B(i\kappa, l)$
$\delta_l^{(1)}(k)$	$-\frac{1}{2}l\pi+(Z/k)\ln 2k+\arg\Gamma(l+1-iZ/k)$	$(\pi/4)+\tan^{-1}[\tanh(\frac{1}{2}\pi\alpha)\tan\{\alpha\ln(2/k)-\chi_\alpha\}]$	$(\pi/4)+\lim_{r\to\infty}\left[\int_{r_1}^{r}k(r')dr'-kr-(Z/k)\ln r\right]$
$\beta(\kappa, l)$	$\pi(-l+Z/\kappa)$	$(\pi/2)+\tan^{-1}[\tanh(\pi\alpha)\tan\{\alpha\ln(2/\kappa)-\chi_\alpha\}]$	$(\pi/2)+\int_{r_1}^{r_2}k(r')dr'$
$D(\kappa, l)$	$\pi^{1/2}(2\kappa)^{Z/\kappa}[\Gamma(l+1+Z/\kappa)\Gamma(-l+Z/\kappa)]^{-1/2}$	$\cosh^{1/2}(\pi\alpha)$	$\sqrt{\frac{1}{2}}\exp\left[\lim_{r\to\infty}\left\{\kappa r-(Z/\kappa)\ln r-\int_{r_2}^{r}\kappa(r')\,dr'\right\}\right]$

[a] The energy is expressed in terms of wave numbers ($\epsilon = \frac{1}{2}k^2 > 0$ or $\epsilon = -\frac{1}{2}\kappa^2 < 0$) in atomic units.

[b] The relationships between B and A and between δ and β reflect the analytic behavior of Jost functions $J^{\pm}(k, l)$; additional relations are $\beta(\kappa, l) = 2\arg[\exp\{i\delta_l^{(1)}(i\kappa)\}]$ and $D^2(\kappa, l)/\sin\beta(\kappa, l) = \lvert\exp\{i\delta_l^{(1)}(i\kappa)\}\rvert^2$.

[c] $Z = -z_1 z_2$, where z_1 and z_2 are the charges of interacting particles.

[d] This column pertains to an attractive $1/r^2$ potential whose strength is represented by the index α that occurs in Eq. (5.29) with $\chi_\alpha = \arg\Gamma(1 - i\alpha)$.

[e] $k^2(r) = -\kappa^2(r) = k^2 - 2v(r) - (l + \frac{1}{2})^2/r^2$ is assumed to have a single maximum corresponding to the bottom of a single well of the combined potential $2v(r) + (l + \frac{1}{2})^2/r^2$; the range of classical motion in this well is limited by two turning points, $r_1 \le r \le r_2$, with $r_2 \to \infty$ for $\epsilon > 0$.

B in Table 5.1. In terms of these, (5.4) takes the alternative form

$$f^{\circ} = [B(k,l)]^{-1/2}(2/\pi k)^{1/2}\{f^{+}\exp[i\delta_l^{(l)}] - f^{-}\exp[-i\delta_l^{(l)}]\}/2i \quad (5.6a)$$

$$\underset{r\to\infty}{\longrightarrow} [B(k,l)]^{-1/2}(2/\pi k)^{1/2}\sin(kr + \zeta\ln r + \delta_l^{(l)}), \qquad \epsilon > 0. \quad (5.6b)$$

The parameter B, therefore, represents the normalization constant that relates regular solutions with the energy-independent normalization in (5.3) to those that are normalized per unit energy range. Comparison of (5.6b) and (4.10) shows that the function

$$f(\epsilon,l,r) = [B(k,l)]^{1/2}f^{\circ}(\epsilon,l,r) \quad (5.7)$$

is the regular solution of (5.1) normalized per unit energy ϵ; it is thus the counterpart of $(2k/\pi)^{1/2}rj_l(kr)$ in Section 4.3. In the important special case of a long-range Coulomb field, we have $f(\epsilon,l,r) = (2/\pi k)^{1/2}F_L(\eta,kr)$ with $L = l$, $\eta = -\zeta = z_1z_2/k$, and F_L the Coulomb function defined in NBS† Chapter 14.

5.3.2 Bound States

We consider here bound states in the long-range potential $v(r)$. Of particular interest to QDT are potentials that fall off at infinity slower than $1/r^2$ or, in the case of an attractive $-1/r^2$ potential, of sufficient strength to bind an infinite number of discrete states. In these cases the systematics of bound states are mostly controlled by the long-range field even in the presence of additional short-range interactions. The modifications due to the short-range terms will be discussed in Section 5.6.

The regular solution in (5.3) is defined irrespective of the energy. In particular, f° is also appropriate to the description of the discrete spectrum when $\epsilon < 0$ and $k = i\kappa$ with κ real and by convention positive. The pair of independent solutions satisfying boundary conditions at infinity $\{f^{-}, f^{+}\}$ are, however, no longer oscillatory functions but rise and fall exponentially at infinity. The presence of a function with singular behavior (the exponentially rising term) causes some difficulties in the proper definition of $\{f^{-}, f^{+}\}$ and (5.4) and, thereby, in the definition of $J^{\pm}$. This is a generic difficulty in defining a singular solution of a differential equation at its point of singularity, which will be encountered again in Section 5.4 and will be discussed in Section 5.11. That discussion will lead to a definite prescription for extrapolating the definitions of $\{f^{-}, f^{+}\}$ and $\{J^{-}, J^{+}\}$ from positive to negative values of the energy ϵ.

Equation (5.4) and the $J^{\pm}$ so obtained thus represent the small-r to large-r connection at $\epsilon < 0$. Bound-state energy eigenvalues are then given by the

† Reference [1] of Chapter 4.

boundary condition that the coefficient of the exponentially rising term vanish, that is,

$$J^-(\kappa, l) = 0. \tag{5.8}$$

A particular illustration of this result was provided in the simple model problem considered in Section 4.9, albeit with a short-range potential. In that problem the bound states in the square well were obtained by equating to zero the coefficient of the rising exponential $e^{\kappa r}$. Another example in the present context is for the Coulomb potential, where the expression for $J^-(\kappa, l)$ involves a gamma function $\Gamma(l + 1 - Z/\kappa)$ in the denominator [2]. The zeroes of J^- lie where the argument of the gamma function becomes zero or a negative integer,

$$l + 1 - (Z/\kappa) = -n_r, \qquad n_r = 0, 1, 2, \ldots. \tag{5.9}$$

This implies the familiar Bohr–Rydberg formula for bound states in a Coulomb potential with n_r the radial quantum number.

As in the case of $\epsilon > 0$, it is customary in QDT to parametrize J^+ and J^- in terms of their various combinations when $\epsilon < 0$. The very large difference in amplitude of f^- and f^+ at large r introduces a new element in this case. At $\epsilon > 0$, it is clear from (5.1), (5.4), and (5.6) that f^- and f^+ have equal amplitude so that the ratio J^+/J^- is a pure phase,

$$J^+/J^- = -\exp(2i\delta_l^{(1)}), \qquad \epsilon > 0. \tag{5.10a}$$

At negative energy, however, the difference in amplitudes of f^+ and f^- translates into a corresponding difference in J^+ and J^-; their ratio (which is now real) is more conveniently parametrized as the product of a smooth factor D^2 and a periodic function of a phase variable β,

$$J^+/J^- = -D^2 \cot \beta, \qquad \epsilon < 0. \tag{5.10b}$$

The parameter D—a smooth positive function of energy—serves to rescale the amplitude of the base pair $\{f^-, f^+)$, so that we only encounter $\{D^{-1}f^-, Df^+\}$, that is, $\{D^{-1}e^{\kappa r}, De^{-\kappa r}\}$, and correspondingly $\{DJ^-, D^{-1}J^+\}$. The parameter β at negative energies is the counterpart of the phase shift $\delta_l^{(1)}$ at $\epsilon > 0$. It represents the number of half wavelengths of the solution to (5.1) between $r = 0$ and $r = \infty$. The definition of bound states in (5.8) as the zeroes of J^- is translated into

$$\sin \beta(\epsilon, l) = 0. \tag{5.11}$$

The product J^+J^-, which is $(2\pi kB)^{-1}$ at $\epsilon > 0$, is written instead as

$-(2\pi\kappa A)^{-1}\sin 2\beta$ at $\epsilon < 0$. As a result of these definitions (5.4) takes the form

$$f^{\circ}(\kappa, l, r) = [A(\kappa, l)]^{-1/2}(1/\pi\kappa)^{1/2}[\sin\beta\, D^{-1}f^{-} - \cos\beta\, Df^{+}], \qquad \epsilon < 0. \tag{5.12}$$

Explicit formulas for the five QDT parameters B, δ, A, β, and D for some standard fields are given in Table 5.1 together with more general WKB expressions drawn from Greene *et al.* [2]. The extension of this treatment to include the effect of short-range forces will be given in Section 5.6.

5.4 PARAMETERS OF SHORT-RANGE INTERACTIONS

The previous sections have described spectral and scattering parameters for the long-range potential $v(r)$ alone. In atomic physics they suffice to give a complete description of the $e + H^{+}$ system. The general problem of scattering and bound states of an electron with an atomic scatterer consisting of the nucleus and all the other electrons involves additional interactions to be represented by a short-range field $v_s(r)$. This interaction is generally nonlocal because of electron exchange. We turn now to a study of the modifications in the scattering parameters due to a $v_s(r)$ that vanishes outside the scatterer's radius r_0.

When the potential $v_s(r)$ is added to the potential $v(r)$ of (5.1), the radial eigenfunction of relevance is the regular solution $F(\epsilon, l, r)$ of

$$\{\tfrac{1}{2}(d^2/dr^2) - \tfrac{1}{2}[l(l+1)/r^2] - v(r) - v_s(r) + \epsilon\}F(\epsilon, l, r) = 0. \tag{5.13}$$

Such a function, regular at the origin, must be integrated outward to the boundary of the scatterer and there expressed in terms of solutions for the field $v(r)$ alone. To define a short-range phase shift shorn of complications due to the effects of $v(r)$ at large distances, we need a second independent solution of $v(r)$ defined at small r. This requirement arises because the function f° alone is not the most general solution of (5.1) and hence is inadequate to the task of describing $F(\epsilon, l, r)$ at the edge of the scatterer.

The general solution includes (besides f°) an irregular companion with which it forms a base pair $\{f^{\circ}, g^{\circ}\}$. The second independent solution g° of the same radial equation (5.1) is irregular at the origin. In the main, g° should lag 90° in phase behind f° for the purpose of defining a phase shift $\delta_l^{(so)}$ that depends only on forces at $r < r_0$. A 90° phase lag would result by defining g° in terms of the $f^{\pm}$, analogously to (5.4), with coefficients differing from (J^{+}, J^{-}) by phase factors $\mp i$. However, this procedure would incorporate in g° dependences on large-r phenomena that are extraneous to the short-range interaction $v_s(r)$ and nonanalytic in the energy ϵ at the threshold $\epsilon = 0$. The elimination of this dependence poses the same problem encountered in Section 5.3.2 when defining f^- at $\epsilon < 0$, whose discussion is deferred to Section 5.11.

The discussion in Section 5.11 leads us to construct the solution $g^o(r)$ of Eq. (5.1) in these steps:

(1) A function $g(\epsilon, l, r)$ is defined that lags 90° in phase at large r behind the solution $f(\epsilon, l, r)$ of (5.7) and is also normalized per unit energy. This function is represented by

$$g(\epsilon, l, r) = -(2/\pi k)^{1/2}\{f^+ \exp[i\delta_l^{(1)}] + f^- \exp[-i\delta_l^{(1)}]\}/2$$
$$\underset{r\to\infty}{\longrightarrow} -(2/\pi k)^{1/2} \cos[kr + \zeta \ln r + \delta_l^{(1)}]. \quad (5.14)$$

(2) The function (5.14) is multiplied by the normalization factor $[B(k, l)]^{1/2}$ of (5.7) so as to obtain the energy-independent irregular behavior at small r,

$$g^o(\epsilon, l, r) \underset{r\to\infty}{\longrightarrow} -[\pi(l + \tfrac{1}{2})]^{-1}\, r^{-l}, \quad (5.15)$$

and to preserve the Wronskian relation

$$W(f^o, g^o) = W(f, g) = 2/\pi. \quad (5.16)$$

(3) Any residual nonanalytic behavior of $g(\epsilon, l, r)$ is eliminated by an operation that affects neither (5.15) nor (5.16), namely, by subtracting from $B^{1/2}g$, a term proportional to the regular solution f^o, thus setting

$$g^o(\epsilon, l, r) = [B(k, l)]^{1/2}\, g(\epsilon, l, r) - \mathscr{G}(\epsilon, l) f^o(\epsilon, l, r). \quad (5.17)$$

The coefficient $\mathscr{G}$ serves as a sixth parameter of QDT, to supplement the ones already discussed and is given at the end of Section 5.11. The full connection between the base sets $\{f, g\}$ and $\{f^o, g^o\}$ is thus represented by

$$\begin{pmatrix} f \\ g \end{pmatrix} = \begin{pmatrix} B^{1/2} & 0 \\ B^{-1/2}\mathscr{G} & B^{-1/2} \end{pmatrix} \begin{pmatrix} f^o \\ g^o \end{pmatrix}. \quad (5.18)$$

(4) The passage between the base sets when $\epsilon < 0$ is given by the same relation (5.18) with A in place of B. The definition of the irregular functions for $\epsilon < 0$ is completed by the relation analogous to (5.12),

$$g^o(\kappa, l, r) = -[A(\kappa, l)]^{1/2}(\pi\kappa)^{-1/2}[\cos\beta\, D^{-1} f^- + \sin\beta\, D f^+] - \mathscr{G} f^o(\kappa, l, r). \quad (5.12')$$

5.4.1 Basic Phase Shift $\delta^{(so)}$

The regular solution $F(\epsilon, l, r)$ of (5.13) is now expressed outside the boundary of the scatterer in terms of the base pair $\{f^o, g^o)$ by

$$F(\epsilon, l, r) = \mathscr{N}(\epsilon) \cos \delta_l^{(so)}[f^o(\epsilon, l, r) - g^o(\epsilon, l, r) \tan \delta_l^{(so)}]. \quad (5.19)$$

The amplitude $\mathcal{N}(\epsilon)$ and the phase $\delta_l^{(\text{so})}$ are determined below by fitting the value and derivative of the radial function at the core boundary; $\mathcal{N}(\epsilon)$ is a normalization factor to be discussed further in Section 5.7.3 and usually chosen to normalize $F(\epsilon, l, r)$ per unit energy range. The superscript $^{\circ}$ on the phase shift indicates that it is shorn of influences of the long-range field by virtue of being referred to the analytic base pair $\{f^{\circ}, g^{\circ}\}$. In the absence of the short-range scatterer this phase shift vanishes and $F(\epsilon, l, r)$ reduces to a regular solution of the long-range field alone, which differs from f° only in its normalization.

5.4.2 Contribution to the Total Phase Shift

Our next task is to translate the short-range shift at the core boundary, $\delta_l^{(\text{so})}$, into a phase shift $\delta_l^{(\text{s})}$, that applies to the base pair $\{f, g\}$ at large r and which will eventually be added to the long-range phase shift $\delta_l^{(\text{l})}$ of (5.6). The translation is accomplished by comparing the expression (5.19) of $F(\epsilon, l, r)$ with the corresponding expression in terms of $\{f, g\}$, namely,

$$F(\epsilon, l, r) = f(\epsilon, l, r) \cos \delta_l^{(\text{s})} - g(\epsilon, l, r) \sin \delta_l^{(\text{s})}, \tag{5.20}$$

which has the standard form (4.3) for a scattering eigenfunction. Substitution of (5.18) into (5.20) and comparison with (5.19) now yields

$$\mathcal{N}(\epsilon) \cos \delta_l^{(\text{so})} = B^{1/2} \cos \delta_l^{(\text{s})} - B^{-1/2}\mathcal{G} \sin \delta_l^{(\text{s})} \tag{5.21a}$$

$$\mathcal{N}(\epsilon) \sin \delta_l^{(\text{so})} = B^{-1/2} \sin \delta_l^{(\text{s})}. \tag{5.21b}$$

The function F in (5.20) is normalized per unit energy, just as are $\{f, g\}$. The value of $\mathcal{N}(\epsilon)$ for energy normalization follows then by squaring and summing (5.21a) and (5.21b),

$$\mathcal{N}^2(\epsilon) = B \cos^2 \delta_l^{(\text{s})} + B^{-1}(1 + \mathcal{G}^2) \sin^2 \delta_l^{(\text{s})} - \mathcal{G} \sin 2\delta_l^{(\text{s})}. \tag{5.22}$$

The reciprocal relations between $\delta_l^{(\text{so})}$ and $\delta_l^{(\text{s})}$ alone,

$$\tan \delta_l^{(\text{so})} = B^{-1} \tan \delta_l^{(\text{s})}/[B^{1/2} - B^{-1/2}\mathcal{G} \tan \delta_l^{(\text{s})}], \tag{5.23a}$$

$$\tan \delta_l^{(\text{s})} = B^{1/2} \tan \delta_l^{(\text{so})}/[B^{-1/2} + B^{-1/2}\mathcal{G} \tan \delta_l^{(\text{so})}], \tag{5.23b}$$

follow instead by eliminating $\mathcal{N}(\epsilon)$ from the pair (5.21). The results (5.23) are therefore the actual analogs of (4.32), to which they reduce on substitution of $\mathcal{G} = 0$ and $B \propto k^{2l+1}$, which are appropriate to zero field as seen by setting $Z = 0$ in the first column of Table 5.1. The expression for the normalization constant in (5.22) can also be cast in terms of the smooth short-range phase shift through (5.23b) as

$$\mathcal{N}^2(\epsilon) = B/[(\cos \delta_l^{(\text{so})} + \mathcal{G} \sin \delta_l^{(\text{so})})^2 + B^2 \sin^2 \delta_l^{(\text{so})}]. \tag{5.22$'$}$$

The full phase shift follows from (5.20) when expressed in terms of the pair

$\{f^-, f^+\}$. Using (5.6a), (5.7), (5.12), (5.12′), (5.14), and (5.18) to write $\{f, g\}$ in terms of $\{f^-, f^+\}$, we have

$$F(\epsilon, l, r) = (2/\pi k)^{1/2}[f^+ e^{i(\delta_l^{(1)} + \delta_l^{(s)})} - f^- e^{-i(\delta_l^{(1)} + \delta_l^{(s)})}]/2i, \qquad \epsilon > 0 \tag{5.24a}$$

$$= (1/\pi\kappa)^{1/2}[D^{-1} f^- \sin(\delta_l^{(s)} + \beta) - D f^+ \cos(\delta_l^{(s)} + \beta)], \qquad \epsilon < 0. \tag{5.24b}$$

The short-range phase shift $\delta_l^{(s)}$ is thus clearly additive to the long-range phases $\delta_l^{(1)}$ when $\epsilon > 0$ or to β when $\epsilon < 0$. Equations (5.24) express the regular solution of the Schrödinger equation (5.13) with the full potential $v(r) + v_s(r)$, in terms of $\{f^-, f^+\}$ in the outer region. Comparison with (5.4) then allows the coefficients of f^- and f^+ to be interpreted as Jost functions of the combined potential. The ratio of these full Jost functions is again exactly as in (5.10) with the addition of $\delta_l^{(s)}$ to $\delta_l^{(1)}$ or β. The resulting modifications to the bound state energy eigenvalues will be taken up in Section 5.6.

5.5 THRESHOLD BEHAVIOR: RESONANCES

The nonanalytic energy dependence of the parameters A, B and $\mathscr{G}$ for long-range fields leads to drastic alterations of the singularities at $\epsilon = 0$ and of the threshold laws for scattering processes in the presence of long-range forces. As discussed earlier following (5.23), $\tan \delta_l^{(s)}$ would be proportional to k^{2l+1} in the absence of long-range forces. Similarly, an inelastic scattering cross section for a final state described by (5.19) has its leading energy dependence contained in $\mathcal{N}^2(\epsilon)$ and is also proportional to k^{2l+1} according to (5.22′). On the other hand, if $v(r)$ contains a Coulomb field, the coefficient B of (5.23) or (5.22′) reduces to a constant as $k \to 0$ for all l and so do the phase shift and the inelastic scattering cross section. The algebra involved in this reduction is complicated, but an essential point of physics is worth noting.

In the presence of long-range potentials, the zone of large r is appropriately viewed as splitting into two. A truly asymptotic zone lies, for any energy ϵ, beyond a radius r_w defined by $|v(r_w)| = \epsilon$. In this zone the familiar asymptotic behavior of base pairs obtains, such as the one in (5.2), with characteristic oscillations controlled by the wavelength $2\pi/k$. A second zone (called the Coulomb zone for problems with the Coulomb potential) lies at r values less than r_w. Here the long-range potential is itself dominant. Characteristic oscillations of the wave function are then controlled by the potential rather than by ϵ. This Coulomb zone stretches to infinity at zero energy.

The energy-normalized zero-energy Coulomb functions take the form (see

Rau [5] for the role played by a pair of branch points and saddle points)

$$f(0, l, r) = (2mr/\hbar^2)^{1/2} J_{2l+1}[(8Zr/a_0)^{1/2}]. \tag{5.25}$$

Thereby f differs from f^0 at the origin only by a constant multiplier, that is, B is independent of ϵ as noted. At finite energy, $\epsilon \gtrsim 0$, Eq. (5.25) constitutes the leading term in an expansion of $f(\epsilon, l, r)$ in powers of ϵ. The occurrence of a potential-dominated zone at large r lies at the heart of the changes in threshold singularities caused by long-range potentials. The Coulomb potential provides one of the most familiar illustrations: The photoionization cross section of a neutral atom remains finite and smooth in the vicinity of the threshold for each partial wave of the outgoing electron. This behavior contrasts with the strong ϵ and l dependence observed for the process of photodetachment from a negative ion, where the Wigner laws apply and the cross section rises from zero at threshold in proportion to $\epsilon^{l+(1/2)}$.

The feature that all the threshold nonanalyticities arise from the parameters B and $\mathcal{G}$ in (5.23) and (5.22′) emphasizes the chief characteristic of threshold behavior, that the sensitive energy dependences are due only to the long-range potential $v(r)$ beyond r_0. The nature or strength of the short-range interactions $v_s(r)$ confined to $r < r_0$ play no role here. The effects of the purely small-r behavior can be described entirely in terms of $\delta_l^{(so)}$, which is smooth in energy. Conversely, the interest in threshold nonanalyticities stems from the information they provide on the long-range interactions between the collision partners, shorn of the complexities that arise at short ranges.

The structure of the equations (5.23) also shows how the renormalization of the short-range pair $\{f^o, g^o\}$ to $\{f, g\}$ gives rise to resonances when the denominator in (5.23b) experiences a zero and a change in sign whereby the phase shift $\delta_l^{(s)}$ rises by π. More precisely, the time delay (Section 4.4) due to the short-range potential, $d\delta_l^{(s)}/dE$, can become large over a small energy range as a result of the energy dependence of B and $\mathcal{G}$.

A resonance also occurs then in the scattering cross section. Since the variation with energy of $\delta_l^{(s)}$ is superposed on the other contributions to the total phase shift such as $\delta_l^{(l)}$, the width and shape of the resonance will depend on all these quantities. This subject will be discussed fully in Chapter 8.

5.6 NEGATIVE ENERGIES: SPECTRUM OF BOUND-STATE LEVELS

The eigenvalues $\varepsilon < 0$ have been expressed in Section 5.3.2 as zeroes of the Jost function J^- or, correspondingly, in (5.11) as the zeroes of sin β. The analysis in Section 5.4 of the effects of the short-range potential on these functions now permits us to study the effects of the short-range $v_s(r)$ on the discrete states. In this manner we get a description of the bound states of a particle held by a long-range potential around a general atomic core.

The key result is (5.24b), stating that the effect of the short-range potential on the eigenfunction at $\epsilon < 0$ is to add $\delta_l^{(s)}$ to the QDT phase parameter β. Therefore, the eigenvalue condition (5.11), stating that the coefficient of the rising exponential f^- in (5.24b) vanishes, now takes the form

$$\sin[\beta(\epsilon) + \delta_l^{(s)}(\epsilon)] = 0, \tag{5.26}$$

that is,

$$\beta(\epsilon) + \delta_l^{(s)}(\epsilon) = \delta_l(\epsilon) = (n_r + 1)\pi, \qquad n_r = 0, 1, 2, \ldots. \tag{5.27}$$

Note that $\delta_l(\epsilon)$ represents the total phase accumulated by $F(\epsilon, l, r)$ from $r = 0$ to $r = \infty$ at negative energies. Of this total $\beta(\epsilon)$ is contributed by the long-range potential, including a fixed contribution of $\pi/4$ from the exponential tail of the wave function at large r to be discussed in Section 5.8. The statement in (5.27) about the total phase accumulation of F means simply that an integral number of half wavelengths of F must fit within the range $0 < r < \infty$. The lowest state of given l has $n_r = 0$ and its radial wave function is nodeless with a single half wavelength between the origin and infinity.

The separation of δ_l into the two contributions $\delta_l^{(s)}$ and β proves important. In the absence of resonances in the valence shell, the short-range contribution can vary appreciably as a function of ϵ only on a scale comparable to the potentials prevailing within the atom, that is, $\gtrsim 5$ eV. The long-range contribution, on the other hand, depends on the weaker portions of $v(r)$ and thus can vary rapidly with ϵ. This is especially so as ϵ approaches zero and the long-range part of $v(r)$ exerts a dominant influence on β. For this reason it is important that this part, which is more sensitive to energy, be handled in a standard manner, as detailed in the previous sections. Asymptotic trends are evaluated analytically for $v(r)$ proportional to $-1/r$ and $-1/r^2$, thus yielding the expressions for β given in Table 5.1. Contributions from other potential terms that converge faster than $1/r^2$ at infinity can be handled typically in terms of WKB phase integrals [2] or numerically if greater accuracy is required.

The distribution of levels within the spectrum depends then primarily on $v(r)$ and can be studied with little reference to short-range interactions. The separation between successive levels $(\epsilon_n, \epsilon_{n+1})$ is such that $\beta(\epsilon_{n+1}) - \beta(\epsilon_n) = \pi$. Each value of $\beta(\epsilon_n)/\pi$ departs from an integer by the nearly constant amount $\delta_l^{(s)}(\epsilon_n)/\pi$, which is accordingly called a quantum defect and generally indicated by μ. The term arose historically in the most important case of a Coulomb field $v(r) = -Z/r$. The dependence of β on ϵ is expressed analytically in Table 5.1, and (5.9) has already recorded the bound-state condition for a pure Coulomb field obtained by equating β to $(n_r + 1)\pi$. The addition in

(5.27) of the term $\delta_l^{(s)}$ due to short-range interactions now yields

$$\epsilon_n = -\tfrac{1}{2}[1/(n - \delta^{(s)}/\pi)^2] = -\tfrac{1}{2}[1/(n - \mu)^2], \tag{5.28}$$

with $n = n_r + l + 1$ the principal quantum number. This result has had countless applications, particularly to the spectra of alkali atoms whose ionic cores have the properties assumed in this chapter. Extension of (5.28) to other atoms will be a main objective of Part C.

Analytical expressions are also available in Table 5.1 for the interaction of an electron with the electric dipole moment of a polar molecule or of an excited H atom in which a permanent dipole moment is allowed by the degeneracy between states of opposite parity. The orbital angular momentum is no longer conserved in such $-1/r^2$ potentials, but the centrifugal and dipole potentials combine to yield $v(r) + l(l+1)/r^2 = \lambda(\lambda+1)/r^2$ with a constant and complex value† of $\lambda = -\frac{1}{2} + i\alpha$. From the value of β in the second column of Table 5.1, the vanishing of $\sin\beta$ now requires

$$\alpha \ln(2/\kappa) - \arg\Gamma(1 - i\alpha) = (n + \tfrac{1}{2})\pi \tag{5.29}$$

so that the series of levels is described by

$$\epsilon_n = \epsilon_0 e^{-2n\pi/\alpha}. \tag{5.30}$$

Here, as for the Coulomb field, the series is infinite and converges to the threshold $\epsilon = 0$ beyond which the particle escapes from the potential field. Potentials $v(r)$ converging at large r faster than $1/r^2$ yields only a finite set of discrete levels ϵ_n and a milder singularity of all QDT parameters at $\epsilon = 0$.

Theoretical knowledge of the long-range contribution to the phases $\beta(\epsilon)$ can be combined with experimental data on ϵ_n to yield a semiempirical determination of the short-range contribution $\delta_l^{(s)}$. This contribution constitutes basic information on the particle–core interaction and determines the influence of this interaction on scattering at $\epsilon > 0$. For purposes of this extrapolation the appropriate parameter is $\delta_l^{(so)}$, the phase shift is smooth in energy and obtained from $\delta_l^{(s)}$ through (5.23a). Since $\delta_l^{(so)}$ does not vary rapidly with energy, changes in energy of even $\simeq 10$ eV around $\epsilon = 0$ cause only a small variation, thus permitting easy and reliable extrapolation into the continuum.

The energy insensitivity of parameters that depend solely on short-range interactions is illustrated in Table 5.2, which gives radial wave functions in uranium calculated in a standard central atomic potential. The positions of

† The definition of regular and irregular functions when the effective angular momentum is complex requires some modification and is developed by Greene *et al.* [2]. Thus, when $f^\circ(\lambda)$ in (5.3) is complex, a real superposition is given by $[f^\circ(\lambda) + f^\circ(\lambda^*)]/2$. This function oscillates rapidly near the origin; the companion irregular function is then simply identified with the linearly independent combination $[f^\circ(\lambda) - f^\circ(\lambda^*)]/2i$.

Table 5.2

Behavior of np Radial Wave Functions r^{-1} $P_{np}(r)$ for Uranium at Small Radial Distances r^a

Orbital	Energy (a.u.)[a]	r_{max}[b]	r_{node}[b]	r_{min}[b]	$P(r_{max})/P(r_{min})$
2p	−624.781	0.0456	—	—	—
3p	−153.292	0.0347	0.0707	0.1490	−0.593
4p	−36.910	0.0334	0.0666	0.1218	−0.692
5p	−7.422	0.0330	0.0657	0.1181	−0.714
6p	−0.959	0.0329	0.0655	0.1173	−0.717
7p	−0.104	0.0329	0.0655	0.1173	−0.718
8p	−0.048	0.0329	0.0655	0.1173	−0.718
9p	−0.028	0.0329	0.0655	0.1173	−0.718
10p	−0.018	0.0329	0.0655	0.1173	−0.718
11p	−0.013	0.0329	0.0655	0.1173	−0.718

[a] From Starace [6]. Wavefunctions and energies computed using the central potential of Herman and Skillman [7].

[b] r_{max}, r_{node}, and r_{min} are the positions (in atomic units) of the first maximum, first node, and first minimum of the radial wavefunction $P_{np}(r)$, respectively.

the first maximum, first node, and first minimum as well as the ratios of the amplitudes of the wave functions at these points are seen to coincide to several decimal places for orbitals from 6p to 11p, even though their energies differ by almost one atomic unit. This weak energy dependence of short-range contributions translates into a similar insensitivity to energy of matrix elements representing other physical quantities, permitting extrapolations over a wide range of continuum and discrete states (Section 5.7.3).

5.7 NORMALIZATION OF WAVE FUNCTIONS

The formulation developed in this chapter allows us to normalize wave functions by a flexible and transparent procedure equally applicable to discrete and continuum levels. The procedure utilizes the continuity equation

$$\partial\rho/\partial t + \operatorname{div}\mathbf{j} = 0, \tag{5.31}$$

for the probability density ρ and flux $\mathbf{j}$ of a particle. The integral of the probability within a large volume is expressed in terms of the flux $\mathbf{j}$ through that volume's outer surface where the wave function takes its asymptotic form. This approach serves two purposes. First, it emphasizes that the normalization of the probability hinges on a basic conservation law, the equation of continuity. Second, it emphasizes that the behavior of an eigenfunction at large r suffices to determine its normalization.

Normalization of an eigenfunction $\psi_\epsilon \exp(-i\epsilon t)$ is properly approached in the context of the orthonormalization of a complete set of eigenfunctions with different energies, ϵ and ϵ'. Normalization proper emerges from an eventual limiting process $\epsilon' \to \epsilon$ [8]. Thus, we consider initially a whole matrix of wave function products $\rho_{\epsilon'\epsilon}(t) = \psi_{\epsilon'}^* \psi_\epsilon \exp[i(\epsilon' - \epsilon)t]$, with rows ϵ' and columns ϵ. We then write (5.31) for a general element $\rho_{\epsilon'\epsilon}$ and integrate it over a spherical volume with very large radius R. Applying Green's theorem to div $\mathbf{j}$ then yields

$$\left[i(\epsilon' - \epsilon)\int_{r \le R} \psi_{\epsilon'}^* \psi_\epsilon \, d\mathbf{r} + \int_{4\pi} d\hat{R}\, R^2 \mathbf{j}_{\epsilon'\epsilon} \cdot \hat{R}\right] \exp[i(\epsilon' - \epsilon)t] = 0. \quad (5.32)$$

Integration over the surface of the sphere yields $\delta_{l'l}$ whereby $R^2 \mathbf{j}_{\epsilon'\epsilon} \cdot \hat{R}$ reduces to a radial Wronskian $-\frac{1}{2} i W[F(\epsilon', l, R)^*, F(\epsilon, l, R)]\delta_{ll'}$. We may now factor $i(\epsilon' - \epsilon)\exp[i(\epsilon' - \epsilon)t]$ out of (5.32), thus obtaining

$$\int_{r \le R} \psi_{\epsilon'}^* \psi_\epsilon \, d\mathbf{r} = \frac{W[F(\epsilon', l, R)^*, F(\epsilon, l, R)]}{2(\epsilon' - \epsilon)}. \quad (5.33)$$

The key point is to choose R so large that the evaluation of the right-hand side is simplified, requiring only a knowledge of the asymptotic behavior of the functions. Thus, the value of the normalization integral on the left is obtained irrespective of complications in the inner volume. This is an important point in atomic scattering where the inner zone has a complicated atomic potential. It also permits application to situations involving many particles provided only one of them reaches large R. Normalization will thus amount to evaluating the right-hand side of (5.33)—a ratio of two quantities that vanish at $\epsilon' = \epsilon$—in the double limit $\epsilon' \to \epsilon$ and $R \to \infty$.

Expressing F in terms of $\{f, g\}$ through (5.20), and those in turn as $\{f^+, f^-\}$, resolves the Wronskian in (5.33) into the four simpler Wronskians $W(f_{\epsilon'}^{\pm *}, f_\epsilon^\pm)$, which are complex conjugate in pairs: $W(f_{\epsilon'}^{+*}, f_\epsilon^+) = W(f_{\epsilon'}^{-*}, f_\epsilon^-)^*$ and $W(f_{\epsilon'}^{+*}, f_\epsilon^-) = W(f_{\epsilon'}^{-*}, f_\epsilon^+)^*$. Further analysis is done separately for $\epsilon > 0$ and $\epsilon < 0$.

5.7.1 Continuum States

The value of the orthonormalization integral for continuum states is an improper function whose significance is established through integration over an energy interval $\epsilon' \sim \epsilon$. The Wronskian in (5.33) will give a nonzero contribution to an integral only through terms that are odd in $\epsilon' - \epsilon$ and thus compensate the odd parity of the denominator. The odd parts of $W(f_{\epsilon'}^{+*}, f_\epsilon^-)$ cancel with their complex conjugates, while those of $W(f_{\epsilon'}^{+*}, f_\epsilon^+)$ and

$W(f_{\epsilon'}^{-*}, f_{\epsilon}^{-})$ coincide. We find that the terms on the right-hand side of (5.33) with a nonzero integral reduce to

$$\sin(k' - k)R/\pi(\epsilon' - \epsilon). \tag{5.34}$$

Since this function diverges for $R \to \infty$ at $\epsilon' = \epsilon$ while its integral over a range $\epsilon' \sim \epsilon$ remains equal to unity, provided only R is sufficiently large, we conclude that (5.34) tends to $\delta(\epsilon' - \epsilon)$ as $R \to \infty$, in accordance with our conventions on the normalization of $\{f, g\}$ in Section 4.3 [9].

5.7.2 Discrete Spectrum

Here we are interested in $\epsilon = \epsilon_n$, where ϵ_n is a root of (5.27). In this event the coefficient of $W(f_{\epsilon'}^{\pm *}, f_{\epsilon_n}^{-})$ vanishes, and we need consider only the contribution of $W(f_{\epsilon'}^{\pm *}, f_{\epsilon_n}^{+})$ to (5.33). The value of $W(f_{\epsilon'}^{+*}, f_{\epsilon_n}^{+})$ vanishes exponentially as $R \to \infty$, while the $W(f_{\epsilon'}^{-*}, f_{\epsilon_n}^{+})$ is readily seen to equal $-(\kappa_n + \kappa')$ with the nonzero value $-2\kappa_n$ for $\epsilon' = \epsilon_n$. On the other hand, the coefficient of $f_{\epsilon'}^{-*}$ in (5.33) includes a factor $\sin \delta_l(\epsilon')$, which vanishes at $\epsilon' = \epsilon_n$, owing to (5.26). The limit of the right-hand side of (5.33) for $R \to \infty$ and $\epsilon' \to \epsilon = \epsilon_n$ must accordingly be evaluated by de l'Hospital's theorem and is found to be

$$\pi^{-1}\, d\delta_l(\epsilon')/d\epsilon' \Big|_{\epsilon' = \epsilon_n} \simeq \pi^{-1}\, d\beta(\epsilon')/d\epsilon' \Big|_{\epsilon' = \epsilon_n}, \tag{5.35}$$

where the right-hand side holds only for a nonzero long-range potential.

The result (5.35) means that discrete-level eigenfunctions can be constructed with proper normalization by first extrapolating continuum eigenfunctions $F(\epsilon, l, r)$, normalized per unit energy, to the range $\epsilon < 0$, by next setting $\epsilon = \epsilon_n$, and finally by multiplying them by

$$N_n = [\pi^{-1}\, d\delta_l/d\epsilon]^{-1/2} \Big|_{\epsilon = \epsilon_n}. \tag{5.36}$$

This procedure puts discrete and continuum states on the same footing and is central to QDT.The normalization constant can be evaluated analytically for fields $v(r) \propto -1/r$ or $-1/r^2$ utilizing Eqs. (5.27–5.28) or (5.29–5.30), respectively. The results are

$$(n - \delta_l^{(s)}/\pi)^{-3/2}[1 + (n - \delta_l^{(s)}/\pi)^{-3}\pi^{-1}\, d\delta_l^{(s)}/d\epsilon]^{-1/2}, \qquad v(r) \propto -1/r, \tag{5.37}$$

which is rather well known and was derived by Seaton [1] in 1958, and

$$\left[-\frac{\alpha}{2\pi\epsilon_0}\frac{\exp(2n\pi/\alpha)}{\tanh \pi\alpha} + \frac{1}{\pi}\frac{d\delta_l^{(s)}}{d\epsilon}\right]^{-1/2}, \qquad v(r) \propto -1/r^2. \tag{5.38}$$

The term with the factor $d\delta_l^{(s)}/d\epsilon$ usually amounts only to a small correction.

The result (5.37) provides the basis for the presentation of data on discrete levels utilized in Fig. 2.1. A broader application is presented next.

5.7.3 Level Dependence of Observable Properties

The calculation of the oscillator strength defined by (2.17), or (2.23) for discrete levels, reduces to the evaluation of a radial integral of the type $\int_0^\infty dr\, F(\epsilon, l, r) r F_0(r)$, where $F_0(r)$ indicates the radial eigenfunction for the ground state. In practice only a limited range of values of r contributes to this integral, namely, the "short" range $0 \leq r \lesssim r_0$ since r_0 is so defined that $F_0(r)$ effectively vanishes at larger radial distances. Accordingly, the same considerations apply to this integral as to the evaluation of the short-range phase shifts $\delta_l^{(s)}$. A smoother and slower dependence on the energy ϵ will be obtained by replacing the excited-state eigenfunction $F(\epsilon, l, r)$ with a renormalized function $F^o(\epsilon, l, r)$ given by dropping the normalization factor $\mathscr{N}(\epsilon)$ in (5.19). The renormalized F^o, just as the pair $\{f^o, g^o\}$, varies smoothly with energy and depends only on the short-range potential, all the sensitive dependences on ϵ due to the long-range $v(r)$ being incorporated into $\mathscr{N}(\epsilon)$ as given in (5.22′).

We set then

$$F(\epsilon, l, r) = F^o(\epsilon, l, r)\mathscr{N}(\epsilon) \qquad \text{for} \quad \epsilon > 0. \tag{5.39a}$$

For discrete eigenfunctions, which are now normalized in accordance with (5.35) and (5.36), the corresponding formula is

$$F_n(l, r) = F(\epsilon_n, l, r)N_n = F^o(\epsilon_n, l, r)\mathscr{N}(\epsilon_n)N_n \qquad \text{for} \quad \epsilon < 0. \tag{5.39b}$$

The renormalization factor $\mathscr{N}(\epsilon)$ in (5.22′) depends on B and $\mathscr{G}$, but the important dependence on ϵ usually stems from B.

On this basis normalization factors are conveniently factored out of transition matrix elements between a ground state and excited states, as indicated by writing

$$(\epsilon|x|0) = \mathscr{N}(\epsilon)(\epsilon^o|x|0) \qquad \text{for} \quad \epsilon > 0, \tag{5.40a}$$

$$(\epsilon_n|x|0) = N_n\mathscr{N}(\epsilon)(\epsilon_n^o\,|x|0) \qquad \text{for} \quad \epsilon_n < 0, \tag{5.40b}$$

for the matrix elements that define the oscillator strengths (2.17) and (2.23). [Here $(\epsilon^o|$ and $(\epsilon_n^o|$ indicate excited states of energy ϵ represented by renormalized radial eigenfunctions F^o.] Renormalized matrix elements such as $(\epsilon^o|x|0)$ can be evaluated adequately, whether by experiment or theory, on a *coarse energy mesh* only. The finer variations of the actual matrix elements $(\epsilon|x|0)$, which are often quite conspicuous, are represented by the *standard factors* $\mathscr{N}(\epsilon)$ and N_n. The evaluation of matrix elements with smoothly varying functions confined to a small range of radial distances will also prove decisive in the multichannel generalization to be considered later in Chapters 7 and 8.

The factoring out of N_n has been utilized in Fig. 2.1 to represent discrete and continuum oscillator strengths on a common basis. This particular analysis has been carried out often—but not universally—in diverse atomic applications [10]. The factoring out of $\mathcal{N}(\epsilon)$ goes further by isolating (particularly at low energy) the influence of energy and centrifugal effects on the dynamics at large r, which only affect the probability of the excited electron penetrating within the atomic core. [Recall that the main coefficient B in (5.22′) is proportional to the probability of finding the electron at the nucleus in a state normalized per unit energy.] Indeed, $\mathcal{N}(\epsilon)^2$ represents the generalization of Wigner's threshold law [Eq. (4.33)] in the presence of the nonzero long-range potential $v(r)$.

This factorization applies to the evaluation of all interaction phenomena that take place at *short ranges* only, even though they involve excited states. These include all the effects of exchange with unexcited electrons, the spin-orbit coupling, and other fine- and hyperfine-structure interactions. Their treatment involves the matrices of operators that we denote here as $T^{(s)}$, meaning that they act on an electron at $r > r_0$ only. The matrix element of such an operator between any two excited states (ϵ', ϵ) is then usefully factored as

$$(\epsilon' | T^{(s)} | \epsilon) = \mathcal{N}(\epsilon')(\epsilon'^{o} | T^{(s)} | \epsilon^{o})\mathcal{N}(\epsilon) \qquad \text{for} \quad (\epsilon, \epsilon') > 0, \tag{5.41}$$

with additional factors $N_{n'}$ and/or N_n when ϵ' and/or ϵ belongs to the discrete spectrum.

5.8 THE PHASE PARAMETER β(ε) in (5.27)

We noted that β in (5.11) and (5.27) is to be interpreted as the phase contribution due to the long-range potential. For $\epsilon < 0$, a general potential (say, with one minimum) has a classically allowed region between turning points r_1 and r_2 and classically forbidden regions for $r \leq r_1$ and $r \geq r_2$. The wave function then oscillates only in the range $r_1 < r < r_2$, but the classically forbidden regions contribute nevertheless to the total phase an amount $\pi/4$ each. In fact, the WKB expression for β is given by (Section 34 of Schiff [4])

$$\beta = \int_{r_1}^{r_2} k(r)\, dr + \pi/2. \tag{5.42}$$

The first term, the integral over the Langer-modified wave number in the oscillatory region [11],

$$k(r) = [2\epsilon - 2v(r) - (l + \tfrac{1}{2})^2/r^2]^{1/2}, \tag{5.42'}$$

represents the phase contribution of that region and the second is the sum of two $\pi/4$ contributions from the exponentially decreasing sections of the wave function.

The origin of this fixed contribution is also most clearly seen in the WKB formalism. It arises in connecting an exponentially damped function across a classical turning point to the oscillatory functions on the other side. The connection is made through the solutions of a linear potential around the turning point with $k^2 \propto (r - r_2)$. There arises, therefore, from the amplitude factor in the WKB wave function $k^{-1/2} \exp[i \int k\, dr]$ a contribution of $\pi/4$ to the phase since $k^{1/2} \propto (r - r_2)^{-1/4} \to (r_2 - r)^{-1/4} \exp[-i\pi/4]$. This result is, of course, not specific to the WKB model because the essential point of the argument is the passage from $k^{-1/2}$ to $(i\kappa)^{-1/2}$ on going from real to imaginary wave numbers; the factor $i^{-1/2}$ adds $\pi/4$ to the phase. Recall also that the WKB (Bohr–Sommerfeld) quantization formula is modified by dropping a $\pi/4$ contribution from β when the wave function vanishes at a turning point without further propagation, i.e., at an infinite wall.

5.9 SPIN-ORBIT COUPLING AND RELATIVISTIC EFFECTS

A specific example of a short-range interaction is the spin-orbit interaction potential, $(r^{-1}\, dv/dr)(2mc^2)^{-1}\mathbf{l} \cdot \mathbf{s}$, which affects the radial motion of an electron in a centrally symmetric potential $v(r)$, depending on the mutual orientation of its orbital and spin momenta (Section 14.2.2 of Fano and Fano [9]). This effect is represented by an additional dependence of the phase shift $\delta_l^{(s)}$ on the quantum number j, because the interaction is sharply peaked at small r, approximately as r^{-3}. When the effect is appreciable we shall indicate $\delta_l^{(s)}$ more precisely by $\delta_{lj}^{(s)}$, with $j = l \pm \frac{1}{2}$. The effect is represented specifically by the difference $\Delta_j\delta = \delta_{l,j=l-(1/2)}^{(s)} - \delta_{l,j=l+(1/2)}^{(s)}$, which vanishes of course for $l = 0$ but attains values of the order of $\pi/10$ for the p electrons of heavy elements, as seen in Table 5.3. The sign of $\Delta_j\delta$ is reversed when the orbital momentum pertains to a vacancy in a shell.

The effect decreases again rapidly with increasing l because the centrifugal force hinders the electron's penetration to the region of high coupling. Under this circumstance $\Delta_j\delta$ is strongly enhanced by any weak interaction that allows the electron to transfer part of its orbital momentum to another so that high and low values are mixed together; the experimental values of $\Delta_j\delta$ then follow no simple pattern. The $\delta_{lj}^{(s)}$ values are related to the $\delta_l^{(s)}$—calculated without considering spin-orbit coupling—by

$$\delta_{l,j}^{(s)} = \delta_l^{(s)} + \Delta_j\delta \times \begin{cases} (l+1)/(2l+1), & j = l - \frac{1}{2}, \\ -l/(2l+1), & j = l + \frac{1}{2}, \end{cases} \tag{5.43}$$

in the weak coupling approximation $\Delta_j\delta/\pi \ll 1$.

The treatment of spin-orbit coupling as a perturbation of the nonrelativistic Hamiltonian suffices for most purposes without replacing the Schrödinger equation by the Dirac equation [12]. A more significant role of relativity in heavy atoms is the change in the connection between momentum and energy of an electron introduced by relativistic kinematics. The resulting increase in binding and the contraction of the

Table 5.3

Sample Values of $\Delta_j\delta$ for $l = 1$[a]

Parameter	Element					
	Li	Na	K	Rb	Cs	Hg
Z	3	11	19	37	55	80
$\Delta_j\delta$	9×10^{-5}	1.9×10^{-3}	6.7×10^{-3}	2.6×10^{-2}	7.6×10^{-2}	0.19

[a] From Fano *et al.* [12].

inner shells affects the phase shifts of all the outer electrons. Self-consistent calculations including all the electrons have shown the net effect to be of opposite sign for lower- and higher-l electrons [13].

5.10 PHASE-AMPLITUDE TREATMENTS

This chapter has discussed the QDT parameters B, A, β_l, and $\delta_l^{(1)}$ which characterize the amplitudes and phases of base pairs such as $\{f, g\}$. There exists a class of scattering theories that also have as a central element the separation of long-range and short-range contributions to the scattering, handling the former through standard analytical expressions and the latter by numerical integrations that extend only over a small range of r. These theories go under different names, but we choose to call them phase-amplitude methods, following the earliest such usage for them [14]. The terminology also reflects the crucial characteristic that these methods concentrate on the nonlinear equations for the phases and amplitudes of these functions instead of dealing with the Schrödinger equation for the functions f and g. (The separation into amplitude and phase is a nonlinear operation, and it is no surprise that the Schrödinger equation is thereby transformed into a nonlinear form.)

The various phase–amplitude methods fall into two classes. One is characterized by a formulation that parallels throughout the WKB approximation and has, in fact, been exploited to develop improvements on the WKB approximation. In this class we have typically a *second-order* nonlinear differential equation for an amplitude function from whose solutions the phase function is obtained by a simple quadrature. The second class constructs first a *first-order* nonlinear differential equation for the phase function and then obtains the amplitude function through a quadrature over the phase.

5.10.1 The Milne–Young–Wheeler Method: Maintaining the WKB Structure

We begin by rewriting the radial Schrödinger equation (5.1) as

$$[d^2/dr^2 + k^2(r)]u(r) = 0, \tag{5.44}$$

where $u(r)$ is a general solution of this equation and $k(r) = [2\epsilon - 2v(r) - l(l+1)/r^2]^{1/2}$ is a local wave number so that $\hbar k(r)$ is the local radial momentum at any point r. The WKB approximation gives the basic approximate solutions of (5.44) in the form

$$u_{\mathrm{WKB}}(r) \propto k^{-1/2}(r) \exp\left[\pm i \int^r k(r')\,dr'\right]. \tag{5.45}$$

The lower limit of the phase integral and an overall normalization factor are the two arbitrary constants in a general solution of (5.44). The solutions (5.45) do not, of course, satisfy the Schrödinger equation (5.44). In fact, they satisfy instead, as can be easily verified by substitution, the equation

$$d^2u_{\mathrm{WKB}}/dr^2 = -[k^2(r) - k^{1/2}(r)(d^2/dr^2)k^{-1/2}(r)]u_{\mathrm{WKB}}. \tag{5.46}$$

The second term in brackets is sometimes called the Schwarzian derivative of a function, [in this case $k(r)$].

A principal characteristic of WKB wave functions is that the derivative of the phase is proportional to the inverse square of the amplitude. It is this characteristic that is maintained in the Milne–Young–Wheeler method by writing a general solution of (5.44), *this time exact*, as

$$u(r) = c\alpha(r) \sin\left[\int^r K(r')\,dr' + b\right], \tag{5.47}$$

where the amplitude and phase functions are related by

$$K(r) = \alpha^{-2}(r), \tag{5.48}$$

while c and b are the integration constants. Instead of stating that the solutions in (5.45) and (5.47) are similar in form, we start here from the alternative statement that the exact solutions in (5.47) obey an equation similar in form to (5.46):

$$d^2u/dr^2 = -[K^2(r) - K^{1/2}(r)(d^2/dr^2)K^{-1/2}(r)]u. \tag{5.49}$$

For this to coincide with the radial equation (5.44), $K(r)$ must be defined by the nonlinear equation

$$K^2(r) - K^{1/2}(r)(d^2/dr^2)K^{-1/2}(r) = k^2(r). \tag{5.50}$$

The function $K(r)$ has been called a "quantal momentum." A first approximation to it is the local wave number $k(r)$ and (5.50) can serve as the starting point for generating further approximations and corrections to quasi-classical procedures such as, for example, the Bohr–Sommerfeld quantization conditions [15].

Our subsequent discussion here focuses on utilizing exact solutions provided by (5.47), (5.48), and (5.50) rather than on generating a sequence of approximate solutions based on (5.50). The original Schrödinger equation is replaced here by the nonlinear equation (5.50) for $K(r)$ or equivalently, through (5.48), by the nonlinear equation for the amplitude $\alpha(r)$:

$$[d^2/dr^2 + k^2(r)]\alpha(r) = \alpha^{-3}(r). \tag{5.51}$$

A single solution of this equation can generate through (5.47) any general solution of the original linear Schrödinger equation (5.44). This method, therefore, also provides a resolution of the problem posed in Section 5.2, of defining uniquely a pair of linearly independent solutions $\{f, g\}$. Starting with a solution $\alpha(r)$, the choice of zero for b and for the lower limit of the integral in (5.47) will define a regular solution f that vanishes at the origin. Setting $b = -\pi/2$ and the lower limit equal to zero provides the companion irregular function g (more precisely $B^{-1/2}g^{o}$).

The Milne–Young–Wheeler method ties together the different choices that are possible for the base pair $\{f, g\}$. Note, first of all, that the choice for f and g is not unique insofar as the construction may start from an arbitrary solution $\alpha(r)$ of (5.51). In most cases a good prescription is that $K(r)$ coincides with the local $k(r)$ at its maximum value, that is, at a potential minimum, where dk/dr vanishes.

The relation between different pairs $\{f, g\}$ and the amplitude and phase functions can be visualized through a geometrical picture. Both f and g satisfy the radial equation (5.44) and, being 90° out of phase, they can be viewed as a pair of Cartesian axes in the two-dimensional space of solutions of the second-order differential equation. Their expression in polar coordinates for this space through

$$f(r) = (2/\pi)^{1/2}\alpha(r)\sin\phi(r), \qquad g(r) = -(2/\pi)^{1/2}\alpha(r)\cos\phi(r), \tag{5.52}$$

yields, on substituting into (5.44) and equating terms in $\sin\phi$ and $\cos\phi$ to zero,

$$d\phi/dr = \alpha^{-2}(r), \tag{5.53}$$

while $\alpha(r)$ satisfies (5.51). These are precisely the Milne–Young–Wheeler equations. Thus, we note from (5.53) or from the equivalent (5.48) that the vector $\boldsymbol{\alpha}(r)$, with Cartesian components f and g, rotates with a speed proportional to the inverse square of its length. Once the length of this vector is known, any mutually orthogonal pair of axes can serve as a base pair $\{f, g\}$. Note also on evaluating the Wronskian $W(f, g)$ that (5.53) amounts to $W(f, g) = 2/\pi$.

5.10.2 The Dashen–Babikov–Calogero Method: The Schrödinger Equation in Riccati Form

The separation of the amplitude and phase of the solutions of the radial equation can be accomplished by an alternative procedure. It begins again by writing

$$f(r) = (2/\pi)^{1/2}\alpha(r)\sin\phi(r), \tag{5.54}$$

but proceeds by adding a second constraint on the functions α and ϕ through the derivative of $f(r)$. Various choices are possible here. A standard choice is to write

$$f'(r) = (2/\pi)^{1/2}\alpha(r)k(r)\cos\phi(r). \tag{5.55}$$

These equations now lead to separate equations for the amplitude and phase functions. This time it is the equation for $\phi(r)$ that occurs in the self-contained form of a first-order nonlinear equation

$$\phi'(r) = k(r) + [k'(r)/2k(r)]\sin 2\phi(r). \tag{5.56}$$

The amplitude function is then given by

$$\alpha'(r) = -[k'(r)/k(r)]\alpha(r)\cos^2\phi(r), \tag{5.57}$$

so that ln $\alpha(r)$ can be obtained by a simple quadrature, once $\phi(r)$ is known. In solving (5.56) a suitable boundary condition is once again $\phi(0) = 0$.

Equivalently, we could proceed by starting with $g(r)$ in the form $-(2/\pi)^{1/2}\alpha \cos \phi$ and enforce a constraint $g' = (2/\pi)^{1/2}\alpha k \sin \phi$. The phase and amplitude functions so obtained will provide an alternative, but different, set of solutions. In particular, this function g and the f obtained earlier will not, in general, oscillate 90° out of phase at all r. Constructing a suitable base pair $\{f, g\}$ is now more complicated than in the Milne–Young–Wheeler method, in which the linear independence is built in by definition. On the other hand, the particular merit of the Dashen–Babikov–Calogero approach is that we need only to solve first-order differential equations, which is more convenient for practical numerical calculations [12].

A variant of this procedure, developed particularly by Calogero, has especially attractive features when applied to the full potential $v(r) + v_s(r)$. It utilizes a base pair of solutions $\{f, g\}$ of (5.1) with the potential $v(r)$, whereby the amplitude and phase to be calculated reflect only the effect of the short-range potential $v_s(r)$. A solution of the radial equation with the combined potential takes the form

$$F(r) = (2/\pi)^{1/2}\alpha(r)[f(r)\cos\delta^{(s)}(r) - g(r)\sin\delta^{(s)}(r)]. \tag{5.58}$$

Here we have changed the generic symbol for the phase ϕ into $\delta^{(s)}(r)$ to emphasize that this function depends on the short-range interaction, vanishing in its absence. In fact, comparison of (5.58) with the solution (5.20) at the boundary r_0 of the short-range potential shows that $\delta^{(s)}(r_0)$ coincides with the short-range phase shift $\delta_l^{(s)}$, while $\alpha(r_0) = 1$. It must be noted that $F(r)$ is defined for all r in (5.58), in contrast to (5.19) and (5.20), which are only valid for $r \geq r_0$.

The phase function $\delta^{(s)}(r)$ takes on the constant value $\delta_l^{(s)}$ once r exceeds r_0. Because F and $\{f, g\}$ are both taken to be energy normalized, the amplitude function $\alpha(r)$ reduces to unity asymptotically—in fact, once $r \geq r_0$. However, $\alpha(r)$ may and generally will depart from unity at small r, thus expressing a relative difference in the amplitudes of the full eigenfunction and of $f(r)$, caused by the presence of the short-range potential. Clearly, since the solution in (5.58) has to be regular at the origin, $\delta^{(s)}(r)$ must satisfy the boundary condition.

$$\delta^{(s)}(0) = 0. \tag{5.59}$$

The ratio $F(0)/f(0)$, i.e., $\alpha(0)$, is therefore the enhancement factor discussed in Problem 4.1 and is of considerable physical interest. This phase–amplitude treatment, therefore, has the merit of separating out the modification of the wave function represented by the phase shift and the amplitude α, both of them being due to the short-range potential.

Once again, Eq. (5.58), wherein one function $F(r)$ has been replaced by two [$\alpha(r)$ and $\delta^{(s)}(r)$], has a large amount of arbitrariness. As before, this arbitrariness is fixed by a condition on $F'(r)$ that is now chosen to be

$$\frac{dF}{dr} = \left(\frac{2}{\pi}\right)^{1/2}\alpha(r)\left[\frac{df}{dr}\cos\delta^{(s)}(r) - \frac{dg}{dr}\sin\delta^{(s)}(r)\right]. \tag{5.60}$$

On combining with the radial equation for $F(r)$, we now get the following phase amplitude equations:

$$\frac{d\delta^{(s)}(r)}{dr} = -\frac{2}{W(f,g)} v_s(r)[f(r)\cos\delta^{(s)}(r) - g(r)\sin\delta^{(s)}(r)]^2 \tag{5.61}$$

and

$$\ln\alpha(r) = -\frac{2}{W(f,g)}\int_r^\infty dr' v_s(r')[f(r')\cos\delta^{(s)}(r') - g(r')\sin\delta^{(s)}(r')] \times [f(r')\sin\delta^{(s)}(r') + g(r')\cos\delta^{(s)}(r')]. \tag{5.62}$$

In these equations $W(f,g)$ is the Wronskian and we have used the boundary condition $\alpha(\infty) = 1$, although, once again, it is clear that $\alpha(r)$ remains at unity throughout the range $r_0 \le r \le \infty$ if $v_s(r)$ vanishes for $r \ge r_0$.

The pair of equations (5.61) and (5.62) has the same structure as (5.56) and (5.57). The phase function is described completely by a first-order nonlinear equation together with the boundary condition (5.59). The phase equation (5.61) also exemplifies the step-by-step accumulation of the phase shift as we proceed from the origin outward, sampling successive bits of $v_s(r)$. At $r = r_0$, $\delta^{(s)}(r)$ reaches the steady value $\delta_l^{(s)}$, which is the full short-range phase shift.

At any intermediate point r, $\delta^{(s)}(r)$ represents the phase shift that would obtain were the potential $v_s(r)$ to be truncated at that point. This feature is an example of the general mathematical technique called invariant imbedding. The problem of scattering by a potential $v_s(r)$ has been imbedded into a family of problems with truncated potentials $\tilde{v}_s(r)$ by viewing r as a variable, $\tilde{v}_s(r') = v_s(r')$ for $r' \le r$, and $\tilde{v}_s(r') = 0$ for $r' \ge r$. In fact, this statement of the principle of invariant imbedding is in itself sufficient to derive (5.56) and (5.61) (Dashen [14] and Rau [16]). It is a general feature of the method of invariant imbedding that it turns a boundary value problem—namely, solving the Schrödinger equation with boundary conditions at the origin and at infinity to extract the phase shift—into an initial-value problem wherein $\delta^{(s)}(\infty)$ [or, for that matter any $\delta^{(s)}(r)$] is evaluated starting from the initial condition $\delta^{(s)}(0) = 0$.

Among other useful features of the Dashen–Babikov–Calogero method is the accumulation of the phase shift $\delta^{(s)}(r)$ which removes the usual $n\pi$ ambiguities of other procedures that deal directly with second-order differential equations. Also, a relation between $\delta_l^{(s)}$ and the sign of $v_s(r)$ follows immediately from (5.61), thus providing the well-known result that attractive potentials lead to positive phase shifts, and purely repulsive potentials give negative $\delta_l^{(s)}$. Phase shifts due to different short-range interactions can also be readily compared; as an example, if $v_{s1}(r)$ is everywhere greater than or equal to $v_{s2}(r)$, then necessarily $\delta_1^{(s)} \le \delta_2^{(s)}$. Similarly, from (5.62) we can make direct correlations between the enhancement factor $\alpha(0)$ and the nature of the potential; as a simple example, the wave function is increased or decreased at the origin relative to f according to whether the potential is attractive or repulsive, respectively.

The simple first-order equation for the phase shift can also serve as a natural starting point for approximation methods. In fact, the first Born approximation to the phase shift results by setting $\delta^{(s)}(r)$ to zero in (5.61). Finally, these first-order equations for the amplitude and phase functions in this scheme have a natural connection to variational formulations. The functions $\alpha(r)$ and $\delta^{(s)}(r)$ are mathematical adjoints in a well-defined sense (Calogero [14] and Rau [16]). Indeed, the boundary conditions $\delta^{(s)}(0) = 0$ and $\alpha(r_0) = 1$ are conjugate and the integrand of (5.62) is formally the derivative of the right-hand side of (5.61) with respect to $\delta^{(s)}(r)$.

5.11 IDENTIFICATION OF IRREGULAR WAVE FUNCTIONS

Solutions of a wave equation that are irregular at a singular point of the equation are characterized primarily by their behavior near that point. This is the case for the wave function $f^-(\epsilon, \lambda, r) = \exp(\kappa r) r^{-i\zeta}$ at $r \to \infty$ for negative energy considered in Section 5.3.2 and for the function $g^0(\epsilon, l, r) \propto r^{-l}$ at $r \to 0$ introduced in Section 5.4. However, this characterization does not suffice to identify a function because it is unaffected by the addition of any regular function that vanishes at the singular point (for example, by superposing cf^+ to f^- at large r or by superposing cf^0 to g^0 at $r \sim 0$).

This subject has been discussed in Greene *et al.* [2], where the following identification procedure was developed. Recall that the problem arises in connection with the desired use of a base pair of independent solutions, one regular and one irregular, which oscillate 90° out of phase. However, none of the solutions of immediate interest oscillates near the singularity at $r \to 0$, or—for negative energy—near $r \to \infty$. The phase relations between the solutions must then be specified in the range where they do oscillate, that is, away from the singular point. To this end the *regular* solution is identified first near the singularity and then integrated away into the region where it oscillates at $\epsilon > v(r)$. Here a companion function is identified which oscillates, for example, with equal amplitude and 90° out of phase with the regular one. Finally, this companion function is integrated back to the singular point.

Two remarks are relevant to this procedure:

(1) The phase difference of any pair of oscillating solutions does not generally remain constant in the course of propagation, particularly in the presence of intervening potential barriers. This dephasing may then be corrected by adjusting the superposition of a base pair as in (5.17).

(2) In a WKB approximation the integration through a range in which $\epsilon < v(r)$ leads to the construction of a "tunneling integral" which constitutes the main ingredient of the QDT parameters (see Section IV of Greene *et al.* [2]. In this approximation the "Langer correction" [11]† required at $r \sim 0$ introduces an effective centrifugal field at $r \sim 0$ that prevents oscillations for any ϵ and l when combined with $v(r)$.

Detailed application of the procedure described earlier may be bypassed at $r \to \infty$ when the wave equation (5.1) has analytical solutions for the potential of interest,

† This correction arises from a procedure that avoids the breakdown of the WKB method at $r \to 0$ by mapping the variable r onto $t = \ln r \to -\infty$. It eventuates in an additional effective centrifugal potential $1/4r^2$ that replaces the $l(l+1)$ coefficient of (5.1) by $(l + \frac{1}{2})^2$.

typically for a Coulomb field. However, the analytical continuation of the solutions and of their Jost functions from positive to negative ϵ needs to be examined and interpreted in detail. Similarly, at $r \to 0$, an analytical solution independent of $f^{\circ}(\epsilon, l, r)$ exists, namely, $f^{\circ}(\epsilon, -l-1, r)$, but the independence fails just for the integer values of l of greatest interest; identification by a limiting procedure is then required. These matters are discussed by Greene *et al.* [2], together with WKB and other alternative methods that reflect the general procedure more directly.

Consideration of the solution $f^{\circ}(\epsilon, -l-1, r)$ serves also to determine the coefficient $\mathscr{G}(\epsilon, l)$ introduced in (5.17). It is convenient here to replace the quantum number l by an index λ, which is generally noninteger, as was done to obtain (5.29). For generic values of λ Appendix B of Greene *et al.* [2] derives the expression

$$\mathscr{G}(\epsilon, \lambda) = -B(k, \lambda) \operatorname{cotan}(\delta_{\lambda}^{(1)} - \delta_{-\lambda-1}^{(1)}), \qquad \epsilon > 0 \tag{5.63}$$

with the replacements $B \to A$ and $\delta_{\lambda}^{(1)} \to \beta_{\lambda}$ for $\epsilon < 0$. This result diverges at the critical values $\lambda = \lambda_c$ where $W(f_{\lambda}^{\circ}, f_{-\lambda-1}^{\circ})$ and $\delta_{\lambda}^{(1)} - \delta_{-\lambda-1}^{(1)}$ vanish. A more detailed analysis in Greene *et al.* [2] then leads to an application of de l'Hospital's theorem, which yields

$$\mathscr{G}(\epsilon, \lambda_c) = \frac{1}{2\pi}\left[\frac{\partial A(\kappa, \lambda)}{\partial \lambda}\right]_{\lambda_c} \qquad \epsilon < 0 \tag{5.64a}$$

$$= -\frac{1}{2\pi}\left\{\partial\left[\frac{B \sin \pi(2\lambda + 1)}{\sin(\delta_{\lambda}^{(1)} - \delta_{-\lambda-1}^{(1)})}\right]\bigg/\partial\lambda\right\}_{\lambda_c} \qquad \epsilon > 0. \tag{5.64b}$$

The set of values $\{\lambda_c\}$ consists generally of all half-integers but, in the presence of a Coulomb field, includes all integers as well.

Problems

5.1 The following absorption lines are observed in potassium vapor (in cm^{-1})

$$\nu_3 = 1.301 \times 10^4, \qquad \nu_5 = 2.900 \times 10^4,$$

$$\nu_4 = 2.471 \times 10^4, \qquad \nu_6 = 3.107 \times 10^4.$$

By fitting these wave numbers to the Rydberg formula (5.28) evaluate the quantum defect μ and locate the position of the ionization threshold with respect to the ground state. The Rydberg constant is $1.097 \times 10^5\ \mathrm{cm}^{-1}$.

5.2 Set $Z = 0$ in the first column of Table 5.1 and simplify to obtain the QDT parameters for zero long-range field (which is applicable, for instance, to the motion of an electron in the field of a neutral atom). With these parameters establish through (5.23) the connection between short-range phase shifts and their renormalized "smooth" forms. Note that the results coincide with (4.32) to within $(2/\pi)$ factors, because of the discrepancy in the definitions of $\{j^{\circ}, y^{\circ}\}$ in Chapter 4 and $\{f^{\circ}, g^{\circ}\}$ in Chapter 5.

5.3 Use the WKB expressions in Table 5.1 to evaluate the QDT parameters for a Coulomb field. Compare with the exact expressions in the table. Note that apart from

β, the WKB approximation is correct only in the limit of $\kappa \to 0$. This exemplifies that the quasi-classical approximation is exact at threshold for the Coulomb field.

5.4 The spectrum of sodium vapor has the characteristic doublet, the so-called D lines at 589.0 and 589.6 nm, corresponding to transitions between the 3s and 3p states. From the level splitting, verify the value of $\Delta_j \delta$ as quoted in Table 5.3. [The ionization potential of Na is 5.13 eV.]

5.5 Evaluate the matrix elements $\langle ns|r^2|1s\rangle$ between the ground state and excited states $n = 2$, 3, and 4 in hydrogen. The radial wave functions may be obtained, for example, from Schiff [4]. Follow the procedure of Section 5.7.3 to obtain the renormalized matrix elements and verify that they depend smoothly on the energy.

5.6 (a) A short-ranged perturbation in the hydrogen atom is described by $V(r) = e^{-\lambda r}/r$. Evaluate its matrix elements between the ground state and the ns states with $n = 2$, 3, and 4. Renormalize the matrix elements according to Section 5.7.3 to exhibit their smooth dependence on the energy.

(b) Calculate the matrix element of $V(r)$ in (a) to a state in the continuum with energy ϵ, once again renormalizing it according to (5.40a). Compare with the discrete matrix elements obtained in (a). [Continuum wave functions are given in standard references such as Schiff [4]. In carrying out the integrals, expand the confluent hypergeometric function in terms of its argument. The expansion converges rapidly in ϵ.]

5.7 *A Variant of the Phase Equation* (5.56). For the Schrödinger equation (5.44) with boundary conditions $u(r) = r^{l+1}$ as $r \to 0$ and $u(r) \sim \sin(kr - \frac{1}{2}l\pi + \delta_l)$ as $r \to \infty$, consider the function

$$\phi(r) = [k/k(r)] \tan^{-1}[k(r)u(r)/u'(r)].$$

Consider also the similar function $\phi_0(r)$ defined with $u_0(r) = rj_l(kr)$ in place of $u(r)$. Set up the nonlinear differential equations obeyed by these functions and interpret them. Contrast them with (5.56).

References

1. M. J. Seaton, *Rep. Prog. Phys.* **46**, 167 (1983).
2. C. H. Greene, A. R. P. Rau, and U. Fano, *Phys. Rev. A* **26**, 2441 (1982); C. H. Greene, U. Fano, and G. Strinati, *Phys. Rev. A* **19**, 1485 (1979).
3. V. A. Zilitis, *Opt. Spectrosc.* **43**, 603 (1977); W. R. Johnson and K.T. Cheng, *J. Phys. B* **12**, 863 (1979).
4. L. I. Schiff, "Quantum Mechanics," 3rd ed., Section 39. McGraw-Hill, New York, 1968; R. G. Newton, "Scattering Theory of Waves and Particles," Section 12.1.2. Springer-Verlag, Berlin and New York, 1982.
5. A. R. P. Rau, *Phys. Rev. A* **4**, 207, Appendix A (1971).
6. A. F. Starace *in* "Encyclopedia of Physics," Vol. 31, p. 1. Springer-Verlag, Berlin and New York, 1982.
7. F. Herman and S. Skillman, "Atomic Structure Calculations." Prentice-Hall, Engelwood Cliffs, New Jersey, 1963.
8. E. Fues, *Ann. Phys. (Leipzig)* **81**, 281 (1926); A. Sommerfeld and G. Schur, *Ann. Phys. (Leipzig)* **4**, 409 (1930). Fues credits H. Weyl for directing him to Weyl's own work [*Math. Ann.* **68**, 220 (1910)] "from which all that is necessary for wave mechanics comes forth."

9. U. Fano and L. Fano, "Physics of Atoms and Molecules," Appendix A 3.5. Univ. of Chicago Press, Chicago, Illinois, 1972.
10. H. A. Bethe and R. W. Jackiw, "Intermediate Quantum Mechanics," Chapter 4. Benjamin, New York, 1968.
11. R. E. Langer, *Phys. Rev.* **51**, 669 (1937).
12. U. Fano, C. E. Theodosiou, and J. L. Dehmer, *Rev. Mod. Phys.* **48**, 49 (1976).
13. J. P. Desclaux and Y. K. Kim, *J. Phys. B* **8**, 1177 (1975).
14. F. C. Calogero, "Variable Phase Approach to Potential Scattering." Academic, New York, 1967; V. V. Babikov, *Sov. Phys. Usp.* **10**, 271 (1967); J. C. Wheeler, *Phys. Rev.* **52**, 1123 (1937); R. Dashen, *Nuovo Cimento* **28**, 229 (1963).
15. E. Young, *Phys. Rev.* **38**, 1613 (1931), *Phys. Rev.* **39**, 445 (1932); C. E. Hecht and J. E. Mayer, *Phys. Rev.* **106**, 1156 (1957).
16. A. R. P. Rau, *J. Math. Phys.* **17**, 1340 (1976).

6 CALCULATION OF THE PHASE SHIFTS

6.1 INTRODUCTION

The phase shifts $\delta_l^{(so)}$, defined by (5.19), represent the effect of short-range interactions between an electron and an atomic core. Their calculation is reasonably straightforward within our present single-channel framework. We consider here different approaches to this problem, namely, the direct solution of the Schrödinger equation in its usual differential form, the configuration mixing approach (which amounts to solving the equation in integral form), and variational approaches that are particularly flexible. We have in mind, of course, the eventual adaptation of these procedures to the much wider range of multichannel problems, a subject deferred to Chapter 9.

6.2 DIRECT INTEGRATION

The radial Schrödinger equation for an electron (or other particle) with orbital quantum number l in the range of radial distances $r < r_0$ occupied by an atomic core with spherical symmetry has the same form as (5.13)

$$\left[\frac{1}{2}\frac{d^2}{dr^2} - \frac{1}{2}\frac{l(l+1)}{r^2} + \epsilon - v(r) - v_s(r)\right]F_l(r) = 0. \tag{6.1}$$

Here, however, the short-range potential $v_s(r)$ generally represents a nonlocal operator, as described later, owing to the exchange of the electron under consideration with the atomic core electrons. The long-range potential $v(r)$ will continue to be a local potential. We proceed initially in a "frozen core approximation" that disregards the perturbation of the core itself by the electron that penetrates it; this perturbation could be included in $v_s(r)$ at this point, but we shall consider it only in the latter part of this chapter.

The direct, or "static," contribution to the potential $v(r)$ consists of the nuclear potential $-Ze^2/r$ and the repulsive potential $\Sigma_i\, e^2/|\mathbf{r} - \mathbf{r}_i|$ averaged over the positions $\mathbf{r}_i$ of all the core electrons. The "exchange" contribution arises from the modification of this averaging resulting from the antisymmetrization of the electron coordinate $\mathbf{r}$ (and of its spin) with the corresponding coordinates of the core electrons. Indicating the radial wave functions of the core electrons by $u_n(r_i)$, the exchange contribution consists of terms of the form

$$\sum_n u_n(r) \int_0^\infty dr_i\, w_{l_n l}(r, r_i) u_n(r_i) F_l(r_i). \tag{6.2}$$

The potential function $w_{l_n l}(r, r_i)$ results from averaging the interaction energy $1/|\mathbf{r} - \mathbf{r}_i|$ over products of the orbital wave function $Y_{l_n m_n}$ and Y_{lm} of the two electrons as detailed later for a special case. The explicit appearance of the core wave functions $u_n(r)$ as a factor external to the integral in (6.2) shows that this exchange term is confined to $r < r_0$ and is, therefore, to be lumped into $v_s(r)$.

As an illustration we work out the calculation of the potential for the interaction with the simple core of a H atom in its ls ground state, that is, with a single electron wave function $(4\pi)^{-1/2}u(r_1)$. The antisymmetrized joint wave function of the two electrons is then

$$\psi(\mathbf{r}, \mathbf{r}_1) = \sqrt{\tfrac{1}{2}}\,[F_l(r)Y_{lm}(\hat{r})u(r_1) \pm F_l(r_1)Y_{lm}(\hat{r}_1)u(r)](4\pi)^{-1/2}, \tag{6.3}$$

where the $-(+)$ sign corresponds to parallel (antiparallel) spins of the two electrons. The potential terms of (6.1) are now constructed as follows. We apply to $\psi(\mathbf{r}, \mathbf{r}_1)$ the *two-electron* Schrödinger operator, with the exact potential $-(e^2/r) - (e^2/r_1) + (e^2/|\mathbf{r} - \mathbf{r}_1|)$, multiply it by $u(r_1)(4\pi)^{-1/2}Y^*_{lm}(\hat{r})$ and integrate the resulting expression over the entire range of $\mathbf{r}_1$ and over the angular variables of $\mathbf{r}$ to get the full one-electron potential. Considering that $u(r_1)$ obeys the radial equation for the H atom, we find

$$[v(r) + v_s(r)]F_l(r) = -\left(\frac{e^2}{r}\right)F_l(r)$$
$$+ \int_{4\pi} d\hat{r} \int d\mathbf{r}_1 (2\pi)^{-1/2} Y^*_{lm}(\hat{r}_1)u(r_1)\left[\frac{e^2}{|\mathbf{r} - \mathbf{r}_1|}\right]\psi(\mathbf{r}, \mathbf{r}_1). \tag{6.4}$$

The potential factor $|\mathbf{r} - \mathbf{r}_1|^{-1}$ is expanded in spherical harmonics

$$\frac{1}{|\mathbf{r} - \mathbf{r}_1|} = \sum_k \frac{4\pi}{2k+1} \frac{r_<^k}{r_>^{k+1}} \sum_q Y^*_{kq}(\hat{r})Y_{kq}(\hat{r}_1), \tag{6.5}$$

where $r_<$ ($r_>$) is the smaller (larger) of r and r_1. The angular integrations in (6.4) reduce to the orthonormality of spherical harmonics, recalling that $(4\pi)^{-1/2} = Y_{00}$.† The two terms of $\psi(\mathbf{r}, \mathbf{r}_1)$ in (6.4) contribute the "direct" and "exchange" terms, respectively, and (6.4) reduces to

$$v(r)F_l(r) = \left[\frac{-e^2}{r} + \int_0^\infty dr_1 u^2(r_1)\frac{e^2}{r_>}\right]F_l(r) \tag{6.6a}$$

$$v_s(r)F_l(r) = \pm e^2(2l+1)^{-1}u(r)\int_0^\infty dr_1 u(r_1)F_l(r_1)\frac{r_<^l}{r_>^{l+1}}. \tag{6.6b}$$

We make a few remarks about each of these terms.

The local potential $v(r)$, which in this example takes the explicit form (see Problem 3.1)

$$v(r) = -e^2[(1/r) + (1/a_0)]\exp(-2r/a_0), \tag{6.6c}$$

is the sum of the nuclear potential and the "direct" electrostatic potential v_d due to the electronic charge distribution, $\rho(r_1) = e^2u^2(r_1)/4\pi$, as given by the solution of Poisson's equation $\nabla^2 v_d = -4\pi\rho$. Retention of this potential alone gives the "static" or "Hartree" approximation for electron scattering from the screened Coulomb potential of the atomic core.

The exchange term in (6.6b), however, is nonlocal. Electron exchange thus causes the solution $F_l(r)$ of the radial equation to appear in the integrand of (6.2) or (6.6b), that is, it converts the Schrödinger equation to integrodifferential form. Direct numerical solution of (6.1) is still possible, if laborious. However, it is much preferable to transform (6.1) into a system of coupled differential equations, by regarding the integral in (6.2), or (6.6b), as a separate dependent variable $\chi_n(r)/r$. The additional equations are obtained by showing that each $\chi_n(r)$ obeys in turn a differential equation that relates it linearly to $F_l(r)$. This procedure is described in a monograph [1] and illustrated in Section 6.7. Smith [1] also provides guidance to the procedures for solving the resulting system numerically, as is normally required in practical applications. The solution is to start from $r = 0$ with $F_l(r) \propto r^{l+1}$ as in Chapter 5.

Smith [1] deals with the calculation of $F_l(r)$ throughout the range of r, $0 \le r < \infty$, whereas we intend here to confine the calculation to $r \lesssim r_0$. At $r = r_0$, once the potential coincides with the standard long-range $v(r)$, F_l reduces to its form (5.19) with a coefficient $\tan\delta_l^{(\mathrm{so})}$ to be determined by matching logarithmic derivatives, as in (4.23). We give here an equivalent

† In the case of atoms other than H the orbital wave functions $Y_{lm}(\theta_i, \varphi_i)$ of most electrons have quantum numbers $l \neq 0$. The angular integrations in (6.4) then involve products of three rather than two spherical harmonics. These integrals reduce nevertheless to algebraic functions of quantum numbers familiar in the theory of angular momentum. See Eq. (2.65) of Smith [1].

form of that matching equation that involves Wronskians of $F_l(r)$ with the base pair of radial functions for the long-range field $\{f^\circ, g^\circ\}$ defined in Chapter 5,

$$\tan \delta_l^{(\mathrm{so})} = W(f^\circ, F_l)/W(g^\circ, F_l). \tag{6.7}$$

Recall now that in (6.1) for F_l, the exchange contribution to the potential vanishes as the electron emerges from the atomic core. Indeed, as noted already, the expressions of the contribution in (6.2) and (6.6b) include as a factor the bound-state radial function $u(r)$, which vanishes exponentially at large r; the radius r_0 of the core boundary is to be chosen so that this factor $u(r)$ is negligible at $r > r_0$. In other words, the potential in the equation (6.1) for $F_l(r)$ must approach the long-range potential $v(r)$ in the corresponding equation (5.1) for the base pair $\{f^\circ, g^\circ\}$, coinciding with it for $r \geq r_0$.

The use of Wronskians in (6.7) serves as a relevant index of the coincidence of the two potential functions. We may evaluate the ratio of Wronskians in (6.7) at successively increasing values of $r \lesssim r_0$. Each Wronskian must approach a constant as (6.1) and (5.1) become equivalent. This convergence is to be regarded as practically adequate whenever the value of $\tan \delta_l^{(\mathrm{so})}$ has become independent of r at any prescribed level of accuracy. Once $\delta_l^{(\mathrm{so})}$ has been obtained, results such as (5.23b) can be applied to obtain the phase shifts $\delta_l^{(\mathrm{s})}$ at infinity and thereby the effect of the atomic core on bound states and scattering.

6.3 CONFIGURATION MIXING

We consider here the electron–core interaction within the framework of the independent particle approach to atomic structure theory. That is, we regard the wave function of the electron–core system—called the central complex in Chapter 1—as a superposition of Slater determinants constructed with radial and orbital functions of its several electrons. Each of these radial functions, $f(\epsilon, l, r)$, is an eigenfunction of (5.1) with an ordinary *local* potential $v(r)$ that is equal for all electrons and that represents the field of the nucleus with an average screening by all electrons. This potential, traditionally called the *self-consistent field*, controls the motion of each electron in the complex. This single-particle description aids in the qualitative understanding of atomic dynamics and serves as an initial, useful starting point for quantitative calculations. The self-consistent local potential $v(r)$ may sometimes include, besides the direct screening term v_d of the previous section, simplified local approximations to the exchange potential.† The important consideration is that $v(r)$ include all the long-range behavior.

† See, for example, Herman and Skillman [2].

Within such an independent particle approach, a configuration of the complex is characterized by the set of numbers $\{N_{nl}\}$ denoting the occupancy of each of the (nl) orbitals consistent with the Pauli exclusion principle. The radial functions of the orbitals correspond to eigenvalues ϵ_{nl} of the discrete or continuum of (5.1).† The Schrödinger equation for the entire complex takes the form

$$H\Psi = (H_0 + V)\Psi = E\Psi, \tag{6.8}$$

where H_0 is the sum of the Hamiltonians of (5.1) for the individual electrons—with the eigenvalue $l(l+1)$ replaced by the operator $\mathbf{l}^2$—and where

$$V = -\sum_i \frac{Ze^2}{r_i} + \sum_{i>j} \frac{e^2}{|\mathbf{r}_i - \mathbf{r}_j|} - \sum_i v(r_i) \tag{6.8$'$}$$

is the difference of the exact and independent-electron potentials.

The eigenfunctions of the independent-electron Hamiltonian H_0 serve as the basis states for expressing the full wave function Ψ in (6.8). These eigenfunctions Ψ_{0i} are antisymmetrized products of the (nl) orbitals expressed most conveniently in determinant form. Each such Slater determinant [3] $\Psi_{0i}(i \equiv \{N_{nl}\})$ is an eigenfunction of H_0 with eigenvalue $E_{0i} = \Sigma_{nl} N_{nl}\epsilon_{nl}$. [To be exact, the index i should specify not only the configuration of the electrons (and hence their energy levels) but also their magnetic and spin quantum numbers.]

The residual potential V and, therefore, the full atomic Hamiltonian H, is not diagonal with respect to the Ψ_{0i}. The actual eigenfunction Ψ is a superposition ("mixing") of Slater determinants

$$\Psi = \sum_i a_i \Psi_{0i}, \tag{6.9}$$

where the sum extends over configurations of the discrete and continuous spectrum. Entering this representation of Ψ into (6.8), multiplying by Ψ_{0i}^*, and integrating over all coordinates $\mathbf{r}_i$ reduces (6.8) to the algebraic form analogous to (3.3′),

$$E_{0i}a_i + \sum_{i'} (i|V|i')a_{i'} = Ea_i. \tag{6.10}$$

This equation is exact but is designed to introduce approximations by restricting the Σ_i in (6.9). Familiar applications to spectroscopy confine the Σ_i to selected configurations of the discrete spectrum, but in collision problems the excited electron belongs, of course, to the continuum.

Here we consider configurations of $N + 1$ electrons, N of which belong to

† See, for example, Fano and Fano [3].

the core and are restricted to discrete levels while the remaining one may belong to the continuous or discrete spectrum. The coefficients a_i will be *restricted* further to yielding a *fixed* superposition of the core configurations, which represents realistically the N-electron core ground state with energy E_c. [Alternative sets of configurations could be included here to represent virtually excited states of the core, whose superposition would allow for perturbation of the core by the $(N + 1)$th electron. We defer considering such additional channels to later chapters.] Thus, we set $a_i = Q_\alpha q_h$, where α labels a discrete configuration of the core and h labels an eigenvalue ϵ_h of (5.13) for the excited electron. The fixed Q_α represents an eigenvector of the analog of (6.10) for the core alone with eigenvalue E_c. The Schrödinger equation in (6.8) or (6.10) for $N + 1$ electrons thus reduces to an equation for the one excited electron alone,

$$\epsilon_h q_h + \sum_{h'} (h|V|h')q_{h'} = (E - E_c)q_h, \tag{6.11}$$

where

$$(h|V|h') = \sum_{\alpha\alpha'} Q_\alpha(\alpha h|V|\alpha' h')Q_{\alpha'} \tag{6.11'}$$

represents a specific average over core configurations. The amplitudes Q_α have served to define this average and will play no further role. The radial function of the excited $(N + 1)$th electron will now be represented as $F_l(r) = \Sigma_h q_h f(\epsilon_h, l, r)$.

6.3.1 Reaction Matrix

When solving (6.11) one should consider that this equation is *singular* at $\epsilon_h = E - E_c - \langle h|V|h\rangle$, where the coefficient of q_h vanishes; the same singularity occurred in (3.3′) and (3.5). This singularity is made explicit by a formula analogous to (3.6), but several differences should be noted. First, Chapter 3 dealt with a perturbation problem whereby (3.5) and, therefore, (3.6) already solved that problem; here, (6.11) is an integral equation for q_h, which still remains to be solved. The analog of (3.6) merely replaces q_h by a new variable $\bar{q}_h$ that will obey a nonsingular equation. We set

$$q_h = P[1/(E - E_c - \epsilon_h)]\bar{q}_h + \delta(E - E_c - \epsilon_h)b, \tag{6.12}$$

where P indicates "principal part", and b is an integration parameter, as in (3.6).

Substitution of (6.12) in (6.11) now yields the nonsingular equation for $\bar{q}_h$, i.e.,

$$\bar{q}_h = \sum_{h'} (h|V|h')\left[\frac{P}{E - E_c - \epsilon_{h'}}\bar{q}_{h'} + \delta(E - E_c - \epsilon_{h'})b\right]. \tag{6.13}$$

Two features of this equation are essential:

(1) The equation is *inhomogeneous*, which makes $\bar{q}_h$ a linear function of the parameter b.

(2) The kernel $(h|V|h')$ depends on interactions within the atomic core and hence varies with the energies $(\epsilon_h, \epsilon_{h'})$ only on a scale of ~ 10 eV; accordingly, $(h|V|h')$ may be evaluated adequately on a *coarse mesh* of these energies.

Equation (6.13) reduces thus to a reasonably sized algebraic inhomogeneous system, with its principal part sum evaluated numerically. This system has been solved with modest effort in several atomic and nuclear applications.†

On substitution in the wave function expression (6.9) the two terms of the expression (6.12) are shown to contribute components that oscillate 90° out of phase at large r, as did the analogous terms of (3.6). Here, however, our purpose is no longer to combine the two terms into an outgoing wave, but to express them as a real standing wave so as to extract the phase-shift $\delta_l^{(s)}$ that will be given in essence by the ratio of coefficients of (6.12), $\bar{q}_h/b$. Since the coefficients a_i in (6.9) factor into $Q_\alpha q_h$, the Σ_i also factors into $\Sigma_\alpha \Sigma_h$; the Σ_α has already been discussed, and the Σ_h is to run over a complete set of radial eigenfunctions of the $(N+1)$th electron.

The latter sum reduces to $\int_0^\infty d\epsilon_h$—plus a sum over the discrete spectrum of $\epsilon_n < 0$—if the set consists of the energy-normalized functions $f(\epsilon, l, r)$. The analysis of the contributions of the two terms in the expression of q_h [(6.12)] is done conveniently by expanding f at large r into exponential components, according to (5.4) and as was done in (3.8). The same considerations that led from (3.8) to (3.10) show that

$$\begin{aligned}\int_0^\infty d\epsilon_h\, f(\epsilon_h, l, r) q_h &= \int_0^\infty d\epsilon_h\, B^{1/2}[J^+ f^+ + J^- f^-] \\ &\quad \times \left[\frac{P}{E - E_c - \epsilon_h}\bar{q}_h + \delta(E - E_c - \epsilon_h) b\right] \\ &= [B^{1/2} J^+ f^+(-i\pi\bar{q}_h + b) + B^{1/2} J^- f^-(i\pi\bar{q}_h + b)]_{\epsilon_h = E - E_c} \\ &\qquad \text{for} \quad E - E_c > 0, \end{aligned} \tag{6.14}$$

while the sum over discrete states defined by $J^- = 0$ as in (5.8) vanishes at large r because of the dying exponential dependence of f^+. The factors $\mp i\pi\bar{q}_h + b$ that multiply $f^\pm$ in this last expression can be recast as

$$\mp i\pi\bar{q}_h + b = (\pi^2\bar{q}_h^2 + b^2)^{1/2} \exp[\pm i\delta_l^{(s)}]|_{\epsilon_h = E - E_c}, \tag{6.15}$$

where

$$\tan \delta_l^{(s)} = (-\pi\bar{q}_h/b)_{\epsilon_h = E - E_c}. \tag{6.16}$$

† Some of the earlier applications are described in Section 6 of Fano and Cooper [4].

Thus, we see how the solution of (6.13) for $\bar{q}_h$ actually amounts to determining the phase shift $\delta_l^{(s)}$ contributed by the core interaction. Notice also that the modulus of the coefficient in (6.15) would have the effect of renormalizing the eigenfunction Ψ of the electron + core complex. To preserve the initial normalization we reduce this coefficient to unity by fixing the integration parameter at

$$b = \cos \delta_l^{(s)}, \tag{6.16'}$$

whereby $-\pi\bar{q}_h = \sin \delta_l^{(s)}$ for $\epsilon_h = E - E_c$; the coefficients $\pm i\pi\bar{q}_h + b$ are thus expressed as pure phase factors with unit modulus. Recalling now, from (4.16c), that the tangent of a phase shift is regarded as a diagonal element of the reaction matrix K, we are led to regard the ratio $-\pi\bar{q}_h/b$ for $\epsilon_h \neq E - E_c$ as an off-diagonal element of an extended matrix $K^{(s)}$ and thus to define

$$(\epsilon_h|K^s|E - E_c) = -\pi\bar{q}_h/b. \tag{6.17}$$

Equation (6.13) then takes the form

$$\begin{aligned}(\epsilon_h|K^{(s)}|E - E_c) = \sum_{h'} (h|V|h') \frac{P}{E - E_c - \epsilon_{h'}} (\epsilon_{h'}|K^{(s)}|E - E_c)& \\ - \pi(h|V|E - E_c),& \end{aligned} \tag{6.18}$$

an equation of the Lippmann–Schwinger type.† Further aspects of this calculation of the phase shift through a Lippmann–Schwinger equation (6.18) for the K-matrix will be discussed later.

6.3.2 Born Approximation to the K Matrix

The structure of (6.18), whose right-hand side explicitly involves matrix elements of V as coefficients, suggests immediately a procedure for solving it when these matrix elements are small. Recall in this connection that the matrices of $K^{(s)}$ and V are dimensionless in the representation with energy-normalized eigenstates h. If, then, the matrix elements of $-\pi V$ are much less than unity, (6.18) gives directly

$$(\epsilon_h|K^{(s)}|E - E_c) = -\pi(h|V|E - E_c) + O(V^2). \tag{6.19}$$

The expansion of $K^{(s)}$ as a Born series in ascending powers of V is obtained by iterating (6.18), that is, by substituting repeatedly the entire right-hand side of (6.18) for $(\epsilon_{h'}|K^{(s)}|E - E_c)$ in its first term.

† See, for example, Rodberg and Thaler [5], where the factor $-\pi$ in (6.17) is not included in the definition of $K^{(s)}$. This type of equation has been generally used to determine the complete K matrix in the wave vector representation (4.16c), rather than the more restricted contribution $K^{(s)}$ of short-range particle–core interactions. Even more frequently we encounter analogous equations in a modified form with the singularity handled not as in (6.12) with real coefficients but replaced by a complex variable pole $(E - E_c - \epsilon_h + i\eta)^{-1}$ to yield a transition matrix (see Section 6.3.4).

6.3.3 Factorized Interaction

Even when the interaction matrix $-\pi(h|V|h')$ is not small, it has often been found in important problems (Section 6.3 of Fano and Cooper [4]) to be represented approximately as the product of two functions

$$-\pi(h|V|h') \sim g(\epsilon_h)g(\epsilon_{h'}). \tag{6.20}$$

Such a factorization of the interaction also permits a ready solution of (6.18), because on substitution of (6.20) into (6.18) each matrix element $(\epsilon_h|K^{(s)}|E - E_c)$ is seen to depend on its row index ϵ_h only through a factor $g(\epsilon_h)$. The full equation is then solved by

$$(\epsilon_h|K^{(s)}|E - E_c) = [g(\epsilon_h)(g(E - E_c)] \bigg/ \left[1 - P\sum_{h'} \frac{g^2(\epsilon_{h'})}{(E - E_c - \epsilon_{h'})}\right], \tag{6.21}$$

where only the "dispersion integral" in the denominator needs to be calculated. This type of solution applies in many linear problems. Inaccuracies of (6.20) can often be taken into account by perturbation methods as in Section 6.3.2.

6.3.4 Outgoing-Wave Boundary Conditions: Transition Matrix

A boundary condition of interest [other than the one in (6.15) and (6.16)] is one that gives a superposition of unperturbed and outgoing waves in (6.12). This is accomplished by replacing $P[1/(E - E_c - \epsilon_h)]$ by $[1/(E - E_c - \epsilon_h + i\eta)]$, where η is an infinitesimal positive quantity.† Further replacement of (6.17) by

$$(\epsilon_h|T^{(s)}|E - E_c) = -\pi\bar{q}_h/b \tag{6.17$'$}$$

yields a complex transition matrix (T-matrix) in place of the real Hermitian K-matrix. We thus have in place of (6.18) the alternative Lippmann–Schwinger equation

$$(\epsilon_h|T^{(s)}|E - E_c) = \sum_{h'}(h|V|h')\frac{1}{E - E_c - \epsilon_{h'} + i\eta}(\epsilon_{h'}|T^{(s)}|E - E_c) - \pi(h|V|E - E_c) \tag{6.18$'$}$$

$$\begin{aligned} &= \sum_{h'}(h|V|h')\frac{P}{E - E_c - \epsilon_{h'}}(\epsilon_{h'}|T^{(s)}|E - E_c) \\ &\quad - i\pi(h|V|E - E_c)(E - E_c|T^{(s)}|E - E_c) \\ &\quad - \pi(h|V|E - E_c). \end{aligned} \tag{6.18$''$}$$

The passage from (6.18′) to (6.18″) involves the operation in the footnote below. Since (6.18″) differs from (6.18) only by a constant factor $[1 + i(E - E_c|T^{(s)}|E - E_c)]$ in the inhomogeneous term, the important relationship follows:

$$(\epsilon_h|T^{(s)}|E - E_c) = \frac{(\epsilon_h|K^{(s)}|E - E_c)}{1 - i(E - E_c|K^{(s)}|E - E_c)},$$

which is implied by (4.16).

† The frequently used expression $(x + i\eta)^{-1}$ with $0 < \eta \ll 1$ is equivalent to $P(1/x) - i\pi\delta(x)$ as can be verified by integration over x. See Eq. (3.106) of Rodberg and Thaler [5].

6.3.5 Extension to Discrete Spectrum

The characterization of the coefficients $\bar{q}_h$ in terms of a reaction matrix has emerged from (6.14), which was designed for the energy range $E - E_c > 0$. Extrapolation to the range $E - E_c < 0$ is required for the applications to bound states described in Chapter 5. However, the preceding equation (6.12) that defines $\bar{q}_h$ no longer applies in this range, since none of the discrete eigenvalues $\epsilon_{h'} < 0$ generally coincides with $E - E_c$. Indeed, (6.11) is now generally nonsingular; it becomes nevertheless quasi-singular by yielding a large value of q_β whenever $E - E_c$ approaches a discrete eigenvalue $\epsilon_{h'} < 0$.

This matter has been studied by Greene *et al.* [6] with the following result. The quasi-singularity in the discrete spectrum is to be eliminated by subtracting an appropriate term, an operation equivalent to the removal of an infinitesimal interval by the principal part integration over the continuum. The K-matrix equation (6.18) thereby takes the form

$$\begin{aligned}(\epsilon_h|K^{(s)}|E - E_c) = {} & \sum_n (\epsilon_h|V|\epsilon_n) \frac{N_n^2}{E - E_c - \epsilon_n} (\epsilon_n|K^{(s)}|E - E_c) \\ & - (\epsilon_h|V|E - E_c) \cot \beta\, (E_c - E)(E - E_c|K^{(s)}|E - E_c)\, \mathrm{St}(E_c - E) \\ & + P \int_0^\infty d\epsilon' (\epsilon_h|V|\epsilon') \frac{1}{E - E_c - \epsilon'} (\epsilon'|K^{(s)}|E - E_c) \\ & - \pi(\epsilon_h|V|E - E_c). \end{aligned} \tag{6.22}$$

Here the first, third, and fourth lines on the right-hand side are as in (6.18) except that in the summation $\Sigma_{h'}$ the discrete and continuum ranges have been separated. The discrete wave functions in $|\epsilon_n)$ are normalized on the energy scale as in Section 5.7.2 with N_n defined in (5.36). The new element in (6.22)—namely, the second term on the right-hand side, not present in (6.18)—serves to eliminate the quasi-singularity. The step function $\mathrm{St}(E_c - E)$, which vanishes for $E - E_c > 0$ and equals unity otherwise, emphasizes that this contribution only arises when $E - E_c$ falls in the discrete part of the spectrum. The P symbol, on the other hand, becomes inoperative for $E - E_c < 0$.

6.3.6 Wave Function Outside the Core

The connection established by (6.16) and (6.17) between q_h [or the reaction matrix $K^{(s)}$] and the phase shift $\delta^{(s)}$ manifests itself in the representation of the full wave function Ψ outside the core. By substituting the expressions (6.16′) and (6.17) of b and $\bar{q}_h$ in the representation (6.12) of q_h and the latter in the coefficients $a_i \equiv Q_\alpha q_h$ of the expansion of Ψ [Eq. (6.9)] we get

$$\Psi = \sum_{\alpha h} \Psi_{0i} Q_\alpha \left[-\frac{1}{\pi} \frac{P}{E - E_c - \epsilon} (\epsilon_h|K^{(s)}|E - E_c) + \delta(E - E_c - \epsilon_h) \right] \cos \delta^{(s)}. \tag{6.23}$$

The Σ_h over the second term in the brackets implies that the wave function $f(\epsilon_h, l, r)$ of the outer electron, included as a factor in the Slater determinant Ψ_{0i}, is to be evaluated at the energy $\epsilon_h = E - E_c$. The Σ_h over the first term, with principal part integration over the singular factor $(E - E_c - \epsilon_h)^{-1}$ is decisive for our purpose and of general significance.

The parameter ϵ_h to be summed over is included in three factors, namely, the function $f(\epsilon_h, l, r)$ in Ψ_{0i}, the singular $(E - E_c - \epsilon_h)^{-1}$, and the function $f(\epsilon_h, l, r)$ represented by the symbol $(\epsilon_h|$ in the matrix element $(\epsilon_h|K^{(s)}|E - E_c)$. The sum over these three factors constitutes the singular Green's function

$$G_{E-E_c}(r, r') = \sum_h f(\epsilon_h, l, r) \frac{P}{E - E_c - \epsilon_h} f(\epsilon_h, l, r') \tag{6.24}$$

for the electron motion in the potential $v(r)$. There are, of course, alternative Green's functions for this potential classified by their boundary conditions at $r \to 0$ and $r \to \infty$. At positive energies $(E - E_c > 0)$, the integration over ϵ_h in (6.24) proceeds readily using the simple asymptotic form of $f(\epsilon_h, l, r)$ for large r and is seen to introduce a 90° phase lag in this function evaluated at $\epsilon_h = E - E_c$, which is thereby changed into $-\pi g(E - E_c, l, r)$. For negative energies (6.24) is to be complemented by a term $-f(E - E_c, l, r)\pi \cot \delta^{(s)} f(E - E_c, l, r)$ according to the analysis that led to (6.22). This complement amounts to replacing the Green's function (6.24) by the "smooth" Green's function

$$G^{(s)}_{E-E_c}(r, r') = \pi g(E - E_c, l, r) f(E - E_c, l, r'), \qquad r > r', \tag{6.24'}$$

which is free of poles and coincides with (6.24) for $E - E_c > 0$. Altogether the Σ_h in (6.23) is expressed entirely in terms of the standard pair $\{f, g\}$ evaluated at $\epsilon_h = E - E_c$, yielding

$$\Psi = \sum_\alpha \Psi_{0i} \Big|_{\epsilon_h = E - E_c} Q_\alpha [-g(E - E_c, l, r) K^{(s)}(E) + f(E - E_c, l, r)] \cos \delta^{(s)}, \tag{6.25}$$

where

$$K^{(s)}(E) = (E - E_c|K^{(s)}|E - E_c) = \tan \delta^{(s)} \tag{6.26}$$

is a diagonal element of the matrix (6.16). The wave function outside the core thus takes its more familiar form (5.20),

$$\Psi = \left[\sum_\alpha \Psi_{0i} Q_\alpha\right] [f(\epsilon, l, r) \cos \delta^{(s)} - g(\epsilon, l, r) \sin \delta^{(s)}] \Big|_{\epsilon = E - E_c}. \tag{6.27}$$

6.3.7 Calculation of $\delta^{(so)}$

Finally, the actual objective of this section is the calculation of $\delta_l^{(so)}$ rather than of $\delta_l^{(s)}$. Equations (5.23) accomplish this goal, of course, once $\delta_l^{(s)}$ is given by (6.16) and the parameters B and $\mathscr{G}$ of the long-range field are known. A procedure that yields the core interaction parameter $\delta_l^{(so)}$ without resorting to an analysis of the large-r behavior of the energy normalized functions $f(\epsilon, l, r)$ would have been more self-contained. Yet use of the $f(\epsilon, l, r)$ was implicit in the configuration mixing approach that expands eigenfunctions Ψ into a set of Ψ_{0i} complete over the entire space. The use of such a set is somewhat foreign to our main purpose of dealing with short-range interactions within the core. The following section will focus on the core exclusively.

6.4 *R*-MATRIX TREATMENT

The preceding remarks lead us to develop procedures to calculate the phase shift $\delta_l^{(\mathrm{so})}$ considering only the electron + core system within a volume of radius r_0, without any reference to the electron's wave function in the outer region $r > r_0$. We remain initially within the frame of the independent-particle configuration-mixing approach.

The key point is the following. The construction of a base set of wave functions Ψ_{0i} for the expansion (6.9), *complete within a finite volume*, requires only a *discrete* spectrum of independent particle radial eigenfunctions $\phi(\epsilon_n, l, r)$ of (6.1) in contrast to the continuous spectrum of Ψ_{0i} in Section 6.3. A discrete set of eigenfunctions and the corresponding set of eigenvalues ϵ_n can be selected by imposing a boundary condition at r_0 that contains an adjustable parameter σ_0,

$$[r\, d\phi/dr + \sigma_0 \phi]_{r=r_0} = 0. \tag{6.28}$$

Wave functions for the N core electrons are to be chosen separately as in Section 6.3 so as to be negligibly small for $r \gtrsim r_0$. They are to be combined with each of the $\phi(\epsilon_n, l, r)$ to construct a set of Slater determinants Ψ_{0i} for the $N + 1$ electrons as in Section 6.3. The procedure of the last section can then be applied, setting $a_i = Q_\alpha q_n$ in (6.9), to obtain (6.11) for the coefficients q_n of the superposition $F_l(r) = \Sigma_n q_n \phi(\epsilon_n, l, r)$.

The calculation of tan $\delta_l^{(\mathrm{so})}$ utilizing the discrete base set $\phi(\epsilon_n, l, r)$ now proceeds in the following context. Recall from Section 6.3 that (6.11) is actually inhomogeneous, owing to its singularity, even though it appears formally homogeneous; its inhomogeneous form [(6.13)] results from the variable change in (6.12) from q_h to $\bar{q}_h$. Physically, the effect of the interaction of the $(N + 1)$th electron with the core may be viewed as the *response of the core* to an incident electron; quantum mechanical response functions generally obey linear *inhomogeneous* equations. Thus, our problem consists, in essence, of solving an inhomogeneous system by constructing its resolvent, that is, the inverse of the matrix of its coefficients. The discrete set of eigenfunctions $\phi(\epsilon_n, l, r)$ of the Hamiltonian for $0 \le r \le r_0$ will be utilized to construct the eigenvectors of this matrix and hence the resolvent of the inhomogeneous system by standard procedures of linear algebra.

The construction of the eigenvectors of the Hamiltonian within the finite range $0 \le r \le r_0$ and with the boundary condition (6.28) on its eigenfunctions Ψ_ν no longer involves a singularity of (6.11) because none of the discrete eigenvalues E_ν generally coincides with E_c. These eigenfunctions will be represented in the form (6.9), $\Psi_\nu = \Sigma_i$

$a_i^{(\nu)}\Psi_{0i}$, requiring now each of the determinants Ψ_{0i} to obey the boundary conditions (6.28). Each Ψ_{0i} will thus consist of a core configuration α as in Section 6.3 and a state of the $(N + 1)$th electron with the radial eigenfunction $\phi(\epsilon_n, l, r)$ of (5.1) that obeys (6.8). This requirement amounts to setting $i \equiv (\alpha, n)$ and $a_i^{(\nu)} = Q_\alpha q_n^{(\nu)}$, that is,

$$\Psi_\nu = \sum_\alpha \sum_n Q_\alpha q_n^{(\nu)} \Psi_{0\alpha n}. \tag{6.29}$$

This structure of Ψ_ν leads to an equation (6.11) with a discrete spectrum of eigenvalues $E = E_\nu$ and with eigenvectors $q_n = q_n^{(\nu)}$. These eigenvectors also identify the radial function of the $(N + 1)$th electron in (6.29) as the superposition

$$\phi_\nu(r) = \sum_n q_n^{(\nu)} \phi(\epsilon_n, l, r). \tag{6.29'}$$

To determine the desired parameter $\delta_l^{(so)}$ utilizing the eigenfunctions Ψ_ν we must solve (6.11) once again, this time for a desired energy E, different from the E_ν, but releasing the radial function of the $(N + 1)$th electron from the boundary condition (6.28). The resulting eigenfunction for the $N + 1$ electrons will be indicated as a superposition

$$\Psi = \sum_\nu c_\nu \Psi_\nu, \tag{6.30}$$

and the radial component of its $(N + 1)$th electron will be represented by

$$F_l(r) = \sum_\nu c_\nu \phi_\nu(r) = \sum_\nu \sum_n c_\nu q_n^{(\nu)} \phi(\epsilon_n, l, r). \tag{6.31}$$

The value of $\delta_l^{(so)}$ will emerge from the boundary value of the logarithmic derivative of $F_l(r)$.

It is unnecessary at this point to solve (6.11) explicitly because the expansion coefficient c_ν in (6.31) can be represented as the overlap integral of $F_l(r)$ and $\phi_\nu(r)$. Through the Schrödinger equation for $F_l(r)$ (with energy E) and $\phi_\nu(r)$ (with energy E_ν), the overlap integral is related in turn to boundary parameters at r_0 of these functions and their first derivatives. These relationships yield

$$\begin{aligned} c_\nu &= \int_0^{r_0} \phi_\nu(r) F_l(r)\, dr \\ &= \frac{[\phi_\nu(r) r\, (dF_l/dr) + \sigma_0 \phi_\nu(r) F_l(r)]_{r=r_0}}{2r_0(E_\nu - E)}, \end{aligned} \tag{6.32}$$

where σ_0 represents the logarithmic derivative of $\phi_\nu(r)$ according to (6.28). The content of (6.32) is made explicit by expressing the logarithmic derivative of $F_l(r)$ at r_0 through a parameter σ analogous to σ_0, thus setting

$$[r\,(dF_l/dr) + \sigma F_l(r)]_{r=r_0} = 0, \tag{6.33}$$

and bringing (6.32) to the more symmetric form

$$c_\nu = -(\sigma - \sigma_0) F_l(r_0) [\phi_\nu(r_0)/2r_0(E - E_\nu)]. \tag{6.32'}$$

The essential information to be extracted from this result is the dependence of the boundary parameter σ on the energy E and on the sets of eigenvalues E_ν and eigenfunctions $\phi_\nu(r)$. To this end we multiply (6.32′) by $\phi_\nu(r_0)$, sum over ν, and introduce the symbol

$$R_l(E) = \sum_\nu \frac{[\phi_\nu(r_0)]^2}{2r_0(E - E_\nu)} \tag{6.34}$$

for the relevant combination of E_ν and $\phi_\nu(r_0)$. The result,

$$\begin{aligned}\sum_\nu c_\nu \phi_\nu(r_0) &= F_l(r_0) \\ &= -(\sigma - \sigma_0)F_l(r_0)R_l(E),\end{aligned} \tag{6.35}$$

allows $F_l(r_0)$ to factor out, thereby reducing to

$$\sigma(E) = \sigma_0 - [1/R_l(E)]. \tag{6.36}$$

The desired phase shift is finally given by (6.7) in terms of $\sigma(E)$,

$$\begin{aligned}\tan \delta_l^{(\mathrm{so})}(E) &= W(f^\circ, F_l)/W(g^\circ, F_l)|_{r=r_0} \\ &= (f^\circ/g^\circ)\left|\frac{\sigma + r\, d \ln f^\circ/dr}{\sigma + r\, d \ln g^\circ/dr}\right|_{r=r_0}.\end{aligned} \tag{6.37}$$

The set of parameters $R_l(E)$ thus represents the whole contribution of core interactions to the collision parameters described in Section 4.5. It constitutes another quantity of the same family, called the R-matrix, which is equivalent to the reaction matrix $K^{(\mathrm{s})}(E)$ of (6.26). The meromorphic dependence of R_l on the energy E greatly reduces the work required to calculate the electron–core interactions at different energies because the surface parameters $\phi_\nu(r_0)$ and σ_0 are independent of E. On the other hand, the values of $R_l(E)$ are not expected to vary greatly over a range of a few electron volts; calculations extending over a broader energy range might well require alterations of the base set of wave functions and hence more numerical work.

A basic shortcoming of the procedure outlined in this section lies in the poor convergence of the expansion $F_l(r) = \Sigma_\nu c_\nu \phi_\nu(r)$ at $r \sim r_0$, rooted in the attempt to reproduce a function $F_l(r)$ with boundary parameter σ as a superposition of $\phi_\nu(r)$ with a different parameter σ_0. This difficulty, familiar in the theory of Fourier analysis, has been minimized by a procedure called the Buttle correction and applied extensively in numerical constructions of R matrices [7]. This correction amounts to obtaining as a preliminary the R matrix for a model potential $\bar{V}$ that is equivalent to (6.8′) at $r \sim r_0$ but allows analytic construction of $R_l(E)$. The numerical calculation of the actual $R_l(E)$ then deals only with the potential difference $V - \bar{V}$, which is large only at a distance from the boundary where the convergence is satisfactory.

Procedures designed to calculate the parameter σ or the equivalent $\delta_l^{(\mathrm{so})}$ for core interactions irrespective of phenomena external to the core are loosely referred to as R-matrix procedures and are discussed further in Section 6.5

from a variational point of view. They were introduced initially by Wigner and Eisenbud in nuclear physics, where, the forces being of short range, r_0 has a natural, well-defined meaning. Their use in atomic physics stems from the two features discussed in this chapter, namely, that the wave functions of many of the electrons fall off exponentially near the core radius and that the most complicated interactions, the exchange terms, are also confined to $r < r_0$.† The inner region is, therefore, the region of strong correlations. The complete set of eigenfunctions $\phi(\epsilon_n, l, r)$ in this region provides a natural basis for expanding the final state wave function $F_l(r)$. This set of eigenfunctions is necessarily discrete so long as r_0 is finite, but the spacing between the eigenvalues decreases as r_0 gets larger. The considerations surrounding a judicious choice of r_0, neither too large nor too small, have been discussed at the end of Section 6.2.

Note also that Wigner originated both the R-matrix procedure and the threshold laws introduced in Section 4.8, thus separating effects stemming from the inner and outer ranges of a radial variable. The very same idea has enabled us in Chapter 5 to isolate the influence of the potential in the outer range that is responsible for threshold singularities in a cross section and to pursue the effect of short-range interactions separately in the present chapter.

6.5 VARIATIONAL APPROACHES

The possibility of representing the state of an electron + core complex within a finite volume by a discrete base set of wave functions, regardless of its excitation level, permits full utilization of the concepts and techniques for calculating the discrete levels of atoms and molecules. Most important among these are the minimum principle for approximating energy levels and the great flexibility in selecting for this purpose any type of wave function, subject to orthogonality to lower-level eigenfunctions. We are thus freed from the rigidity inherent in starting from a systematic independent-particle description of atomic states.

Variational principles have been introduced long ago in scattering theory for the direct determination of the phase shifts of continuum wave functions. However, their application has been beset by difficulties, conceptual and technical; for example, alternative procedures were available with different but not very transparent implications [8]. The interpretation and interconnections of different procedures will be stressed in this section. Further details are given in Section 6.8.

† In the present context, where only a single electron reaches asymptotically large distances, the confinement applies to all but one of the $N + 1$ electrons. Extension to problems with more than one electron reaching large radial distances will be considered in Part D.

The developments described in Chapter 5 show that $\delta_l^{(l)}$, the contribution of long-range atomic fields to a phase shift, is known analytically for the longest-range field components and is evaluated by WKB or direct numerical procedures otherwise (Table 5.1). It is the short-range component $\delta_l^{(s)}$ whose calculation requires elaborate analysis. We see now, particularly in the R-matrix procedures of Section 6.4, that this problem is reduced in essence to the problem of calculating energy eigenvalues for discrete levels. Ordinary variational procedures apply directly to the calculation of the eigenvalue E_ν and its corresponding eigenfunctions in the R-matrix element (6.34). It can be seen from (6.34) that the minimum properties of the E_ν—chiefly that a variational estimate of E_ν always exceeds its actual value—imply corresponding minimum properties of $R_l(E)$; thereby the parameter σ and the phase shifts $\delta_l^{(so)}$ obtained from R_l by (6.36) and (6.37) exhibit maximum properties.

The equivalence between the variational properties of energy levels and phase shifts rests on a fundamental *monotonic* functional relationship between the eigenfrequencies of any wave in a bounded region and the boundary parameter σ of (6.33). The parameter σ represents, in turn, the tangent of the oscillation phase of $F_l(r)$ and is accordingly monotonically related to $\delta_l^{(so)}$. A prototype of this monotonic relationship is the functional dependence between the pitch of a musical string and the tension applied to it. Any *overestimate* of the pitch achieved with a given stress implies an *underestimate* of the stress required to achieve a given pitch. A classical treatment of this relationship [9] stresses its extremely broad range of validity, which will again prove essential in Chapter 9.

We now utilize this relationship between energy eigenvalues of an atom and boundary parameter σ directly to calculate the phase shift $\delta_l^{(so)}$ and the corresponding wave function $F_l(r)$ variationally for any desired energy E within the range of our single-channel representation. This approach bypasses the eigenstate expansion discussed in the previous section and the need for representing the correct σ by superposition of eigenfunctions with an incorrect σ_0. The procedure approximates the radial function $F_l(r)$ by a trial function $F_{lt}(r)$ represented as a superposition of a suitable finite set of analytical or numerical expressions $w_n(r)$,

$$F_{lt}(r) = \sum_n q_{nt} w_n(r). \tag{6.38}$$

The set of base radial functions $w_n(r)$ thus replaces the set of $f(\epsilon_n, l, r)$ utilized in the configuration mixing procedure of Section 6.3. Note, however, that each $w_n(r)$ is *not* viewed here as representing the wave function of one electron in an initial approximation; this role will be filled by the full expression (6.38) after optimization of its coefficients. The boundary parameter σ is similarly represented as an explicit function of the coefficients q_{nt},

$$\sigma = -r_0 \left[\sum_n q_{nt} \frac{dw_n}{dr} \right]_{r=r_0} \Bigg/ \sum_n q_{nt} w_n(r_0). \tag{6.39}$$

The next step consists of combining the trial radial function $F_{lt}(r)$ of the excited electron with the wave functions of core electrons and with the orbital functions to construct the Slater determinants Ψ_{0i} and the approximation to the complete wave function Ψ of (6.9). The resulting functions Ψ_t depend linearly on the coefficients q_{nt} only, since the other factors Q_α of a_i are fixed by the identification of the core ground state. The variational procedure now consists of seeking a minimum of the expectation value of the energy

$$\langle E \rangle = (\Psi_t | H | \Psi_t)/(\Psi_t | \Psi_t), \tag{6.40}$$

with H as defined by (6.8).†

Expressing F_{lt} in terms of the known functions w_n, as in (6.38), and varying the q_{nt} subject to the subsidiary condition (6.39) with a given value of σ, we obtain the functional relationship

$$\langle E \rangle_{\min} = \mathscr{E}(\sigma) \tag{6.41}$$

and also determine the corresponding optimum values of the coefficients, $q_n^{\text{opt}}(\sigma)$. This relationship may be mapped by evaluating numerically $\langle E \rangle_{\min}$ and the q_n^{opt} for a set of values of σ; a coarse mesh of values may often be adequate. Graphical or numerical interpolation of $\langle E \rangle_{\min}$ then serves to determine the value of σ for which $\langle E \rangle_{\min}$ takes any preestablished value E. This value constitutes a variational lower bound estimate σ^{opt} of the value of σ at the given energy E. The corresponding estimate of $\delta_l^{(\text{so})}$ is then obtained from σ^{opt} through (6.37). The variationally optimized wave function for this energy is finally obtained by setting $a_i = Q_\alpha q_n^{\text{opt}}(\sigma_{\text{opt}})$ in the expansion $\Sigma_i\, a_i \Psi_{0i}$ of Ψ_t.

† That (6.40) is indeed the variational functional leading to (6.1) and (6.6) can be demonstrated as follows. Using (6.3) and carrying out the integral over $d\mathbf{r}_1$, we have

$$\begin{aligned}(\Psi_t | H - E | \Psi_t) = &-\frac{1}{2}\left(F_{lt}\left|\frac{d^2}{dr^2}\right|F_{lt}\right) - \frac{1}{2}\left(F_{lt}\left|\frac{l(l+1)}{r^2}\right|F_{lt}\right) \\ &- \left(F_{lt}\left|\frac{e^2}{r}\right|F_{lt}\right) - \left[\frac{1}{2}\left(u\left|\frac{d^2}{dr^2}\right|u\right) + \left(u\left|\frac{e^2}{r}\right|u\right) + E\right] \\ &\times (F_{lt}|F_{lt}) + e^2\left(\Psi_t\left|\frac{1}{|\mathbf{r}-\mathbf{r}_1|}\right|\Psi_t\right).\end{aligned}$$

The expression in square brackets is $E - E_c = \epsilon$. By using (6.5) and taking the functional derivative $\delta(\Psi_t | H - E | \Psi_t)/\delta F_{lt}$, we have (6.1) with $v(r)$ as given in (6.6).

Alternatively and more effectively, we may determine σ directly for a desired energy E by interchanging the role of the energy functional (6.40) and the subsidiary condition (6.39). The boundary parameter σ thus takes the role of the eigenvalue. This approach is discussed in Section 6.8 and will take a central role in Chapter 9.

A further extension of this variational procedure may be envisaged to remove the frozen core approximation, that is, to take into account that penetration of the $(N + 1)$th electron into the core contributes an additional screening to the nuclear attraction experienced by the N core electrons. Recall that the coefficients Q_α of the Slater determinants of the core had been chosen so as to minimize the energy of the core by itself. This choice is appropriate only for positions of the $(N + 1)$th electron at $r \gtrsim r_0$. A correlation between the radial coordinate of this electron and the core wave function can be introduced by adding to the representation of Ψ [(6.9)] further terms $\Sigma_\rho[\Sigma_{\alpha n} Q_\alpha^\rho q_n^\rho \Psi_{0\alpha n}^{(\rho)}]$. In a first step, the $\Psi_{0\alpha n}^{(\rho)}$ term may be kept identical to the initial Slater determinants $\Psi_{0i} \equiv \Psi_{0\alpha n}$, thus merely introducing alternative sets of coefficients $\{Q_\alpha^\rho\}$ to change the mixing of core configurations. Additional flexibility results by constructing Slater determinants with radial eigenfunctions $u_n^\rho(r)$ of (6.1) with different potentials $v^{(\rho)}(r)$ or even by introducing zero-order functions $\Psi_{0\alpha n}^{(\rho)}$ with entirely different structure.

In any event, the correlations thus introduced should disappear when the $(N + 1)$th electron emerges from the core; this requirement is met by setting the boundary parameter of (6.39) at $\sigma^{(\rho)} = \infty$ for the correlation terms. With these stipulations we may proceed again to seek a minimum of $\langle E\rangle$ in (6.33). The values of $\langle E\rangle_{\min}$ thus obtained will be certainly lower than those found in the absence of the additional parameters $q_n^{(\rho)}$. The influence of the electron–core correlations will be deemed significant to the extent that the resulting reduction of $\langle E_{\min}\rangle$ and the corresponding increase in the estimate of the phase shift $\delta_l^{(\mathrm{so})}$ are themselves statistically significant from the point of view of analysis of variance. This procedure will be considered further in Chapter 9 together with its extension to multichannel problems.

6.6 CLOSE-COUPLING EQUATIONS

We return now to the direct determination of collision wave functions through the Schrödinger differential—or, more accurately, integrodifferential—equation by extending the procedure of Section 6.2 by removing the frozen core approximation. For simplicity we refer only to the $e + H$ collision and two alternative stationary states of the core, namely, 1s and 2s, with eigenfunctions $u_1(r)(4\pi)^{-1/2}$ and $u_2(r)(4\pi)^{-1/2}$. Extension to atoms with many electrons and many alternative states of an atom would merely involve writing down additional factors and terms without changing the structure of the equations.

The wave function for the $e + H$ complex consists of two terms with the structure of (6.3),

$$\begin{aligned}\Psi_1(\mathbf{r}, \mathbf{r}') &+ \Psi_2(\mathbf{r}, \mathbf{r}') \\ &= \sqrt{\tfrac{1}{2}}[F_l^1(r) Y_{lm}(\hat{r}) u_1(r') \pm F_l^1(r') Y_{lm}(\hat{r}') u_1(r)](4\pi)^{-1/2} \\ &\quad + \sqrt{\tfrac{1}{2}}[F_l^2(r) Y_{lm}(\hat{r}) u_2(r') \pm F_l^2(r') Y_{lm}(\hat{r}') u_2(r)](4\pi)^{-1/2}, \quad (6.42)\end{aligned}$$

with the alternative signs for triplet or singlet states. The two-electron equation is

$$\begin{aligned}&\left\{\frac{1}{2}\left(\frac{d^2}{dr^2} + \frac{d^2}{dr'^2}\right) - \frac{1}{2}\left(\frac{\mathbf{l}^2}{r^2} + \frac{\mathbf{l}'^2}{r'^2}\right) + \epsilon + \frac{e^2}{r} + \frac{e^2}{r'} - \frac{e^2}{|\mathbf{r} - \mathbf{r}'|}\right\} \\ &\quad \times [\Psi_1(\mathbf{r}, \mathbf{r}') + \Psi_2(\mathbf{r}, \mathbf{r}')] = 0. \quad (6.43)\end{aligned}$$

In Section 6.2 the potential terms of a single-variable radial equation for $F_l(r)$ were obtained from those of the two-electron equation through multiplication by $u(r_1)(4\pi)^{-1/2} Y_{lm}(\hat{r})$ followed by integration over $d\mathbf{r}_1$ and $d\hat{r}$. Similarly, here we obtain a system of two radial equations, for $F_l^{(1)}(r)$ and $F_l^{(2)}(r)$, by multiplying (6.43) by $u_i(r')(4\pi)^{-1/2} Y_{lm}(\hat{r})$, setting alternatively $i = 1$ or $i = 2$, and integrating over $d\mathbf{r}'$ and $d\hat{r}$. Proceeding further as in Section 6.2 we recover, for $i = 1$, *two* potential terms, each of them with the structure of (6.6), which are contributed by the terms of (6.43) with Ψ_1 and Ψ_2, respectively, and are indicated by

$$v_{11}(r) F_l^1(r) + v_{12}(r) F_l^2(r). \quad (6.44)$$

By setting $i = 2$ we recover instead another pair of analogous terms indicated by

$$v_{21}(r) F_l^1(r) + v_{22}(r) F_l^2(r). \quad (6.44')$$

In this procedure, as in Section 6.2, we consider also that $u_i(r')$, with $i = 1$ or 2, obeys the radial equation for the H atom

$$\left[\frac{1}{2}\frac{d^2}{dr'^2} + \frac{e^2}{r'}\right] u_i(r') = -\epsilon_i u_i(r'). \quad (6.45)$$

Thus, we obtain the system of two radial equations†

$$\begin{aligned}&\left[\frac{1}{2}\frac{d^2}{dr^2} - \frac{1}{2}\frac{l(l+1)}{r^2} + \epsilon - \epsilon_1 - v_{11}(r)\right] F_l^1(r) - v_{12}(r) F_l^2(r) = 0, \\ &\left[\frac{1}{2}\frac{d^2}{dr^2} - \frac{1}{2}\frac{l(l+1)}{r^2} + \epsilon - \epsilon_2 - v_{22}(r)\right] F_l^2(r) - v_{21}(r) F_l^1(r) = 0.\end{aligned} \quad (6.46)$$

† These equations can also be derived alternatively by the variational procedure of Section 6.5. Construct the expression $(\Psi|H - \epsilon|\Psi) = 0$ with (6.42) as a "trial form" for Ψ, that is, multiply (6.43) by (6.42) and integrate over $\mathbf{r}_1$ and $\mathbf{r}_2$, then set to zero the two functional derivatives of this expression with respect to $F_l^1(r)$ and $F_l^2(r)$.

The two equations (6.46) are said to be "close coupled" because it is not intended to treat the coupling potentials v_{12} and v_{21} as small perturbations. The radial functions $F_l^1(r)$ and $F_l^2(r)$ represent, of course, the probability amplitudes of finding simultaneously the colliding electron at a radial distance r and the atom in its state 1s or 2s, respectively. Using two such radial functions serves to represent a correlation between the states of the colliding electron and the atom; using a larger number of functions $F_l(r)$ would represent a more detailed correlation. Because each of the potentials $v_{ij}(r)$ in (6.46) includes an exchange term, the system is integrodifferential but can be converted into a system of four coupled, purely differential, equations by the procedure of Section 6.7. Use of close-coupling equations with numerical solution of rather large systems has been developed extensively to study electron–atom collisions [1, 10].

Many, or even most, of the successful applications of close-coupling procedures have dealt with inelastic processes to be discussed in later chapters. In this section we still deal with the single-channel situation in which the total energy is insufficient to excite the target, that is, where $\epsilon - \epsilon_2 < 0$. The wave function $F_l^2(r)$ will then decay with increasing r outside the extension of the excited-state eigenfunction $u_2(r)$. In this context let us return now to the effective core radius r_0, beyond which the effect of core interaction is represented entirely by the phase shift $\delta_l^{(\mathrm{so})}$ of $F_l^1(r)$. In the present application the value of r_0 will have to be large enough to enclose the whole extension of the excited-state wave function $u_2(r)$ and hence exceed by far the extension of the ground state's $u_1(r)$, which had served earlier as reference for the value of r_0.

This consideration points up an inherent limitation of the close-coupling procedure, common to all methods that describe the effects of electron–core interaction by expansion into a complete set of core eigenstates. (Configuration mixing is included among these methods if its Slater determinants are constructed with a complete set of single-electron eigenfunctions as envisaged in Section 6.3.) Such sets of eigenstates are complete over the entire space and hence overcomplete in the relevant volume of the core in its ground state. Two consequences follow: Irrelevant information is processed needlessly when using these expansions, and the addition of each successive eigenstate contributes less and less to the description of particle–core interactions. Avoiding these consequences requires the R-matrix or the related variational procedures introduced in the preceding sections.

6.6.1 Optical Potentials

In the system of coupled equations (6.46) the radial function $F_l^2(r)$ plays the subsidiary role of representing the correlation of the colliding electron with the core electron. However, the key collision parameter is still the phase shift $\delta_l^{(\mathrm{so})}$ of the other radial function $F_l^1(r)$. For the purpose of calculating $\delta_l^{(\mathrm{so})}$ we need not know $F_l^2(r)$, and it may be useful to eliminate this radial function from the problem. The elimination proceeds readily—if only formally at the outset—by a straightforward application of operator algebra.

The system (6.46) can be written more compactly as

$$L_1 F_l^1(r) - v_{12} F_l^2(r) = 0, \tag{6.47}$$

$$L_2 F_l^2(r) - v_{21} F_l^1(r) = 0, \tag{6.48}$$

where L_1 and L_2 represent the operators in the first bracket of the two equations (6.46). Symbolic multiplication of (6.48) by L_2^{-1} yields the formal expression of F_l^2

$$F_l^2(r) = L_2^{-1} v_{21} F_l^1(r). \tag{6.49}$$

[This formal operation amounts to treating the term $v_{21}F_l^1$ of (6.48) as the inhomogeneous term of an equation for F_l^2 and solving this equation in terms of its Green's function represented by L_2^{-1}.] Substitution of F_l^2 in (6.47) by its expression (6.49) yields a single equation for F_l^1 only,

$$[L_1 - v_{12} L_2^{-1} v_{21}] F_l^1(r) = 0. \tag{6.50}$$

This procedure is, of course, formally equivalent to the elementary method of solving an algebraic system by elimination of the variable $F_l^2(r)$; it can be extended to eliminating all but one variable of a larger multichannel system.

The term $-v_{12}L_2^{-1}v_{21}$, which has been added to L_1 in (6.50), serves to represent compactly the effect of correlations of the colliding electron with the core; it is called an "optical potential." Note that this effective potential is not only nonlocal but also energy dependent because L_2 itself contains the energy ϵ as a parameter. Practical applications of (6.50) hinge on constructing an analytically or numerically usable representation of the Green's function L_2^{-1} or rather of the entire expression $v_{12}L_2^{-1}v_{21}$. An essential point is that such a representation need incorporate only those elements of (6.48) that are relevant to the solution of (6.47); in particular, it can avoid representing the electron–core correlation by expansion into an overcomplete system. A specific illustration is given in Section 6.9. We will also return to such eliminations of unobserved variables in Chapter 8.

Other notable effects of the elimination of unobserved channels are the occurrences of "resonances," sharp enhancements of $F_l^1(r)$ over a small energy range because of zeroes in the operator L_2 in (6.50). Such effects will be discussed extensively in Chapter 8, but we conclude this section with an allied approximation called the distorted wave approximation. In this method coupled equations such as (6.46) are treated by setting v_{12} to zero (or, more generally, v_{1n} when there are many channels). This makes the first equation in (6.46) homogeneous, permitting a solution by one of the methods of the earlier sections in this chapter. The logic behind this approximation is that inelastic cross sections are often much smaller than the elastic ones, so that the elastic scattering channel is only slightly perturbed by the coupling to the other states. (Clearly, this will be inapplicable near resonances.) Neglect of v_{1n} without disregarding v_{n1}, which represents the influence of the elastic channel on the others, enables us to solve for F_l^1 independently and then obtain each F_l^n in turn from (6.48) and analogous equations, if any.

A general shortcoming of distorted-wave and close-coupling approximations is that they often overestimate inelastic cross sections. This seems paradoxical in that an approximation initially guided by the principle of weak coupling between states, or by truncating some coupling altogether, gives rise to an overestimate of the coupling that is retained, namely, the effect of channel 1 on the others. The point is that complete neglect of coupling acts in effect as a rigid boundary condition that imposes a strong constraint on the form of the wave function. Note that truncation of a discrete sequence of equations, however weakly coupled, amounts in effect to the introduction of a sharp potential barrier, thus overestimating the effect of the truncated channels.

6.7 REDUCTION OF INTEGRODIFFERENTIAL EQUATIONS TO A COUPLED DIFFERENTIAL SYSTEM

The conversion of the integrodifferential equation defined by (6.1) and (6.6) to a coupled differential system starts by rewriting (6.6b) in the form

$$v_s(r)F_l(r) = \pm \frac{e^2}{2l+1}\frac{u(r)}{r}\chi_l(r) \tag{6.51}$$

with

$$\chi_l(r) = r^{-l}\int_0^r dr_1 u(r_1)F_l(r_1)r_1^l + r^{l+1}\int_r^\infty dr_1 u(r_1)F_l(r_1)r_1^{-l-1}. \tag{6.52}$$

A differential equation for $\chi_l(r)$ is now constructed by differentiating (6.52)

$$\frac{d\chi_l(r)}{dr} = -lr^{-l-1}\int_0^r dr_1 u(r_1)F_l(r_1)r_1^l + (l+1)r^l\int_r^\infty dr_1 u(r_1)F(r_1)r_1^{-l-1},$$

$$\frac{d^2\chi_l}{dr^2} = \frac{l(l+1)}{r^2}\chi_l(r) - (2l+1)\frac{u(r)}{r}F_l(r). \tag{6.53}$$

Substitution of (6.51) into (6.1) gives

$$\left[\frac{1}{2}\frac{d^2}{dr^2} - \frac{1}{2}\frac{l(l+1)}{r^2} + \epsilon - v(r)\right]F_l(r) = \pm\frac{e^2}{2l+1}\frac{u(r)}{r}\chi_l(r), \tag{6.54}$$

which combines with (6.53) to form a system of two equations for the pair (F_l, χ_l). The $\pm$ signs correspond to the triplet–singlet alternative in (6.3).

This procedure is generalized in Smith [1] to multichannel equations involving M different functions F and N different $u(r)$; N different functions $\chi(r)$ are introduced to yield a system of $M + N$ differential equations.

6.8 VARIATIONAL PRINCIPLES FOR SCATTERING

The formulation of variational principles in Section 6.4, with functions normalizable within the range $r < r_0$, will be contrasted here with earlier approaches [8] that proceed without explicit separation of short- and long-range contributions to the potential in (6.1). Instead of imposing a boundary condition on $F_l(r)$ at $r = r_0$ as in (6.33), we represent the radial function for $\epsilon > 0$ and zero-field at infinity in the form

$$F_l(r) \underset{r\to\infty}{\longrightarrow} (2/\pi k)^{1/2}[\sin(kr - \tfrac{1}{2}l\pi) + \tan\delta_l(\epsilon)\cos(kr - \tfrac{1}{2}l\pi)]. \tag{6.55}$$

Here δ_l stands for the full phase shift, including the contributions from both the long-range $v(r)$ and the short-range $v_s(r)$.

As a trial approximation to the solution of (6.1), we choose a function $F_{lt}(r)$ which is known for all r, satisfies the boundary conditions $F_{lt}(0) = 0$, and takes the asymptotic form

$$F_{lt}(r) \underset{r\to\infty}{\longrightarrow} (2/\pi k)^{1/2}[\sin(kr - \tfrac{1}{2}l\pi) + \tan\delta_{lt}(\epsilon)\cos(kr - \tfrac{1}{2}l\pi)], \tag{6.56}$$

where δ_{lt} serves as a trial approximation to the phase shift. Consider the expression

$$\int_0^\infty dr\{F_l(r)[(h-\epsilon)F_{lt}(r)] - F_{lt}(r)[(h-\epsilon)F_l(r)]\}, \tag{6.57}$$

where $(\epsilon - h)$ is the operator in (6.1). Application of Green's theorem with the boundary condition $F_{lt}(0) = F_l(0) = 0$ reduces the derivative terms in (6.57) to

$$\tfrac{1}{2}W[F_{lt}(r), F_l(r)]_{r\to\infty} = \pi^{-1}(\tan\delta_{lt} - \tan\delta_l); \tag{6.58}$$

all the other terms cancel in pairs. Since the second term in (6.57) vanishes by virtue of (6.1), Eq. (6.57) can be written as

$$\tan\delta_l = \tan\delta_{lt} - \pi\int_0^\infty dr F_l(r)[(h-\epsilon)F_{lt}(r)], \tag{6.59}$$

a formula known as the "Kato identity" and derived similarly to the normalization equation (5.33).

The Kato identity is an integral expression for the exact phase shift δ_l in terms of trial quantities δ_{lt} and F_{lt} and the exact solution F_l. It amounts to a formal consequence of (6.1), since F_l is not known, but it casts (6.1) into an integral form suitable for approximations. Insofar as F_{lt} is an approximate solution, $(h - \epsilon)F_{lt}$ vanishes to within a first-order error; further replacement of F_l by F_{lt} in (6.59) should, therefore, only introduce errors of second order.

This replacement now gives the Kohn variational principle for the phase shift

$$\tan \delta_{lv} = \tan \delta_{lt} - \pi \int_0^\infty dr F_{lt}(r)[(h - \epsilon)F_{lt}(r)]. \tag{6.60}$$

We have denoted here the variational approximation to δ_l by δ_{lv}, the difference $\delta_{lv} - \delta_l$ being formally of order $(F_{lt} - F_l)^2$.

As an example, consider the scattering of electrons from hydrogen for $l = 0$ with neglect of exchange. The resulting static or Hartree potential is given in (6.6a), and its explicit form for hydrogen is given in (6.6c). With a simple trial function of the form

$$F_{0t} = (2/\pi k)^{1/2}[\sin kr + (c_1 + c_2 e^{-r})(1 - e^{-r}) \cos kr], \tag{6.61}$$

where c_1 and c_2 are variational parameters, the following tabulation gives the s-wave phase shifts δ_{0v} computed from (6.60), compared with the exact results for δ_0 [11].

k (a.u.)	Born	Kohn principle	Exact
0.1	0.099	0.721	0.730
0.2	0.194	0.972	0.973
0.3	0.281	1.045	1.046
0.5	0.423	1.044	1.045
1.0	0.597	0.904	0.906

The Born approximation is also shown because it results as a special case of the Kohn principle with the extreme choice of a free wave for F_{lt} in (6.60), so that $\delta_{lt} = 0$ and $c_1 = c_2 = 0$ in (6.61). Contrast these results with those of Problem 6.3, which applies the procedure of Section 6.5 to the same problem.

For the electron–hydrogen scattering with inclusion of exchange, that is, with the combined potentials (6.6a) and (6.6b), the most accurate calculations with the Kohn variational principle are due to Schwartz, who used the trial functions

$$\begin{aligned}\Psi_t(\mathbf{r}_1, \mathbf{r}) = {} & \frac{u(r_1)}{r_1}[\sin kr + \tan \delta_{lt}(1 - e^{-\lambda r}) \cos kr] \\ & \times \sum_{pqs} [C_{pqs}|\mathbf{r} - \mathbf{r}_1|^p r^q r_1^s \pm (r \leftrightarrow r_1)],\end{aligned} \tag{6.62}$$

with up to 50 variational parameters $\{\lambda, C_{pqs}\}$, the alternative signs denoting triplet and singlet scattering as in (6.3). The results are shown in Fig. 7.1.

The connection between the Kohn variational principle and the one in Section 6.5 emerges by replacing the asymptotic boundary condition (6.55) on $F_l(r)$ with a scattering boundary condition of the form (5.20) at the boundary $r = r_0$ of the short-range potential $v_s(r)$:

$$F_l(r) = (f(\epsilon, l, r) - \tan \delta_l^{(s)} g(\epsilon, l, r))|_{r=r_0}. \tag{6.63}$$

The functions f and g are defined in Chapter 5 for the relevant long-range field $v(r)$; the full phase shift has been separated explicitly into its short- and long-range pieces, the latter being included in the base functions $\{f, g\}$. A similar replacement is made in (6.56) for $F_{lt}(r)$ and the upper limit of the integral in the expression (6.57) is set at r_0. Thereby (6.58) is replaced by the corresponding expression evaluated at r_0, which reduces to $\frac{1}{2}W(f, g)$ $(\tan \delta_{lt}^{(s)} - \tan \delta_l^{(s)})$. With the Wronskian normalized to $2/\pi$ (as in Chapter 5), there results the form of the Kato identity appropriate to the contribution of the short-range interaction,

$$\tan \delta_l^{(s)} = \tan \delta_{lt}^{(s)} - \pi \int_0^{r_0} F_l(r)[(h - \epsilon)F_{lt}(r)]. \tag{6.64}$$

The variational estimate, $\delta_{lv}^{(s)}$ follows on replacing F_l by F_{lt}. Use of the wave functions $\{f, g\}$ of the long-range potential in (6.63) has thus compensated the omission of the integration from r_0 to ∞ in equation (6.64).

6.9 POLARIZATION FIELD

Section 6.6 dealt with the elimination of unobserved variables or "closed channels" in evaluating the phase shift in an open channel. A canonical example is provided by the influence of excited states on elastic scattering of electrons by an atom. At low energies the dominant effect is the polarization induced in the ground state by the electric field of the electron that is being scattered.

This effect can be evaluated in terms of virtual excitations of the atom to higher states, that is, by extending the optical potential $v_{12}L_2^{-1}v_{21}$ of equation (6.50) into a sum over all excited and ionized states and by representing L_n^{-1} explicitly by its perturbation theory form $(\epsilon_1 - \epsilon_n)^{-1}$,

$$-v_{12}L_2^{-1}v_{21} \rightarrow -\sum_n \frac{|v_{1n}(\mathbf{r})|^2}{\epsilon_1 - \epsilon_n}. \tag{6.65}$$

The coefficients $v_{1n}(\mathbf{r})$ are matrix elements of the Coulomb potential (6.5) between the ground state 1 and the excited states n of the atom. Since the term of (6.5) with $k = 0$ is already included in L_1, the usual polarization potential is obtained by truncating the sum over k in (6.5) to the next leading term, the dipole term with $k = 1$. This gives

$$-\sum_n \frac{|v_{1n}(\mathbf{r})|^2}{\epsilon_1 - \epsilon_n} \sim v_{\text{pol}}(\mathbf{r}) = -\left(\frac{1}{r^2}\frac{4\pi}{3}\right)^2 \sum_q |Y_{1q}(\hat{r})|^2$$
$$\times \sum_n [|(1|\sum_i r_i Y_{1q}(\hat{r}_i)|n)|^2/(\epsilon_1 - \epsilon_n)], \qquad r > r_0, \tag{6.66}$$

where the index i pertains to the several atomic electrons. This expression is simplified further by laying the polar axis along $\hat{r}$—whereby $(4\pi/3)|Y_{1q}(\hat{r})|^2 = \delta_{q0}$ and averaging over the orientation of the atom (which is unnecessary for $L = 0$ atoms) thus yielding

$$v_{\text{pol}}(r) = -\frac{1}{r^4} \sum_n [(1|\sum_i r_i \cos \theta_i|n)^2/(\epsilon_1 - \epsilon_n)], \qquad r > r_i. \tag{6.67}$$

The potential thus obtained is attractive and is responsible for a dramatic increase in phase shift illustrated in Fig. 7.1. It is called the static polarization potential and is represented in ordinary units by

$$v_{\text{pol}} = -\alpha e^2/2r^4, \tag{6.68}$$

where α is the static dipole polarizability of the atom. The term static relates here to the interaction of the atom with the electron at a fixed distance r with no allowance made for its motion. The exact value of α for hydrogen is $\alpha = 4.5a_0^3$. Higher multipole contributions, with $k > 1$, leading to the quadrupole polarizability, may be added to (6.68). Additional high-order terms would result from nonadiabatic contributions due to the motion of the scattered electron.

The static polarizability can become very large for excited states of atoms; for example, $\alpha \simeq 600a_0^3$ for the electron–potassium system. The polarization potential v_{pol} is then a particularly significant piece of the long-range interaction and is best incorporated into the base functions $\{f, g\}$ and $\{f^{\circ}, g^{\circ}\}$ for the purpose of calculating the effects of additional short-range forces in the system. The radial Schrödinger equation with v_{pol},

$$\left[\frac{1}{2}\frac{d^2}{dr^2} - \frac{1}{2}\frac{l(l+1)}{r^2} + \frac{\alpha}{2r^4} + \frac{1}{2}k^2\right]F_l(r) = 0, \tag{6.69}$$

has solutions expressible in terms of known special functions, the Mathieu functions. Regular and irregular solutions are, therefore, available in analytical form just as in the cases of the Coulomb and dipole potentials discussed in Chapter 5 [12]. However, Mathieu functions are somewhat more complicated in their algebraic handling and it proves easier in many applications to follow the procedure given in Chapter 5 for fields that fall off faster than $1/r^2$. The QDT parameters appropriate to the long-range field in (6.69) can be evaluated as WKB phase integrals. Particularly at low energy, which is also the region where v_{pol} is most important, solutions in the $1/r^4$ field can be evaluated to any desired accuracy as a series in k^2 by elementary integrations that involve Bessel functions [13].

Problems

6.1 Evaluate the Hartree potential for electron–H($2p$) scattering by following the procedure in Section 6.2 but considering the hydrogen atom to remain throughout in the $2p$ state.

6.2 Calculate a variational approximation to the phase shift curves of Fig. 4.1 by using as a trial wave function the superposition of three terms

$$u(r) = A\sin(\pi r/r_0) + B\sin(3\pi r/2r_0) + C\sin(2\pi r/r_0),$$

and treating A, B, and C as variational parameters. *Suggested procedure* (Not much work is required, but departures from simple procedure may increase it by orders of magnitude): Introduce a boundary condition (6.33) at r_0, with σ as a boundary parameter, and use it to eliminate one of the parameters (i.e., B). Use r_0/π as a unit of r. Calculate the variational expression

$$\int_0^{r_0} dr\, u(r)[H - E]u(r) = \text{min}.$$

Set to zero its derivatives with respect to remaining parameters, thus obtaining a compatibility equation quadratic in E and σ. Plot this result as a curve $F(E, \sigma) = 0$ from which you read out $E(\sigma)$ or $\sigma(E)$. [This plot will have unrealistic features resulting from inadequacy of your trial function to represent *large* negative or positive values of $(u|H|u)$.] Handle calculation expeditiously, paying attention to analysis of final results. How do they compare with the solution in Fig. 4.1? Why?

6.3 For electron–hydrogen scattering calculate a variational approximation to the elastic scattering phase shift for $l = 0$ from the static potential given in (6.6c) by the same procedure as that in the previous problem. Choose $r_0 = a_0$.

6.4 Repeat Problem 6.2 with the Kohn variational principle in (6.60) and the trial function given in (6.61) with c_2 set equal to zero. Vary with respect to the one parameter c_1 to get the stationary results. Compare with the results in the table in Section 6.8. Note that setting c_1 equal to zero gives the Born approximation.

6.5 Obtain the close-coupled equations in (6.46) by the variational procedure sketched in the footnote that accompanies (6.46).

References

1. K. Smith, "The Calculation of Atomic Collision Processes." Wiley, New York, 1971.
2. F. Herman and S. Skillman, "Atomic Structure Calculations." Prentice-Hall, Engelwood Cliffs, New Jersey, 1963.
3. U. Fano and L. Fano, "Physics of Atoms and Molecules," Section 16.2. Univ. of Chicago Press, Chicago, Illinois, 1972.
4. U. Fano and J. W. Cooper, *Rev. Mod. Phys* **40**, 441 (1968).
5. L. S. Rodberg and R. M. Thaler, "The Quantum Theory of Scattering," p. 239. Academic, New York, 1967.
6. C. H. Greene, U. Fano, and G. Strinati, *Phys. Rev. A* **19**, 1485 (1979).
7. P. G. Burke and D. Robb, *Adv. At. Mol. Phys.* **11**, 144 (1975) and "Invited Papers to X ICPEAC," (G. Watel, ed.). North-Holland Publ., Amsterdam,
8. E. Gerjuoy, A. R. P. Rau, and L. Spruch, *Rev. Mod. Phys.* **55**, 725 (1983); R. K. Nesbet, "Variational Methods in Electron–Atom Scattering Theory." Plenum, New York, 1980.
9. R. Courant and D. Hilbert, "Methods of Mathematical Physics," Vol. 1, Chapter 6. Wiley, New York, 1953. See, in particular, Theorem 6 on page 410.
10. N. F. Mott and H. S. W. Massey, "The Theory of Atomic Collisions." Oxford Univ. Press, London and New York, 1965.
11. B. H. Bransden, "Atomic Collision Theory," p. 57. Benjamin, New York, 1970.
12. T. F. O'Malley, L. Spruch, and L. Rosenberg, *J. Math. Phys.* **2**, 491 (1961); S. Watanabe and C. H. Greene, *Phys. Rev. A* **22**, 158 (1980).
13. O. Hinckelmann and L. Spruch, *Phys. Rev. A* **3**, 642 (1971).

Part C

MULTICHANNEL PHENOMENA

7 NONRESONANT CONTINUA AND DISCRETE SPECTRA

7.1 INTRODUCTION

Most collision processes involve exchanges of energy and/or angular momentum between a particle and an atomic core, a class of phenomena that we had not considered in Part B. These exchanges are described here as transitions of the particle–core system from one "channel" to another. The transition parameters depend on short-range interactions; their calculation will be dealt with in Chapter 9.

We first consider "fragmentation channels," which classify the states of a particle–core system in the limit where these constituents are widely separated. States containing one or more photons that interact weakly with the rest of a system also constitute a channel. In this chapter *particle* usually means an electron or photon, and *core* refers to an atom or positive ion.

States of a particle–core system belonging to its ith fragmentation channel are normally characterized as consisting of an atomic (or molecular) core in a specific energy eigenstate—with energy $\bar{E}_i$, angular momentum quantum number $\bar{J}_i$, and additional constants, if any, indicated by α_i—and of a particle with orbital quantum number l_i, total j_i, and energy $\epsilon_i = E - \bar{E}_i$. Here E represents the total energy of the system and takes different values for different states of the same channel. The total angular momentum J of the system is instead a fixed label of a channel (in the absence of external fields). A state of this channel is thus represented by

$$|(\bar{E}_i\bar{J}_i, \epsilon_i l_i j_i)EJM). \tag{7.1}$$

A stationary state of the system is represented in the fragmentation limit as a superposition of different channel states (7.1), with the same E and J.

As an illustration consider the fragmentation channels in the photodetachment of a negative ion such as F^-. The initial state of the negative ion is $^1S_0$ and, therefore, the total J equals 1 in the final state because of dipole selection rules. Low-energy photoelectrons have $l = 0, j_l = \frac{1}{2}$, since the Wigner threshold law of Chapter 4 restricts the outgoing electron to its lowest partial wave. The neutral F atom that is left behind can be in one of the two states $\bar{J}_i = \frac{1}{2}, \frac{3}{2}$. This problem, therefore, involves two fragmentation channels distinguished by the alternative values of $\bar{J}_i$. At higher photoelectron energies the other l value allowed by angular momentum and parity selection rules, namely, $l = 2$, is no longer suppressed by the Wigner threshold law (4.33). Five channels with the common value $J = 1$ are then available, namely, $(\bar{J} = \frac{1}{2}, l = 0, j_l = \frac{1}{2})$, $(\bar{J} = \frac{1}{2}, l = 2, j_l = \frac{3}{2})$, $(\bar{J} = \frac{3}{2}, l = 0, j_l = \frac{1}{2})$, $(\bar{J} = \frac{3}{2}, l = 2, j_l = \frac{3}{2})$, and $(\bar{J} = \frac{3}{2}, l = 2, j_l = \frac{5}{2})$.

The total angular momentum quantum numbers J and M in the channel symbol are, of course, collision constants. However, the circumstances of observation may permit the identification of a fragmentation channel through the separate magnetic quantum numbers of particle and core, or even of the particle's spin and orbit. Thus for instance, in the preceding example of the photodetachment of F^-, a spin analysis of either the outgoing electron or the residual neutral atom may permit sorting out different $\bar{M}_i$ compatible with a given $\bar{J}_i$. In this event the ket (7.1) is replaced by

$$|\bar{E}_i \bar{J}_i \bar{M}_i, \epsilon_i l_i j_i m_i) \tag{7.1'}$$

or

$$|\bar{E}_i \bar{J}_i M_i, \epsilon_i m_{si} l_i m_{li}), \tag{7.1''}$$

where m_{si} is the magnetic quantum number of the photoelectron's spin. These alternative channel representations are related simply by angular momentum coupling transformations, irrespective of collision dynamics.

The concept of channel has been introduced here with reference to a particle–core system, the core being viewed as very massive, but is actually more general. It applies particularly to molecular dissociation or indeed to the separation of any atomic or nuclear system into two (or more) fragments. For the purpose of general applications the orbital momentum l of the particle must be reinterpreted as the orbital momentum of the fragment pair about its center of mass and the particle spin as the internal angular momentum $\tilde{J}$ of the second fragment. Vector addition of $\bar{\mathbf{J}}$, $\tilde{\mathbf{J}}$, and $\mathbf{l}$ to yield the total $\mathbf{J}$ may proceed by alternative couplings. For simplicity we shall deal explicitly with a particle–core system.

The representation of a stationary state as a superposition of fragmentation channel states could be utilized for the entire range of particle–core distances. This is done, for example, in the close-coupling treatments of collisions. However, a different representation of channels may be relevant for the complex formed on close approach of particle and core. In the preceding example of F^-, where $\bar{J}_i$ and j_i characterize the $(F + e)$ complex in the fragmentation limit, alternative quantum numbers, i.e., $\bar{L}_i = 1$, $\bar{S}_i = \frac{1}{2}$ of the ^2P residual atom and $L_i = 1$, $S_i = 0, 1$ for the $F + e(l = 0)$ complex, afford a more suitable description when the electron is close to the F atom. These representations will be introduced in the course of treating specific systems of increasing complexity.

Transformations from one to another representation of the state of a system are required for a complete description of collision processes. The initial state of a collision process is, in fact, an eigenstate of the relative momentum $\mathbf{p} = \hbar\mathbf{k}$ of particle and core. This state is then to be resolved into eigenstates of angular momentum with different l_i and J belonging to a specific fragmentation channel—the entrance channel—in the manner discussed in Chapter 4 for elastic single-channel processes. The entrance channel itself is only one among alternative channels that are superposed in an energy eigenstate. The treatment of collisions thus necessarily consists of an extended sequence of steps. Familiarity with such procedures is developed through study of successive examples.

Fragmentation of a central complex directly into three fragments, typically two electrons and an ionic core, also occurs at sufficiently high excitation levels. The study of these channels will be outlined in Part D, but its development remains incomplete. There results a restriction to the range of excitations to be considered in this part, which reflects a serious limitation of current theory.

Elaboration of the concepts of fragmentation and central complex channels and the quantitative relationships between channel properties will proceed in successive steps. Initially we consider the simple exchange of spin angular momentum between an incident electron and a H atom which involves only the coupling and decoupling of spins in addition to the single-channel procedures of Part B (Section 7.2). The following section will formulate the connection between eigenfunctions F_α of a central complex and eigenfunctions of fragmentation channels pertaining to the initial and final states of reactants. These connections, represented by matrices and patterned after those introduced in Section 4.5 for a simple case, serve in turn to formulate cross sections and all other observable parameters of interest. The results will be expressed analytically in terms of two separate classes of basic parameters, dynamical and geometrical.

Implementation of the numerous details of the general formulas will be illustrated through examples that prove laborious even under restrictive conditions. Even in the simple case (Section 7.4) of $e + F$ interactions, at low energies that allow only transitions between the doublet ground states of the F atom, several aspects must be worked out: (1) the classification of channels, (2) the connection of channels through angular momentum algebra (Section 7.4.1), (3) the phase shifts of eigenchannels of the F^- complex (Section 7.4.2),

(4) the construction of wave functions, which requires the explicit inversion of a matrix (Section 7.4.3), (5) the possible binding of the electron within F^- (Section 7.4.4), and (6) the photodetachment of an electron from F^- (Section 7.4.6). Further and more complex examples, to be treated in lesser detail, include the photoabsorption of molecular hydrogen (Section 7.5.1) which has served as a prototype for the development of the procedures developed in this chapter and in Chapter 8.

7.2 TWO DEGENERATE CHANNELS: SPIN EXCHANGE IN e + H COLLISIONS

The elastic scattering of electrons by H atoms has been considered in Chapter 6 as an example for the calculation of phase shifts in a single-channel setting. The alternative occurrence of wave functions symmetric or antisymmetric in the position coordinates of the incident and atomic electron was seen to determine the sign of the exchange term of the potential (6.6b). However, this process actually involves two channels when the spin orientation of the two electrons is considered explicitly, because the collision may involve a "spin flip." This is the process in which the spin orientation of the electron emerging from the elastic collision is opposite that of the incident electron. We reconsider the problem here in this broader setting.[†]

We proceed to write a two-electron wave function that represents the initial state as consisting (at large distance) of an incident electron with momentum parallel to the z axis and an atomic electron in a 1s state with wave function u_0. The final state consists instead of a scattered electron at $r > r_0$ [as in the last expression of (4.6+)] and the atom remaining in its ground state. The spin orientations of the colliding and atomic electrons in the initial state will be represented by $|s_{\rm pi})$ and $|s_{\rm ai})$ for particle and atom, respectively, with $s = \pm\frac{1}{2}$. The wave function should be antisymmetric in the combined position and spin coordinates of the two electrons as indicated by

$$\Psi = \sqrt{\tfrac{1}{2}}[e^{ikz_1}(1|s_{\rm pi})u_0(r_2)(2|s_{\rm ai}) - e^{ikz_2}(2|s_{\rm pi})u_0(r_1)(1|s_{\rm ai})] + \text{outgoing scattered wave.} \tag{7.2}$$

The scattered part of Ψ is to be calculated and its amplitude will provide the cross sections for scattering with and without spin flip. Spin-orbit coupling is very weak in H and will be disregarded for most of this section, meaning that the association of particle label (1 or 2) and spin orientation ($s_{\rm pi}$ or $s_{\rm ai}$) remains invariant in the collision.

[†] The elastic scattering of electrons by alkali atoms presents the same problem. For the lower scale of excitation energies of alkalis, virtual excitations of the atomic valence electron may be disregarded realistically only in a much narrower range, ~1 eV instead of ~5 eV.

Two cases may be distinguished at the outset. If the two spins are parallel, $s_{pi} = s_{ai} = s_i$, spin flip is excluded by conservation of the total $S_z = s_{pi} + s_{ai} = \pm 1$ and we deal, in fact, with a single channel. Indeed, the spin representation, $(1|s_i)(2|s_i)$, factors out of (7.2) and the remaining position wave function is antisymmetric, as behooves a triplet state. The outgoing wave is then represented, in analogy to (4.6+), by

$$(1|s_i)(2|s_i) \sum_l \frac{e^{2i\delta_{l-}} - 1}{2i} (2l+1)$$

$$\times \sqrt{\tfrac{1}{2}} \left[\frac{e^{ikr_1}}{kr_1} P_l(\cos\theta_1) u_0(r_2) - \frac{e^{ikr_2}}{kr_2} P_l(\cos\theta_2) u_0(r_1) \right], \tag{7.3}$$

where the subscript " $-$ " has been added to δ_l to specify which sign is to be chosen in (6.6b) for the potential when calculating the phase shift. The scattering cross section is given accordingly by (4.17) with $\delta_l \equiv \delta_{l-}$.

If the two spins are antiparallel in the initial state ($s_{pi} = -s_{ai}$), we deal with a truly two-channel system. Yet we may identify the two fragmentation channels simply by the alternative spin wave functions $(1|s_{pi})(2|s_{ai})$ and $(2|s_{pi})(1|s_{ai})$, in the absence of spin-orbit coupling, without introducing the quantum numbers j_i and J as we did in (7.1). The essential point here is that the collision process at $r < r_0$ is properly represented in terms of singlet and triplet states—that is, of eigenstates of the total spin $\mathbf{S}^2 = |\mathbf{s}_1 + \mathbf{s}_2|^2$. On the other hand, the invariant spin orientations of the two electrons are labeled by the eigenvalues of the observables s_{z1} and s_{z2} of the initial state, which are incompatible with $\mathbf{S}^2$. Accordingly, we must complement the partial wave analysis of the electron's motion with a parallel analysis of their separate spin orientation into eigenstates of their total spin $\mathbf{S}^2$. That is, we resolve the initial state (an eigenstate of $\mathbf{k}$) into partial waves (eigenstates of $\mathbf{l}^2$) for the spatial part of the wave function and simultaneously the pair of spin orientations (s_{pi}, s_{ai}) into singlet and triplet states, labeled by $S = 0$ or 1. This further analysis is performed simply by writing

$$\left.\begin{matrix}(1|s_{pi})(2|s_{ai}) \\ (2|s_{pi})(1|s_{ai})\end{matrix}\right\} = \sqrt{\tfrac{1}{2}}\{\sqrt{\tfrac{1}{2}}[(1|s_{pi})(2|s_{ai}) + (2|s_{pi})(1|s_{ai})]$$

$$\pm \sqrt{\tfrac{1}{2}}[(1|s_{pi})(2|s_{ai}) - (2|s_{pi})(1|s_{ai})]\}$$

$$= \sqrt{\tfrac{1}{2}}\{\chi_{S=1}(1,2) \pm \chi_{S=0}(1,2)\}. \tag{7.4}$$

Substitution of these alternative expansions in the initial-state wave function of (7.2) yields directly the separate expressions of the outgoing spherical waves for $S = 1$ or 0. The expression for $S = 1$ coincides with (7.3) and the one for $S = 0$ differs from it only by replacing the minus sign by plus in the label of δ_l and in the bracket. Thus, we have

$$\Psi = \tfrac{1}{2}\chi_{S=1}\left\{[e^{ikz_1}u_0(r_2) - e^{ikz_2}u_0(r_1)] + \sum_l \frac{e^{2i\delta_{l-}} - 1}{2i}(2l+1)\left[\frac{e^{ikr_1}}{kr_1}P_l(\cos\theta_1)u_0(r_2) - \frac{e^{ikr_2}}{kr_2}P_l(\cos\theta_2)u_0(r_1)\right]\right\}$$
$$+ \tfrac{1}{2}\chi_{S=0}\left\{[e^{ikz_1}u_0(r_2) + e^{ikz_2}u_0(r_1)] + \sum_l \frac{e^{2i\delta_{l+}} - 1}{2i}(2l+1) \times \left[\frac{e^{ikr_1}}{kr_1}P_l(\cos\theta_1)u_0(r_2) + \frac{e^{ikr_2}}{kr_2}P_l(\cos\theta_2)u_0(r_1)\right]\right\}, \qquad r \to \infty. \tag{7.5}$$

7.2.1 Scattering Amplitudes and Cross Sections

The wave function Ψ of (7.5) represents the system completely at $r > r_0$ in terms of the channels $S = 0, 1$ appropriate to the range $r < r_0$. We may now return (7.5) to the representation of fragmentation channels simply by expressing $\chi_{S=1}$ and $\chi_{S=0}$ once again in terms of the left-hand side of (7.4) and factoring out these spin factors. The result is expressed compactly in terms of the permutation operator P_{12} that interchanges the particle labels 1 and 2

$$\Psi = \sqrt{\tfrac{1}{2}}(1 - P_{12})(1|s_{\text{pi}})(2|s_{\text{ai}}) \times \left[e^{ikz_1}u_0(r_2) + \sum_l \frac{e^{2i\delta_{l-}} + e^{2i\delta_{l+}} - 2}{4i}(2l+1)\frac{e^{ikr_1}}{kr_1}P_l(\cos\theta_1)u_0(r_2) + \sum_l \frac{e^{2i\delta_{l+}} - e^{2i\delta_{l-}}}{4i}(2l+1)\frac{e^{ikr_2}}{kr_2}P_l(\cos\theta_2)u_0(r_1)\right]. \tag{7.6}$$

This formula shows two scattered waves with alternative particle labels on the position coordinates. The second of these waves associates the *atomic* electron function $u_0(r_1)$ in the final state with the spin $(1|s_{\text{pi}})$ that belonged to the *incident* electron in the initial state; thus, it represents a spin exchange, or spin-flip process. The amplitudes of the two scattered waves are identified accordingly as the transition matrix elements for no-flip and spin-flip scattering

$$\begin{aligned} T_l^{\text{nf}} &= \frac{e^{2i\delta_{l-}} + e^{2i\delta_{l+}} - 2}{4i} = \frac{1}{2i}[e^{i(\delta_{l+}+\delta_{l-})}\cos(\delta_{l+} - \delta_{l-}) - 1], \\ T_l^{\text{sf}} &= \frac{e^{2i\delta_{l+}} - e^{2i\delta_{l-}}}{4i} = \frac{1}{2}e^{i(\delta_{l+}+\delta_{l-})}\sin(\delta_{l+} - \delta_{l-}), \end{aligned} \tag{7.7}$$

which bear an interesting relation to (4.16*b*). The spin-flip amplitude would vanish, of course, in the absence of a difference between δ_{l+} and δ_{l-}. Further reference to (4.17) leads to the two separate cross sections

$$\begin{aligned} d\sigma^{\mathrm{nf}}/d\Omega &= k^{-2}\left|\sum_l T_l^{\mathrm{nf}}(2l+1)P_l(\cos\theta)\right|^2 \\ d\sigma^{\mathrm{sf}}/d\Omega &= k^{-2}\left|\sum_l T_l^{\mathrm{sf}}(2l+1)P_l(\cos\theta)\right|^2. \end{aligned} \tag{7.8}$$

7.2.2 Results of Calculations

The calculation of the phase shifts δ_{0+} and δ_{0-} for $e + \mathrm{H}$ has long been a testing ground for computational techniques. High-accuracy results obtained by Schwartz in 1961 by a variational technique (Section 6.8) are shown in Fig. 7.1 under the label of singlet and triplet. Note how the curves approach a multiple of π in the zero-energy limit in accordance with (4.26). A negative

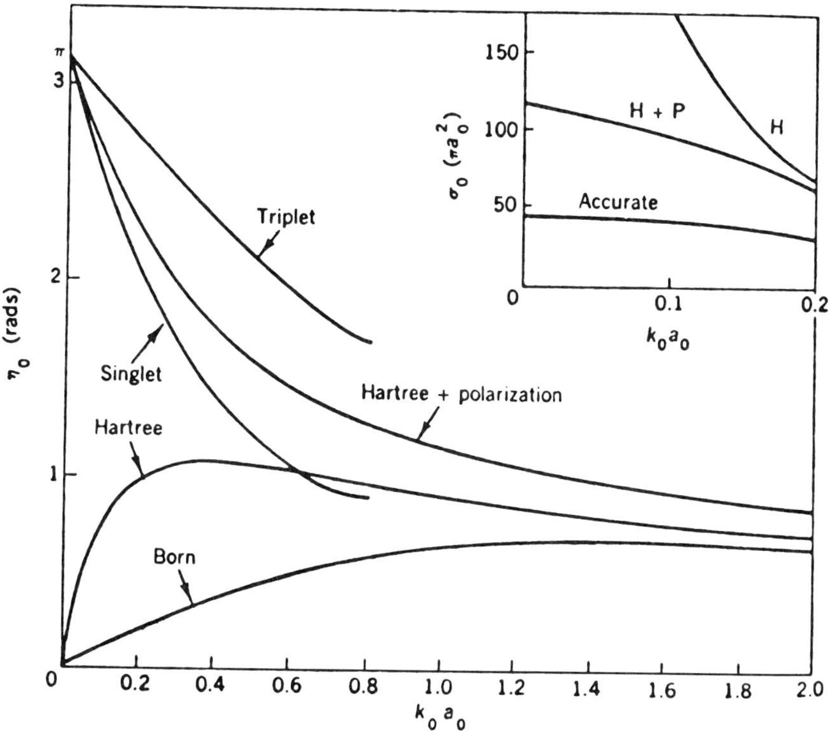

Fig. 7.1 The s-wave phase shifts for e–H elastic scattering in successive approximations (cross section shown in inset): Born approximation to static potential [Eq. (6.6a)]; Hartree, exact for static potential [Eq. (6.6a)]; Hartree + polarization [Eq. (6.6a) + Eq. (6.68)]; Singlet and triplet, accurate calculation through wave function (6.62) for potential (6.6a) + (6.68) $\pm$ (6.6b). (From S. Geltman [1].)

slope of $\delta_{0\pm}(E)$ near this limit would also result from the action of a repulsive local potential. This interpretation does not apply strictly here, in the presence of a correlated motion of two electrons. In the singlet state the apparent repulsion relates to the occurrence of a bound singlet state of H^-, with the low binding energy of 0.75 eV, which completes the filling of the 1s shell. The binding energy of H^- is related to the scattering length for singlets at $E = 0$, as explained in Section 4.9. For triplets, on the other hand, the negative slope of $\delta_{0-}(E)$ reflects the actual hindrance to penetration of the incident electron into the H atom, which results from the mutual exclusion of electrons with parallel spin. The scattering lengths for the two channels obtained by Schwartz are

$$a_+ = 5.965 \quad \text{a.u.}, \quad S = 0 \qquad \text{and} \qquad a_- = 1.769 \quad \text{a.u.}, \quad S = 1. \tag{7.9}$$

These values represent upper bounds to the actual values as explained in Section 6.5. (Note that lower bounds on the phase shift translate into upper bounds on the scattering lengths.) These values of the scattering lengths suffice to determine the e + H scattering only over a narrow range of energies because the potential field of an atom is not limited as sharply as that of a spherical well. For example, to obtain the binding energy of H^- from (4.35) we should not simply set $\tan \delta_0^\circ = -a_+/r_0$ as in (4.36), which would predict a binding energy of $\simeq$0.4 eV; the higher approximation (4.29) is required.

Schwartz's results take into account the attractive long-range field that results from the atomic polarization induced by the incident electron; this field decays as r^{-4} for $r > r_0$ (Section 6.9). The other values of δ_0 shown in Fig. 7.1 result from less refined calculations than Schwartz's. In Fig. 7.1 a comparison of the curves labeled H and H + P demonstrates the very substantial influence of the long-range potential $-\alpha/r^4$ due to the polarization of the H atom by the electron even at $r \gg r_0$. The effect of this polarization is represented by a long-range phase shift contribution $\delta_0^{(1)}$, which depends on the energy and thus restricts sharply the range of application of the threshold laws discussed in Section 4.8.2. This restriction was studied long ago, and quite recently in greater detail, by evaluating $\delta_l^{(1)}$ as a function of the energy and the core polarizability. The energy dependence of $\delta_l^{(1)}$ and other QDT parameters can be worked out for the $1/r^4$ field by the procedure of Chapter 5. The expansion of $\delta_l^{(1)}$ in terms of k has also been given earlier and is known in the literature as the modified effective range expansion of O'Malley, Spruch, and Rosenberg.†

7.2.3 The e + He^+ System

The e + He^+ system is isoelectronic to e + H and hence has the same set of channels. The e + He^+ system differs, of course, by its long-range

† See Section 6.9 and references therein.

Coulomb field about the He^+ core at $r > r_0$, which generates a long-range phase shift $\delta_l^{(l)}$ far in excess of that generated in $e + H$ by the polarization of the H atom. This phase shift is to be added to the short-range phase shift $\delta_{l\pm}^{(s)}$ contributed by the interaction within the atomic core. Here, as in $e + H$, the long-range phase shift is equal for the singlet and triplet channels, and hence is equal for all their superpositions, including the fragmentation channels. This degeneracy of $\delta_l^{(l)}$ rests on the degeneracy of spin orientations in the hydrogen atom.

The bound-state levels of the neutral He spectrum are represented by (5.28). Departures from the hydrogenic Bohr expression $-1/n^2$ stem from the short-range forces between e–He^+ and are expressed in terms of the short-range phase shifts $\delta_{l\pm}^{(s)}$ that operate in e–He^+ collisions, with appropriate extrapolation of $\delta_{l\pm}^{(s)}$ to negative electron energies. For triplets the spin-orbit coupling introduces a small dependence of $\delta_l^{(s)}$ on the total quantum number J; thus, we actually have three slightly different phase shifts $\delta_{l,-,J}^{(s)}$ for the alternative values $J = l - 1, l, l + 1$. This difference is hardly observable in ordinary optical spectra but has been studied in great detail by radio-frequency techniques.

The small effect of spin-orbit coupling also manifests itself more readily in $e + He^+$ than in $e + H$ because the centrifugal repulsion keeps the electron away from the stronger field in the He^+ core, provided the orbital momentum is appreciable ($l > 2$) and the electron energy is low. In this event, the entire short-range phase shift vanishes and with it the singlet–triplet difference $\delta_{l+}^{(s)} - \delta_{l-}^{(s)}$. For $e + H$ there results simply an absence of scattering, to within the small effect of polarization. For $e + He^+$, on the other hand, there remains the Coulomb field contribution $\delta_l^{(l)}$, which also depends—albeit weakly—on spin-orbit coupling and hence splits into distinct phase shifts $\delta_{lj}^{(l)}$ with $j = l \pm \frac{1}{2}$ (Section 5.9). This effect is important because it qualitatively alters the structure of scattering at low energies and for $l > 2$ and especially the structure of the discrete spectrum. This phenomenon need not be discussed in detail here, but it points to the opportunity of introducing a description of multichannel processes more flexible than the elementary method utilized for the $e + H$ system.

7.2.4 Photodetachment of H^-

We have seen how the scattering length of $e + H$ collisions at very low energy relates to the binding of an electron and a H atom to form a stable H^- anion. This connection also enables us to calculate the photoabsorption spectrum of H^- in terms of the scattering length—or rather in terms of the bound-state energy—owing to a special feature of negative ions. The weak binding of the second electron in H^-, $\frac{1}{2}\kappa_B^2 = 0.028$ a.u ($= 0.75$ eV) implies that its scattering length exceeds the atomic radius substantially. Most of the electron distribution is thus contributed by the tail of its wave function $\psi \propto e^{-\kappa_B r}$, according to Section 4.9. Even though photoabsorption by atoms and molecules is generally localized within their multielectron core, negative ions are exceptional in this respect. The probability of detachment of the excess electron by incident light (visible or even infrared) can indeed be determined to a very good approximation by elementary procedures. This approach, called the weak binding or zero-range field approximation, originates from the well-known treatment of the deuteron in nuclear physics.†

The wave function of H^- may be represented accordingly by

$$\psi_i(\mathbf{r}_1, \mathbf{r}_2) \sim \sqrt{\tfrac{1}{2}}(1 + P_{12})u_0(r_1)Ce^{-\kappa_B r_2}/r_2, \tag{7.10}$$

for r_2 values in excess of the effective range r_e of the e–H interaction. [The permutation operator P_{12} and the H wave function u_0 are the same here as in (7.6), with the understanding that P_{12} acts only on the spatial coordinates. The sign in front of P_{12} corresponds to the singlet character of the bound state of H^-.] The normalization constant C can be taken from a very accurate calculation of the complete wave function [3], or it can be adjusted by fitting the scale of the cross-section spectrum. Photoabsorption lifts an electron from its s orbital in H^- to a p orbital with energy $\frac{1}{2}k^2$ in a.u. The complete wave function of the e + H system in its final $^1P^o$ state, normalized per unit energy according to Section 4.3, is represented simply by

$$\psi_f(\mathbf{r}_1, \mathbf{r}_2) = \sqrt{\tfrac{1}{2}}(1 + P_{12})u_0(r_1)(2k/\pi)^{1/2}j_1(kr_2)\cos\theta_2\,[3/(4\pi)]^{1/2}, \tag{7.11}$$

disregarding the phase shift of the detached electron which hardly approaches the H atom at low energy owing to the centrifugal barrier (Section 4.8).

On entering these wave functions in the dipole transition matrix element of (2.9), its evaluation reduces to a standard integral [4],

$$(\epsilon|x|0) \to (\psi_f|z_1 + z_2|\psi_i) = (\tfrac{2}{3})^{1/2}C\frac{(2k)^{3/2}}{(\kappa_B^2 + k^2)^2} \quad \text{a.u.} \tag{7.12}$$

Allowance for the fact that (7.10) holds only for $r_2 > r_e$ is included here in the factor C. The resulting absorption cross section for photons of energy $\omega = \frac{1}{2}(\kappa_B^2 + k^2)$ a.u. is then

$$\sigma = \frac{32\pi^2}{3(137)}C^2\frac{k^3}{(\kappa_B^2 + k^2)^3}. \tag{7.13}$$

† See, for example, Segrè [2].

Ohmura and Ohmura [3], who followed this treatment, showed this approximate cross section to agree with experiment better than earlier, more elaborate ones up to $\hbar\omega \sim 2$ eV (Fig. 7.2). Note the k^3 dependence of the cross section in (7.13) for small k, characteristic of a p-wave threshold as in (4.33).

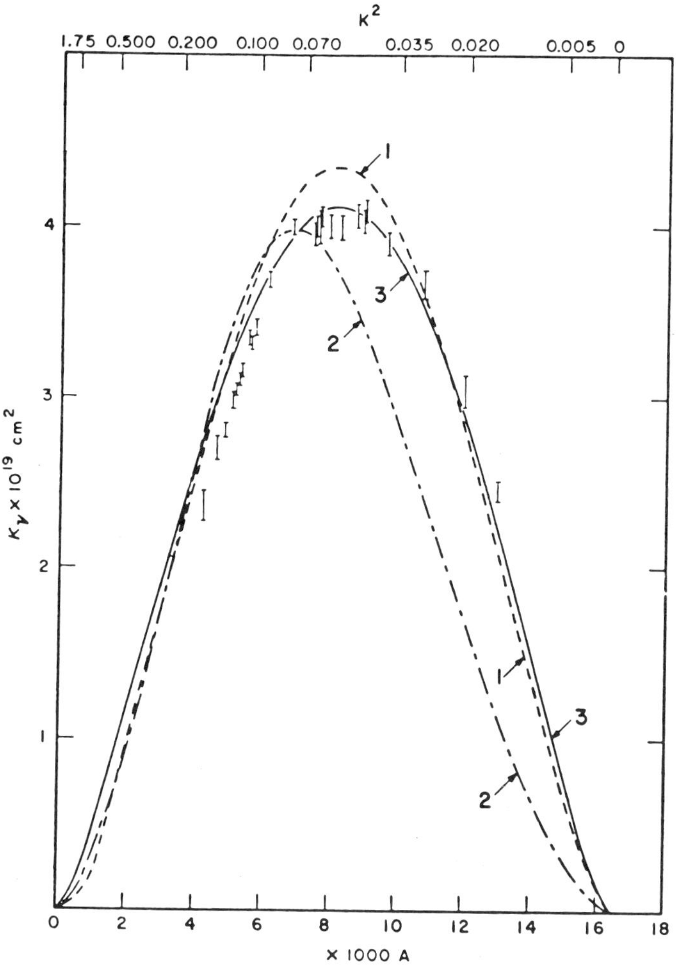

Fig. 7.2. Continuous absorption coefficient for the photodetachment of H^-. Results from 20-parameter numerical wave functions in two different calculational schemes [(1) dipole velocity and (2) dipole length] are compared with the scattering length approximation Eq. (7.13), [curve (3)] and with data from experiment. For curve (3) $K_\nu = (6.8475\kappa_B k^3 \times 10^{-18})/[(1 - \kappa_B r_e)(\kappa_B^2 + k^2)^3]$ cm^2. (From Ohmura and Ohmura [3].)

7.3 COLLISION PARAMETERS

The essential point emerging from the treatment of spin exchange in $e + \mathrm{H}$ collisions is that the effect of short-range electron–core interactions is properly described by two different phase shifts δ_{l+} and δ_{l-} pertaining to the singlet and triplet channels, whereas the spin orientations of e and H are identified in the pair of fragmentation channels. The connection between the two sets of channels is represented by the simple transformation (7.4) in the case of $e + \mathrm{H}$, but it will be less trivial in other cases.

Recall also that we stressed in Chapter 5 the separation of short- and long-range contributions to the phase shifts, $\delta^{(s)}$ and $\delta^{(l)}$, of which $\delta^{(l)}$ is sensitive to the particle energy for $\epsilon \sim 0$, that is, near the fragmentation threshold. The contribution $\delta^{(l)}$ is expected to depend in general on the fragmentation channel energy ϵ_i, even though its value is the same for all channels in the special cases of $e + \mathrm{H}$ and $e + \mathrm{He}^+$. Thus, we require a parametrization that combines short-range and long-range contributions to the phase shifts even though these contributions are properly identified with reference to different base sets of functions.

To this end we generalize to a multichannel situation the concept of Jost functions introduced in Chapter 5. The alternative pairs of independent radial functions $\{f^\circ, g^\circ\}$ and $\{f^-, f^+\}$ considered in Chapter 5 are replaced here by alternative *maximal sets* of independent solutions of the Schrödinger equation for the entire complex of particle–core with specified values of E, J, and M. One of the sets consists of *standing waves*, such as $\{f^\circ, g^\circ\}$, identified by properties that are characteristic of the core region at $r \lesssim r_0$. Here r denotes the radial coordinate of the short-range interaction in which the particle emerges at r_0 with an eigenphase shift $\delta_\alpha^{(s)}$. We indicate this set by $\{F_{\alpha JM}(\omega, r), G_{\alpha JM}(\omega, r)\}$, where ω stands for all coordinates other than r. The second set is characterized instead by boundary properties at large r that specify the particle motion as represented by an ingoing (or outgoing) wave in a *single fragmentation channel* (7.1), $\Phi_{iJM}(\omega) f^{\mp}(\epsilon_i, l_i, r)$, with f^- (f^+) identified by (5.2).

The wave function in all the other coordinates is represented by Φ_{iJM}. This function consists of a properly antisymmetrized product of the complete N-electron wave function of the core with the angular and spin functions of the incident or outgoing particle, together with angular momentum coupling coefficients appropriate to a prescribed coupling scheme. It thus contains key information on the state of interest, but it may be indicated simply by $\Phi_{iJM}(\omega)$ for most of our applications.

Any function belonging to one of the independent sets (F, G) or $f^\pm$ with different boundary conditions can be expanded into a superposition of solutions of the other set. Specifically, we expand

$$F_\alpha(\omega, r) = \frac{1}{2i}\sum_i \left(\frac{2}{\pi k_i}\right)^{1/2} \Phi_i(\omega)[(-i)^{l_i} f^+(\epsilon_i, l_i, r) J^+_{i\alpha}(k_i) - (i)^{l_i} f^-(\epsilon_i, l_i, r) J^-_{i\alpha}(k_i)], \tag{7.14}$$

where the coefficients $J^+_{i\alpha}(k)$ and $J^-_{i\alpha}(k)$ will be regarded as elements of two Jost function matrices J^+ and J^-, and the indices JM have been dropped for simplicity. A channel-dependent normalization coefficient $\pm(\mp i)^{l_i}(1/2i)(2/\pi k_i)^{1/2}$ has been factored out of the Jost matrices in (7.14) at variance with (5.4) for the following reason. With this choice each channel function in (7.14) goes asymptotically to the energy-normalized form $(2/\pi k_i)^{1/2} \sin(k_i r - \frac{1}{2} l_i \pi + \cdots)$, thereby simplifying† the later definition of a unitary scattering matrix from the ratios of coefficients of f^+ and f^-.

The coefficients $J^\pm_{i\alpha}$ will serve as collision parameters. Their complex phases combine contributions from short-range and long-range interactions, as shown by specific examples in the following sections. The magnitude of $J^\pm_{i\alpha}$ represents instead the admixture of the fragmentation channel i in the short-range eigenchannel α.

7.3.1 Scattering Matrix

The construction of a scattering wave function with an incident particle in channel i and outgoing wave boundary conditions follows the procedure described for the elementary case in Section 4.3. The coefficients c_l were selected there to cancel the phase shift $e^{-i\delta_l}$ of the ingoing wave; here they will have to cancel the $J^-_{i\alpha}$, which is achieved by elements of the matrix reciprocal to J^-. Accordingly, we represent the desired wave, with a given JM, by the superposition

$$\sum_\alpha F_\alpha(\omega, r)(J^{-\,-1})_{\alpha i} = \frac{1}{2i}\sum_j \left(\frac{2}{\pi k_j}\right)^{1/2} \Phi_j(\omega)\Big\{-\delta_{ji}(i)^{l_i} f^-(\epsilon_i, l_i, r) + (-i)^{l_j} f^+(\epsilon_j, l_j, r)\Big[\sum_\alpha J^+_{j\alpha}(J^{-\,-1})_{\alpha i}\Big]\Big\}. \tag{7.15+}$$

Here the amplitudes of the outgoing waves in the various fragmentation channels

$$S_{ji} = \sum_\alpha J^+_{j\alpha}(J^{-\,-1})_{\alpha i} = [J^+(J^-)^{-1}]_{ji} \tag{7.16}$$

† Returning from atomic units to ordinary units, we obtain with this choice spherical waves defined by $(\mu_i/2\pi\hbar^2 k_i)^{1/2} e^{\pm i k_i r}/r$, where μ_i is the reduced mass of the particle in the channel i. The velocity dependent factor $(\mu_i/\hbar k_i)^{1/2} = v^{-(1/2)}$ normalizes the flux density to unity, as required by the definition of unitary scattering matrices [5].

represent elements of the scattering matrix, as $e^{2i\delta_l}$ does in (4.6+). More specifically, they constitute a diagonal block (with fixed EJM) of the entire unitary scattering matrix [5] in the representation of fragmentation channels (7.1).

The arrangement of terms in (7.16) has a direct interpretation. Note from (7.14) that the $J^{\pm}_{i\alpha}$ represent projections of the α eigenchannel onto the i fragmentation channels. Thus, reading (7.16) from right to left, the inverse matrix of the ingoing part, $(J^-)^{-1}$, carries a fragmentation channel i into alternative eigenchannels α; the outgoing wave matrix J^+ connects these α channels to the fragmentation channel j to give the scattering matrix element for going from i to j. A summation over all α is, of course, involved. The complete wave function with an incident plane wave is to be constructed in a final step by superposition of wave functions (7.15+) with different values of J, in the manner of (4.6+). This procedure will be described in Section 7.9.

7.3.2 Bound States

Our earlier study of a single channel in Section 5.4 pointed out that the equation for the energy eigenvalues of bound states can be expressed as requiring the Jost function $J^-(\kappa, l)$ to vanish. In this chapter we consider for simplicity only two extreme cases where the fragmentation channels are either all open ($\epsilon_i > 0$) or all closed ($\epsilon_i < 0$); the intermediate cases will be the subject of Chapter 8. In the case of closed channels the functions $f^-(\epsilon_i, l_i, r)$ diverge as $e^{\kappa_i r}$. Nondiverging bound-state eigenfunctions are then constructed as superpositions $\Sigma_\alpha F_\alpha(\omega, r) A_\alpha$ with the requirement that the coefficient of *each* $f^-(\epsilon_i, l_i, r)$ vanishes,

$$\sum_\alpha J^-_{i\alpha}(i\kappa_i) A_\alpha = 0. \tag{7.17}$$

The existence of roots A_α of the homogeneous system (7.17) requires in turn that

$$\mathrm{Det}|J^-_{i\alpha}(i\kappa_i)| = 0. \tag{7.18}$$

This is the multichannel form of (5.8).

The components $Z_i^{(n)}$ of an eigenvector of the system (7.17) normalized to unity and corresponding to an energy level $E = E_n$ serve in the spectroscopy literature to characterize the structure of each eigenstate.

7.3.3 Photoionization

The final state wave function of photoionization in which an electron travels away to infinity is constructed in analogy to (7.15+). A superposition of the eigenfunctions $F_\alpha(\omega, r)$ in (7.14) is again involved, but the ingoing wave boundary conditions now require a cancellation of $J_{i\alpha}^{+}$ in accordance with Section 4.3. The cancellation is achieved by the reciprocal of *this* Jost matrix and the desired wave is given by

$$\sum_{\alpha} F_{\alpha}(\omega, r)(J^{+^{-1}})_{\alpha i} = \frac{1}{2i}\sum_{j}\left(\frac{2}{\pi k_j}\right)^{1/2} \Phi_j(\omega)$$

$$\times \left\{\delta_{ji}(-i)^{l_i} f^{+}(\epsilon_i, l_i, r) - (i)^{l_j} f^{-}(\epsilon_j, l_j, r)\left[\sum_{\alpha} J_{j\alpha}^{-}(J^{+^{-1}})_{\alpha i}\right]\right\}. \quad (7.15-)$$

This wave is characterized by the relevant values of J and M that are common to the i and α channels.

The ingoing wave function (7.15−) corresponds to the outgoing wave function (7.15+) much as the free particle wave function (4.6−) corresponds to (4.6+). However, the wave functions of Chapter 4 had the ingoing or outgoing component at large r in the form of a plane wave constructed by superposing components with different l, to give, for example in (4.6−), the form of a plane wave at infinity and of purely ingoing waves in all the separate l channels. Here a further superposition of wave functions (7.15−) with different JM will describe the ejection of a particle into a plane wave of the ith channel (Section 7.9).

Note the amplitude of the ingoing waves in (7.15−) involves the reciprocal of the S matrix defined in (7.16). This is the multichannel generalization of the factor $e^{-2i\delta_l}$ in (4.6−). A reading of the left-hand side of (7.15−) also proves instructive. In a process such as photoionization, we can view the electron as excited at small r into one or more of the eigenchannels α. The electron then escapes to infinity along the open fragmentation channels i. According to the basic expansion (7.14) of eigenchannels in terms of fragmentation channels, the coefficients $J_{i\alpha}^{\pm}$ represent the projection of the channel α onto the channel i for outgoing and ingoing waves, respectively. The probability amplitude of detecting the photoelectron at $r \to \infty$ in any particular channel i is thus to be obtained by tracing the ith outgoing wave back into its excitation channel α; this operation is accomplished by inserting in (7.15−) the reciprocal matrix element $(J^{+^{-1}})_{\alpha i}$. Each of these matrix elements is weighted in (7.15−) by the wave function F_{α} for the corresponding αth eigenchannel. The product, summed over all α on the left-hand side of (7.15−), represents the desired amplitude for receiving the photoelectron at infinity in the channel i.

Such a description of photoionization amounts to "half of a scattering" from the point of view outlined after (7.16), which regards each scattering event as broken into two parts. This picture of half-scattering, with photoionization described in terms of input strengths at small r and half the scattering matrix—specifically, $J^{+^{-1}}$—playing the key role, proves useful in establishing connections between photoionization and scattering cross sections. The discussion in Section 7.2.4 of the photodetachment of H^- contained this idea implicitly in that the parameter κ_B in (7.13) is intimately linked to the scattering length a_+. The further discussion in Sections 7.4 and 7.5 will exploit and illustrate the close connection between photoionization and scattering.

7.3.4 Remarks on the Jost Matrices

Phenomena in which the Jost matrices can be constructed readily by a combination of theory and experimental data are described through a sequence of examples in the next sections. Even simple examples involve numerous details. However, each step of the procedure deals only with concepts developed in this or earlier sections, complemented by applications of angular momentum algebra such as occur in the theory of line spectra. Remarkably, the procedure will yield *analytical expressions* of observable cross sections, in terms of quantum numbers and a few dynamical parameters embodied in Jost matrices. The procedure will be described step by step and in considerable detail for the initial example in Section 7.4 but will be abbreviated in the following ones.

We conclude here with some remarks on the formulas presented in this section:

(1) The introduction in (7.14) of the Jost matrices $J^{\pm}$, which are complex conjugate at $\epsilon > 0$, has the effect of splitting the scattering matrix (7.16) into two factors $(J^-)^{-1}$ and J^+. Whereas S_{ji} connects an ingoing wave directly to outgoing waves, $J^+(J^-)$ connects the standing wave $F_\alpha(\omega, r)$, which is real and hence even under time reversal, to the outgoing (ingoing) wave separately. This process of analysis fills the prescription outlined in Section 1.4. It will be sufficient to calculate a complete set of F_α for any one system and obtain their expansions (7.14). The rest will follow by a standard procedure using *only Jost matrix elements* rather than explicit wave functions, as well as the *coefficients of* (*4.6*) that connect plane waves and spherical waves at $r \to \infty$. The Jost matrices depend on atomic dynamics and on channel quantum numbers: directions of incidence and ejection and other experimental specifications will appear instead only in the coefficients of (4.6) or in the analogous expressions to be described in Section 7.9.

(2) The factorization $S = J^+J^{-^{-1}}$ is not only independent of the normalization of the standing waves F_α but also invariant when the F_α experience any linear transformation $T_{\beta\alpha}$ with nonzero determinant. However, it is important in our applications to restrict such transformations by requiring the F_α's to be eigenfunctions of the short-range particle–core interactions, that is, to remain uncoupled to one another by this interaction. These F_α's, which are themselves invariants of the collision but are phase-shifted by

different amounts, are thus seen to *mediate the transitions* from one fragmentation channel to another.

(3) At negative energies ($\epsilon_i < 0$) the radial functions f^+ and f^- are no longer complex conjugate, but each of them is, in fact, even under time reversal. The boundedness condition, that the coefficient of each f^- vanishes, affects each channel independently only in the absence of channel coupling, as in the familiar spectra of alkalis to which (5.8) and (5.27) are applicable. In general, the phase shifts of all fragmentation channel waves are involved in the more complicated (7.18), which replaces (5.8). The study of (7.18) will replace the level-by-level calculations of intermediate coupling effects that are otherwise required to predict or interpret the Rydberg spectra of nonalkaline elements within the frame of independent particle approaches.

(4) Like (7.16), Eq. (7.18) is also invariant under a linear transformation $T_{\beta\alpha}$ of the base set F_α, because this transformation would merely muliply (7.18) by the factor $\mathrm{Det}|T_{\beta\alpha}|$, which is defined to be nonzero.

7.4 ELECTRON-FLUORINE INTERACTION AT LOW ENERGY

The elastic collisions $e + \mathrm{H}$, described in Section 7.2, yield exchanges of spin orientation without any energy transfer because the alternative spin orientations are degenerate in ground state H. We consider here $e + \mathrm{F}$ collisions in which a change of mutual orientation of spin and orbit in the ground state doublet of F requires energy to be exchanged between e and F. We deal for simplicity with very low-energy collisions ($E \lesssim 0.2$ eV), such that only $l = 0$ channels need be considered, contributions of other channels being suppressed by centrifugal barriers (Section 4.8.2). This simplification does not apply equally to the analogous collisions with Cl, Br, or I, which are better known experimentally, because their doublet levels have a large separation.

7.4.1 Channels and Their Interconnections

Earlier in this chapter we have seen how to describe a collision in terms of channels identified by different sets of quantum numbers at short and large radial distances. The connections between such channels were very simple in Section 7.2, but the case of $e + \mathrm{F}$ will introduce more general procedures.

An incident electron and the F atom form a complex in a fragmentation channel labeled in the notation of (7.1″) by

$$|\bar{E}_i \bar{J}_i \bar{M}_i, \epsilon_i m_{si}, l_i = 0, m_{li} = 0). \tag{7.19}$$

The energy $\bar{E}_i$ of the F doublet depends only on the value of $\bar{J}_i$ (Fig. 7.3), but alternative choices of the magnetic quantum numbers yield a total of 12

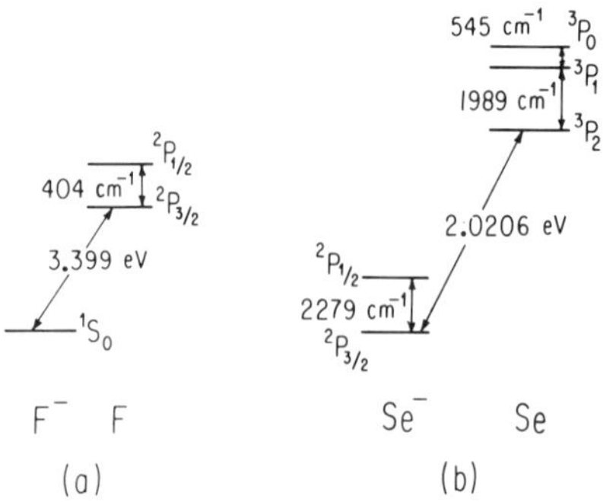

Fig. 7.3. Energy levels for the photodetachment of the negative ion (not to scale): (a) $e + \mathrm{F}$ and (b) $e + \mathrm{Se}$.

channels with $(\bar{J}_i = \frac{3}{2}, \bar{M}_i = \pm\frac{3}{2}, \pm\frac{1}{2})$ or $(\bar{J}_i = \frac{1}{2}, \bar{M}_i = \pm\frac{1}{2})$ and with $m_{si} = \pm\frac{1}{2}$.[†] The doublet separation is $\bar{E}_{3/2} - \bar{E}_{1/2} = 0.0501$ eV. Note that the incidence channel should be properly labeled in terms of the electron's momentum or wave vector, by $|\bar{E}_i\bar{J}_i\bar{M}_i, \mathbf{k}_i m_{si})$; however, the connection of the wave vector's eigenstate with the $l = 0$ partial wave is elementary and is reviewed in Section 7.9.

On the other hand, the collision dynamics should be studied with reference to channels identified in a notation similar to (7.1), with quantum numbers (JM) that are collision constants, namely, the channels labeled $|\alpha JM)$ in Section 7.3 and indicated here explicitly by

$$|E(\bar{S}_i s_i)S(\bar{L}_i, l_i = 0)LJM), \tag{7.19'}$$

with $\bar{S}_i = s_i = \frac{1}{2}$ and $\bar{L}_i = L = 1$. In this channel the (SL) coupling of the colliding spins and orbits has replaced the $(\bar{J}_i j_i)$ coupling of atom and electron utilized in (7.19) because S and L are approximate collision constants. (The spin-orbit coupling in the complex $e + \mathrm{F} = \mathrm{F}^-$ is indeed much weaker than the exchange interaction, which depends on S.)

The connection between the states (7.19) and (7.19′) is described by representing the fragmentation channel (7.19) as a superposition of states (7.19′) with coefficients of the appropriate orthogonal transformation,

$$\begin{aligned}&|\bar{E}_i\bar{J}_i\bar{M}_i, \epsilon_i m_{si})\\ &\quad = \sum_{SLJM} |E(\bar{S}_i s_i)S(\bar{L}_i 0)LJM)((\bar{S}_i s_i)SLJM|(\bar{S}_i\bar{L}_i)\bar{J}_i s_i\bar{M}_i m_{si}).\end{aligned} \tag{7.20}$$

[†] Channels that differ only in their magnetic quantum numbers (i.e., in their angular momenta) are, however, simply related through standard coupling coefficients. Thus, the problem under consideration involves in its dynamical aspects essentially two channels, $\bar{J} = \frac{3}{2}$ and $\frac{1}{2}$.

Here we have indicated explicitly the total angular momentum $\bar{J}_i$ of the F atom as the resultant of $\bar{S}_i = \frac{1}{2}$ and $\bar{L}_i = 1$. The transformation coefficients in (7.20) consist of products of three separate additions of angular momenta $\mathbf{J} = \mathbf{S} + \mathbf{L}$, $\mathbf{S} = \bar{\mathbf{S}}_i + \mathbf{s}_i$, and $\bar{\mathbf{S}}_i + \bar{\mathbf{L}}_i = \bar{\mathbf{J}}_i$ indicated, in the notation of Condon–Shortley [6], by

$$((\bar{S}_i s_i)SLJM|(\bar{S}_i\bar{L}_i)\bar{J}_i s_i \bar{M}_i m_{si})$$
$$= \sum_{M_S M_L} (JM|SLM_S M_L)(SM_S|\bar{S}_i s_i \bar{M}_{\bar{S}i} m_{si})(\bar{S}_i \bar{L}_i \bar{M}_{\bar{S}i} \bar{M}_{Li}|\bar{J}_i \bar{M}_i). \qquad (7.21)$$

Here we have $\bar{L}_i = L$ because $l_i = 0$ and hence $\bar{M}_{\bar{L}i} = M_L$, as well as $\bar{M}_i + m_{si} = M$. Algebraic expressions of all coefficients on the right-hand side of (7.21) (Condon and Shortley [6, pp. 76–77]) serve to calculate each element of this matrix.

Alternatively, (7.21) may be reformulated by multiplying it by the respective sides of the orthonormality identity

$$\delta_{\bar{M}_i\bar{M}_i'}\delta_{m_{si}m_{si}'} = \sum_{J'M'} (\bar{J}_i s_i \bar{M}_i m_{si}|J'M')(J'M'|\bar{J}_i s_i \bar{M}_i' m_{si}'),$$

and summing over $\bar{M}_i$ and m_{si}. The new formula is

$$((\bar{S}_i s_i)SLJM|(\bar{S}_i\bar{L}_i)\bar{J}_i s_i \bar{M}_i m_{si}) = ((\bar{S}_i s_i)SLJ|(\bar{S}_i\bar{L}_i)\bar{J}_i s_i J)(JM|\bar{J}_i s_i \bar{M}_i m_{si}), \qquad (7.21')$$

where

$$((\bar{S}_i s_i)SLJ|(\bar{S}_i\bar{L}_i)\bar{J}_i s_i J)\delta_{JJ'}\delta_{MM'} = \sum_{M_S M_L \bar{M}_i m_{si}} (JM|SLM_S M_L)$$
$$\times (SM_S|\bar{S}_i s_i \bar{M}_{\bar{S}i} m_{si})(\bar{S}_i \bar{L}_i \bar{M}_{\bar{S}i} \bar{M}_{\bar{L}i}|\bar{J}_i \bar{M}_i)(\bar{J}_i s_i \bar{M}_i m_{si}|J'M'), \qquad (7.21'')$$

indicates a transformation matrix from LS to jj coupling, diagonal in (JM) and independent of M. Transformations of this type are expressed and tabulated in terms of standard $6j$ coefficients [7].

The orthogonal transformation between fragmentation channels (i channels) and the eigenchannels of total angular momentum and short-range interactions (α channels), introduced here by the matrix (7.21″), will be indicated by $U_{i\alpha}^{(J)}$ or by its reciprocal $\tilde{U}_{\alpha i}^{(J)}$. The superscript J indicates that this matrix is diagonal in total angular momentum. This transformation is generally complemented by a further step, represented here by the last factor of (7.21′), which resolves $\mathbf{J}$ into the sum of momenta $\mathbf{J}_i + \mathbf{s}_i$ of the separate fragments.

In our case of e + F, the numerical values of the matrix elements (7.21″) for $J = 1$ are

$$\tilde{U}_{S\bar{J}_i}^{(1)} = \begin{array}{c|cc} S \backslash \bar{J}_i & \frac{1}{2} & \frac{3}{2} \\ \hline 1 & \sqrt{\frac{2}{3}} & -\sqrt{\frac{1}{3}} \\ 0 & \sqrt{\frac{1}{3}} & \sqrt{\frac{2}{3}} \end{array}. \qquad (7.22)$$

The corresponding matrices for $J = 0, 2$ reduce to a single element equal to unity, because $S = 0$ is not consistent with $L = 1$ and $J \neq 1$. The channel transformations thus have been carried out without reference to radial variables. The orthogonal matrix (7.22) connecting short- and long-range channels is an example of a frame transformation. It is the analog of the one encountered earlier in (7.4).

7.4.2 Eigenphase Shifts

The effect of short-range interactions on collisions has been represented in Section 7.3 by an eigenphase shift $\delta_\alpha^{(s)}$. Here we indicate it more explicitly by δ_{SLJ} omitting the label $^{(s)}$ because no additional phase shift $\delta^{(l)}$ arises from long-range interactions outside the neutral F atom.[†] [The index S of this phase shift replaces the index $\pm$ used for e + H collisions in Section 7.2; $\pm$ served there to indicate the symmetry under permutation of space variables and was thus equal to $(-1)^S$.] The index L may also be omitted here because it has the single value $L = 1$.

The alternative combinations of indices (SLJ) with $L = 1$ may be indicated by the spectroscopic symbol $^{2S+1}\mathrm{P}_J$, further complemented by the indication of odd parity, $^{2S+1}\mathrm{P}_J^\mathrm{o}$. For the singlet $S = 0$ only $J = 1$ occurs, $^1\mathrm{P}_1^\mathrm{o}$; the triplets $S = 1$ yield instead three alternative values of J, $^3\mathrm{P}_{0,1,2}^\mathrm{o}$. Application of vector addition rules to (7.21) shows, as we have seen, that $J = 0$ is consistent only with $\bar{J}_i = \frac{1}{2}$ and that the matrix element (7.21″) with $J = 0$ reduces to $\delta_{\bar{J}_i 1/2}$. Similarly, $J = 2$ is consistent only with $\bar{J}_i = \frac{3}{2}$ and again (7.21″) equals $\delta_{\bar{J}_i 3/2}$ for $J = 2$. These collision channels yield accordingly only elastic scattering from the two alternative levels of the F doublet, respectively. Sign reversal of the spin orientation quantum number of the electron, m_{si}, may, however, occur for $J = 0$ or 2, together with a compensating change of the orientation quantum number of the F atom, $\bar{M}_i$, as it does for e + H collisions in Section 7.2; the last factor of the transformation (7.21″) plays the same role as the coefficients $\pm\sqrt{\frac{1}{2}}$ do in (7.4). Inelastic collisions, with transition of the F atom from one to the other level of the doublet, may instead occur in the $J = 1$ channels, where the transformation (7.21″) is nontrivial and $^3\mathrm{P}_1^\mathrm{o}$ is superposed with $^1\mathrm{P}_1^\mathrm{o}$.

In our range of low energies each phase shift δ_{SLJ} approaches zero according to Section 4.8 and can be represented in terms of a scattering length as in (4.27). Recall in this connection—from (4.32) or (5.23)—that a scattering length is immediately related to a short-range phase shift δ^o rather

[†] In fact, the potential due to the polarization of F by the electron (Section 6.9) extends appreciably beyond the radius of F into the region where the electron motion depends on its energy $\epsilon_i = E - \bar{E}_i$ and hence on the i-channel parameter $\bar{J}_i$. We disregard its effect here for simplicity. Part of the polarization effect may be included in the short-range phase shifts δ_{SLJ}.

than to an observable phase shift δ, the connection of δ^o and δ being provided by formulas that depend on the electron motion at long ranges. In our two-channel problem the long-range parameters depend on a specific fragmentation channel. Accordingly, we must append an additional label i to the phase shift symbol $\delta_{SL\,J}$; Eq. (4.27) leads thus to

$$\delta_{SLJi} = -k_i a_S, \tag{7.23}$$

where $\frac{1}{2}k_i^2 = \epsilon_i = E - \bar{E}_i$. The analog of (4.32) is instead

$$\delta^o_{SLJ} = -a_S/r_0. \tag{7.23$'$}$$

The indices $L = 1$ and J have been omitted on the right of these formulas because L has a single value and the value of J hardly affects the scattering length of the low-Z atom F (see Section 5.9).

7.4.3 Wave Functions

We proceed now to construct the explicit form of the wave function $F_\alpha(\omega, r)$ for the e + F complex and of its expansion (7.14) into wave functions of the fragmentation channels. Here, to "construct" means to describe the radial dependence explicitly at distances larger than the radius of the F atom, $r > r_0$, as was done in Chapter 4 for elastic scattering. The dependence on angular and spin variables, indicated by ω in Section 7.3, need not be indicated explicitly, being implied by the relevant quantum numbers.

The index α of Section 7.3 takes the explicit form $(SLJM)$, as noted earlier, with $S = 0, 1$; $L = 1$; $J = 0, 1, 2$. The dependence of $F_{SLJM}(\omega, r)$ on r at $r > r_0$ is represented by the factor $f^o - g^o \tan \delta^{(so)}$ of (5.19). [The residual factors of (5.19) are normalization coefficients inappropriate to our multichannel context.] Since the wave function pair $\{f^o, g^o\}$ of Chapter 5 corresponds to the pair $\{rj_0, ry_0\}$ of Chapter 4 in the absence of long-range forces, the radial factor of F_{SLJM} coincides with the limit of (4.31), multiplied by r, for $kr \ll 1$ and $k \to 0$. The value of $\tan \delta^{(so)} = (kr_0)^{-1} \tan \delta^{(s)}$ for $l = 0$ coincides here with the scattering length a_S of (7.23), which yields the limit

$$r[j_0(kr) - y_0(kr) \tan \delta_{SLJ}] \underset{k \to 0}{\longrightarrow} r - a_S, \qquad r > r_0. \tag{7.24}$$

The dependence of $F_{SLJM}(\omega, r)$ on r thus factors out. The residual dependence on ω, which will not be used explicitly, will be indicated by $X_{SLJM}(\omega)$. Thus, we set finally

$$F_{SLJM}(\omega, r) = X_{SLJM}(\omega)(r - a_S), \qquad r > r_0, \quad k \to 0. \tag{7.25}$$

The expansion (7.14) of this function into fragmentation channel components involves two separate operations. The angular function X_{SLJM} is first transformed into a superposition of i-channel functions $\Phi_i(\omega) \equiv \Phi_{\bar{J}_i \bar{M}_i m_{si}}(\omega)$

by the reciprocal of the transformation (7.20),

$$X_{SLJM}(\omega) = \sum_i \Phi_{\bar{J}_i \bar{M}_i m_{si}}(\omega)((\bar{S}\bar{L})\bar{J}_i s_i \bar{M}_i m_{si}|(\bar{S}s_i)SLJM), \tag{7.26}$$

whose matrix

$$((\bar{S}\bar{L})\bar{J}_i s_i \bar{M}_i m_{si}|(\bar{S}s_i)SLJM) = (\bar{J}_i s_i \bar{M}_i m_{si}|JM)U^{(J)}_{\bar{J}_i S} \tag{7.26'}$$

is the reciprocal of (7.21′).

The wave function $\Phi_{\bar{J}_i \bar{M}_i m_{si}}$ itself consists of two factors, one representing the state of the F atom and the other the spin of the incident electron. Thus, we write more explicitly

$$\Phi_{\bar{J}_i \bar{M}_i m_{si}} = \mathscr{A}(1, 2, \ldots, 9|\bar{J}_i \bar{M}_i)(10|m_{si}). \tag{7.26''}$$

The symbol $\mathscr{A}$ indicates that the position and spin coordinates of the nine electrons of the F atom must be antisymmetrized with those of the incident electron, indicated here by 10. The symbol $(10|m_{si})$ pertains here only to the spin of the incident electron (because Φ is independent of angular coordinates for $l = 0$), but $\mathscr{A}$ is meant to operate also on the radial function that follows $\Phi(\omega)$, specifically on the factor $(r - a_S)$ in (7.24). In the following functions $\Phi(\omega)$ always will be *implied*, for brevity, to include the symbol $\mathscr{A}$.

The construction of the large-r form of $F_{SLJM}(\omega, r)$ [(7.14)] is now performed by replacing $(r - a_S)$ in (7.25) by the expression on the left-hand side of (7.24) evaluated for $kr \gg 1$, while $ka_S \ll 1$ in $\tan \delta_{SLJ}$. The value of k now depends on the index i in (7.26) and we obtain

$$F_{SLJM}(\omega, r) \xrightarrow[k_i r \gg 1]{} \sum_i \Phi_{\bar{J}_i \bar{M}_i m_{si}}(\omega) \frac{1}{2ik_i} [e^{ik_i r}(1 - ik_i a_S) - e^{-ik_i r}(1 + ik_i a_S)]$$
$$\times (\bar{J}_i s_i \bar{M}_i m_{si}|JM)U^{(J)}_{\bar{J}_i S}. \tag{7.27}$$

Comparison of this expression with (7.14) yields the Jost coefficients

$$J^{(J)\pm}_{iS} = \left(\frac{\pi}{2k_i}\right)^{1/2} (\bar{J}_i s_i \bar{M}_i m_{si}|JM)U^{(J)}_{\bar{J}_i S}(1 \mp ik_i a_S). \tag{7.28}$$

In this equation the factor $U^{(J)}_{\bar{J}_i S}$ is given explicitly by (7.22) for $J = 1$; in combination with the factor $(1 \mp ik_i a_S)$, the dependence of the Jost matrix on the eigenchannel index S takes the explicit form

$$U^{(1)}_{\bar{J}_i S}(1 \mp ik_i a_S) = \begin{array}{c|cc} \bar{J}_i \backslash S & 1 & 0 \\ \hline \frac{1}{2} & \sqrt{\frac{2}{3}}(1 \mp ik_{1/2}a_1) & \sqrt{\frac{1}{3}}(1 \mp ik_{1/2}a_0) \\ \frac{3}{2} & -\sqrt{\frac{1}{3}}(1 \mp ik_{3/2}a_1) & \sqrt{\frac{2}{3}}(1 \mp ik_{3/2}a_0) \end{array}. \tag{7.28'}$$

[For $J = 0$, 2 the matrix $U^{(J)}$ has a single-unit element, as we have seen.]

Inversion of the matrix J^- is required to construct the wave function (7.15+) with outgoing wave boundary conditions and the scattering matrix (7.16). The inversion is simple because J^- is the direct product of separate matrices. Its first factor in (7.28) depends on i only and is inverted trivially. The second factor constitutes a separate orthogonal matrix inverted by simple transposition. The last pair of factors, represented explicitly by (7.28′), also constitutes a separate matrix, but its inversion needs to be worked out explicitly. Note that the quantum number M never departs from its value for the incident channel, $M = \bar{M}_i + m_{si}$. The reciprocal of J^- is then to be multiplied by J^+ to yield the scattering matrix (7.16),

$$S_{ji} = k_j^{-1/2} \sum_J (\bar{J}_j s_j \bar{M}_j m_{sj} | JM) \left\{ \sum_S (1 - ik_j a_S) U^{(J)}_{\bar{J}_j S} [(1 + ika) U^{(J)}]^{-1}_{S\bar{J}_i} \right\} \times (JM | \bar{J}_i s_i \bar{M}_i m_{si}) k_i^{1/2}. \qquad (7.29)$$

Hence the matrix product in the braces has the explicit form

$$\begin{array}{c|cc|c} j \backslash i & \frac{1}{2} & \frac{3}{2} & \\ \frac{1}{2} & \Delta(-k_{1/2}, k_{3/2}) & i\sqrt{2}\frac{2}{3}k_{1/2}(a_1 - a_0) & \dfrac{1}{\Delta(k_{1/2}, k_{3/2})} \\ \frac{3}{2} & i\sqrt{2}\frac{2}{3}k_{3/2}(a_1 - a_0) & \Delta(k_{1/2}, -k_{3/2}) & \end{array} \qquad (7.29')$$

where $\Delta(k_{1/2}, k_{3/2})$ is the determinant of (7.28′), namely,

$$\Delta(k_{1/2}, k_{3/2}) = 1 - k_{1/2}k_{3/2}a_1 a_0 + \tfrac{1}{3}ik_{1/2}(2a_1 + a_0) + \tfrac{1}{3}ik_{3/2}(a_1 + 2a_0). \qquad (7.29'')$$

[We note here a byproduct of the normalization of the asymptotic wave function (7.14), with a factor $(2/\pi k_i)^{1/2}$, that additional factors $k_i^{\pm 1/2}$ appear in the Jost and scattering matrices [(7.28–7.29)]. The same normalization factors, which appear again in the following expressions of wave functions, will generally combine again with an identical factor introduced by a coefficient of the outgoing wave, namely, S_{ji} or an analogous expression.]

The scattering wave function can finally be expressed by (7.15+) in terms of (7.28) and (7.29),

$$\sum_{SJ} F_{SLJM}(\omega, r)([J^{(J)-}]^{-1})_{Si} \underset{kr \gg 1}{\longrightarrow} \frac{1}{2i}\left(\frac{2}{\pi}\right)^{1/2} \left\{ -\Phi_{\bar{J}_i \bar{M}_i m_{si}}(\omega) \frac{e^{-ik_i r}}{k_i^{1/2}} + \sum_j \Phi_{\bar{J}_j \bar{M}_j m_{sj}}(\omega) \frac{e^{ik_j r}}{k_j^{1/2}} S_{ji} \right\}. \qquad (7.30)$$

Here k_i indicates the wave number of the incident electron while k_j indicates either of the alternative values of k for the scattered electron in the two alternative i channels. Energy conservation requires, of course, that

$\frac{1}{2}k_j^2 + \bar{E}_j = \frac{1}{2}k_i^2 + \bar{E}_i$ in atomic units. Note how the scattering matrix (7.29) would become diagonal if the two scattering lengths a_1 and a_0 were to coincide in (7.29′) and how its nonzero elements would depend on $k_{1/2}$ and $k_{3/2}$, respectively. Similarly, the spin-flip amplitude vanished in (7.7) when $\delta_{l+} = \delta_{l-}$. Quite generally, channels remain uncoupled when the short-range phase shifts coincide because of the degeneracy that this represents.

7.4.4 Closed Channel

The e + F collision is necessarily elastic when the F atom is initially in the lowest level of its doublet, $\bar{J}_i = \frac{3}{2}$, and the incident electron's energy $\epsilon_i = \frac{1}{2}k_i^2$ is lower than the doublet's separation in Fig. 7.3, ($\bar{E}_{1/2} - \bar{E}_{3/2} = 0.0501$ eV). Virtual excitation of F to $\bar{J}_i = \frac{1}{2}$ nevertheless plays the role of a closed channel in the collision to be outlined here as a preliminary to the fuller treatment of closed channels.

To this end we return to the representation (7.27) of an eigenchannel function $F_{SLJM}(\omega, r)$ rewriting it explicitly for the case $\epsilon_{1/2} = \frac{1}{2}k_{1/2}^2 < 0$, in which $k_{1/2}$ is replaced by its imaginary form. In accordance with Section 4.9 we write

$$
\begin{aligned}
F_{SLJM}(\omega, r) \xrightarrow[k_{3/2}r \gg 1]{} & [\Phi_1(\tfrac{3}{2}\tfrac{1}{2}\bar{M}_1\tfrac{1}{2}|JM) + \Phi_2(\tfrac{3}{2}\tfrac{1}{2}\bar{M}_2 - \tfrac{1}{2}|JM)] \frac{1}{2ik_{3/2}} \\
& \times [e^{ik_{3/2}r}(1 - ik_{3/2}a_S) - e^{-ik_{3/2}r}(1 + ik_{3/2}a_S)]U^{(J)}_{3/2S} \\
& - [\Phi_3(\tfrac{1}{2}\tfrac{1}{2}\bar{M}_3\tfrac{1}{2}|JM) + \Phi_4(\tfrac{1}{2}\tfrac{1}{2}\bar{M}_4 - \tfrac{1}{2}|JM)] \\
& \times \frac{1}{2\kappa_{1/2}}[e^{-\kappa_{1/2}r}(1 + \kappa_{1/2}a_S) \\
& - e^{\kappa_{1/2}r}(1 - \kappa_{1/2}a_S)]U^{(J)}_{1/2S}
\end{aligned}
\tag{7.27a}
$$

in a notation where the index $i = 1, 2, 3, 4$ of $\Phi_i(\omega)$ and $\bar{M}_i$ stand for the combinations

i	$\bar{J}$	$\bar{M}_i$	$m_{s\bar{J}}$
1	$\frac{3}{2}$	$M - \frac{1}{2}$	$\frac{1}{2}$
2	$\frac{3}{2}$	$M + \frac{1}{2}$	$-\frac{1}{2}$
3	$\frac{1}{2}$	$M - \frac{1}{2}$	$\frac{1}{2}$
4	$\frac{1}{2}$	$M + \frac{1}{2}$	$-\frac{1}{2}$.

Note that only either Φ_3 or Φ_4 exists for $M = \pm 1$.

The factor $\exp(\kappa_{1/2}r)$ in this expression diverges at $r \to \infty$; accordingly (7.27a) does *not* represent a stationary state of the e + F complex whose wave function should remain finite at all r. However, we can construct an eigenfunction of $e + \mathrm{F}$ by *superposing* wave functions (7.27a) with $J = 1$ and $S = 0, 1$ and adjusting their coefficients so as to cancel the term with $\exp(\kappa_{1/2} r)$.

The desired superposition is identified by recalling that the diverging factor $\exp(\kappa_{1/2} r)$ would become $\exp(-ik_{1/2} r)$ for $\epsilon_{1/2} > 0$ and that the factor $\exp(-ik_j r)$ is absent (for $j \neq i$) in the eigenfunction (7.30) with outgoing wave boundary conditions. Accordingly, the scattering wave function (7.30) itself will represent the desired elastic scattering in the single open channel, accompanied by virtual excitation in the closed channel, if we set $(k_i = k_{3/2} = (2\epsilon_{3/2})^{1/2}, \epsilon_{3/2} < \bar{E}_{1/2} - \bar{E}_{3/2}, \bar{J}_i = \frac{3}{2})$ and for $j \neq i$, $(k_j = i\kappa_{1/2} = (2\epsilon_{1/2})^{1/2}, \bar{J}_j = \frac{1}{2})$. These values are compatible with $J = 1$ only. The scattering eigenfunction of the $e + \mathrm{F}$ complex with $J = 1$, $\epsilon_{3/2} < \bar{E}_{1/2} - \bar{E}_{3/2}$ and total energy $E = \bar{E}_{3/2} + \epsilon_{3/2}$ then has the explicit form

$$\begin{aligned}\Psi_E(\omega, r) \xrightarrow[k_{3/2} r \gg 1]{} & -\left(\frac{2}{\pi}\right)^{1/2} \Phi_1(\omega) \frac{1}{2ik_{3/2}^{1/2}} e^{-ik_{3/2} r} \\ & + \left(\frac{2}{\pi}\right)^{1/2} (\Phi_1 S_{11} + \Phi_2 S_{21}) \frac{1}{2ik_{3/2}^{1/2}} e^{ik_{3/2} r} \\ & + \left(\frac{2}{\pi}\right)^{1/2} (\Phi_3 S_{31} + \Phi_4 S_{41}) \frac{k_{1/2}^{1/2}}{2\kappa_{1/2}} e^{-\kappa_{1/2} r}. \end{aligned} \tag{7.31}$$

The first two terms of this wave function correspond to those of $e + \mathrm{H}$ elastic scattering with or without spin flip [(7.5)], these two alternatives being represented by the indexes 1 and 2. The last term, which vanishes at $r \to \infty$, represents the additional virtual excitation of F to the $\bar{J}_j = \frac{1}{2}$ level of the doublet; its electron is seen here to tunnel toward large r as far as allowed by its negative energy $\epsilon_{1/2} = E - \bar{E}_{1/2}$. The presence of this virtual excitation, confined within a radius of the order $\kappa_{1/2}^{-1}$, might seem hardly accessible to experimental detection. In fact, this term manifests itself through sharp changes in the open-channel cross section, as we shall see in Section 8.2. (Note finally that a complex factor $k_{1/2}^{1/2}$ has been combined with the S-matrix element (7.29) in the last term to make it real, its reciprocal having been incorporated into the factor $1/\kappa_{1/2}$.)

7.4.5 Cross Sections

The cross section for $e + \mathrm{F}$ collision at low energy is obtained by adapting the elastic scattering cross section (4.17), whether the collision be elastic, spin flip, or inelastic. For our present case of $l = 0$, (4.17) reduces to $d\sigma/d\Omega = k^{-2}|(e^{2i\delta_0} - 1)/2i|^2$, namely, to the reciprocal of the squared wave number times the squared transition amplitude, which was indicated by $(S - 1)/2i$ in

(4.16b). The expression thus described maintains its structure quite generally; we simply treat S as a scattering matrix and 1 as a unit matrix and recall from earlier formulas that k^{-2} belongs to the incidence channel. The cross section for scattering from an initial fragmentation channel i to a final one j, per unit solid angle, thus takes its general form

$$\left|\frac{d\sigma}{d\Omega}\right|_{ji} = \frac{1}{k_i^2}\left|\frac{S_{ji} - \delta_{ji}}{2i}\right|^2 = \frac{1}{4k_i^2}\,[|S_{ji}|^2 - 2\delta_{ji}\,\mathrm{Re}\,S_{ii} + \delta_{ji}]. \tag{7.32}$$

The values of S_{ji} for $e + \mathrm{F}$ are provided by (7.29) for our example of $e + \mathrm{F}$ and $l = 0$. The total cross section is then equal to $4\pi\,(d\sigma/d\Omega)$. Equation (7.32) also holds for $l \neq 0$, but we must then include in S_{ji} the transformation coefficients analogous to the $(\hat{k}|lm)$ of (4.16a), as detailed in Section 7.9.

Summation of the cross section over alternative exit channels j, with or without spin flip, is made easy by the structure of S_{ji} in (7.29). The final state magnetic spin quantum number m_{sj} appears in (7.29) only in the first Wigner coefficient (together with the complementary quantum number $\bar{M}_j$ of the F atom since $\bar{M}_j + m_{sj} = M$ is conserved). The sum over the product of these coefficients, in the $\Sigma_{m_{sj}}|S_{ji}|^2$, yields a unit matrix.† Hence, the cross section summed over final spins is obtained simply by dropping the first Wigner coefficient in (7.29) and shifting the Σ_J symbol outside the square in $|S_{ji}|^2$. (Explicit expressions of the results appear in Problem 7.4.)

The Σ_S in (7.29) nevertheless remains within S_{ji} even after the sum over final spin orientations, thus yielding interference effects between singlet and triplet states of the $e + \mathrm{F}$ complex. Note how the final result is represented primarily in terms of the determinants $\Delta(\pm k_{3/2}, \pm k_{1/2})$ in (7.29′) and how each of these determinants combines wave numbers k_i of fragmentation channels with scattering lengths a_S of the complex.

7.4.6 Photodetachment of F^-

The bound state of the negative ion F^- has the same quantum numbers as the H^- treated in Section 7.2.4, namely, ${}^1\mathrm{S}_0$. Photoabsorption transfers F^- to a continuum state ${}^1\mathrm{P}_1^{\mathrm{o}}$ (again as for H^-), but the fragmentation of F^- generally yields the alternative doublet states of F, ${}^2\mathrm{P}^{\mathrm{o}}_{1/2,\,3/2}$, thus allowing the electron to detach with $l = 0$ at low energies. The F^- ion also differs from H^- by having a much larger binding energy (3.4 eV instead of 0.75 eV); accordingly, the volume occupied by the excess electron, $\sim\kappa_{\mathrm{B}}^{-3}$ in the notation of Section 7.2.4, is nearly 10 times smaller for F^- than for H^-. The simple treatment of the H^- wave function in Section 7.2.4 is hardly applicable

† The relevant orthogonality formula is

$$\sum_{m_{sj}\bar{M}_j} (JM|\bar{J}_j\bar{M}_j s_j m_{sj})(\bar{J}_j\bar{M}_j s_j m_{sj}|J'M') = \delta_{JJ'}\delta_{MM'}.$$

to F^-. Therefore, we deal here mainly with the multichannel aspect of F^- photodetachment in contrast to the earlier emphasis on the spectrum of $h\nu + H^- \to H + e$.

To represent the cross section for photodetachment into a specific fragmentation channel i of F^- we shall utilize the cross-section formula (2.9), whose final state is indicated by $(\epsilon|$, entering the wave function (7.15−) with ingoing wave boundary condition in place of a generic $|\epsilon_i)$. We take the direction of linear polarization as the z axis, as in Section 7.2.4, so that $M = 0$ in the final state just as in the ground state. Entering (7.15−) with $S = 0$, $L = 1$, $J = 1$, and $M = 0$, in (2.9) then gives

$$\sigma_i = 4\pi^2 e^2(\omega/c)|(J^{+\,-1})^*_{i0}(F_{0110}|z|0)|^2, \tag{7.33}$$

where the matrix elements of J^+ are provided by (7.28) and those of its reciprocal $J^{+\,-1}$ are to be obtained by using (7.29′). Since the ground state $|0)$ only extends over a limited range ($r \lesssim \kappa_B^{-1}$), so does the effective range of integration in the dipole matrix element of (7.33). Accordingly, the calculation of this matrix element involves *no* dependence on the final fragmentation channel i. This dependence is *concentrated* instead in the matrix element of $J^{+\,-1}$.†

The F^- detachment considered here is particularly simple in that a single α channel contributes to it. Indeed, (7.33) could be transcribed as

$$\sigma_i = |(J^{+\,-1})_{i0}|^2\bar{\sigma}, \qquad \bar{\sigma} = 4\pi^2 e^2(\omega/c)|(F_{0110}|z|0)|^2, \tag{7.33′}$$

where $\bar{\sigma}$ is common to all fragmentation channels and may be treated as an empirical parameter for many purposes. In other applications (to be considered in Problem 7.2 and elsewhere in the following), alternative values of the final angular momentum J contribute coherently to photodetachment in any one channel.

The angular distribution of photodetached electrons is isotropic in the present example, whose final state is limited to $l = 0$; a less trivial distribution will be met in Section 7.5.3. Summing the cross section over final spin orientations of the detached electron is as straightforward here as it was for the collisions in Section 7.4.5. The Wigner coefficient in the Jost matrix element (7.28) can be inverted separately as noted in the discussion of that expression and can then be squared and summed over m_{si} to yield unity for

† Numerical calculations require attention to dimensional elements. In (2.9), $e^2(\omega/c)$ represents an energy and $(\epsilon|$ is normalized per unit energy, whereby the dimensions of σ are provided by the square of the dipole operator; accordingly, the wave function (7.15−) must be restored explicitly to normalization per unit energy. Proceeding alternatively in dimensionless atomic units, we set $e^2 \to 1$ and the wavenumber ω/c in reciprocal Bohr radii.

the given value of $J = 1$. The sum over i in (7.33′) thus gives

$$\sum_{m_{si}} |(J^{+^{-1}})_{io}|^2 = k_i |[U^{(1)}(1 - ika)]_{io}^{-1}|^2. \tag{7.34}$$

Note here the proportionality to k_i, which accords with the Wigner threshold law (4.33) for $l_i = 0$.

The matrix element on the right-hand side of (7.34) is to be obtained by inverting the matrix (7.28′), which yields

$$\begin{aligned} \frac{\sigma_{3/2}}{\bar{\sigma}} &= \sum_{m_{s(3/2)}} |(J^{+^{-1}})_{(3/2)0}|^2 = k_{3/2} \frac{2}{3} \frac{1 + k_{3/2}^2 a_1^2}{|\Delta(-k_{1/2}, -k_{3/2})|^2}, \\ \frac{\sigma_{1/2}}{\bar{\sigma}} &= \sum_{m_{s(1/2)}} |(J^{+^{-1}})_{(1/2)0}|^2 = k_{1/2} \frac{1}{3} \frac{1 + k_{1/2}^2 a_1^2}{|\Delta(-k_{1/2}, -k_{3/2})|^2}. \end{aligned} \tag{7.34′}$$

with Δ defined by (7.29″). The cross section depends, in general, on all four parameters $\{k_{1/2}, k_{3/2}, a_0, a_1\}$. The two expressions (7.34′) yield the separate cross sections for the ejection of two groups of electrons with different energies $\frac{1}{2}k_{3/2}^2 = \omega - \bar{E}_{3/2}$ and $\frac{1}{2}k_{1/2}^2 = \omega - \bar{E}_{1/2}$ corresponding to the alternative doublet levels of the F residue. A special case occurs when the $\bar{J}_i = \frac{1}{2}$ channel is closed, that is, when $\frac{1}{2}k_{1/2}^2 < 0$. The treatment of photodetachment must then be adjusted much as the treatment of e + F collisions was adjusted in Section 7.4.4. Here we must replace $k_{1/2}$ by $-i\kappa_{1/2}$ and we find, instead of (7.34′), the single coefficient†

$$\sigma_{3/2}/\bar{\sigma} = \sum_{m_{s(3/2)}} |(J^{+^{-1}})_{(3/2)0}|^2 = k_{3/2} \frac{2}{3} \frac{(1 - \kappa_{1/2} a_1)^2}{|\Delta(i\kappa_{1/2}, -k_{3/2})|^2}. \tag{7.34″}$$

The main energy dependence of (7.34′) and (7.34″) is represented by their factor k_i because their last factors remain ~ 1 in the low-energy range considered here, owing to the small values of k_i in a.u. Figure 7.4 displays data on Se^- photodetachment, which is analogous to F^- but involves a larger number of thresholds.

These cross sections for photodetachment display an important result. Even though photoabsorption conserves the singlet character of the bound F^- (that is, its initial spin quantum number, $S = 0$), the photodetachment cross section also depends on the scattering length a_1 for the unbound triplet channel of e + F, because a_1 occurs in the full matrix $U_{\bar{J}_i S}^{(1)}(1 - ik_i a_S)$ and

† The inversion of J^+ to arrive at (7.34″) involves the determinant $\Delta(-k_{1/2}, -k_{3/2})$ in the denominator. Note the minus signs absent in the inversion of J^- that led in (7.29′) to the results for e + F scattering. Note also that we set $k_{1/2} = -i\kappa_{1/2}$ for a closed channel in (7.34″), namely, the complex conjugate of the convention established for scattering in Section 7.4.4. Complex conjugation derives from the time reversal connection between the ingoing and outgoing wave boundary conditions that, respectively, are involved.

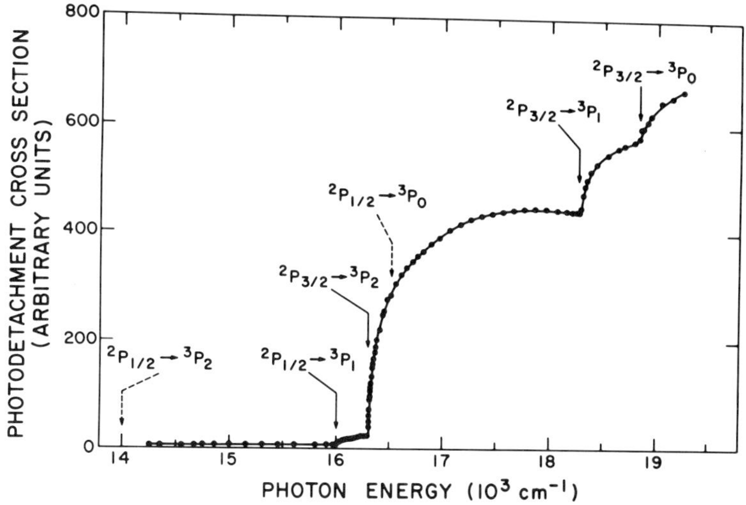

Fig. 7.4. Photodetachment cross section of Se^-. $Se^-(^2P_{3/2,1/2}) + h\nu \to Se(^3P_{2,1,0}) + e^-$. Note the six fine structure thresholds (Fig. 7.3b) and proportionality to (energy − threshold)$^{1/2}$. (From Hotop *et al.* [8].)

remains entangled with the other parameters when this matrix is inverted. The inversion results from the difference in the boundary conditions governing the propagation of the electron in the alternative fragmentation channels. This propagation involves a partial reflection, especially when one channel is closed. The partial reflection back into the small-r region leads to rescattering by *both* alternative eigenchannels of the F^- complex with their parameters a_0 and a_1. The initial singlet character persists, however, through the fragmentation stage of photodetachment in the high-energy limit where the difference $k_{3/2} - k_{1/2} \ll k_{3/2}$ and we may replace both $(k_{3/2}, k_{1/2})$ by a single value k. In this event the simplified expression $(1 - ika_S)$ factors out from the residual matrix $U^{(1)}_{J_i S}$ and can be inverted separately without entangling a_1 with a_0. Only in this limiting case can we visualize the photodetached electron as propagating freely from the range where it forms a complex with F into the fragmentation channels, much as a light beam would do in the limit of geometrical optics.

The final expressions for photodetachment, as well as the earlier ones for $e + F$ scattering cross sections, exemplify how a full analysis of the multichannel problem is developed in terms of a small set of parameters, here four. Three of these—the two scattering lengths and the dipole matrix element contained in $\bar{\sigma}$—are characteristic of the short-range interaction alone. The fourth, representing the numerical coefficients $\frac{2}{3}$ and $\frac{1}{3}$ in (7.34′) and (7.34″), arises from the U matrix in (7.22). It is the single rotation angle that describes the frame transformation connecting the i and α channels. The next two sections of this chapter will illustrate further this viewing of multichannel problems in terms of a few short-range and frame-transformation parameters, the long-range aspects being handled analytically.

7.5 PHOTOABSORPTION OF H_2

The approach to multichannel processes outlined in Section 7.3 and illustrated in Section 7.4 by the example of $e + \mathrm{F}$ processes was, in fact, originally introduced through the interpretation of a photoabsorption spectrum obtained by Herzberg by using a sample of hydrogen molecules kept in their ground state of vibration and rotation ($v = 0$, $J = 0$) at very low temperature. We take up this study next. The Coulomb force in this system e-H_2^+ makes for a far richer spectrum than that in the analysis of a negative ion.

As in the case of F^-, photoabsorption excites, or ionizes, the molecules of H_2 only to states with odd parity and total angular momentum $J = 1$. This momentum can be shared between the orbital motion of the excited electron and the rotational motion of the H nuclei about their center of mass. This sharing is determined by short-range electron–core interactions. The analysis of the spectrum was facilitated by the smallness of the excitation energy that can be taken up by rotation, namely ~ 0.02 eV, which is far less than the binding energy of the lower Rydberg states.

Accordingly, the electron–core interaction forces the electron's orbital motion to remain coupled rigidly to the axis of nuclear rotation in the lower levels of electron excitation. For $J = 1$ this coupling allows two alternative orbital motions, characterized by projections of $\mathbf{l}$ along the nuclear axis with quantum numbers $\Lambda = \mathbf{l} \cdot \hat{R} = 0$ or 1 and spectroscopic labels Σ and Π, respectively. Thus, Σ and Π are the specific labels of the eigenchannels of short-range interactions for this system. The corresponding eigenphase shifts are known to be $\delta_\Sigma^{(s)} = 0.2\pi$ and $\delta_\Pi^{(s)} = -0.08\pi$.

Combination of these phase shifts with long-range phase shifts $\delta^{(l)}$ proceeds quite readily for the lower excitations, because here—as for the $e + \mathrm{He}^+$ problem in Section 7.2.3—$\delta^{(l)}$ is given by the Coulomb field formula $\beta = -l\pi + \pi/\kappa$ from Table 5.1 and is substantially independent of the channel label, whether α ($\equiv \Sigma$ or Π) or i. Thus, we have a separate Rydberg series of levels, $n^1\Sigma_u$ and $n^1\Pi_u$, where u stands for odd parity, determined by the eigenvalue equations

$$\delta_\Sigma^{(s)} + \pi/\kappa = n\pi, \qquad \delta_\Pi^{(s)} + \pi/\kappa = n\pi, \qquad 2 \le n \lesssim 10, \tag{7.35}$$

and the resulting equation (5.28). The values of $\delta_\Sigma^{(s)}$ and $\delta_\Pi^{(s)}$, in fact, have been derived by fitting these results to experimental data and by using very accurate *ab initio* calculation of the levels for $n = 2$. (The considerable influence of vibrational excitation will be mentioned later.)

The approximation involved in regarding $\delta^{(1)}$ as channel-independent breaks down for energies of the excited electron comparable to the difference of core energies $\bar{E}_i$ for different fragmentation channels. Two such channels are relevant here, labeled by rotational quantum numbers of the H_2^+ core $\bar{J}_i \equiv N = (0, 2)$. (These are the only values of N consistent with $J = 1$, with the parahydrogen character of the $J = 0$ initial state and with orbital momentum $l_i = 1$; higher values of l_i of odd parity are unlikely.) Figure 7.5a, a diagram prepared by Herzberg and Jungen [9], illustrates in its upper portion the two fragmentation channels, labeled as npN (with $N = 0, 2$), on a scale of total energies E measured from the ground state and expressed in reciprocal centimeters. The lower part of the diagram indicates instead the Rydberg series corresponding to the α channels, Σ and Π, which are *not* intended to be in one-to-one correspondence with the fragmentation channels $N = 0$ and 2. (This distinction of molecular state channels, characterized as "Hund's cases b and d," has been known since the 1920s.) The actual appearance of the excitation spectrum of the two fragmentation channels (overlaid by broad lines due to vibrational excitation) is shown by the reproduction of Herzberg's spectral plate in Fig. 7.5b.

The analytical connection between the short-range and long-range channels—akin to that of Section 7.4.1—is established by comparing their respective rotational wave functions, which we indicate as $X_M^\Lambda(\hat{r}, \hat{R})$ and $V_M^N(\hat{r}, \hat{R})$. Here $\hat{r}$ is the direction of the excited electron and $\hat{R}$ is that of the molecular axis, $J = 1$, $M = 0$ or ± 1, $\Lambda \equiv \Sigma$ or Π, and $N = 0$ or 2. Both functions are constructed with orbital $l = 1$ wave functions of the electron and rotational functions of the H_2 molecule for X and H_2^+ for V. Their connection is represented by†

$$X_M^\Lambda(\hat{r}, \hat{R}) = \sum_{N=0,2} V_M^N(\hat{r}, \hat{R}) U_{N\Lambda}, \tag{7.36}$$

where

$$U_{N\Lambda} = \begin{array}{c|cc|} N \backslash \Lambda & 0 & 1 \\ \hline 0 & \sqrt{\tfrac{1}{3}} & \sqrt{\tfrac{2}{3}} \\ 2 & -\sqrt{\tfrac{2}{3}} & \sqrt{\tfrac{1}{3}} \end{array} \tag{7.36'}$$

† In this example, J takes only one value (unity). It is, therefore, not explicitly indicated in the quantities X and V and in their following expressions. More generally, alternative J values may be allowed, in which case X and V would carry a further superscript J. The total angular momentum J is conserved throughout and alternative J's will contribute coherently in the final expressions. The frame transformation matrix (7.36′) is essentially similar to (7.22), even though they represent different algebraic functions of small quantum numbers.

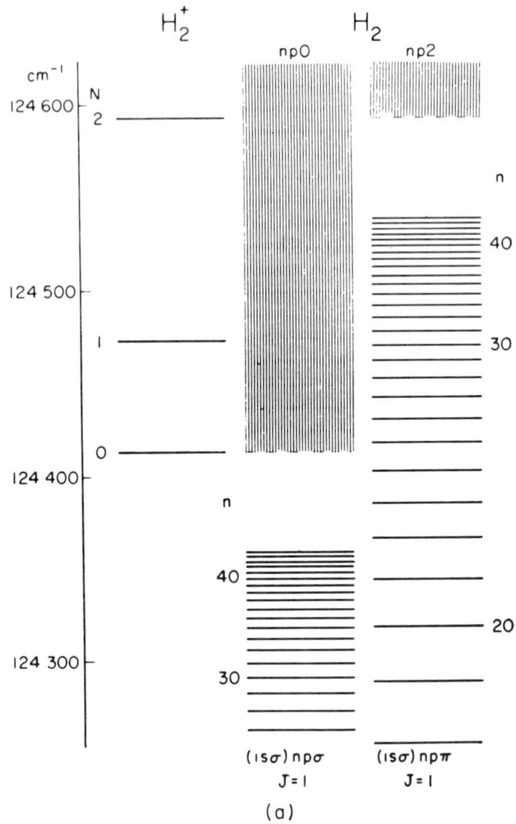

Fig. 7.5. Photoabsorption spectrum of $H_2(1s\sigma^2\ \Sigma_g^+\ v = J = 0)$. (a) Schematic of states converging to the $N = 0$ and 2 rotational levels of H_2^+. Upper area represents fragmentation channels, and lower area represents eigenchannels. (b) Observed spectrum (not to scale). (From Herzberg and Jungen [9].)

is an orthogonal matrix analogous to (7.22). On this basis we can proceed to construct the expansion (7.14) of the wave functions $F_\alpha(\omega, r)$, $\alpha \equiv \Lambda = \{\Sigma, \Pi\}$.

Consider first the range of energies and radial distances where the dependence of $\delta^{(1)}$ on channel indexes can be disregarded. Here we can write [see, for instance, (5.20)]

$$F_\Lambda(\omega, r) = [f(\epsilon, l = 1, r) \cos \delta_\Lambda^{(s)} - g(\epsilon, l = 1, r) \sin \delta_\Lambda^{(s)}] \times X_M^\Lambda(\hat{r}, \hat{R})\Phi(\omega), \qquad r > r_0, \tag{7.37}$$

where $\{f, g\}$ indicate the base pair of functions† of Chapter 5 calculated with

† The multichannel problems considered in this section and subsequent ones involve a long-range Coulomb field. We may then depart from the base pair $\{f^\circ, g^\circ\}$ of Chapter 5 in defining the Jost matrices, as the alternative pair $\{f, g\}$ often proves more suitable. Care is required in applying this alternative, as discussed in Section 7.8.

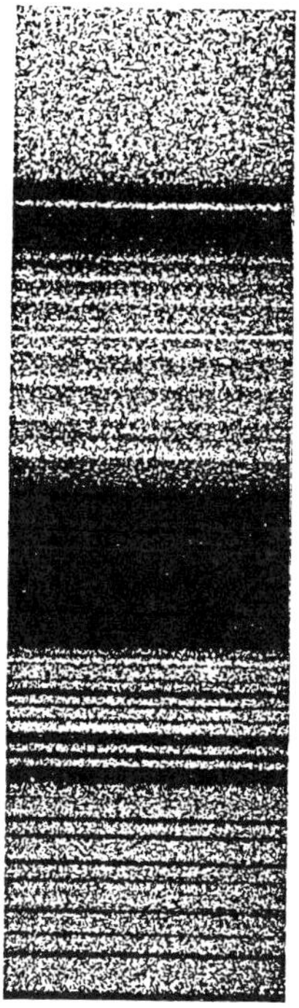

(b)

Fig. 7.5. (*Continued*)

$v(r) = -e^2/r$, and $\Phi(\omega)$ is an antisymmetrized product of the core electron eigenfunctions $1s\sigma$, of the singlet spin wave functions of both electrons, both nuclei, and the nuclear vibration wave function $\chi_{v=0}(R)$. We now enter in (7.37) the expansions of f and g from (5.6a), (5.7), and (5.14) and find

$$F_\Lambda(\omega, r) = \left(\frac{2}{\pi k}\right)^{1/2} \frac{1}{2i} [e^{i\delta_\Lambda^{(s)} + i\delta^{(1)}} f^+(\epsilon, l, r) - e^{-i\delta_\Lambda^{(s)} - i\delta^{(1)}} f^-(\epsilon, l, r)] \times X_M^\Lambda(\hat{r}, \hat{R})\Phi(\omega), \qquad r > r_0, \quad \epsilon > 0. \tag{7.38}$$

This expression is of the form given in (5.24a).

The final step consists of introducing the expansion of X in (7.36) into the two fragmentation channel functions V and specifying that the Coulomb field phase shifts $\delta^{(1)} = k^{-1} \ln 2k - l\frac{1}{2}\pi + \arg \Gamma(l + 1 - i/k)$ and the wave functions $f^\pm$ must be evaluated with the parameters ϵ_N, k_N, and $l = 1$ appropriate to each fragmentation channel. Thus, we find

$$F_\Lambda(\omega, r) = \sum_{N=0,2} \left(\frac{2}{\pi k_N}\right)^{1/2} \frac{1}{2i} [e^{i(\delta_\Lambda^{(s)} + \delta_N^{(1)})} f^+(\epsilon_N, l, r) - e^{-i(\delta_\Lambda^{(s)} + \delta_N^{(1)})} f^-(\epsilon_N, l, r)] \times V_M^N(\hat{r}, \hat{R}) U_{N\Lambda} \Phi(\omega), \qquad r > r_0, \quad \epsilon > 0. \tag{7.39}$$

The product of two factors $V_M^N(\hat{r}, \hat{R})\Phi(\omega)$ constitutes the function indicated in (7.14) by $\Phi_i(\omega)$. Comparison with (7.14) gives the explicit form of the Jost matrix elements for our problem

$$J_{i\alpha}^{\pm} \equiv J_{N\Lambda}^{\pm}(k_N) = (\pm i)^l e^{\pm i\delta_N^{(l)}(k_N)} U_{N\Lambda} e^{\pm i\delta_\Lambda^{(s)}}, \qquad \epsilon_N > 0. \qquad (7.40)$$

Reading from right to left, the connection between α and i channels is explicitly factored into its basic elements. First, the $\exp[\pm i\delta_\Lambda^{(s)}]$ term expresses the phase shift in the α channel, which depends on the short-range interactions. The $U_{N\Lambda}$ term is the geometrical factor, the frame transformation connecting the α and i channels, and finally the other factor $\exp[\pm i\delta^{(l)}(k_N)]$ involves the long-range interaction in the fragmentation channel N. The factor $(\pm i)^l$ in (7.40) cancels the phase $-l\pi/2$ that occurs in the long-range phase shift according to Chapter 5.

For $\epsilon < 0$ we apply the expressions for $\{f, g\}$ from Chapter 5 to express F in (7.37) as in (5.24b) and find for $l = 1$,

$$J_{N\Lambda}^{+} = -DU_{N\Lambda}\cos[\beta(\kappa_N) + \delta_\Lambda^{(s)}],$$
$$J_{N\Lambda}^{-} = D^{-1}U_{N\Lambda}\sin[\beta(\kappa_N) + \delta_\Lambda^{(s)}], \qquad \epsilon < 0. \qquad (7.41a)$$

Accordingly, we have

$$F_\Lambda(\omega, r) = \sum_{N=0,2} (\pi\kappa_N)^{-1/2}\{-D\cos[\beta(\kappa_N) + \delta_\Lambda^{(s)}]f^{+}(\epsilon_N, l, r)$$
$$+ D^{-1}\sin[\beta(\kappa_N) + \delta_\Lambda^{(s)}]f^{-}(\epsilon_N, l, r)\} V_M^N(\hat{r}, \hat{R})U_{N\Lambda}\Phi(\omega), \qquad \epsilon < 0, \qquad (7.41b)$$

where

$$\beta(\kappa_N) = -\pi + (\pi/\kappa_N) = -\pi + \pi\nu_N \qquad (7.41c)$$

from Table 5.1.

7.5.1 General Remarks on Quantum Defect Theory

As already noted, the basic structure of $J_{i\alpha}^{+}$ in (7.41a) consists of an element of an orthogonal transformation matrix $U_{i\alpha}$ and a phase factor that results from the sum of two contributions: a phase shift $\delta_\alpha^{(s)}$ characteristic of a short-range eigenchannel and a $\delta_i^{(l)}$ characteristic of a fragmentation channel. At negative energies the phase factor turns into a cosine for $J_{i\alpha}^{+}$ and into a sine for $J_{i\alpha}^{-}$. Additional coefficients in (7.40) and (7.41a) only reflect conventions in the definition of $J_{i\alpha}^{\pm}$.

The schematic and very logical manner of combining the short- and long-range phase shift contributions displayed in (7.40) and (7.41) hinges on a single feature of the system considered here. This is the occurrence of an intermediate range of radial distances $r > r_0$, where the field acting on the electron has central symmetry and contributes to the phase shift at the *same rate* per unit radial distance for all channels. This circumstance implies that, on the one hand, the anisotropic components of the core field have short range and, on the other hand, the differences between alternative energy levels $\bar{E}_i$ of the core loom important only at larger radial distances.

This factorization of Jost matrices in (7.40) and (7.41a) constitutes a fundamental approximation that disregards the influence of any coupling of α channels because of multipole fields at intermediate distances. It underlies the procedures of multichannel quantum defect theory to be utilized in the rest of this chapter and in much of the following ones.

7.5.2 Photoionization into Specific Channels

We have seen that photoionization of H_2, under the conditions specified previously, can leave the H_2^+ residue in either of two rotational states ($N = 0$ or 2). To calculate the separate probability of either of these events, the calculation of photoabsorption outlined in Chapter 2 must be carried out using a final-state wave function with outgoing wave in one fragmentation channel only, exactly as in Section 7.4.5. This ingoing wave boundary condition specifying photoemission into a channel with given quantum number N is then given by (7.15−) as

$$\sum_{\Lambda} F_{\Lambda}(\omega, r)(J^{+^{-1}})_{\Lambda N} = \sum_{N'} \Phi_{N'}(\omega)\left(\frac{2}{\pi k_{N'}}\right)^{1/2} \frac{1}{2i}\left\{(-i)^l f^+(\epsilon_{N'}, l, r)\delta_{N'N} - (i)^l f^-(\epsilon_{N'}, l, r)\left[\sum_{\Lambda} J^-_{N'\Lambda}(J^{+^{-1}})_{\Lambda N}\right]\right\}. \tag{7.42}$$

(The coefficient of f^- is again an element of the reciprocal scattering matrix $S^{-1} = J^- J^{+^{-1}}$.)

While the right-hand side of (7.42) serves to characterize the fragmentation properties of the state of interest, its left-hand side relates to its properties within the core region in a standing wave representation and is accordingly appropriate to represent the excited state indicated by $(\epsilon|$ in the dipole matrix element of (2.9). If we represent the state $(\epsilon|$ by the left-hand side of (7.42) (or, rather, by its complex conjugate), the dipole matrix element for photoionization to channel N takes the form analogous to (7.33)

$$(\epsilon_N|z|0) = \sum_{\Lambda} (J^{+^{-1}})^*_{\Lambda N}(F_{\Lambda}|z|0). \tag{7.43}$$

This result is fully in line with our approach. Two basic dipole matrix elements are considered, corresponding to transitions to states $\Lambda \equiv \Sigma$ or Π. (The dipole moment is parallel to the molecular axis for transitions to the Σ states, perpendicular to it for transitions to Π.) The calculation of these matrix elements extends only within the core volume, to which the ground state is confined, whereas the Jost functions introduce the data relevant to final state selection. Note that two values of Λ contribute to (7.43), whereas only $S = 0$ contributes to (7.33). [Incidentally, the value of (7.43) is very small for $N = 2$, reflecting the circumstance that the dipole polarizability of the H_2 molecules is nearly isotropic, whereby photoionization is unlikely to excite rotational motion.[†]]

7.5.3 Angular Distribution of Photoelectrons

The cross section for photoabsorption (2.9), at photon energies ϵ sufficient to ionize, may be regarded as the integral over a differential cross section $d\sigma(\theta, \phi)/d\Omega$ for photoemission in a specified direction (θ, ϕ). The differential cross section is obtained by specifying the quantum numbers (l, m) of the photoelectron in the final state $|\epsilon)$, whereby the squared matrix element $|(\epsilon|z|0)|^2$ should actually read $\sum_{lm}|(\epsilon lm|z|0)|^2$. The dipole amplitude $(\epsilon lm|z|0)$ should then be multiplied by the probability amplitude $(\theta\phi|lm) \equiv Y_{lm}(\theta, \phi)$ that the photoelectron will be ejected in the direction (θ, ϕ). The cross section (2.9) thus becomes

$$\frac{d\sigma(\theta, \phi)}{d\Omega} = 4\pi^2 e^2 \frac{\omega}{c} \left| \sum_{lm} Y_{lm}(\theta, \phi)(\epsilon lm|z|0) \right|^2. \tag{7.44}$$

The dipole amplitude $(\epsilon lm|z|0)$ in this formula should, of course, pertain to ionization into a specific channel.

Application of (7.44) to our H_2 process is quite straightforward if the channel under consideration is the one with $N = 0$. In this case the dipole matrix element (7.43) has the selection rule $(\epsilon_0 lm|z|0) = (\epsilon_0 10|z|0)\delta_{l1}\delta_{m0}$ because the ground state has $J = M = 0$. Since $Y_{10} = (3/4\pi)^{1/2} \cos\theta$, (7.44) gives

$$d\sigma_0(\theta, \phi)/d\Omega = 4\pi^2 e^2(\omega/c)|(\epsilon_0 10|z|0)|^2(3/4\pi)\cos^2\theta. \tag{7.45}$$

This very simple result can be written in the standard general form for the angular distribution of photoelectrons ejected by linearly polarized light as

$$d\sigma(\theta, \phi)/d\Omega = \sigma_{tot}[1 + \beta P_2(\cos\theta)]/4\pi, \qquad \beta = 2, \tag{7.46}$$

where θ indicates the angle between the directions of ejection and light

† Alternative couplings of the electron in the H_2^+ residue with the rotation state $N = 2$ of the molecular ion introduce a fine structure of the energy $\bar{E}_2$ that has been disregarded here. This interaction would disturb the spin coupling of the two electrons. The spin of the ionized electron is also weakly coupled to its orbit. Weak phenomena of this type are widespread and have been discussed in the recent literature [10].

polarization and $P_2 = \frac{1}{2}(3\cos^2\theta - 1)$. Only the value of β varies in (7.46), depending on the specific process under consideration, as long as the light acts by the electric dipole mechanism.† For unpolarized light, θ represents the angle between the directions of ejection and light propagation, and the parameter β is replaced by $-\frac{1}{2}\beta$ in (7.46).

The treatment is more complicated for the channel of photoionization that leaves the H_2^+ in its excited rotational state with $N = 2$. An elementary treatment would specify the fragmentation channel $|\epsilon_2)$ in the uncoupled representation (7.1″), which lists the magnetic quantum numbers of molecular rotation and spin and orbital orientation separately. The angular distribution would be calculated for each value of these quantum numbers separately and would then be averaged over all values of the unobserved parameters.

In practice we rely on the following argument of angular momentum recoupling that yields directly the reduction of the value of β ("depolarization") because of this averaging. The angular momentum coupling of core rotation and orbital motion of the photoelectron is represented by $(Nl)J$, with $N = 2$, $l = 1$, and $J = 1$; the electron spins can be ignored because they remain coupled to a singlet, with $S = 0$ [10]. In (7.46) the coefficient β of the harmonic P_2 plays the role of a quadrupole moment of the angular distribution. A general 2^k-pole moment of the fragmenting channel state is represented by the coupling relation $[(Nl)J(Nl)J, k]$, where J appears quadratically as l does in (7.44). The averaging over the orientation of the rotational momentum $\mathbf{N}$ amounts to changing this coupling relation into $[(NN)0(ll)k, k]$, a standard operation of angular momentum algebra. The value of β is accordingly reduced by the ratio of the recoupling coefficients for $k = 2$ and $k = 0$,

$$\frac{((Nl)J(Nl)Jk|(NN)0(ll)kk)_{k=2}}{((Nl)J(Nl)Jk|(NN)0(ll)kk)_{k=0}} = \frac{3.5^{-1/2}\begin{Bmatrix}1 & 1 & 2\\ 1 & 1 & 2\end{Bmatrix}}{3.5^{-1/2}\begin{Bmatrix}1 & 1 & 0\\ 1 & 1 & 2\end{Bmatrix}} = \frac{1}{10}, \tag{7.47}$$

according to Eqs. (3.9) and (3.19) of Rotenberg *et al.* [7]. The value of β is thus reduced from 2 to $\frac{1}{5}$.

Note that the procedure followed here for the calculation of β took advantage of simplifying circumstances of our problem, namely, that H_2 has initially $J = 0$ and that the fragmentation channels are nondegenerate. More elaborate procedures are generally required. Conversely, the experimental study of angular distributions may serve to evaluate collision parameters with particular sensitivity when they reflect interference of alternative fragmentation channels.

† We use the standard notation β for the asymmetry parameter of the angular distribution to be discussed further in Section 7.10. Earlier in this chapter, a quantum defect parameter has been denoted by $\beta(\kappa)$. No confusion need arise between the two usages because the *QDT* parameter is always indicated as the function of an argument κ.

7.5.4 Bound-State Levels

An explicit form of the eigenvalue equation [(7.18)] is obtained by entering in it the matrix elements of $J^-_{N\Lambda}$ from (7.41)—disregarding irrelevant factors—and the values (7.36′) of the $U_{N\Lambda}$, that is,

$$\begin{vmatrix} U_{0\Sigma}\sin[\pi\nu_0+\delta^{(s)}_\Sigma] & U_{0\Pi}\sin[\pi\nu_0+\delta^{(s)}_\Pi] \\ U_{2\Sigma}\sin[\pi\nu_2+\delta^{(s)}_\Sigma] & U_{2\Pi}\sin[\pi\nu_2+\delta^{(s)}_\Pi] \end{vmatrix}$$
$$= \tfrac{1}{3}\sin[\pi\nu_0+\delta^{(s)}_\Sigma]\sin[\pi\nu_2+\delta^{(s)}_\Pi] + \tfrac{2}{3}\sin[\pi\nu_0+\delta^{(s)}_\Pi]\sin[\pi\nu_2+\delta^{(s)}_\Sigma] = 0. \qquad (7.48)$$

This equation establishes an implicit relation between the energy parameters $\nu_0 = (-2\epsilon_0)^{-1/2}$ and $\nu_2 = (-2\epsilon_2)^{-1/2}$, which is plotted in Fig. 7.6 modulo 1.

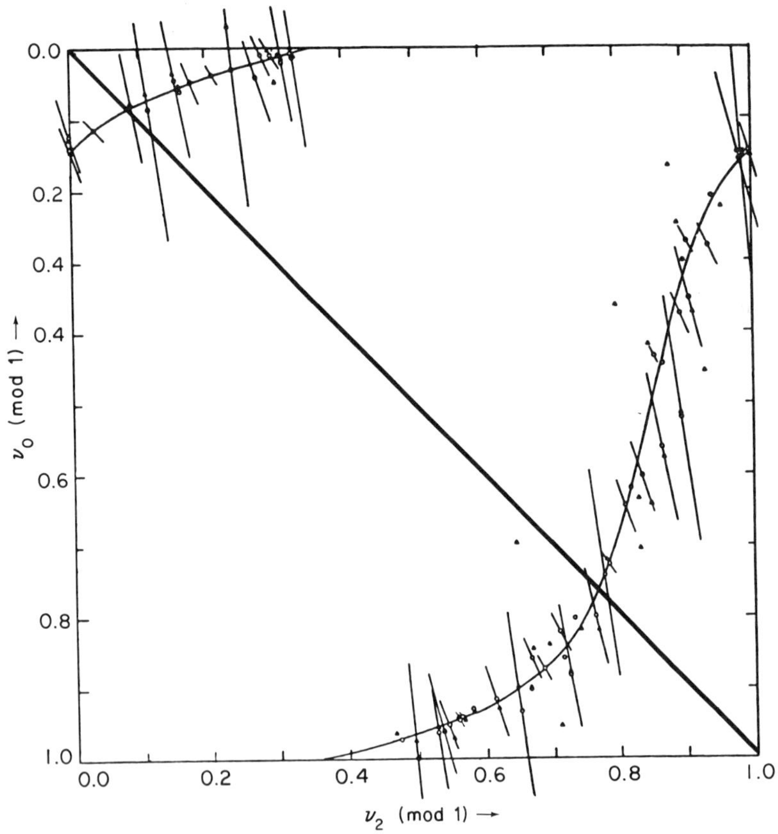

Fig. 7.6. Plot of Eq. (7.48) for the $e + H_2^+$ system, a two-channel quantum defect plot. Here $\nu_i = (-2\epsilon_i)^{-1/2}$, $i \equiv N = 0, 2$.

A second relation, implicit in the definition of ϵ_0 and ϵ_2, is $\epsilon_0 - \epsilon_2 = \bar{E}_2 - \bar{E}_0 = 174\ \text{cm}^{-1} = 8 \times 10^{-4}$ a.u. ~ 0.022 eV. Line segments that represent this second relation,

$$\nu_2^2 = \nu_0^2/(1 + 0.0016\nu_0^2), \tag{7.49}$$

are also plotted in Fig. 7.6. The energy eigenvalues lie accordingly at the intersections of the segments with the main curve.

A large number of experimental points are plotted in the figure, each of them with the coordinates (ν_0, ν_2) corresponding to the energy value of an observed spectral line. These points are clearly clustered along the curve representing (7.48) but show appreciable deviations. These deviations were interpreted in the initial study by Herzberg and Jungen [9] as perturbations by extraneous lines belonging to fragmentation channels with vibrational excitation. The extension of the present treatment to vibrational channels is outlined in Section 7.5.5.

The lower levels of the spectrum, in the range where (7.49) reduces to $\nu_2 \sim \nu_0$, lie on the diagonal line across the plot of Fig. 7.6. This line crosses the line representing (7.48) at the points $\nu_2 = \nu_0 = -\delta_\Sigma^{(s)}/\pi$ and $-\delta_\Pi^{(s)}/\pi$, thus verifying that the system of equations (7.48–7.49) reduces in this case to the Rydberg series formula (5.28) for independent channels Σ and Π. For higher energies the line representing (7.49) becomes increasingly steep and the energy level positions seem to vary erratically when plotted on an energy scale rather than on the mapping of Fig. 7.6.

Although the intensities seem to vary erratically from one line to another, they are actually represented [in analogy to (7.33)] by the analytic formula

$$\left|\sum_\Lambda A_\Lambda^*(F_\Lambda|x|0)\right|^2, \tag{7.50}$$

where the A_Λ are unit eigenvectors of the linear system with the determinant (7.48). The simple law on the distribution of levels in Rydberg series—level positions at unit intervals of $\nu = \delta^{(1)}/\pi$—now takes a much weaker form: one level *on the average* per unit interval of the *eigenphase sum* $\delta_{N=0}^{(1)} + \delta_{N=2}^{(1)}$. Illustrative data on the intensity distribution will be given when we return to this subject in a broader context (Sections 8.4 and 8.5).

7.5.5 Extension to Vibrational Excitation

Vibrational excitation is very likely to result from photoabsorption because the H_2^+ core of the $e + H_2^+$ system has a far larger equilibrium distance of its nuclei than the initial H_2 molecule. The fragmentation channels bear accordingly a double label, Nv, where v is the vibrational quantum number of the H_2^+ core, which may range up to ~ 10.

A second label must also be added to the eigenchannels of electron–core

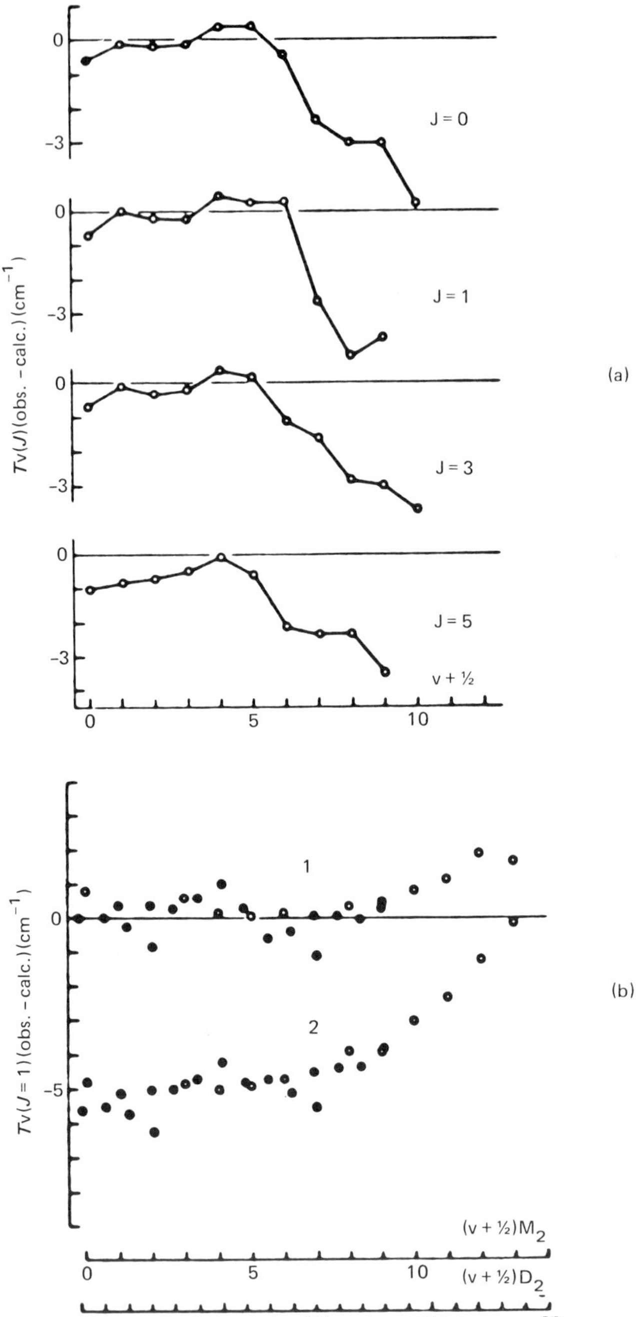

Fig. 7.7. (a) Deviations of the observed levels $T_v(J)$ in the $B^1\,\Sigma_u^+$ state of H_2 from those obtained *ab initio* using quantum defect theory. (b) Deviations of the observed $T_v(J = 1)$ levels of H_2 and D_2 in the $D^1\Pi_u^-$ state from those obtained with quantum defect theory (1) by fitting the potential function to the experimental levels and (2) by using *ab initio* results. (From Jungen *et al.* [11].)

interaction. Since these eigenchannels are properly studied in the Born–Oppenheimer approximation, much as the electronic ground state of H_2, the eigenphase shifts $\delta_\Lambda^{(s)}$ are regarded as functions of the internuclear distance R, just as are the electronic energies. In fact, the phase shifts may be obtained from the Born–Oppenheimer potential energy curves through (5.28). (In a time-dependent view, an excited electron stays within the core too briefly to allow the core to settle in a stationary vibrational state.) The core interaction eigenphase shifts are accordingly labeled $\delta_\Lambda^{(s)}(R)$. One treats R here as the variable conjugate to the vibrational quantum number v just as Λ is conjugate to the rotational quantum number N.

The interconnection between the two sets of channels—short-range and long-range—was represented by the transformation of rotational eigenfunctions (7.36) as long as vibration was being disregarded. This transformation must now be complemented by a transformation that connects the internuclear distance R (a parameter of the short-range interaction) to stationary states of the core vibration; the matrix elements of this transformation are just the vibrational eigenfunctions $\chi_v(R) \equiv (v|R)$ of the H_2^+ core. Extension of our treatment to deal with vibrational excitation consists of substituting

$$U_{N\Lambda} \to U_{N\Lambda}(v|R) \tag{7.51}$$

in the expansion (7.39) of $F_\Lambda(\omega, r)$ and in the following expressions of the Jost matrices. These may now be represented as

$$(Nv|J^\pm(k_{Nv})|\Lambda R) = (\pm i)^l e^{\pm i\delta_{Nv}^{(l)}} U_{N\Lambda}(v|R) e^{\pm i\delta_\Lambda^{(s)}(R)}. \tag{7.52}$$

The application of this procedure to photoionization and to the study of spectral levels is now far more involved, of course. This application has been conducted by Jungen and collaborators, accounting for extensive experimental data in astonishing detail [11]. Figure 7.7 shows samples of these results, cast as differences between observed and calculated levels; notice the expanded scale of ordinates (1 $cm^{-1} \simeq 1.25 \times 10^{-4}$ eV).

7.6 SEMIEMPIRICAL ANALYSIS OF e + Ne^+

As a next example we describe here results from a study of excited states of the Ne atom [12]. The Ne^+ core has the same valence shell configuration $2p^5$ as the F atom, with two fine structure levels $^2P_{1/2}$ and $^2P_{3/2}$ separated by $\bar{E}_{1/2} - \bar{E}_{3/2} = 0.1$ eV, which amounts to $\sim 3\%$ of the binding energy of the lowest Rydberg levels and, thus, is not altogether negligible. Hence, the short-range eigenchannels cannot be safely identified through the character of such lower states as in the previous examples of the H_2 spectrum and the detachment of F^- even though they are likely to be classed appropriately as singlets and triplets. The study of discrete states thus takes the form of semiempirical analysis, based on the presumable physicomathematical similarity between this system and that of H_2 excitations described in Section 7.5.

Two sets of excited states will be considered: (1) with an s electron, $J = 1$, and odd parity, and (2) with a p excited electron, $J = 0$, and even parity. Possible mixing with other channels is disregarded. In both cases, two fragmentation channels occur: (1) ${}^2P_{3/2}$ core + $s_{1/2}$ and ${}^2P_{1/2}$ core + $s_{1/2}$ and (2) ${}^2P_{3/2} + p_{3/2}$ and ${}^2P_{1/2} + p_{1/2}$. The two short-range eigenchannels should be classified approximately as singlet and triplet in either case, but this classification will be verified empirically. This $e + Ne^+$ system differs from $e + H_2^+$ essentially by the doublet split of Ne^+ ($\bar{J}_i = \frac{3}{2}, \frac{1}{2}$) replacing the rotational split ($N = 0, 2$) of H_2^+ and the presumed singlet–triplet ($S = 0, 1$) split of short-range Ne eigenchannels replacing the Σ–Π split in H_2.

We then proceed to an empirical analysis of experimental data on the discrete spectrum assuming the Jost matrix elements to have the structure shown in (7.41a) for H_2,

$$J^{\pm}_{i\alpha}(\kappa_i) = \mp D_i^{\pm 1} U_{i\alpha} \begin{matrix}\cos\\ \sin\end{matrix} [\beta(\kappa_i) + \delta_\alpha^{(s)}], \qquad \epsilon < 0. \tag{7.53}$$

Here the index i is identified by the alternative values of the Ne^+ core quantum number $\bar{J}_i = \frac{3}{2}$ or $\frac{1}{2}$ and will, in fact, be set equal to these values, both for (1) and (2). Imaginary values of $k_i = i\kappa_i$ are given by

$$-\tfrac{1}{2}\kappa_{3/2}^2 + \bar{E}_{3/2} = -\tfrac{1}{2}\kappa_{1/2}^2 + \bar{E}_{1/2} = E \tag{7.54}$$

for each discrete level. These determine the values of the parameter $\beta(\kappa_i) + l\pi = \pi/\kappa_i = \pi\nu_i$ from Table 5.1. The matrix $U_{i\alpha}$ is regarded here as unknown, but we set $\alpha \equiv S = (0, 1)$ for labeling purposes and express $U_{i\alpha}$ as a general 2×2 orthogonal matrix

$$U_{i\alpha} = \begin{array}{c|cc|} i \backslash \alpha & 0 & 1 \\ \hline \frac{3}{2} & \cos\phi & \sin\phi \\ \frac{1}{2} & -\sin\phi & \cos\phi \end{array}, \tag{7.55}$$

thus leaving the angle ϕ to be determined. We note, however, from (7.22) that $\phi = \arctan(1/\sqrt{2}) = 0.6155$ rad corresponds to singlet–triplet short-range channels. The values of $\delta_\alpha^{(s)}$ are also unknown and not readily estimated, though they could be calculated *ab initio* by an extension of the methods of Chapter 6 to be discussed in Chapter 9.

The semiempirical procedure runs now as follows:

(1) Read out from standard tables all the energy levels of configurations $2p^5ns$ with $J = 1$ or $2p^5np$ with $J = 0$, and convert each of them into a *pair* of parameters ($\nu_{1/2} = \kappa_{1/2}^{-1}$, $\nu_{3/2} = \kappa_{3/2}^{-1}$) through (7.54).

(2) Plot each of these levels as a point with coordinates ($\nu_{1/2}, -\nu_{3/2}$) on a graph analogous to Fig. 7.6. The results are shown in Fig. 7.8 for cases (1) and (2).

(3) Draw through the points a curve (or curves) that obeys the analog of (7.48),

$$\cos^2\phi \sin[\pi\nu_{3/2} + \delta_0^{(s)}] \sin[\pi\nu_{1/2} + \delta_1^{(s)}] + \sin^2\phi \sin[\pi\nu_{3/2} + \delta_1^{(s)}] \sin[\pi\nu_{1/2} + \delta_0^{(s)}] = 0, \tag{7.56}$$

adjusting the values of ϕ, $\delta_0^{(s)}$, and $\delta_1^{(s)}$ as necessary. [Recall that $\delta_\alpha^{(s)} = -\pi\nu_{3/2} = -\pi\nu_{1/2} \pmod{\pi}$ at the intercepts of the curve with the diagonal.] The availability of 15–20 points per graph was sufficient to determine accurate values of these three parameters, thus allowing each of them to depend linearly on the energy E. The values of ϕ thus determined turned out to be 0.6150 for $2p^5ns$ and 0.6097 for $2p^5np$, with no significant dependence on E; both of them lie very close to the value 0.6155 expected for $\alpha \equiv S$. The values of $\delta_\alpha^{(s)}$ turned out instead to depend on E appreciably, as evidenced by the fact that all points representing lower levels lie progressively higher and to the left than the others. The values of $\delta_\alpha^{(so)}$ thus obtained, related to the $\delta_\alpha^{(s)}$ by (5.23), are

$$\delta_0^{(so)} = 0.2832 + 0.1372\,\tfrac{1}{2}\kappa_{3/2}^2, \qquad \delta_1^{(so)} = 0.3151 + 0.1348\,\tfrac{1}{2}\kappa_{3/2}^2, \qquad \text{for } 2p^5ns,$$

$$\delta_0^{(so)} = 0.6783 + 0.0375\,\tfrac{1}{2}\kappa_{3/2}^2, \qquad \delta_1^{(so)} = 0.8091 - 0.1237\,\tfrac{1}{2}\kappa_{3/2}^2, \qquad \text{for } 2p^5np. \tag{7.57}$$

The values of $\delta_1^{(so)}$ for np^5ns could be double-checked by fitting the levels of single-channel series of the same configurations with $J = 0$ and $J = 2$, which gave $0.3142 + 0.1369\,\tfrac{1}{2}\kappa_{3/2}^2$ and $0.3149 + 0.1359\,\tfrac{1}{2}\kappa_{3/2}^2$, respectively.

Note that the position of each experimental point on the curves of Fig. 7.8 characterizes the eigenvector $\{A_\alpha\}$ of the homogeneous system with the determinant (7.56), which represents the coupling of the core and particle angular momenta for the particular level. Thus, the fact that the points labeled 3s and 3s′ in Fig. 7.8a lie exactly on the diagonal where $\nu_{1/2} = \nu_{3/2} = -\delta_\alpha^{(s)}/\pi \pmod{1}$ confirms that these levels are indeed LS-coupled with $S = 1$ and 0, respectively; conversely the points 3p and 3p′ lie a little off the diagonal, showing some departure from LS. In the opposite limit of jj coupling, points lying on a horizontal branch of a curve would have uniform values of $\nu_{3/2} \pmod{1}$ corresponding to jj coupling with the core in its ${}^2P_{3/2}$ state; this is indeed the meaning of the conventional unaccented label ns or np. Similarly, points lying on a vertical branch would also be jj-coupled with the core in ${}^2P_{1/2}$, this being the meaning of the accented label. The actual shape of the curves indicates predominance of jj coupling except in the most curved sections near the diagonal. Notice how the 10p point happens to lie on the diagonal reflecting singlet coupling, in spite of its high excitation.

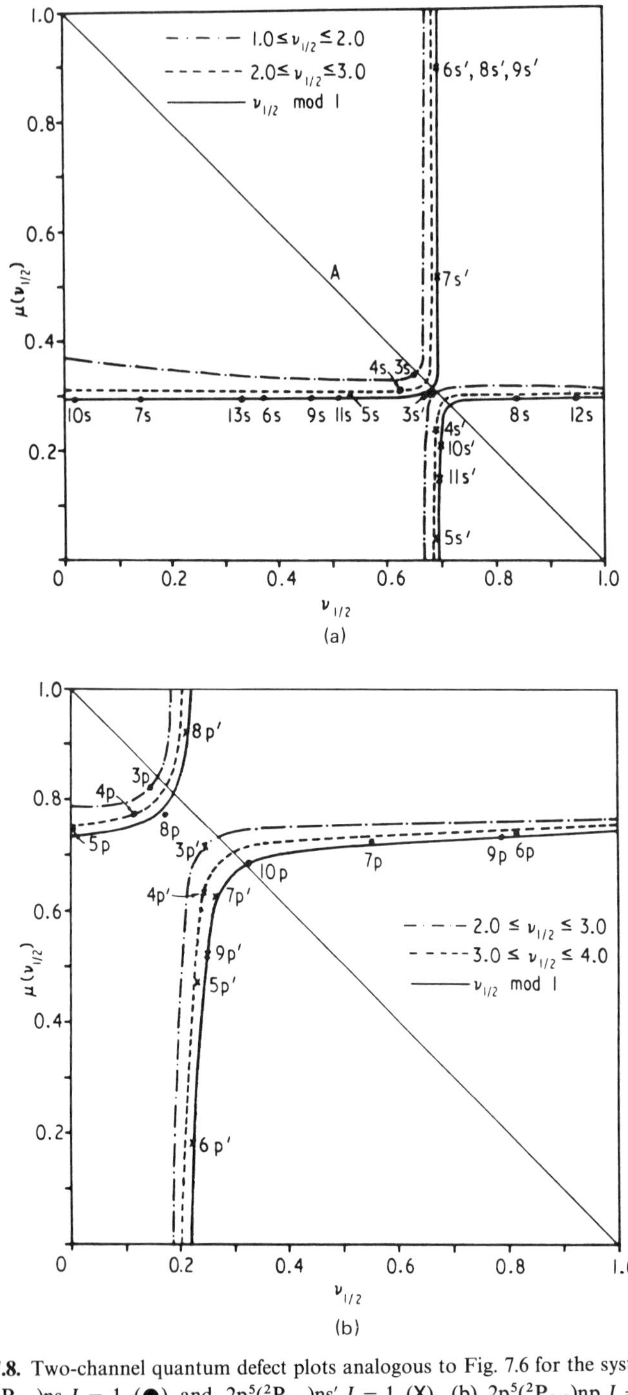

Fig. 7.8. Two-channel quantum defect plots analogous to Fig. 7.6 for the system $e + Ne^+$. (a) $2p^5(^2P_{3/2})ns\ J = 1$ (●) and $2p^5(^2P_{1/2})ns'\ J = 1$ (X). (b) $2p^5(^2P_{3/2})np\ J = 0$ (●) and $2p^5(^2P_{1/2})np'\ J = 0$ (X). (From Starace [12].)

The small difference between the values of $\delta_0^{(so)}$ and $\delta_1^{(so)}$ in (7.57) implies a small value of the exchange energy that separates singlets from triplets; a small value is known to occur frequently for interactions of p and s electrons, thus reflecting cancellations in the exchange integral. This close approach of short-range interaction eigenphases contrasts with the large difference of $\delta_\Sigma^{(s)}$ and $\delta_\Pi^{(s)}$ in Fig. 7.6. Prevalence of *jj* coupling, and more generally of fragmentation channel labeling, is thus favored by near degeneracy of the $\delta_\alpha^{(s)}$'s (see Problem 7.4). Similarly, it could be favored by a different mechanism, namely, near coincidence of alternative channels, that is, by values of ϕ near 0 or $\frac{1}{2}\pi$; in this event the effects of channel coupling would amount to a small perturbation.

Semiempirical analysis by the procedure of this section, extended or adjusted as appropriate, has been applied to a number of discrete spectra, providing considerable new insight on the effects of intermediate coupling and interpretations of previously unexplained anomalies. Further examples will be shown in the next chapter, in a broader setting.

7.7 SPIN POLARIZATION OF PHOTOELECTRONS

Anisotropies in the angular distribution of photoelectrons have been considered in Section 7.5.3 for the example of $h\nu + H_2$. The experimental study of such anisotropies has provided much information on electron interactions with atomic or molecular cores, specifically on the l quantum numbers of electrons emerging from collisions or photoprocesses. The orbital motion is, in fact, generally coupled to spin orientation in the fragmentation channels.

Measurements of the spin orientation of the ejected electron, the residual atomic system, or both, provide information on the contributions to the transition amplitude from specific states $\bar{M}_i, m_{si}$, the magnetic quantum numbers of the residual ion and the photoelectron, respectively. Detailed information on the spin aspects of the collision thus makes it possible to disentangle from experimental data the full set of short-range dynamical parameters, scattering lengths or eigenphase shifts $\delta_\alpha^{(s)}$, and especially the matrix $U_{i\alpha}$, under circumstances more complicated than those that prevail in the examples discussed in this chapter [13].

The photodetachment of F^- in Section 7.4 provides a simple illustration of spin polarization. Here the results follow from angular momentum conservation alone with no reference to other aspects of the dynamics. The amplitude for the absorption of linearly polarized light leading to fragmentation into channel i is given in (7.33). As noted there, the value of M is restricted in this case to 0 by selection rules. If we consider the channel $i = \frac{1}{2}$, that is, the photoelectrons with the energy $\epsilon_{1/2}$ characteristic of leaving behind the F atom in its $^2P_{1/2}$ level, both values of $\bar{M}_i = -m_{si} = \frac{1}{2}$ and $-\frac{1}{2}$ contribute to photodetachment equally, the recoupling coefficient in the Jost matrix in (7.29) being equal to $\sqrt{1/2}$ in both cases. Both of these alternatives, therefore, are present with equal strength, and there is no net preference for the spin orientation of the photoelectrons (or of the F atom).

On the other hand, selection rules constrain M to ± 1, respectively, if we consider photoabsorption of circularly polarized light, where the operator z in the matrix element of (7.33) is replaced by $(\pm x + iy)/\sqrt{2}$, for right- and left-hand polarizations. In this case it follows unambiguously that the only allowed combinations are $\bar{M}_i = m_{si} = \pm\frac{1}{2}$ for the photoelectrons of energy $\epsilon_{1/2}$. Thus, this group of photoelectrons ejected by circularly polarized light is 100% spin polarized, as the Jost matrix in (7.28) vanishes for one set of $(\bar{M}_i, m_{si})$ by conservation of the angular momentum component J_z. Similar consideration shows that the group of photoelectrons leaving F in its $^2P_{3/2}$ level is 50% spin polarized if the incident light is circularly polarized.

The preceding example represents a particularly simple, and trivial, example of spin effects because a single value of J is involved and the Jost functions in (7.28) involve a complete decoupling of dependences on $\bar{M}_i$ and m_{si} from all the dynamical aspects. In a more general situation, when several values of J contribute and the short-range parameters such as a_S in (7.28) also depend on J, the azimuthal dependences are entangled with dynamical aspects of the radial wave function. The study of spin polarizations then provides nontrivial information on dynamics.

A canonical example is afforded by the spin polarization of photoelectrons following ionization of alkali metal vapors by circularly polarized light. Here the spin-orbit coupling leads to slightly different quantum defects for the outgoing $p_{1/2}$ and $p_{3/2}$ electrons (see Section 5.9) and thereby to a relative shift in their radial wave functions. These features give rise to a large electron spin polarization over a narrow range of photoelectron energies [13]. The study of this polarization serves to probe spin-orbit interactions in the continuum, which are less accessible to study by other means because they are generally weak.

7.8 ALTERNATIVE BASE PAIRS IN MULTICHANNEL PROBLEMS

Chapter 5 introduced $\{f^\circ, g^\circ\}$ and $\{f, g\}$, two alternative base pairs of standing wave solutions of the radial Schrödinger equation. The former are characterized by energy-independent normalization at the origin, whereas the latter are energy normalized, thereby having comparable amplitude at large kr. The two pairs are connected through the QDT parameters B and $\mathcal{G}$ as in (5.18).

Effects arising purely from short-range interactions are expected to be insensitive to the energy of the particle at large r and are accordingly viewed suitably in terms of the pair $\{f^\circ, g^\circ\}$. Also, Jost functions are properly defined in terms of this pair expanded at large r in terms of $\{f^+, f^-\}$ as in (5.4). [The expansion should be done in terms of the energy normalized $k_i^{-1/2} f^{\pm}(\epsilon_i)$ for multichannel problems as noted in the discussion of (7.14).]

However, one feature often makes it more convenient to use the pair $\{f, g\}$ in practical calculations, particularly at large k, namely, that f and g are of comparable amplitude for large values of their argument kr, unlike f° and g°, which can be very unequal. Use of the latter to define phase shifts in superpositions such as (5.19) can then prove awkward. In short, both pairs have their own merits and disadvantages. No single choice between them seems appropriate for all circumstances.

For single-channel problems, both pairs can be used with equal facility in defining phase shifts and Chapter 5 has considered the connections between them. Likewise, the definition of Jost functions could also have been done with f in (5.4) and would differ then only by an overall factor $B^{1/2}$. This factor, even though energy dependent (sometimes strongly), would essentially be irrelevant because it drops out of physically significant quantities such as the S matrix that are defined through ratios of Jost functions. The same is no longer true in multichannel problems, in which the energy-dependent parameter B carries a channel index and varies from one fragmentation channel to the other.

The proper definition of Jost matrices must involve $\{f^{\text{o}}, g^{\text{o}}\}$, as in the general expression (7.14) and its specific realization in (7.27–7.28) for zero long-range field. Sections 7.5 and 7.6, nevertheless, followed the common practice of using $\{f, g\}$ in the presence of a long-range Coulomb field and low channel energies, for which the QDT parameters in Table 5.1 reduce to $\mathscr{G} \sim 0$ and $B(\epsilon) \simeq B(0)$.

For larger values of $|\epsilon)$, in which the energy dependence of B and $\mathscr{G}$ is nonnegligible, the Jost functions should be defined with the base pair $\{f^{\text{o}}, g^{\text{o}}\}$. In this event, the Jost matrices differ from the form in Sections 7.5 and 7.6 by energy-dependent factors that are derived explicitly here.

The use of $\{f^{\text{o}}, g^{\text{o}}\}$ in place of $\{f, g\}$ replaces (7.37) with

$$F^{\text{o}}_{\Lambda}(\omega, r) = [f^{\text{o}}(\epsilon, l=1, r) - g^{\text{o}}(\epsilon, l=1, r) \tan \delta^{(\text{so})}] X^{\Lambda}_{M}(r, R)\Phi(\omega), \tag{7.58}$$

where both $\{f^{\text{o}}, g^{\text{o}}\}$ and the phase $\delta^{(\text{so})}$ are smooth in energy as discussed in Section 5.5. Drawing now on Section 5.5 for the connection between the alternatives (5.19) and (5.20), we find that

$$F_{\Lambda}(\omega, r) = [\mathscr{N}(\epsilon) \cos \delta^{(\text{so})}] F^{\text{o}}_{\Lambda}(\omega, r). \tag{7.59}$$

Their replacement in (7.38) and (7.39), with ϵ in place of ϵ_N, yields instead of (7.40)

$$\begin{aligned} J^{\text{o}\pm}_{i\alpha} \equiv J^{\text{o}\pm}_{N\Lambda}(k_N) &= [\mathscr{N}(\epsilon_N) \cos \delta^{(\text{so})}]^{-1} (\pm i)^l e^{\pm i\delta^{(1)}_N(k_N)} U_{N\Lambda} e^{\pm i\delta^{(\text{s})}_{\Lambda}} \\ &= [\mathscr{N}(\epsilon_N) \cos \delta^{(\text{so})}]^{-1} J^{\pm}_{N\Lambda}(k_N), \qquad \epsilon_N > 0, \end{aligned} \tag{7.40'}$$

where $J^{\pm}_{N\Lambda}(k_N)$ are the matrices in (7.40). The normalization factor in (7.40′) is obtained from (5.22′)

$$[\mathscr{N}(\epsilon_N) \cos \delta^{(\text{so})}]^{-1} = B^{-1/2}(k_N)[\{1 + \mathscr{G}(k_N) \tan \delta^{(\text{so})}\}^2 + \{B(k_N) \tan \delta^{(\text{so})}\}^2]^{1/2}. \tag{7.60}$$

The energy-dependent factor in (7.60), which should multiply the Jost functions in (7.40), (7.41), (7.52), and (7.53), remains constant in the presence of a long-range attractive Coulomb field as long as $\mathscr{G}$ and B and $\delta^{(\text{so})}$ are independent of energy. On the other hand, for zero long-range field as in Section 7.4, Table 5.1 gives $\mathscr{G} = 0$ and $B(k, l=0) = (2k/\pi)$. With $\cos \delta^{(\text{so})} = 1$, $\delta^{(l)} = 0$, and $\exp(i\delta^{(\text{s})}) = (1 - ika_S)$ from (7.23), (7.40′) reduces properly to the result (7.28) in Section 7.4.

7.9 CROSS SECTIONS IN THE LABORATORY FRAME

Collision experiments usually deal with beams of incident particles or light that are monoenergetic and monodirectional, that is, in momentum eigenstates. Collision products are similarly detected in eigenstates of momentum (that is, of wave vectors). In this chapter, on the other hand, scattering matrices have been constructed for the sake of simplicity with reference to eigenstates of angular momentum, that is, with reference to transitions between partial waves. These alternative formulations nearly coincide in the examples of this chapter. General applications require two additional steps, namely, the superposition of alternative J values and the conversion of matrices $S_{ll'}$ for transitions between partial waves to a base of wavevectors $S_{\mathbf{kk}'}$ utilizing the transformation matrices $(lm|\hat{k})$ of (4.16).

The transformation from partial waves to collision momenta proceeds along the line followed in Sections 4.2 and 4.3, namely, by considering initially an uncoupled state of projectile and target approaching each other with momentum $\hbar k_i$ along the z axis. Indicating this state by $|\bar{E}_i \bar{J}_i \bar{M}_i, m_{si}\epsilon_i \mathbf{k}_i)$ in this chapter's notation, its expansion corresponding to (4.2) takes the form

$$(1, 2, \ldots, N|\bar{E}_i \bar{J}_i \bar{M}_i)\chi_{m_{si}} e^{ik_i z} \xrightarrow[k_i r \to \infty]{}$$

$$(1, 2, \ldots, N|\bar{E}_i \bar{J}_i \bar{M}_i)\chi_{m_{si}} \sum_{l_i=0}^{\infty} \frac{1}{2ik_i r} [4\pi(2l_i + 1)]^{1/2} Y_{l_i 0}(\theta, \phi)[-(-1)^{l_i} e^{-ik_i r_i} + e^{ik_i r_i}]. \tag{7.61}$$

Here the quantum number m_{l_i} is restricted to zero by our choice of $\mathbf{k}_i$ parallel to $\hat{z}$; $\chi_{m_{si}}$ represents the normalized spin state of the projectile if it is a single particle or its complete state if the projectile has internal structure.

Recasting the right-hand side of (7.61) in the mold of the incident channel wave function in (7.15+) requires the following adaptations:

(1) Replacement of the asymptotic factors $\exp(\pm ik_i r)$ by the corresponding radial functions

$$e^{\pm ik_i r} \to f_i^{\pm}(r). \tag{7.62a}$$

(2) Replacement of the product of separate factors $Y_{l_i 0}$, $\chi_{m_{si}}$ and $(1, 2, \ldots, N|\bar{E}_i \bar{J}_i \bar{M}_i)$ by its expansion into the functions Φ_{iJM} that represent eigenstates of $\mathbf{J} = \bar{\mathbf{J}}_i + \mathbf{j}_i$ and of $\mathbf{j}_i = \boldsymbol{l}_i + \mathbf{s}_i$,

$$(1, 2, \ldots, N|\bar{E}_i \bar{J}_i \bar{M}_i)\chi_{m_{si}} Y_{l_i 0}(\theta, \phi) \to \sum_J \sum_{j_i} \Phi_{iJM}(JM|j_i \bar{J}_i m_{si} \bar{M}_i)(j_i m_{si}|s_i l_i m_{si} 0). \tag{7.62b}$$

Recall that the detailed description of the functions $\Phi_i(\omega)$—or rather X_{SLJM}—in Eqs. (7.26ff.) specified them to include antisymmetrization of all electron coordinates (incident and target), which was not intended on the left-hand side of (7.62b). We understand here the symbol → in (7.62b) to add any appropriate symmetry operation.

Our objective is now to construct the superposition of scattering wave functions (7.15+) whose incident wave part coincides with (7.61) at $k_i r \to \infty$. The treatment will be extended to reactive collisions by inserting a reduced mass factor μ_i (in atomic units) in the coefficient of each channel, in accordance with Landau and Lifshitz [5]. Comparison with (7.62a, b) shows the desired wave function to be

$$\frac{\pi}{r}\sum_i\sum_{\alpha JM} \mathrm{F}_{\alpha JM}(\omega, r)(J^{--1})_{\alpha i}\left[\frac{2(2l_i+1)}{\mu_i k_i}\right]^{1/2} i^{l_i}(JM|j_i\bar{J}_i m_{si}\bar{M}_i)(s_i l_i m_{si}0|j_i m_{si})$$

$$= \sum_{l_i j_i JM}\frac{1}{2ir}\left\{\Phi_{iJM}\frac{-(-1)^{l_i}f_i^-(r)+f^+(r)}{k_i}\right.$$

$$\left.+\sum_j \Phi_{jJM}\frac{f_j^+(r)}{k_j}(S_{ji}-\delta_{ji})\left(\frac{\mu_j k_j}{\mu_i k_i}\right)^{1/2} i^{l_i}(-i)^{l_j}\right\}$$

$$\times [4\pi(2l_i+1)]^{1/2}(JM|j_i\bar{J}_i m_{si}\bar{M}_i)(j_i m_{si}|s_i l_i m_{si}0). \tag{7.63}$$

Note that the summation Σ_j in (7.63) actually runs over all indexes so labeled (i.e., $\bar{J}_j$, $\bar{M}_j$, j_j, l_j, m_{sj}, m_{l_j}). Furthermore, Φ_j in (7.63) represents the coupled state

$$\sum_{\bar{M}_j m_{lj} m_{sj}} Y_{l_j m_{l_j}}(\hat{r})\chi_{m_{sj}}(1, 2, \ldots, N|\bar{J}_j\bar{M}_j)(s_j l_j m_{sj} m_{lj}|j_j m_j)(j_j\bar{J}_j m_j\bar{M}_j|JM). \tag{7.64}$$

The various magnetic quantum numbers are constrained by the usual requirements of addition of angular momenta, particularly $M = \bar{M}_i + m_{si} = \bar{M}_j + m_{lj} + m_{sj}$.

The two expressions in the braces of (7.63) replace, respectively, the two terms in the single-channel wave function (4.6+). Just as the elastic collision cross section in (4.17) includes the squared modulus of the coefficient of (e^{ikr}/r) in (4.6+), the differential cross section can now be read off from the coefficient of (f_j^+/r) in (7.63). Recalling that a cross section is a ratio of fluxes, we also need now to include kinematical factors because $e^{ik_i z}$ and $(e^{ik_j r}/r)$ represent fluxes of different magnitude, (k_i/μ_i) and (k_j/μ_j), respectively. Taking the spin-wave functions $\chi_{m_{si}}$ and atomic functions $(1, 2, \ldots, N|\bar{J}_i\bar{M}_i)$ to be normalized to unity, we thus obtain the cross-section formula

$$\frac{d\sigma}{d\Omega}(\bar{J}_i\bar{M}_i k_i m_{si}\to\bar{J}_j\bar{M}_j m_{sj}) = \frac{k_j/\mu_j}{k_i/\mu_i}\left|\sum_{l_i j_i J j_j l_j}\left(\frac{\mu_j}{\mu_i}\right)^{1/2}\left[\frac{\pi(2l_i+1)}{k_i k_j}\right]^{1/2} i^{l_i}(-i)^{l_j}\right.$$

$$\times (JM|j_j\bar{J}_j m_j\bar{M}_j)(j_j m_j|s_j l_j m_{sj} m_{lj})(S_{ji}-\delta_{ji})$$

$$\left.\times (s_i l_i m_{si}0|j_i m_{si})(j_i\bar{J}_i m_{si}\bar{M}_i|JM)Y_{l_j m_{lj}}(\hat{r})\right|^2. \tag{7.65}$$

This expression represents the cross section for a "complete" experiment that measures all spin projections, initial and final, of the fragments. The quantity m_{lj} takes the single value $\bar{M}_i + m_{si} - \bar{M}_j - m_{sj}$. The five indices summed over in the amplitude in (7.65) take values ranging as follows: $0 \le l_i \le \infty$, j_i as determined by the angular momentum addition $\boldsymbol{l}_i + \mathbf{s}_i = \mathbf{j}_i$ and J by the addition $\bar{\mathbf{J}}_i + \mathbf{j}_i = \mathbf{J}$, j_j as restricted by $\mathbf{J} = \mathbf{j}_j + \bar{\mathbf{J}}_j$, and l_j as restricted by $\mathbf{j}_j = \mathbf{l}_j + \mathbf{s}_j$. The cross section for scattering without any spin measurement on the outgoing particles and with unpolarized incident beams is obtained by summing (7.65) over $\bar{M}_j$ and m_{sj}, and averaging over $\bar{M}_i$ and m_{si}:

$$\frac{d\sigma}{d\Omega}(\bar{J}_i \to \bar{J}_j) = \frac{1}{2(2\bar{J}_i + 1)} \sum_{\bar{M}_i m_{si} \bar{M}_j m_{sj}} \frac{d\sigma}{d\Omega}(\bar{J}_i \bar{M}_i m_{si} \to \bar{J}_j \bar{M}_j m_{sj}). \tag{7.66}$$

Note how the general cross-section formula in (7.65) reduces to the result in Section 7.4.5 for $e + \mathrm{F}$ scattering at low energy. By setting $l_i = l_j = 0, j_i = j_j = \frac{1}{2}$, and $J = 1$, only one term remains in the summation. The various angular momentum coupling coefficients reduce to unity, the kinematical factors (with $\mu_j = \mu_i$) give $(1/k_i^2)$, the factors $\pi |Y_{00}(\hat{r})|^2$ give the $\frac{1}{4}$ in (7.32). As a result, (7.65) reduces to (7.32).

The construction of photoionization cross sections from the fragmentation channel wave function in (7.15−) is carried out in an analogous manner. A superposition of the functions in (7.15−) with alternative J and M is required to represent the ejection of the photoelectron in terms of a plane wave in the i-channel, with $|\bar{E}_i \bar{J}_i \bar{M}_i, m_{si} \epsilon_i \mathbf{k}_i)$ as given in (7.61), together with purely ingoing waves $f^-(\epsilon_j, l_j, r)$ in all channels $j \neq i$. Since the quantization axis is customarily taken along the photon polarization, the momentum $\mathbf{k}_i$ of the ejected electron is unspecified. In place of (7.61), therefore, we write the alternative form drawn from (4.1),

$$(1, 2, \ldots, N | \bar{E}_i \bar{J}_i \bar{M}_i) \chi_{m_{si}} e^{i\mathbf{k}_i \cdot \mathbf{r}} \xrightarrow[k_i r \to \infty]{} (1, 2, \ldots, N | \bar{E}_i \bar{J}_i \bar{M}_i) \chi_{m_{si}}$$

$$\times \sum_{l_i = 0}^{\infty} \sum_{m_i = -l_i}^{l_i} \frac{1}{2ik_i r} 4\pi Y^*_{l_i m_i}(\theta_k, \phi_k) Y_{l_i m_i}(\theta, \phi)[-(-1)^{l_i} e^{-ik_i r} + e^{ik_i r}]. \tag{7.67}$$

To match the coefficient of f_j^+ on the right-hand side of (7.15−) with that in (7.67) requires a superposition of α channels according to

$$\frac{4\pi}{r} \sum_{l_i m_i j_i m_{ji} J M} \left(\frac{\pi}{2\mu_i k_i}\right)^{1/2} i^{l_i} Y^*_{l_i m_i}(\theta_k, \phi_k)(j_i m_{ji} | \tfrac{1}{2} l_i m_{si} m_i)(JM | j_i \bar{J}_i m_{ji} \bar{M}_i) \sum_{\alpha} F_\alpha (J^{+-1})_{\alpha i}. \tag{7.68}$$

The wave function in (7.68) is normalized to a plane wave with unit amplitude. Its use in the photoabsorption matrix element would give the cross-section differential in $d\mathbf{k}_i/(2\pi)^3$ according to (4.8). By writing $k_i^2 \, dk_i \, d\Omega$ as $(\mu_i k_i) \, d\epsilon_i \, d\Omega$, we obtain the wave function to be entered in (2.9),

$$(1, 2, \ldots, N + 1 | \epsilon_i) = \frac{1}{r} \sum_{l_i m_i j_i m_{ji} J M} i^{l_i} Y^*_{l_i m_i}(\theta, \phi)$$

$$\times (j_i m_{ji} | \tfrac{1}{2} l_i m_{si} m_i)(JM | j_i \bar{J}_i m_{ji} \bar{M}_i) \sum_{\alpha} F_\alpha(\omega, r)(J^{+-1})_{\alpha i}. \tag{7.69}$$

Equation (2.9) gives then the desired photoabsorption cross section for ejection of photoelectrons with momentum $\mathbf{k}_i$ in a solid angle $d\Omega$.

7.10 ANGULAR DISTRIBUTION OF COLLISION FRAGMENTS

The treatment of the direction of photoemission in Section 7.5.3 provides a simple example of the theory of the distribution of electrons or other fragments emerging from a collision. Theory relates the detection of fragments in particular directions to parameters of the collision dynamics. It then serves conspicuously in the twin capacities of predicting the detection of fragments from assumed knowledge of the dynamics and of determining the dynamics from observation of the fragment distribution.

A basic feature of theory lies here in sorting out dynamical from geometrical elements. Dynamics is, of course, independent of the orientation of collimators and detectors and is accordingly represented by scalar parameters such as the coefficient β in (7.46). The geometrical aspects depend instead on such orientations and relate to the dynamics only through the values of angular momentum quantum numbers, which may be viewed as Fourier transforms of angular coordinates. Angular momentum algebra, the main tool of geometrical analysis, is well developed, standardized, and very flexible, but effective utilization of its flexibility requires some skill.

We outline here two elements of the theory of angular distributions: (1) the general procedure for separating dynamical from geometrical elements, and (2) a special example, namely, the geometrical analysis in terms of angular momentum transfer between observed and unobserved collision variables.

7.10.1 Disentangling Dynamical Parameters

The dipole matrix element $(\epsilon lm|z|0)$ in the photoabsorption equation [(7.44)] combines parameters of the incident radiation and the target molecule asymmetrically, the former being treated as an operator z, the latter as row and column indexes of the matrix. A symmetric treatment of the reactants labels the columns of a transition matrix by angular momentum quantum numbers of both the incident projectile (p) and the target (T), and its rows by corresponding indexes of the ejected fragment (f) and the target residue (R). The dipole operator z may then be indicated in the scalar form $\mathbf{r}\cdot\hat{e}_{\mathrm{p}}$, where $\mathbf{r}$ represents the position of a target electron and $\hat{e}_{\mathrm{p}}$ the light polarization,

$$(\epsilon lm|z|0) \to (j_{\mathrm{R}}m_{\mathrm{R}}, \epsilon j_{\mathrm{f}}m_{\mathrm{f}}|\mathbf{r}\cdot\hat{\epsilon}_{\mathrm{p}}|j_{\mathrm{T}}m_{\mathrm{T}}, j_{\mathrm{p}}m_{\mathrm{p}}). \tag{7.70}$$

In a general process the transition operator would be indicated by a scalar symbol T rather than by its form $\mathbf{r}\cdot\hat{\epsilon}$ specific to dipole interaction with light.

The main consideration now stems from the rotational invariance of the Hamiltonian of the whole system and its resulting transition operator T. This invariance is reflected in the matrix of T, specifically of $\mathbf{r}\cdot\hat{\epsilon}_{\mathrm{p}}$, being *diagonal* in the total angular momentum number J and *independent* of the projection M of the vector $\mathbf{J}$ on an arbitrarily oriented axis z. These properties are manifested by separately adding the angular momenta of the initial and those of the final state reactants to yield the total momenta of the combined complex. Each matrix element (7.70) is thus expanded into diagonal elements with vector addition (Wigner) coefficients,

$$(j_R m_R, \epsilon j_f m_f | \mathbf{r} \cdot \hat{\epsilon}_p | j_T m_T, j_p m_p) = \sum_{JM} (j_R m_R, j_f m_f | JM) \times [(j_R, \epsilon j_f) J | \mathbf{r} \cdot \hat{\epsilon}_p | (j_T j_p) J] (JM | j_T m_T, j_p m_p). \quad (7.71)$$

Each of the matrix elements diagonal in J is thus independent of the choice of coordinate axes and hence of collimators or detectors in an experiment. It is thus classed as a purely "dynamical parameter." Geometrical elements consist instead of the combination of relevant Wigner coefficients and angular wave functions. The standard passage from plane waves representing the incident beam to spherical waves of the partial wave expansion (Section 4.2) constitutes the first step in disentangling geometry from dynamics. As seen in Sections 4.5 and 7.9, the two aspects are combined through the relevant Wigner and $(lm|\hat{k})$ coefficients.

In the example of Section 7.5.3 each matrix element equivalent to (7.70) was multiplied by the photoelectron's orbital wave function $Y_{l_f m_f}(\theta, \phi)$, summed over alternative values of m_f and squared to obtain the probability of ejection in the direction (θ, ϕ). The squared matrix element would then generally be averaged over the alternative values of the unobserved orientation quantum numbers (m_T, m_R) of target and residue, setting $m_p = 0$ for the incident light if the z axis coincides with the light polarization $\hat{e}_p$. These manipulations involve only angular momentum algebra analogous to its applications in Section 7.4; they are generally performed analytically, yielding expansions into spherical functions with standard coefficients.

There remains the summation over alternative values of the quantum number J in (7.71), which, on squaring, takes the form of a *quadratic* expression in the dynamical parameters $[(j_R, \epsilon j_f)J|T|(j_T j_p)J]$. This expression represents the angular distribution of fragments as the average of a number of standard expressions $F_{JJ'}(\theta, \phi)$, each of them labeled by a pair of values of the total J and weighted by the product of the dynamical parameters with that pair of values,

$$P(\theta, \phi) = \sum_{JJ'} F_{JJ'}(\theta, \phi)[(j_R, \epsilon j_f)J|T|(j_T j_p)J'][(j_R, \epsilon j_f)J|T|(j_T j_p)J']. \quad (7.72)$$

This is the general form of angular distribution of collision fragments.

The initial example in Section 7.5.3 was simplified greatly by the specification of zero angular momenta of both the target and the residual ion following photoemission, $j_T = j_R = 0$. It follows that $m_T = m_R = 0$ and $J = j_p = j_f$, whereby the angular distribution parameter β reaches its maximum value of 2. The second example dealt with the alternative value $j_R = 2$, whereby the averaging over m_R reduces the value of β to $\frac{1}{5}$ [Eq. (7.47)].

On the other hand, much more elaborate circumstances may occur, in which the target is prepared initially with nonrandom distribution of its quantum number m_T. Averaging over this quantum number can nevertheless be performed analytically by angular momentum algebra. Similarly, we can predict the outcome of experiments that bear on the state of the collision residue after ejection of a fragment.

7.10.2 Angular Momentum Transfer to the Ionized Fragment

The averaging process in (7.72), bearing on standard distributions $F_{JJ'}(\theta, \phi)$ weighted by a two-index quadratic expression, obscures the contribution of its individual terms $F_{JJ'}$. The structure of the angular distribution formula can be made more transparent by reducing it to single-index-averaging when no information is available on the orientation of the target or of its residue, that is, when both indices (m_T, m_R) contribute with equal weight to the sum over the product of (7.71) and its Hermitian conjugate.

To this end we transform the pair of Wigner coefficients in (7.71) by a "recoupling" procedure analogous to that which changed the *jj*-coupled representation (7.19) of F^- into the *LS*-coupled representation (7.19′). In our case the pair of Wigner coefficients in (7.71) represents the additions of angular momenta

$$\mathbf{J} = \mathbf{j}_T + \mathbf{j}_p, \qquad \mathbf{J} = \mathbf{j}_R + \mathbf{j}_f, \tag{7.73}$$

where $\mathbf{J}$ indicates the total angular momentum of the collision complex. Alternative combinations of the same angular momenta represent instead the angular momentum $\mathbf{J}_t$ *transferred* from the observed to the unobserved reactants,

$$\mathbf{J}_t = \mathbf{j}_p - \mathbf{j}_f, \qquad \mathbf{J}_t = \mathbf{j}_R - \mathbf{j}_T. \tag{7.74}$$

The pair of Wigner coefficients in (7.71), corresponding to the additions in (7.73), is expanded into alternative pairs corresponding to the additions in (7.74) by a standard formula [7, Eqs. (1.1) and (2.19)].

$$\sum_M (j_R m_R, j_f m_f | JM)(JM | j_T m_T, j_p m_p)$$

$$= (2J+1) \sum_{J_t M_t} (-1)^{J+J_t+m_T-m_f} \begin{Bmatrix} j_p j_T J \\ j_R j_f J_t \end{Bmatrix} (j_R m_R, j_T - m_T | J_t M_t)(J_t M_t | j_f - m_f, j_p m_p). \tag{7.75}$$

Applying this transformation to the expansion (7.71) of the transition matrix elements yields an alternative expansion, in terms of a different set of dynamical parameters,

$$(j_R m_R, \epsilon j_f m_f | T | j_T m_T, j_p m_p)$$

$$= \sum_{J_t M_t} (-1)^{J_t+m_T-m_f} (j_R m_R, j_T - m_T | J_t M_t)[(j_R j_T)J_t | T | (\epsilon j_f j_p)J_t](J_t M_t | j_f - m_f, j_p m_p), \tag{7.76}$$

where

$$[(j_R j_T)J_t | T | (\epsilon j_f j_p)J_t] = \sum_J (-1)^J (2J+1) \begin{Bmatrix} j_p j_T J \\ j_R j_f J_t \end{Bmatrix} [(j_R, \epsilon j_f)J | T | (j_T j_p)J], \tag{7.76′}$$

and the operator $\mathbf{r} \cdot \hat{e}_p$ in (7.71) has been replaced by the general symbol T. The simplification achieved by this alternative expansion becomes apparent on calculating the squared modulus of the matrix element (7.76) averaged over (m_T, m_R) to obtain the probability of fragment ejection in any given direction. The orthonormality of the Wigner coefficients $(j_R m_R, j_T - m_T | J_t M_t)$ causes cross terms with different values of $J_t M_t$ to vanish in the sum over squares of (7.76). Accordingly, the angular distribution is weighted only by the set of squared dynamical parameters $|[(j_R j_T)J_t | T | (\epsilon j_f j_p)J_t]|^2$.

Further essential steps remain to complete the construction of the cross section for ejection of a fragment in a direction (θ, ϕ). The first of them consists of multiplying the transition matrix element (7.76) by the fragment's wave function $(\theta, \phi | \epsilon j_f m_f)$. In (7.44) this wave function is indicated simply by $Y_{lm}(\theta, \phi)$, but in general it may involve a dependence on the fragment's own spin. In that case additional information on spin and orbital correlation may become accessible, which is more detailed than that contained in the angular distribution [13]. Similarly, Section 7.5.3 represented the state of the incident light simply by its polarization direction, but the general state of the incident projectile will be represented here as a superposition $\Sigma_{m_p} | j_p m_p) a_{m_p}$. Accordingly, the complete transition amplitude for detection of fragments in the direction (θ, ϕ) is represented by complementing the matrix element (7.76) in the form

$$\begin{aligned}
&\sum_{m_f m_p} (\theta\phi | \epsilon j_f m_f)(j_R m_R, \epsilon j_f m_f | T | j_T m_T, j_p m_p) a_{m_p} \\
&\quad = \sum_{J_t} (-1)^{J_t} [(j_R j_T) J_t | T | (\epsilon j_f j_p) J_t) \sum_{M_t} (-1)^{m_T} (j_R m_R, j_T - m_T | J_t M_t) \\
&\quad \times \sum_{m_f m_p} (-1)^{m_f} (\theta\phi | \epsilon j_f m_f)(J_t M_t | j_f - m_f, j_p m_p) a_{m_p}.
\end{aligned} \tag{7.77}$$

On squaring and summing over the unobserved variables (m_R, m_T), as noted earlier,

$$\begin{aligned}
&\sum_{m_R m_T} \left| \sum_{m_f m_p} (\theta\phi | \epsilon j_f m_f)(j_R m_R, \epsilon j_f m_f | T | j_T m_T, j_p m_p) a_{m_p} \right|^2 \\
&\quad = \sum_{J_t} |(j_R j_T) J_t | T | (\epsilon j_f j_p) J_t|^2 \sum_{M_t} \left| \sum_{m_f m_p} (-1)^{m_f} (\theta\phi | \epsilon j_f m_f)(J_t M_t | j_f - m_f, j_p m_p) a_{m_p} \right|^2 .
\end{aligned} \tag{7.78}$$

The last factor of (7.78), $\Sigma_{M_t} |\Box|^2$, consists of geometrical elements only; it thus corresponds to the factor of (7.72) indicated by the generic symbol $F_{JJ'}(\theta, \phi)$, but is to be averaged only over the single index J_t rather than over the pair JJ'. The weight of each value of J_t is given by a dynamical factor, the squared matrix element of T. The occurrence of a *single* sum over J_t is an important distinguishing element of the analysis in terms of the angular momentum transfer. Unlike the treatment in terms of J, amplitudes with different J_t do not interfere.

The geometrical factor, which is now separated in the second factor of (7.78), may be transformed once again to articulate its dependence on the separate factors, i.e., motions of the incident projectile and the ejected fragment. Expansion of the squared polynomial in this factor yields the partial sum

$$\sum_{M_t} (j'_f - m'_f, j'_p m'_p | J_t M_t)(J_t M_t | j_f - m_f, j_p m_p),$$

where we inserted primes to distinguish indexes to be treated as independent. This expression is now recast by the identity (7.75) into

$$(2J_t + 1)\sum_{kq}(-1)^{J_t+k-m_f+m'_p}\begin{Bmatrix} j_p j_f J_t \\ j'_f j'_p k \end{Bmatrix}(j'_f - m'_f, j_f m_f | kq)(kq | j'_p - m'_p, j_p m_p).$$

The new index k in this expression represents formally the vector difference of angular momenta of a wave function and its complex conjugate,

$$\mathbf{k} = \mathbf{j}'_f - \mathbf{j}_f = \mathbf{j}_p - \mathbf{j}'_p, \tag{7.79}$$

but it is more properly described as the degree of 2^k-pole components of the probability distributions of projectile and fragment. This final transformation thus recasts the geometrical factor of (7.78) into a sum of scalar products of the multipole expansions of particle densities,

$$\sum_{M_t}\left|\sum_{m_f m_p}(-1)^{m_f}(\theta\phi|\epsilon j_f m_f)(J_t M_t | j_f - m_f, j_p m_p) a_{m_p}\right|^2$$

$$= (-1)^{J_t}(2J_t + 1)\sum_{kq}(-1)^k\left[\sum_{m_f m'_f}(\theta\phi|\epsilon j_f m_f)(\theta\phi|\epsilon j'_f m'_f)^*(-1)^{m'_f}(j'_f - m'_f, j_f m_f | kq)\right]$$

$$\times \begin{Bmatrix} j_p j_f J_t \\ j'_f j'_p k \end{Bmatrix}\left[\sum_{m_p m'_p} a_{m_p} a_{m'_p}{}^*(-1)^{m_p}(j_p - m'_p, j_p m_p | kq)\right]. \tag{7.80}$$

When the wave functions $(\theta\phi|\epsilon j_f m_f)$ are spherical harmonics, the first factor on the right of (7.80) is automatically represented in terms of single harmonics $Y_{kq}(\theta, \phi)$.

The preceding discussion for a general process of a projectile plus target leading to a residue and an ejected fragment may be slightly modified for a process such as photoejection of electrons. Here the photon in the electric dipole approximation couples only to the orbital angular momentum of the electron, so that only the orbital part of $\mathbf{j}_f$ is considered in defining the angular momentum transfer. Thus, with $\mathbf{j}_f = \boldsymbol{l}_f + \mathbf{s}_f$, we replace (7.74) by

$$\mathbf{J}_t = \mathbf{j}_p - \boldsymbol{l}_f, \qquad \mathbf{J}_t = \mathbf{j}_R + \mathbf{s}_f - \mathbf{j}_T = \mathbf{j}'_R - \mathbf{j}_T, \tag{7.81}$$

where $\mathbf{j}'_R$ is an intermediate combination of the spin of the fragment with total angular momentum of the residue. All the development and formulas subsequent to (7.74) remain valid but for the replacement of j_R and its projection m_R by the corresponding primed quantities. As an illustration, photoionization of a noble gas atom such as Ne or Xe has $j_R = \frac{3}{2}$ and $\frac{1}{2}$, $l_f = 0$ and 2, and $J_t = 1, 2$.

7.10.3 Parity Transfer in the Collision

The analysis of angular distributions in terms of angular momentum transfer is supplemented by considering the attendant parity changes in the reaction. Treating once again the process of photoabsorption in the electric dipole approximation, it is clear from (7.81) that l_f is restricted to the values J_t and $J_t \pm 1$. Overall conservation of parity can always be ensured given the parities of the target and the residue but by focusing only on the incident photon and the photoelectron that receives the angular momentum J_t, we can classify the reactions into two categories. Processes with $l_f = J_t \pm 1$ yield simultaneous transfers of angular momentum and parity from the photon to the photoelectron and are accordingly labeled "parity favored." By contrast transitions with $l_f = J_t$ are "parity unfavored." In the illustration at the end of the last paragraph, transitions with $l_f = 2$ receive both parity-favored and parity-unfavored contributions, depending on whether $J_t = 1$ or 2.

The classification into parity-favored and parity-unfavored processes takes on particular significance because of a general result that the asymmetry parameter $\beta(J_t)$ of the angular distribution is uniquely -1 for the unfavored transitions, yielding an angular distribution proportional to $\sin^2\theta$. This distribution is, of course, opposite to the $\cos^2\theta$ (i.e., $\beta = 2$) distribution expected from a classical model of electric dipole interactions. For parity-favored transitions, general expressions for $\beta(J_t)$ have been developed in terms of two basic complex amplitudes, $[(j_R' j_T)J_t|\mathbf{r}\cdot\hat{e}_p|(\epsilon l_f j_p)J_t]$ with $l_f = J_t \pm 1$ [14]. For such transitions the value of β ranges, in general, over its entire interval $-1 \leq \beta \leq 2$, but it frequently remains positive. Thereby the considerations of angular momentum transfer and parity favoredness give together a compact accounting of the angular distribution in terms of a few basic dynamical parameters.

The occurrence of parity-unfavored elements in a process such as photoabsorption fits naturally into our approach, which separates the initial stage of absorption of the photon in the α channels from the subsequent propagation of the electron into the i channels. The initial stage of direct interaction between the photon and a particular electron is necessarily parity favored. If $\boldsymbol{l}_0$ denotes the initial orbital angular momentum of this electron and $\boldsymbol{l}_\alpha$ its momentum immediately after absorption, we have $\boldsymbol{l}_\alpha = \mathbf{j}_p + \boldsymbol{l}_0$. The angular momentum transfer at this stage takes a single value, $\mathbf{j}_p - \boldsymbol{l}_\alpha = -\boldsymbol{l}_0$. Other values of $J_t \neq \boldsymbol{l}_0$ realized at infinity can be viewed as arising from torques exerted on the outgoing electron by anisotropic interactions, whether electron–electron or spin–orbit, during the escape.

Problems

7.1 Consider the system e + N at 5 eV above the ground state of the nitrogen atom. Using standard tables of atomic energy levels, list all the open fragmentation channels with $J = 0$, classifying them according to their quantum numbers and parities. Also list the alternative set of short-range eigenchannels of the N^- complex.

7.2 Consider the photodetachment of the negative ion S^- which has two fine structure states $p^5\,{}^2P^o_{1/2,3/2}$. At low detachment energies, the outgoing s wave is dominant and the neutral atom is left in its lowest states $p^4\,{}^3P_{2,1,0}$. Determine

(a) the alternative values of the total angular momentum J of the (e + S) complex;

(b) the quantum numbers characterizing the eigenchannels at small r. Note that the fragmentation channels are indexed by $\bar{J}_i = 2, 1, 0$;

(c) the transformation matrices $U^{(J)}_{i\alpha}$ between fragmentation and eigenchannels for each value of J.

7.3 Analyze the photodetachment of the negative ion of thallium, Tl^-, as in Problem

7.2. Note that the two problems are complementary with regard to the quantum numbers of the neutral atom and negative ion.

7.4 Consider the scattering of electrons from the F atom in its ground state $\bar{J} = \frac{3}{2}$ for energies insufficient to excite the atom to $\bar{J} = \frac{1}{2}$, such that only elastic scattering is possible as in Section 7.4.2. Show, according to Sections 7.4.3–7.4.5, that the S matrix elements for scattering with ($S_{\rm sf}$) and without ($S_{\rm nf}$) spin flip are

$$S_{\rm nf} = \frac{3}{4}\frac{\Delta(i\kappa_{1/2}, -k_{3/2})}{\Delta(i\kappa_{1/2}, k_{3/2})} + \frac{1}{4}\frac{1 - ik_{3/2}a_1}{1 + ik_{3/2}a_1}$$

$$S_{\rm sf} = \frac{\sqrt{3}}{4}\left\{-\frac{\Delta(i\kappa_{1/2}, -k_{3/2})}{\Delta(i\kappa_{1/2}, k_{3/2})} + \frac{1 - ik_{3/2}a_1}{1 + ik_{3/2}a_1}\right\},$$

and that the corresponding cross sections are

$$\sigma_{\rm nf} = 4\pi\frac{d\sigma_{\rm nf}}{d\Omega} = \frac{\pi|S_{\rm nf} - 1|^2}{k_{3/2}^2}$$

$$\sigma_{\rm sf} = 4\pi\frac{d\sigma_{\rm sf}}{d\Omega} = \frac{\pi|S_{\rm sf} - 1|^2}{k_{3/2}^2}.$$

Discuss your results for equal values of triplet and singlet scattering lengths $a_1 = a_0$.

7.5 For photoabsorption in noble gases just above the $^2P_{1/2}$ and $^2P_{3/2}$ thresholds, determine the allowed values of J_t and classify them according to whether they are parity favored or parity unfavored.

References

1. S. Geltman, "Topics in Atomic Collision Theory." Academic, New York, 1969.
2. E. Segrè, "Nuclei and Particles," Section 10.1. Benjamin, New York, 1964.
3. T. Ohmura and H. Ohmura, *Phys. Rev.* **118**, 154 (1960).
4. I. S. Gradshteyn and I. M. Ryzhik, "Tables of Integrals," Eq. 6.623.1. Academic, New York, 1980.
5. L. D. Landau and I. M. Lifshitz, "Quantum Mechanics," 3rd ed., Section 144. Pergamon, Oxford, 1977.
6. E. U. Condon and G. H. Shortley, "The Theory of Atomic Spectra," Chapters 3 and 12. Cambridge Univ. Press, London and New York, 1959.
7. M. Rotenberg, R. Bivins, N. Metropolis, and J. K. Wooten, Jr., "The 3-j and 6-j Symbols." MIT Press, Cambridge, Massachusetts, 1959.
8. H. Hotop, T. A. Patterson, and W. C. Lineberger, *Phys. Rev. A* **8**, 762 (1973).
9. G. Herzberg and Ch. Jungen, *J. Mol. Spectrosc.* **41**, 425 (1972).
10. S. Watanabe, U. Fano, and C. H. Greene, *Phys. Rev. A* **29**, 177 (1984).
11. Ch. Jungen and O. Atabek, *J. Chem. Phys.* **66**, 5584 (1977); Ch. Jungen and D. Dill, *J. Chem. Phys.* **73**, 3338 (1980); M. Raoult and Ch. Jungen, *J. Chem. Phys.* **74**, 3388 (1981).
12. A. F. Starace, *J. Phys. B* **6**, 76 (1973).
13. J. Kessler, "Polarized Electrons." Springer-Verlag, Berlin and New York, 1976.
14. D. Dill, *Phys. Rev. A* **7**, 1976 (1973).

8 RESONANCES AND THE ELIMINATION OF CHANNELS

8.1 INTRODUCTION

Chapter 7 considered for the most part ranges of energy in which all channels are either closed with a discrete spectrum or open with alternative continua for the scattered or photoelectron to escape to infinity. We consider now multichannel phenomena in the range of energies E where some, but not all, of the fragmentation channel energies $\epsilon_i = E - \bar{E}_i$ are positive, that is, where some of these channels are open and the rest are closed. This is, in fact, the normal situation for most physical systems below the energy threshold for complete disintegration.

If excitations into different fragmentation channels were independent of one another, there would occur a spectrum of discrete levels ϵ_{in} in each of the closed channels. However, the channels are, in fact, coupled: A particle bound in a closed channel can be scattered by collision with the core into an open channel whereby it becomes permanently detached from the core. The energy for the detachment is provided by the core dropping from a level $\bar{E}_i > E$ to a lower one, $\bar{E}_j < E$. This phenomenon is called autodetachment for atomic systems with negative charge and autoionization (preionization) for neutral atoms or positive ions.

In the prototype two-electron atoms H^- and He, such autodetaching or autoionizing states exist for energies $E \lesssim \bar{E}_i$, where $\bar{E}_i (i \geq 2)$ is the energy of the ith level of H and He^+, respectively, with respect to the ground state. For example, there are two dominant groups of channels at $E \lesssim \bar{E}_2$. The channels with $i = 1$ are open, the H or He^+ atom being in its ground state and the second electron free with kinetic energy E. The channels with $i = 2$ are closed, the second electron being bound with respect to the energy position $\bar{E}_2$ of the

$i = 2$ state of H or He^+. The sequence of thresholds and channels for He is diagrammed in Fig. 2.2.

In our usual time-independent representation, stationary states have a continuous spectrum of energies E but are represented by superpositions of wave functions of open and closed channels. The admixture of closed-channel components depends, often sensitively, on the energy E as the particle energy in a closed channel, ϵ_i, sweeps through a quasi-eigenvalue ϵ_{in}. The existence of such eigenvalues ϵ_{in} is evidenced by the occurrence of resonances in the spectra of collisions or photoabsorption. The term *resonance* means that the total energy E resonates with the quasi-eigenvalue $\bar{E}_i + \epsilon_{in}$ of a closed channel.

In the description of stationary states as superpositions of independent particle wave functions belonging to different configurations, a resonance is represented by a rising admixture of a configuration of the discrete spectrum of a closed channel when E approaches its quasi-eigenvalue $\bar{E}_i + \epsilon_{in}$. In the examples of H^- or He, the continuum configuration in the $i = 1$ channel, $1s\epsilon L$, mixes with two-electron bound state configurations $2\mathbf{l}_1 n\mathbf{l}_2$ with $n \geq 2$ and $\mathbf{l}_1 + \mathbf{l}_2 = \mathbf{L}$ at a sequence of resonance energies E below $\bar{E}_2$. The admixture and the transitions between configurations of different channels are then ascribed to configuration interaction. The concept of scattering by the core stresses that this interaction actually operates at $r < r_0$, where the particle penetrates the core.

Since direct observation bears mainly on the open channels, while the closed channels influence the observations indirectly through resonances, we are led to focus the theory on the open channels. The base set of independent wave functions at any energy equals the number of open channels. The closed channels are eliminated from direct representation in this set, but their influence is represented by additional short-range terms of each independent wave function or by corresponding parameters in cross sections. This procedure of channel elimination is useful not only for its stated immediate purpose but serves as a prototype for very broad applications to other phenomena considered in this book and throughout physics as well.

The number of channels to be considered may seem extremely large at first sight, when closed channels are included, because the number of fragmentation channels of a many-particle complex becomes larger and larger as the spectrum of core levels becomes richer at higher energies $\bar{E}_i$. However, this number is sometimes limited effectively by the following consideration. For any given energy E, the energy of the particle leaving the core in a fragmentation channel, $\epsilon_i = E - \bar{E}_i$, becomes increasingly negative as $\bar{E}_i$ becomes larger and larger; the radial wave function of this particle decays faster and faster, as $\exp(-\kappa_i r)$ with $\kappa_i = [2(\bar{E}_i - E)]^{1/2}$, and its value becomes negligible for $r > r_0$. Such channels give rise to no resonance in the range of interest and need not be considered explicitly in the treatment of phenomena in which one particle dissociates out of the complex to the extent of reaching radial distances $r > r_0$ (see Section 6.6). On the other hand, these "strongly closed" channels must be considered, if only indirectly, in the study of the complex itself and its short-range eigenchannels, which is to be resumed in Chapter 9. The main practical limitation of our present treatment lies, as noted earlier, in excluding fragmentation channels in which two or more particles escape from the core. This limitation will be partially lifted in Part D.

We shall start again here with a simple example and extend its treatment later.

8.2 RESONANCES IN ELASTIC SCATTERING

In Section 7.4 we considered elastic scattering of electrons with energy $\epsilon_{3/2} < \bar{E}_{1/2} - \bar{E}_{3/2}$ from F in its fine-structure ground state with $\bar{J} = \frac{3}{2}$. The closed channel with $\bar{J} = \frac{1}{2}$ adds to the scattering wave function a term with the factor $e^{-\kappa_{1/2}r}$ in (7.31). This decaying term characterizes any channel that is closed, but it does not contribute directly to the wave function at large r and hence to the cross-section expressions in (7.32) and (7.33). However, as $\kappa_{1/2}$ approaches zero, that is, as the electron begins to have an energy almost sufficient to excite the F atom to $\bar{J} = \frac{1}{2}$, the effect of this term is no longer negligible and may give rise to a resonance in the elastic scattering cross section. No experimental data exist for e + F, partly because of the difficulty of obtaining monoenergetic electrons with the required energies of $\simeq \bar{E}_{1/2} - \bar{E}_{3/2} = 0.05$ eV.

To study this effect we turn to a different system of electron + atom, where resonances in elastic scattering have been studied extensively both in experiment and in theory. Instead of a spin flip, we consider an electron excitation whose energy range is more accessible to experimental investigation. Consider then the total cross section (4.19) for the elastic scattering of electrons on He in its ground state, $1s^2\ {}^1S_0$. As the electron energy rises to approach the threshold for the lowest excitation to He($1s2s\ {}^3S_1$) at $\bar{E}_1 = 19.82$ eV, the $l = 0$ channel of the incident electron, $|1s^2\ {}^1S_0, \bar{E}_0, l = 0, j = \frac{1}{2}, \epsilon_0, J = \frac{1}{2})$, becomes significantly coupled to the closed channel $|1s2s\ {}^3S_1, \bar{E}_1, l = 0, j = \frac{1}{2}, \epsilon_1, J = \frac{1}{2})$.

Regardless of details of the short-range interaction with the core, the presence of a closed channel with $0 > \epsilon_1 \gtrsim -1$ eV suggests that we represent the phase shift $\delta_{l=0}$ considered in Chapter 4 in the form $\delta_0^{(l)} + \Delta$. Here $\delta_0^{(l)}$ is the contribution of the long-range potential as in Chapter 5 and Section 7.5, and Δ represents the full effect of the short-range interaction, including the effect of the closed channel.

The detailed analytical treatment of the contribution from the closed channel, such as the last term in (7.31) for $e + \mathrm{F}$ scattering in the energy range where $\bar{J}_i = \frac{1}{2}$ is closed, will be taken up in Section 8.3. However, we can anticipate that its effect on the phase shift δ for the channel that is open will cause it to rise rapidly by π at an energy corresponding to a bound-state eigenvalue in the closed channel. In the example of $e + \mathrm{He}$ just below the 1s2s $^3\mathrm{S}_1$ threshold of He, such a rise in Δ results in a resonance in the elastic scattering channel as shown in Fig. 8.1. The occurrence of this resonance was, in fact, anticipated before its observation and associated with the existence of an independent particle configuration $1s2s^2\ ^2\mathrm{S}_{1/2}$ formed by the attachment of an electron to the 1s2s $^3\mathrm{S}_1$ state of He.

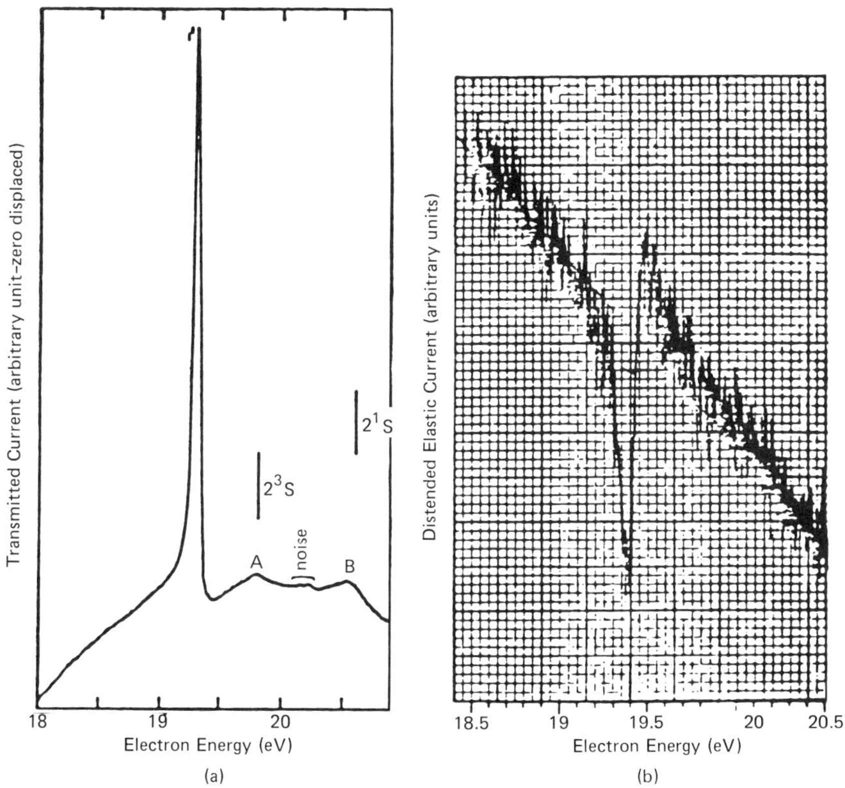

Fig. 8.1. Resonances in e–He elastic scattering. (a) Scattered intensity as a function of incident energy. (From Kuyatt *et al.* [1].) (b) Original observation of the $(1s2s^2)^2\mathrm{S}$ resonance in e–He elastic scattering at 72°. See Fig. 1.5a for later data at different angles. (From Schulz [2].)

Accordingly, we rewrite the total elastic scattering cross section (4.19) as

$$\sigma = \frac{4\pi}{k^2} \sum_l (2l+1) \sin^2 \delta_l \to \left(\frac{4\pi}{k^2}\right) \left\{ \sin^2(\delta_0^{(1)} + \Delta) + \sum_{l=1} (2l+1) \sin^2 \delta_l \right\}. \quad (8.1)$$

As a next step, anticipating a rapid variation of Δ in a narrow energy range, we break up the $\sin^2(\delta_0^{(1)} + \Delta)$ in the form

$$\sin^2(\delta_0^{(1)} + \Delta) = \sin^2(\delta_0^{(1)} + \Delta_a) \frac{[\cot(\Delta - \Delta_a) + \cot(\delta_0^{(1)} + \Delta_a)]^2}{1 + \cot^2(\Delta - \Delta_a)}, \quad (8.2)$$

where Δ_a indicates the steady value of Δ outside of a resonance range. A rapid rise of Δ by π yields a characteristic resonance structure in the cross section σ in (8.1), namely, a rapid alternation of a maximum and a zero of the expression (8.2), the maximum at $\Delta = \Delta_a + \frac{1}{2}\pi$ and the zero at $\Delta = -\delta_0^{(1)}$ (mod π). The details of this (Beutler–Fano) resonance profile are identified by the single parameter

$$q = -\cot(\delta_0^{(1)} + \Delta_a). \quad (8.3)$$

Figure 8.1a shows a classic example of resonance observed in the spectrum of unscattered electrons with intensity $I_0 \exp(-\sigma t)$, where σ is given by (8.1) and (8.2) and t is the sample thickness in atoms per square centimeter; the peak of the observed current corresponds to the zero of (8.2). Figure 8.1b shows the same resonance as observed in the electron current scattered by 72°; here the zero of (8.2) coincides with the minimum of the observed current, the remaining current being contributed by terms of (8.1) with $l \neq 0$. The very weak increase of scattered current from its steady value $\sin^2(\delta_0^{(1)} + \Delta_a)$ to the peak shows that the value of $\delta_0^{(1)} + \Delta_a$ is not very different from $\frac{1}{2}\pi$(mod π). The fact that σ drops instead of rising in Fig. 8.1b, as the energy ϵ_0 rises to the resonance, shows more specifically that $\delta_0^{(1)} + \Delta_a > \frac{1}{2}\pi$; its value lies, in fact, between 100 and 110° (see Problem 8.2). Figure 1.5a shows how the resonance profile alters in the differential cross section at different angles depending on the interference of terms of (4.17) with different l values. Resonances analogous to those shown in Fig. 8.1 have been observed for elastic scattering of electrons by all noble gases at energies ~0.5 eV below the first excitation threshold.

The elastic scattering of electrons from a positively charged ion can be expected to show resonances at energies corresponding to the bound states. In particular, the elastic scattering $e + H_2^+(v = N = 0)$, the system considered in Section 7.5, should display a whole series of resonances in its $l = 1$ channel as Δ rises by π again and again. However, it is difficult to prepare samples of H_2^+ in this particular state and monoenergetic electrons with the required positive energy $\epsilon_0 < \bar{E}_2 - \bar{E}_0 \sim 0.02$ eV. Instead, we will return in the next section to the photoabsorption of H_2 treated in Section 7.5, where the resolutions attainable in optical spectroscopy permit the detailed study of these resonances. In Chapter 7 we had considered the bound-state region. This study will now be extended into the resonance region.

8.3 RESONANCES IN THE TWO-CHANNEL PHOTOABSORPTION OF H_2

We now complete the analysis of the photoabsorption of H_2 in its ground state, $1s\sigma^2\ {}^1\Sigma_g^+$, $v = J = 0$. This study was developed in Section 7.5 excluding the energy range $\bar{E}_0 < E < \bar{E}_2$. Photoionization is achieved in this range only if the H_2^+ core remains in its ground rotational state ($N = 0$), whereas the excited electron remains bound when H_2^+ is raised to the excited rotational state $N = 2$. Under these circumstances, the state of the two-channel system $H_2^+ + e$ can be represented by a standing-wave superposition $\Sigma_\Lambda F_\Lambda(\omega, r)A_\Lambda$, as in Section 7.5, but its real coefficients A_Λ must now satisfy the bound-state condition (7.17) for the $N = 2$ channel *only*,

$$\sum_\Lambda J_{2\Lambda}^- A_\Lambda = D^{-1} \sum_\Lambda \sin(\beta_2 + \delta_\Lambda^{(s)}) U_{2\Lambda} A_\Lambda = 0. \tag{8.4}$$

Recall from Chapter 5 that the long-range phase is represented by β_2 for $\epsilon_2 < 0$. Analogy with the construction of (7.15) shows that (8.4) is satisfied by

$$A_\Lambda \propto (J^{- -1})_{\Lambda 0}. \tag{8.5}$$

Equation (8.5) constitutes, in essence, the solution of our problem, but its implications are brought out by writing the wave function for $r > r_0$ explicitly. For this purpose, $F_\Lambda(\omega, r)$ is drawn from (7.39) and (7.41b) for ϵ_0 and ϵ_2, respectively, intending to adapt these equations, designed for both channels open or both closed, to the present situation where $N = 0$ is open and $N = 2$ closed. To this end we replace the summation over N by writing its two terms separately and setting the coefficient of $f^-(\epsilon_2, l, r)$ equal to zero according to (8.4), which yields

$$\begin{aligned}\sum_\Lambda F_\Lambda(\omega, r)A_\Lambda = \sum_\Lambda \Bigg\{ & \Phi_0(\omega)\left(\frac{2}{\pi k_0}\right)^{1/2} \\ & \times \frac{1}{2i}[f^+(\epsilon_0, l, r)e^{i(\delta_0^{(l)} + \delta_\Lambda^{(s)})} - f^-(\epsilon_0, l, r)e^{-i(\delta_0^{(l)} + \delta_\Lambda^{(s)})}]U_{0\Lambda} \\ & - \Phi_2(\omega)(\pi\kappa_2)^{-1/2} D f^+(\epsilon_2, l, r)\cos(\beta_2 + \delta_\Lambda^{(s)})U_{2\Lambda}\Bigg\} A_\Lambda, \quad r > r_0.\end{aligned} \tag{8.6}$$

Equation (8.6) for one open and one closed channel displays the generic structure of a multichannel wave function in a region when some of the channels are open and others closed. Scattering wave functions extend out to large r in each of the open channels, with phase shifts that represent the combined effects of short- and long-range interactions, but also include exponentially decaying components in the closed channels. Note that (8.6) coincides in form with a similar expression derived for the $e + \mathrm{F}$ system in (7.31). At large r, where $f^+(\epsilon_2, l, r)$ becomes negligible, the wave function reduces to its component with $N = 0$, which represents a standing wave in a single channel. This function should accordingly take the standard form normalized per unit energy,

$$\sum_\Lambda F_\Lambda(\omega, r) A_\Lambda \xrightarrow[r\to\infty]{} \Phi_0(\omega)\left(\frac{2}{\pi k_0}\right)^{1/2} \sin(k_0 r + \zeta \ln r + \delta_0^{(l)} + \Delta), \qquad (8.7)$$

where the additional phase Δ will represent the effect of the resonance as in Section 8.2.

This asymptotic form removes from (8.6) any explicit consideration of the closed channel $N = 2$ and the short-range decomposition into Λ channels. All these effects are lumped into the effective phase shift Δ, which accordingly represents

(1) The combined effect of the short-range interaction eigenphase shifts $\delta_\Sigma^{(s)}$ and $\delta_\Pi^{(s)}$. These are slowly varying functions of energy.

(2) The effect of coupling with the closed channel, mediated by short-range scattering on the H_2^+ core, which depends sensitively on the energy through $\beta_2 = -\pi + \pi\nu_2$.†

Before taking up the reduction of (8.6) to its large-r form [(8.7)], we pause to note that these are key equations in multichannel scattering in the region between thresholds for the opening of alternative channels. The two-channel example being considered here illustrates the fundamental feature of the replacement of unobserved variables (and channels) by effective parameters, such as Δ in the open channel, which are observable at infinity. Multichannel problems in atomic systems involve an infinity of closed channels lying above the energy range of interest particularly when there is a long-range Coulomb field. The joint occurrence of open and closed channels considered in this chapter thus takes on added significance.

The formal transformation of the wave function (8.6) into (8.7) proceeds at

† In the rest of this section we will set $\beta_N = \pi\nu_N$, dropping the term $-\pi$, because the possible overall factor of (-1) thus introduced in trigonometric functions will be irrelevant to our analysis.

large r by rearranging the coefficients of $f^{\pm}(\epsilon_0, l, r)$ as follows:

$$\Phi_0\left(\frac{2}{\pi k_0}\right)^{1/2}(\pm i)^l \sum_\Lambda J^{\pm}_{0\Lambda} A_\Lambda = \Phi_0(\omega)\left(\frac{2}{\pi k_0}\right)^{1/2} \sum_\Lambda e^{\pm i(\delta_0^{(l)} + \delta_\Lambda^{(s)})} U_{0\Lambda} A_\Lambda$$

$$= \Phi_0(\omega)\left(\frac{2}{\pi k_0}\right)^{1/2} e^{\pm i(\delta_0^{(l)} + \Delta)} \sum_\Lambda e^{\pm i(-\Delta + \delta_\Lambda^{(s)})} U_{0\Lambda} A_\Lambda. \tag{8.8}$$

Comparison with (8.6) and (8.7) shows these two equations to be indeed equivalent, provided the complex Σ_Λ on the right-hand side of (8.8) reduces to unity, that is, provided

$$\sum_\Lambda \sin(-\Delta + \delta_\Lambda^{(s)}) U_{0\Lambda} A_\Lambda = 0, \tag{8.9}$$

$$\sum_\Lambda \cos(-\Delta + \delta_\Lambda^{(s)}) U_{0\Lambda} A_\Lambda = 1. \tag{8.10}$$

Equations (8.9) and (8.4) together form a homogeneous system that determines Δ and A_Λ, to within normalization, while (8.10) normalizes A_Λ.

The value of the effective phase shift Δ is determined by the condition of compatibility of the system of Eqs. (8.4) and (8.9), namely, that the determinant of its coefficient matrix vanish,

$$\begin{vmatrix} U_{0\Sigma}\sin(-\Delta + \delta_\Sigma^{(s)}) & U_{0\Pi}\sin(-\Delta + \delta_\Pi^{(s)}) \\ U_{2\Sigma}\sin(\beta_2 + \delta_\Sigma^{(s)}) & U_{2\Pi}\sin(\beta_2 + \delta_\Pi^{(s)}) \end{vmatrix} = 0. \tag{8.11}$$

This equation differs from (7.48) only by the occurrence of the effective phase shift Δ—actually of $-\Delta$—in place of the phase shift $\beta_0 = \pi\nu_0$, which pertains to the channel $N = 0$ when it is closed. This important connection has the immediate consequence that Fig. 7.6 gives Δ/π directly as a function of $\nu_2 = \beta_2/\pi$; it suffices to replace the downward ordinate scale of ν_0 by an upward scale of Δ/π (mod 1). This parameter is seen to increase by unity whenever β_2/π does so, with

$$d\Delta/d\beta_2 > 0. \tag{8.12}$$

A less immediate consequence, namely, a reinterpretation of the results of Section 7.5 is discussed in Section 8.3.3.

8.3.1 Resonance Parameters

The dependence of the effective phase shift Δ on the closed-channel phase parameter β_2 [(8.11)] is a key feature of the resonance process. We examine it here in some detail, making our notation independent of special features of the H_2 example. To this end we indicate the orthogonal transformation matrix $U_{N\Lambda}$ in (8.11) by

$$\begin{vmatrix} U_{0\Sigma} & U_{0\Pi} \\ U_{2\Sigma} & U_{2\Pi} \end{vmatrix} = \begin{vmatrix} \cos\phi & \sin\phi \\ -\sin\phi & \cos\phi \end{vmatrix} \tag{8.13}$$

as in (7.55). With this notation the determinant in (8.11) expands into

$$\cos^2\phi \sin(\delta_\Sigma^{(s)} - \Delta)\sin(\delta_\Pi^{(s)} + \beta_2) + \sin^2\phi \sin(\delta_\Pi^{(s)} - \Delta)\sin(\delta_\Sigma^{(s)} + \beta_2) = 0. \tag{8.14}$$

Equation (8.14) is periodic (with unit period) in both Δ/π and β_2/π just as (7.48) utilized in Section 7.5.4 for determining the bound-state eigenvalues. Viewing Fig. 7.6 as a plot of Δ/π as a function of β_2/π, the point of maximum slope $d\Delta/d\beta_2$ may be interpreted as the position of the resonance. To exhibit this slope more transparently, it is convenient to reexpress (8.14) in terms of three other parameters (μ_1, μ_2, ξ) alternative to $(\delta_\Sigma^{(s)}, \delta_\Pi^{(s)}, \phi)$ such that the determinantal equation is cast as

$$\begin{vmatrix} \tan(-\Delta + \pi\mu_1) & \xi \\ \xi & \tan(\beta_2 + \pi\mu_2) \end{vmatrix} = 0, \tag{8.15}$$

that is, in the form of an equilateral hyperbola in the variables $\tan(-\Delta + \pi\mu_1)$ and $\tan(\beta_2 + \pi\mu_2)$.

The connection between the alternative parameters is established by expanding both (8.14) and (8.15) in terms of $\tan\Delta$ and $\tan\beta_2$ and equating the coefficients of $\tan\Delta$, $\tan\beta_2$, and $\tan\Delta\tan\beta_2$:

$$\cos^2\phi \tan\delta_\Sigma^{(s)} + \sin^2\phi \tan\delta_\Pi^{(s)} = \frac{\tan\pi\mu_1 + \xi^2\tan\pi\mu_2}{1 - \xi^2\tan\pi\mu_1\tan\pi\mu_2}, \tag{8.16a}$$

$$\sin^2\phi \tan\phi_\Sigma^{(s)} + \cos^2\phi \tan\delta_\Pi^{(s)} = \frac{\tan\pi\mu_2 + \xi^2\tan\pi\mu_1}{1 - \xi^2\tan\pi\mu_1\tan\pi\mu_2}, \tag{8.16b}$$

$$\tan\delta_\Sigma^{(s)}\tan\delta_\Pi^{(s)} = \frac{-\xi^2 + \tan\pi\mu_1\tan\pi\mu_2}{1 - \xi^2\tan\pi\mu_1\tan\pi\mu_2}. \tag{8.16c}$$

Equivalently, these equations can be inverted to give

$$\pi\mu_1 = \tfrac{1}{2}(\delta_\Sigma^{(s)} + \delta_\Pi^{(s)} - \pi\eta), \tag{8.17a}$$

$$\pi\mu_2 = \tfrac{1}{2}(\delta_\Sigma^{(s)} + \delta_\Pi^{(s)} + \pi\eta), \tag{8.17b}$$

$$\xi^2 = \frac{\cos\pi\eta - \cos(\delta_\Pi^{(s)} - \delta_\Sigma^{(s)})}{\cos\pi\eta + \cos(\delta_\Pi^{(s)} - \delta_\Sigma^{(s)})} = \tan\tfrac{1}{2}(\pi\eta + \delta_\Pi^{(s)} - \delta_\Sigma^{(s)})\tan\tfrac{1}{2}(\delta_\Pi^{(s)} - \delta_\Sigma^{(s)} - \pi\eta), \tag{8.17c}$$

where the parameter η has been defined by

$$\tan \pi\eta = \cos 2\phi \tan(\delta_{\Pi}^{(s)} - \delta_{\Sigma}^{(s)}). \tag{8.17d}$$

An alternative relation between ξ^2 and the short-range parameters is

$$\sin^2 2\phi \sin^2(\delta_{\Pi}^{(s)} - \delta_{\Sigma}^{(s)}) = \frac{4\xi^2}{(1+\xi^2)^2} = \frac{4/\xi^2}{(1+1/\xi^2)^2}. \tag{8.17e}$$

Here ξ^2 may be interpreted as a mixing parameter, that is, as an index of the effective interaction between the open and the closed channels. Equations (8.16) or (8.17e) show that ξ^2 vanishes on either of two grounds: degeneracy of the short-range eigenchannels, [i.e., $\delta_{\Sigma}^{(s)} = \delta_{\Pi}^{(s)}$] or coincidence of the eigenchannels with the fragmentation channels [i.e., $\phi = 0$ or $\pi/2$]. In such an instance of decoupled channels, the plot of Δ versus β_2 consists of two intersecting straight lines: $\Delta = \pi\mu_1$ and $\beta_2 = \pi\mu_2$.

For ξ^2 nonzero and small, Δ remains near a plateau value of $\pi\mu_1$ except in the vicinity of $\beta_2 \simeq -\pi\mu_2$, where $-\Delta + \pi\mu_1$ switches rapidly from $-\pi/2$ to $\pi/2$. Such a plot, shown in Fig. 8.2a, resembles Fig. 7.6 except that the plateau value has been centered in the middle of the unit square in diagramming (8.15). The analytical expression for the variation of Δ with β_2, as given by differentiating this equation,

$$d\Delta/d\beta_2 = [\xi^2 \cos^2(\beta_2 + \pi\mu_2) + \xi^{-2} \sin^2(\beta_2 + \pi\mu_2)]^{-1}, \tag{8.18}$$

is displayed in Fig. 8.2b. The slope ranges from its minimum value ξ^2 (when $\Delta \sim \pi\mu_1$) to its maximum value $1/\xi^2$ (at $\beta_2 = -\pi\mu_2$). Resonances result from a rapid change of the short-range phase shift $\pi\Delta$ in the open channel as the closed channel phase parameter $\beta_2(=\pi\nu_2)$ cycles through unity. The infinite sequence of Rydberg states in the closed channel with energies $E = \bar{E}_2 - \mathrm{Ry}/\nu_2^2$ gives rise to a corresponding infinite number of resonances in the open channel over the energy range $\bar{E}_0 < E < \bar{E}_2$. These resonance positions, therefore, follow the Rydberg-like formula

$$E_n = \bar{E}_2 - \mathrm{Ry}/(n - \mu_2)^2. \tag{8.19}$$

The time delay (Section 4.4) due to the short-range interactions,

$$2\hbar \frac{d\Delta}{dE} = h \frac{d\Delta}{d\beta_2} \frac{d\nu_2}{dE} = h \frac{\nu_2^3}{2\mathrm{Ry}} \frac{d\Delta}{d\beta_2}, \tag{8.20}$$

represents the lifetime of a resonant multichannel scattering process and takes the form

$$2\hbar \left(\frac{d\Delta}{dE}\right)_n = \frac{\pi\hbar(n - \mu_2)^3}{\mathrm{Ry}} \frac{1}{2}. \tag{8.21}$$

This time delay for elastic scattering of $e + H_2^+$ in the $N = 0$ channel, arising from the short-range interactions with the $N = 2$ channel, must be combined

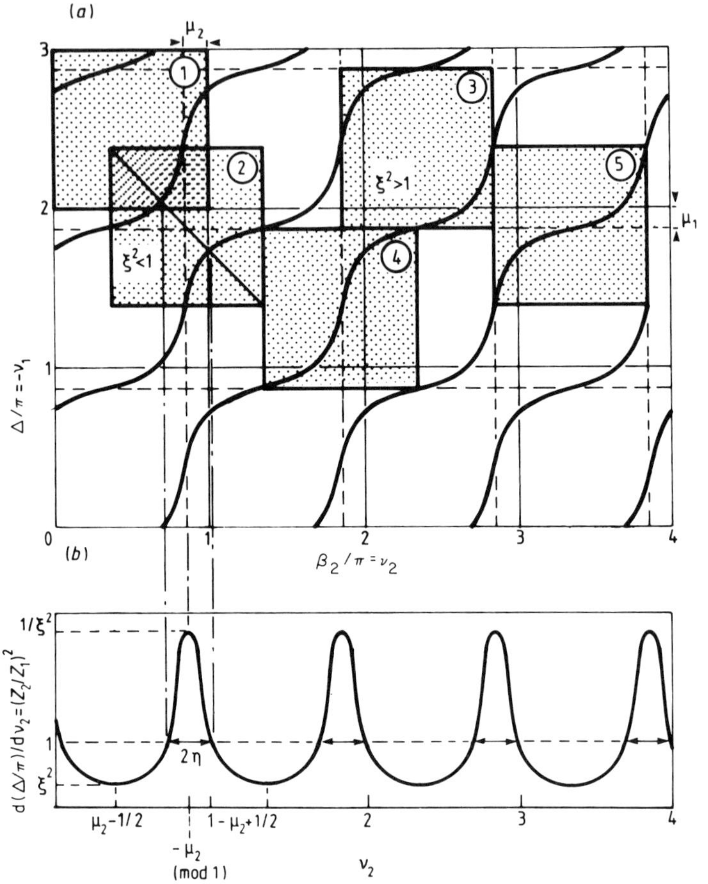

Fig. 8.2. (a) Extended quantum defect plot for two-channel interaction. Full line grid as in Fig. 7.6 with axes at integer values of ν_i (reduced zone 1) describes Eq. (8.11). Dashed line grid corresponds to $\nu_i = -\mu_i$ (mod 1). Alternative reduced zones 2–5 adapt to different symmetries of the plot; zone 2 represents Eq. (8.15). (b) Slope $d\Delta/d\beta_2$, equal to the squared ratio of channel amplitudes [Eq. (8.18)]. (From Giusti-Suzor and Fano [3].)

additively with $2\hbar d\delta_0^{(1)}/dE$, the contribution of the long-range Coulomb interaction in the $N = 0$ channel itself. The time delay in (8.21) is usually identified with $4T_n$, where T_n is the resonance lifetime reciprocal to the resonance width: $T_n = \hbar/\Gamma_n$. This delay is factored according to the middle of (8.20) into $d\Delta/d\beta_2$, a probability for reversible rotational excitation of the closed channel $N = 2$, and $2\hbar d\beta_2/dE = 2\pi\nu_2^3$, which is the orbital period of the incident electron while bound in this channel.

The preceding discussion applies to variations of a mixing parameter in its natural range $0 < \xi^2 < 1$. However, ξ^2, as defined in (8.17c) in terms of the short-range parameters, may in fact assume values greater than unity. In this event, a reparametrization may be applied which replaces $\xi^2 > 1$ by $1/\xi^2 < 1$. [See also (8.17e).] Since (8.15) and (8.16) are invariant under the replacements $\mu_1 \to \mu_1 + \frac{1}{2}$, $\mu_2 \to \mu_2 + \frac{1}{2}$, and $\xi^2 \to 1/\xi^2$, we may interpret $1/\xi^2$ as the mixing parameter while shifting the μ_i by $\frac{1}{2}$. These replacements, which amount to a permutation of the short-range eigenchannels Σ and Π, are to be made throughout, for example, in expressions such as (8.18) and (8.21). As a result, the widths of the resonances as given in (8.21) always remain smaller than the separation in energy of the resonances. The reparametrization is also indicated diagrammatically in Fig. 8.2 and shows in particular that the points of minimum and maximum slope (time delay) interchange roles on passing from $\xi^2 < 1$ to $\xi^2 > 1$.

The wave function in (8.6) can be cast into a simpler form, permitting ready interpretation in terms of the parameters (μ_1, μ_2, ξ). The coefficient of Φ_0 is first reexpressed in terms of the phase shift Δ by using (8.9) and (8.10). Next, the coefficient of the closed-channel factor Φ_2 is recast as

$$\sum_{\Lambda} \cos(\beta_2 + \delta_{\Lambda}^{(s)}) U_{2\Lambda} A_{\Lambda} = \left(\frac{d\Delta}{d\beta_2}\right)^{1/2}, \tag{8.22}$$

where $d\Delta/d\beta_2$ is given in (8.18). The algebra for this recasting involves manipulations of (8.4), (8.9), and (8.10) and passage through (8.16) to the alternative parameters. The final result is just as expected. It accords with the identification made earlier of $d\Delta/d\beta_2$ as the probability for excitation of the $N = 2$ channel. The wave function is thus expressed in the transparent form

$$\begin{aligned}\sum_{\Lambda} F_{\Lambda}(\omega, r) A_{\Lambda} = \Phi_0(\omega)\left(\frac{2}{\pi k_0}\right)^{1/2} & \frac{1}{2i}[f^+(\epsilon_0, l, r)e^{i(\delta_0^{(1)} + \Delta)} \\ & - f^-(\epsilon_0, l, r)e^{-i(\delta_0^{(1)} + \Delta)}] \\ & + \Phi_2(\omega)(\pi\kappa_2)^{-1/2} D(\epsilon_2, l) f^+(\epsilon_2, l, r)\left(\frac{d\Delta}{d\beta_2}\right)^{1/2}, \qquad r > r_0.\end{aligned} \tag{8.23}$$

The parameter $(d\Delta/d\beta_2)^{1/2}$ in (8.22) can be written in a form slightly different from (8.18), namely,

$$\left(\frac{d\Delta}{d\beta_2}\right)^{1/2} = -\frac{\sin(-\Delta + \pi\mu_1)}{\xi \cos(\beta_2 + \pi\mu_2)}. \tag{8.22$'$}$$

The right-hand side of this equation represents the ratio of the components of the eigenvector of (8.15), that is, the ratio of amplitudes of the two channels denoted by Z_2/Z_1 in the spectroscopy notation introduced after (7.18).

Equation (8.23) may be regarded as an extension of (8.7) to include at smaller r the admixture of the closed-channel term of the wave function, whose coefficient has now been cast in terms of the derivative $d\Delta/d\beta_2$. At asymptotically large r, this term involving $f^+(\epsilon_2, l, r)$ vanishes and (8.23) takes the standard form (8.7) of the wave function for elastic scattering of electrons from H_2^+ in the $N = 0$ channel. As discussed earlier, the phase shift Δ remains near a plateau value $\pi\mu_1$ except in the vicinity of the energies given by (8.19). Identifying Δ_a in (8.2) and (8.3) with $\pi\mu_1$, these equations then describe the structure of the resonance profiles in the elastic cross section.

8.3.2 Resonances in Photoabsorption

Resonances in the photoabsorption cross section of H_2 in the range of energies $\bar{E}_0 \leq E \leq \bar{E}_2$ parallel the resonances in elastic scattering. The intensity of photoabsorption in this energy range is proportional to the expression in (7.50), $|\Sigma_\Lambda A_\Lambda^*(F_\Lambda|x|0)|^2$. Unlike the various combinations of A_Λ that have occurred so far, as in (8.22), we now require explicit knowledge of these amplitudes as obtained from (8.4), (8.9), and (8.10). Inspection of the latter two gives

$$A_\Sigma = N \sin\phi \sin(-\Delta + \delta_\Pi^{(s)}) \tag{8.24a}$$

$$A_\Pi = -N \cos\phi \sin(-\Delta + \delta_\Sigma^{(s)}) \tag{8.24b}$$

with the normalization factor

$$N = [\sin\phi \cos\phi \sin(\delta_\Pi^{(s)} - \delta_\Sigma^{(s)})]^{-1}. \tag{8.24c}$$

A convenient recasting follows on elimination of ϕ through (8.17c) and (8.17d) to give

$$\begin{matrix}\cos\\ \sin\end{matrix}\ \phi = [\sin(\delta_\Pi^{(s)} - \delta_\Sigma^{(s)} \pm \pi\eta)/2\cos\pi\eta \sin(\delta_\Pi^{(s)} - \delta_\Sigma^{(s)})]^{1/2}, \tag{8.24d}$$

so that

$$N = 2\left[1 - \frac{\cos^2(\delta_\Pi^{(s)} - \delta_\Sigma^{(s)})}{\cos^2\pi\eta}\right]^{1/2} = \frac{2}{\xi}\left[1 + \frac{\cos(\delta_\Pi^{(s)} - \delta_\Sigma^{(s)})}{\cos\pi\eta}\right]^{-1}. \tag{8.24e}$$

Note that $\pi\eta$ equals $\delta_\Pi^{(s)} - \delta_\Sigma^{(s)}$ and N reduces to $1/\xi$ in the limit of vanishing mixing ($\xi \to 0$).

The photoabsorption intensity is now obtained by expressing the short-range phases in the arguments of the sine functions in (8.24a) and (8.24b) in terms of $\pi\mu_1$ through (8.17a), substituting for ϕ from (8.24d), and collecting terms,

$$I = |\Sigma_\Lambda A_\Lambda^*(F_\Lambda|x|0)|^2 = [\cos(-\Delta + \pi\mu_1)\, d_1 - \sin(-\Delta + \pi\mu_1)\, d_2/\xi]^2. \tag{8.25}$$

Here the dipole amplitudes d_1 and d_1 are combinations of the eigenchannel amplitudes:

$$d_1 = (F_\Sigma|x|0)\left[1 + \frac{\sin \pi\eta}{\sin(\delta_\Pi^{(s)} - \delta_\Sigma^{(s)})}\right]^{1/2} + (F_\Pi|x|0)\left[1 - \frac{\sin \pi\eta}{\sin(\delta_\Pi^{(s)} - \delta_\Sigma^{(s)})}\right]^{1/2} \tag{8.26a}$$

$$d_2 = -(F_\Sigma|x|0)\left[1 - \frac{\sin \pi\eta}{\sin(\delta_\Pi^{(s)} - \delta_\Sigma^{(s)})}\right]^{1/2} + (F_\Pi|x|0)\left[1 + \frac{\sin \pi\eta}{\sin(\delta_\Pi^{(s)} - \delta_\Sigma^{(s)})}\right]^{1/2}, \tag{8.26b}$$

to which they reduce in the limit of zero mixing, $\pi\eta = \delta_\Pi^{(s)} - \delta_\Sigma^{(s)}$.

Equation (8.25) holds, with appropriate changes of notation, for any system with one open and one closed channel of excitation. A key feature is the periodicity of this expression, which reaches its characteristic values d_1^2 and d_2^2/ξ^2 once in each cycle at $\Delta = \pi\mu_1$ and $\Delta = \pi\mu_1 \pm \frac{1}{2}\pi$, respectively, and vanishes at $\Delta = \pi\mu_1 - \operatorname{arccot}(d_2/d_1\xi)$. The photoabsorption cross section (2.9), $\sigma(\omega) = 4\pi^2 e^2(\omega/c)I$, can be written in a form analogous to the elastic scattering formula (8.2),

$$I = d_1^2 \frac{[\cot(-\Delta + \pi\mu_1) - d_2/d_1\xi]^2}{1 + \cot^2(-\Delta + \pi\mu_1)}, \tag{8.27a}$$

or, alternatively, with the aid of (8.15), as a function of β_2

$$I = d_1^2 \frac{[\tan(\beta_2 + \pi\mu_2)/\xi^2 - d_2/d_1\xi]^2}{1 + \tan^2(\beta_2 + \pi\mu_2)/\xi^4}. \tag{8.27b}$$

Once again, as for elastic scattering, the photoabsorption cross section exhibits a series of resonances at the energies (8.19). The resonance widths remain the same as before, but the profile index

$$q = -d_2/d_1\xi \tag{8.28}$$

now depends on the photoabsorption matrix elements in (8.26). The diverse resonance profiles depend on the parameters η and ξ (Fig. 8.3). Their q values can be obtained conveniently through the ratio of the intensity expression in (8.27b) at $\beta_2 = -\pi\mu_2$ and $-\pi\mu_2 \pm \frac{1}{2}\pi$. The sign of q determines whether the zero or the peak occurs at a lower value of β_2 within a resonance. As in the case of elastic scattering (Section 8.2), this single parameter determines the profile of the autoionization resonances. When $\xi^2 > 1$, the replacements $\mu_i \to \mu_i + \frac{1}{2}$ and $\xi^2 \to 1/\xi^2$ lead again to the same form (8.27) of the resonance profiles for photoabsorption with q replaced by its inverse in (8.28).

It should be stressed that these widely observed properties of resonance profiles hinge only on the trigonometric relations that define the effective phase shift Δ and lead to (8.27). They are altogether independent of the

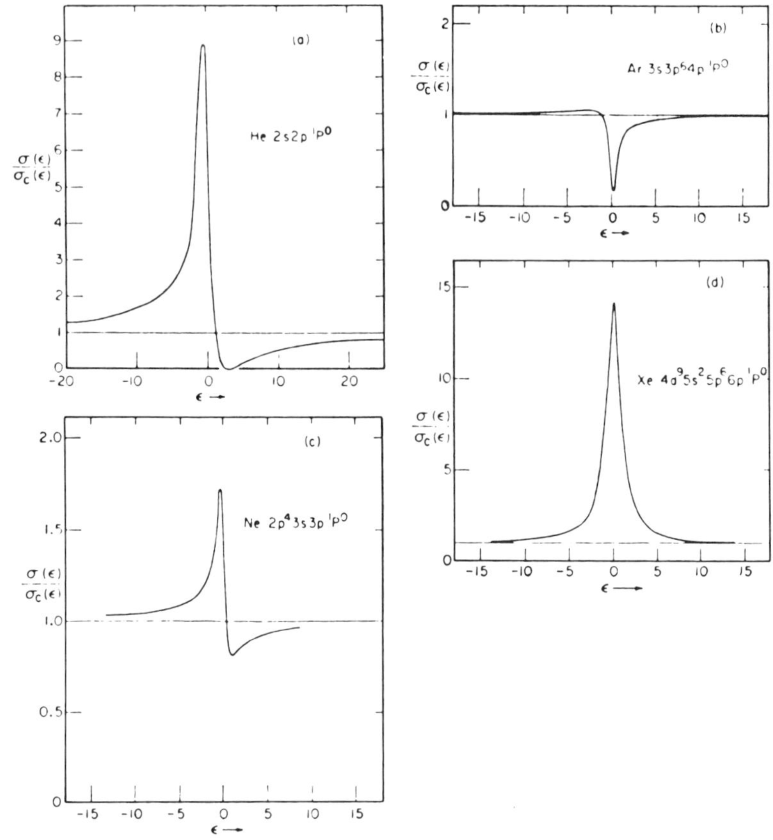

Fig. 8.3. Examples of photoabsorption resonance profiles for (a) helium, (b) argon, (c) neon, and (d) xenon. (From Fano and Cooper [4].)

physical mechanisms of channel interaction. It should then cause no surprise to observe their equal manifestations in quite different processes. The resonances discussed here, arising from coupling to closed channels, are called "Feshbach resonances" to distinguish them from resonances arising from short-range interactions within a single channel. The simple theory presented here accounts only crudely for the resonances in H_2 photoabsorption shown in Fig. 7.5. Very accurate spectral observations in this region have been interpreted in complete detail by including vibrational excitations as outlined in Section 7.5.5 (Fig. 8.4).

8.3.3 Alternative Interpretation of Discrete Spectrum

Having introduced here the concept of an effective phase shift Δ for the *single* channel $N = 0$ with lowest ionization threshold, we may extend the wave function (8.23) below this threshold and consider the entire discrete

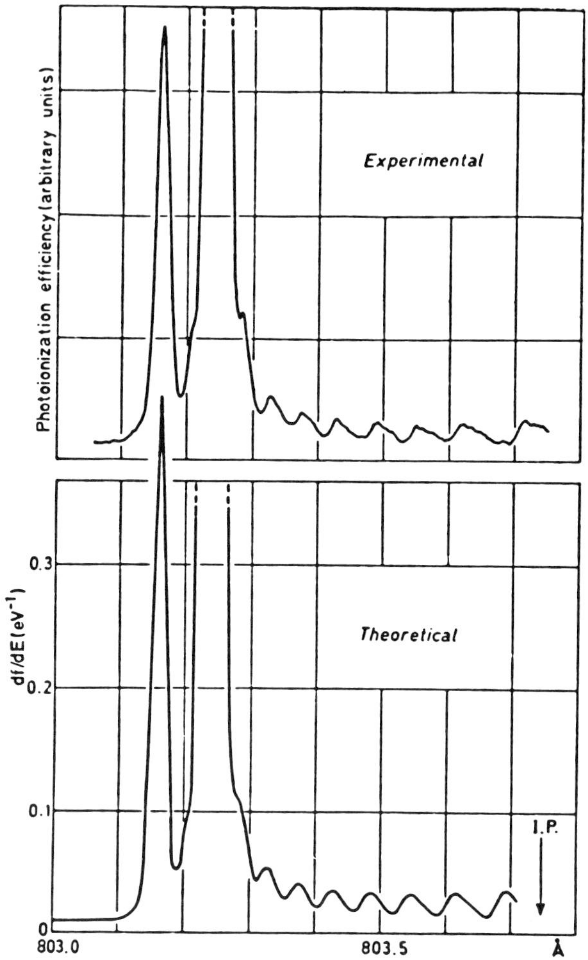

Fig. 8.4. Photoionization yield of $H_2(^1\Sigma_g^+ \, v = J = 0)$ just above threshold. Theoretical results have been folded into experimental resolution. (From Jungen and Dill [5]; experimental data from Dehmer and Chupka [6].)

spectrum as belonging to this single channel. The influence of scattering into the closed channel $N = 2$ has already been fully taken into account through the energy dependence of Δ and of the solutions A_Λ [(8.24)]. Accordingly, we may reclassify the levels as belonging to a single series in accordance with (5.11) and (5.26) but replacing $\delta^{(s)}$ by Δ, i.e.,

$$\beta_{N=0}(\kappa_0) + \Delta(\epsilon_2) = \pi v_0 + \Delta(\epsilon_2) = n\pi, \tag{8.29}$$

$$\epsilon_{0n} = -\frac{1}{2}\frac{1}{[n - \Delta(\epsilon_{2n})/\pi]^2} = -\frac{1}{2}\frac{1}{[n - \Delta(\bar{E}_0 + \epsilon_{0n} - \bar{E}_2)/\pi]^2}. \tag{8.30}$$

The variations in level spacing shown in Fig. 7.5b are thus seen to reflect the dependence of Δ on the energy ϵ_2, imposed by (8.11) between Δ and $\beta_{N=2}$ and by $\beta_{N=2} = \pi/\kappa_2 = \pi(-2\epsilon_2)^{-1/2}$. The variations of intensity similarly follow the photoabsorption profile (8.27), adjusted by the discrete level normalization factor given by (5.37)

$$N_n^2 = \{[n - \Delta(\epsilon_{2n})/\pi]^3 + \pi^{-1}(d\Delta/d\epsilon_2)_{\epsilon_{2n}}\}^{-1}. \tag{8.31}$$

These results have a general implication for the spectra of Rydberg levels perturbed by the levels of closed channels. The closed-channel levels reappear in the combined spectrum as modulations of its intensity that reflect the variations of $d\Delta/d\epsilon$, especially when the intensity of the closed channel is very high. The H_2 spectrum shown in Fig. 7.5b fails to provide a very clear example of this phenomenon, owing to the low intensity of the $N = 2$ channel and intervening vibrational excitations, but striking examples appear in the analogous spectra shown in Fig. 8.5 and plotted as histograms according to the prescription of Fig. 2.1.

The interpretation of $\Delta(\epsilon_{2n})/\pi$ as a quantum defect for the discrete spectrum and modulations (8.31) of the intensity in this region suggest a reinterpretation of the autoionization region. Resonance positions and intensity profiles may be reinterpreted in terms of a complex quantum defect, whose real and imaginary parts provide the position and width of the resonance. This connection will be made in Section 8.6.2.

8.4 MULTICHANNEL TREATMENT

The physicomathematical considerations developed for the two-channel example of $e + H_2^+$ are readily extended to any particle + core system with M closed fragmentation channels, labeled by $j = 1, 2, \ldots, M$, and N open channels labeled by $i = 1, 2, \ldots, N$. We shall develop here the formalism required for this extension following the procedure of Section 8.3.

The quantum defect procedure that was stressed in Section 7.5.1 and also underlies the treatment in Section 8.3 for the case of H_2 will continue to provide the basis for the general multichannel treatment to be developed here. Its key approximation, embodied in (7.40) and (7.41a) for H_2, lies in the factorization of the Jost matrices, reproduced here in the general notation of (7.14),

$$J_{i\alpha}^{\pm}(k_i) = (\pm i)^{l_i} e^{\pm i\delta_i^{(1)}(k_i)} U_{i\alpha} e^{\pm i\delta_\alpha^{(s)}}, \qquad \epsilon_i > 0, \tag{8.32a}$$

$$J_{i\alpha}^{\pm}(\kappa_i) = \mp [D(\kappa_i, l_i)]^{\pm 1} U_{i\alpha} \begin{matrix}\cos\\ \sin\end{matrix} [\beta(\kappa_i) + \delta_\alpha^{(s)}], \qquad \epsilon_i < 0. \tag{8.32b}$$

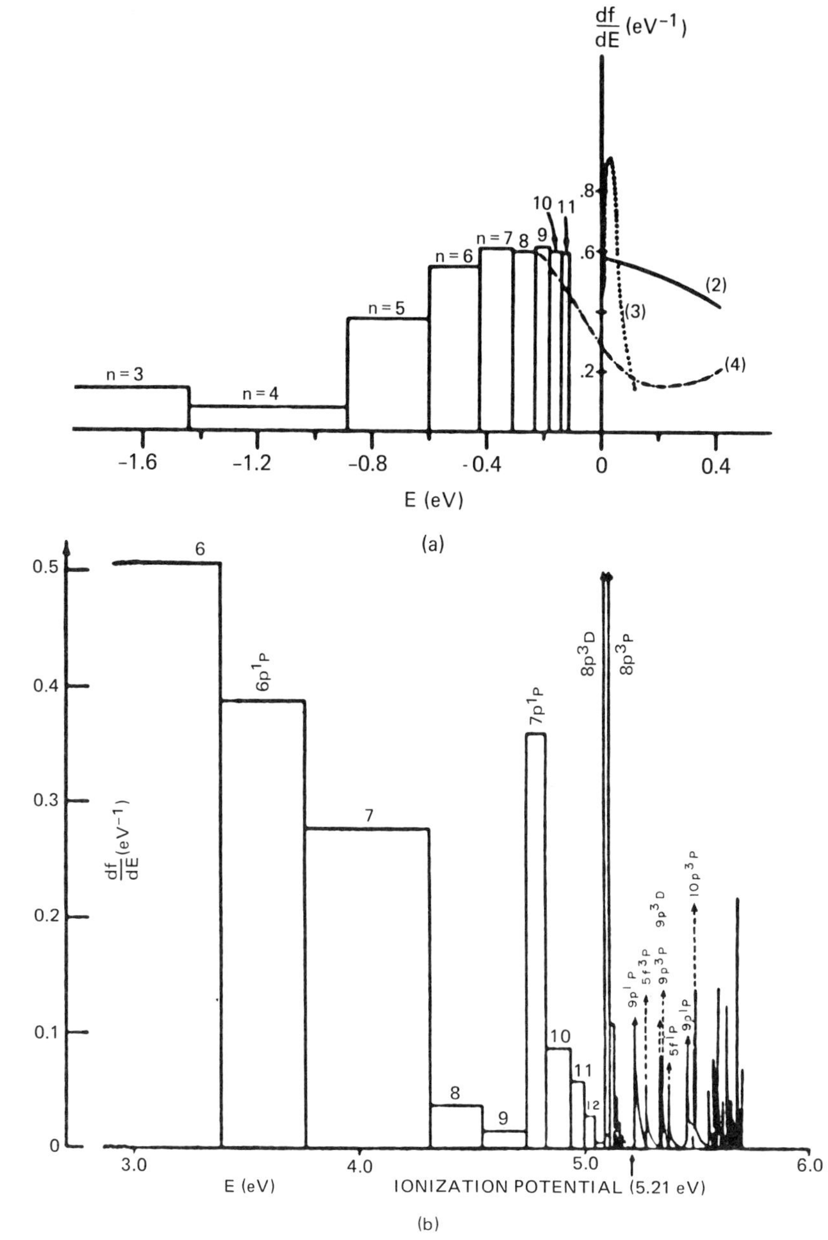

Fig. 8.5. Perturbed absorption series plotted as in Fig. 2.1. (a) Aluminum $3s^23p \to 3s^2nd\ ^1D_2$ perturbed by $3s3p^2\ ^1D_2$. (From Lin [7]. Courtesy of the author and the *Astrophysical Journal*, copyright 1974 by the American Astronomical Society.) (b) Barium $6s^2 \to 6snp\ ^1P_1$ perturbed by $5dnl\ ^1P_1$. (From Fano and Cooper [4].)

The i channels are the fragmentation channels defined in (7.1) and the α channels the eigenchannels of the short-range interaction introduced in Section 7.3.[†]

In accordance with Section 8.3, we apply the bound-state condition (7.17) to each of the M closed channels. This condition takes now the form analogous to (8.4),

$$\sum_\alpha J_{j\alpha}^- A_\alpha = D^{-1}(\kappa_j, l_j) \sum_\alpha \sin[\beta(\kappa_j) + \delta_\alpha^{(s)}] U_{j\alpha} A_\alpha$$

$$= D^{-1}(\kappa_j, l_j) \operatorname{Im} \{e^{-i\beta(\kappa_j)} \sum_\alpha U_{j\alpha} e^{-i\delta_\alpha^{(s)}} A_\alpha\} = 0. \tag{8.33}$$

The representation of $\Sigma_\alpha J_{j\alpha}^- A_\alpha$ on the right-hand side of this equation, as the imaginary part of a complex parameter, has been introduced to underscore that the real part of the same parameter will reappear below as the coefficient of the closed-channel wave function $f^+(\epsilon_j, l_j, r)$, the analog of the coefficient of $f^+(\epsilon_2, l, r)$ in (8.6). [Similarly, Eqs. (8.9) and (8.10) represent the imaginary and real parts of the Σ_Λ expression in (8.8).] The complete wave function, at $r > r_0$, for a standing wave state—even under time reversal—has the form analogous to (8.6)

$$\sum_\alpha F_\alpha(\omega, r) A_\alpha = \sum_i \Phi_i(\omega)(2/\pi k_i)^{1/2} \frac{1}{2i}\left[(-i)^{l_i} f^+(\epsilon_i, l_i, r) \sum_\alpha J_{i\alpha}^+ A_\alpha \right.$$

$$\left. - (i)^{l_i} f^-(\epsilon_i, l_i, r) \sum_\alpha J_{i\alpha}^- A_\alpha \right]$$

$$+ \sum_j \Phi_j(\omega)(\pi\kappa_j)^{-1/2} f^+(\epsilon_j, l_j, r) \sum_\alpha J_{j\alpha}^+ A_\alpha, \qquad r > r_0. \tag{8.34}$$

Here we have $\epsilon_i > 0$ and $\epsilon_j < 0$, by the definition of i and j given earlier. The last parameter $\Sigma_\alpha J_{j\alpha}^+ A_\alpha$ equals the real part of the last factor in (8.33) as anticipated.

The essential aspects of the procedure of Section 8.3, which we should generalize here, are the following. Since collision and photoemission experiments bear directly on the open-channel portions of the wave function (8.34), the amplitudes $\Sigma_\alpha J_{i\alpha}^\pm A_\alpha$ of its outgoing and ingoing waves are conveniently reformulated in terms of effective ("mock") eigenphase shifts Δ, which indirectly reflect the coupling to closed channels. Specifically, this coupling manifests itself through the resonant behavior of the Δ consisting of rapid rises at energies that correspond, at least approximately, to bound-state levels

[†] The procedure to be developed here is equivalent to the multichannel quantum defect theory as developed by Seaton and his collaborators and reviewed by him [8]. Seaton's version makes no substantial use of α channels or Jost functions. Details of the connection between alternative procedures will be indicated in Section 8.6.1, where the α channels are eliminated from explicit consideration.

in the closed channels. The mock phases Δ thus incorporate jointly (1) the effects of the eigenphases $\delta_\alpha^{(s)}$ representing the short-range particle–core interactions, which depend weakly on energy, and (2) the effect of coupling with closed channels, which depends sensitively on the small energies ϵ_j but is nevertheless mediated by the core interaction parameters $\delta_\alpha^{(s)}$ and $U_{i\alpha}$.

Here the presence of more than one open channel requires us to extend the treatment of Section 8.2 by introducing as many mock-eigenchannels as there are open channels, that is, N of them. These eigenchannels will no longer be in one-to-one correspondence with the fragmentation channels i and will thus be labeled by a new index $\rho = 1, 2, \ldots, N$, which replaces $\alpha = 1, 2, \ldots, M + N$. Each eigenfunction's specific amplitude A_α will be labeled $A_\alpha^{(\rho)}$ and will be characterized by an eigenphase shift Δ_ρ. The Jost functions $J_{i\alpha}^\pm$ accordingly will be replaced by new functions $J_{i\rho}^\pm$, which will usually be factored, in analogy to (7.40), as

$$J_{i\rho}^\pm = (\pm i)^{l_i} e^{\pm i\delta_i^{(1)}(k_i)} T_{i\rho} e^{\pm i\Delta_\rho}. \tag{8.35}$$

The determination of Δ_ρ and $T_{i\rho}$ is our next task.

The analogous task was approached in (8.8) by restructuring the expression $\Sigma_\alpha J_{i\alpha}^\pm A_\alpha$. Here we follow the same path, writing

$$e^{\pm i\delta_i^{(1)}} \sum_\alpha U_{i\alpha} e^{\pm i\delta_\alpha^{(s)}} A_\alpha = e^{\pm i(\delta_i^{(1)} + \Delta)} \sum_\alpha e^{\pm i(-\Delta + \delta_\alpha^{(s)})} U_{i\alpha} A_\alpha, \tag{8.36}$$

and requiring the Σ_α on the right to equal a real amplitude coefficient T_i. We introduce the analogs of (8.9) and (8.10),

$$\sum_\alpha \sin(-\Delta + \delta_\alpha^{(s)}) U_{i\alpha} A_\alpha = 0, \tag{8.37a}$$

$$\sum_\alpha \cos(-\Delta + \delta_\alpha^{(s)}) U_{i\alpha} A_\alpha = T_i, \qquad \sum_i T_i^2 = 1. \tag{8.37b}$$

The N equations [(8.37a)] combine with the M equations [(8.33)] to form a homogeneous system of $M + N$ equations in the $M + N$ amplitudes A_α. The consistency of this system requires the vanishing of a determinant with two subsets of M and N rows that depend, respectively, on $\beta(\kappa_j)$ and $-\Delta$,

$$F[\Delta, \beta(\kappa_j)] = \mathrm{Det}\begin{vmatrix} U_{j\alpha} \sin[\beta(\kappa_j) + \delta_\alpha^{(s)}] \\ U_{i\alpha} \sin(-\Delta + \delta_\alpha^{(s)}) \end{vmatrix} = 0. \tag{8.38}$$

Here Δ plays the role of an eigenvalue parameter. Equation (8.38) has N roots Δ because it can be expanded into a polynomial of Nth degree in $\tan(-\Delta)$, much as (8.11) was expanded in its form (8.15). These eigenvalues will be designated as Δ_ρ, the corresponding eigenvectors of systems (8.33) and (8.37a) by $A_\alpha^{(\rho)}$ and their normalization coefficients T_i [(8.37b)] by $T_{i\rho}$. In the absence of open channels ($N = 0$), (8.38) becomes the eigenvalue equation for bound-state levels, namely, the quantum defect form of the multichannel equation (7.18):

$$\mathrm{Det}|U_{j\alpha}\sin[\beta(\kappa_j)+\delta_\alpha^{(s)}]|=0. \tag{8.38$'$}$$

As noted after (7.18), the eigenvectors of the system of equations corresponding to (8.38′) constitute the set of parameters $Z_i^{(n)}$ that characterize the structure of each state.

The mock Jost matrices $J_{i\rho}^{\pm}$ constructed according to this prescription are to be used in the analysis or prediction of experimental observations, typically when combined into scattering or transition matrices, as discussed earlier for the $J_{i\alpha}^{\pm}$. Notice that the $J_{i\rho}^{\pm}$ depend sensitively on the closed channel energies ϵ_j not only through the Δ_ρ in (8.35) but also through the $T_{i\rho}$. The latter dependence is a further, more complicated manifestation of resonances. The examples in Section 8.5 illustrate how to bring out such manifestations in the semiempirical analysis of spectra. The rapidly varying parameters Δ_ρ and $T_{i\rho}$ serve as stepping-stones for the analysis, which aims at the determination of the slowly varying parameters $\delta_\alpha^{(s)}$ and $\{U_{i\alpha}, U_{j\alpha}\}$ on the left-hand side of (8.33) and (8.37).

The mock Jost matrices $J_{i\rho}^{\pm}$ serve, of course, to construct the scattering matrix for the *open channels* as in (7.16),

$$\begin{aligned} S_{ii'} &= \sum_\rho J_{i\rho}^{+}(J^{-\,-1})_{\rho i'} \\ &= i^{l_i}e^{i\delta_i^{(1)}}\sum_\rho T_{i\rho}e^{2i\Delta_\rho}\tilde{T}_{\rho i'}(i)^{l_{i'}}e^{i\delta_{i'}^{(1)}}. \end{aligned} \tag{8.35$'$}$$

This matrix represents observable phenomena at $r\to\infty$. Its analysis in terms of the mock eigenchannels ρ has been introduced here for purposes of convenient classification; alternative representations may be used. The main point is that $S_{ii'}$ embodies the effect of resonances in the closed channels. These effects cause the identification of the ρ channels to depend sensitively on the energy as noted earlier.

8.4.1 Analysis of Resonances

Whereas (8.11), which determines Δ as a function of β_2 for a two-channel system, can be transformed readily into (8.15), which gives Δ explicitly and traces its resonance behavior to a singularity when $\beta_2 \simeq -\pi\mu_2$, no correspondingly simple transformation of (8.38) is available for multichannel systems. An approximate representation of Δ as a function of energy can be obtained in the special, but not infrequent, case when the closed channels j are weakly coupled. In this event the closed channels can be placed into one-to-one correspondence with M of the eigenchannels α; one $M\times M$ submatrix $U_{j\alpha}$ is then nearly diagonal, with diagonal elements ~ 1, and the remaining elements $U_{j\alpha}$ can be treated as weak interaction parameters. Each resonance to be observed in the open channels can then be traced to the vanishing of one of the factors $\sin(\beta_j + \delta_\alpha^{(s)})$ in (8.38) whose coefficient $U_{j\alpha}$ approaches unity. This is illustrated by the example of (8.13–8.17) with a single closed channel. If this closed channel $j = 2$ is in one-to-one correspondence with, say, the eigenchannel $\alpha = \Pi$, so that $U_{2\Pi} = 1$, $U_{2\Sigma} = 0$ (i.e., $\phi = 0$), then the singularity of Δ occurs at $\beta_2 = -\delta_\Pi^{(s)}$.

In the general case of strong coupling of the closed channels, the resonances cannot be assigned to specific (closed) fragmentation channels. However, they correspond to peak values of the time delay parameter introduced in Section 4.4 and in (8.20). An elementary approach considers the energy dependence of the "eigenphase sum" $\Delta_{\text{tot}} = \Sigma_\rho \Delta_\rho$

$$2\hbar \frac{d\Delta_{\text{tot}}}{dE} = 2\hbar \sum_j \frac{\partial \Delta_{\text{tot}}}{\partial \beta_j} \frac{d\beta_j}{dE}. \tag{8.39}$$

Sharp variations of Δ_{tot} mark the positions of the resonances (Fig. 8.6).

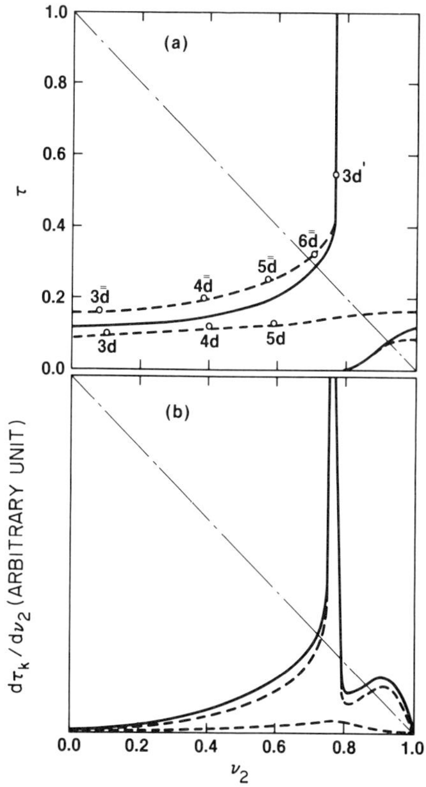

Fig. 8.6. (a) Eigenphase sum and (b) time delay for photoionization of chlorine; — —, experiment; ——, schematized two-channel model. (From Wang and Lu [9].)

More systematically, we may start with the extended definition of the time delay in its matrix form as given in (4.16d). In this expression for the time delay matrix T_d of a multichannel problem the S matrix may be constructed as in (8.35′) by using the Jost functions in (8.35) obtained by the elimination of the closed channels.† Diagonalization of T_d then identifies eigenchannels of time delay, which provide independent contributions to the scattering process; each resonance corresponds to a maximum of one of these contributions. (Note that T_d does *not* have the same eigenvectors as S itself.) This method of analysis serves to predict how each resonance would manifest itself in different open fragmentation channels. No complete study of this manifestation has been carried out yet in multichannel phenomena [10].

We see in the next section an example of analysis of discrete multichannel spectra in which (1) the wealth of experimental data permits a ready identification of the ρ eigenchannels and evaluation of short-range parameters and (2) the weak coupling of closed channels with different orbital numbers l_j preserves the association of resonances with each of these channels. In this event the decay width of the resonance, that is, twice the reciprocal of the corresponding eigenvalue of T_d (see Section 8.3.1), may be incorporated [8] as an imaginary part of a quantum defect, as will be seen in Section 8.6.2. On the other hand, little progress has been achieved on the broader task of identifying resonances of strongly coupled closed channels from their effects on the cross sections for elastic and inelastic collisions in several alternative open channels.

8.5 SEMIEMPIRICAL ANALYSIS OF MULTICHANNEL PHOTOABSORPTION: ARGON AND OTHER SPECTRA [11]

The photoabsorption spectra of the noble gases, Ne through Xe, have served as testing ground for the analysis of the multichannel phenomena outlined in Section 8.4. The example of p → s excitation in Ne introduced in Section 7.6 could be treated approximately on a two-channel basis because fragmentation channels with different l_i remain weakly coupled upon interaction with a Ne^+ core, but this coupling increases in the heavier elements. This section accents a semiempirical analysis of data. A more complete analysis, including the results of *ab initio* calculations, is now available [12].

8.5.1 Identification of Channels

The first step of the analysis consists of listing and characterizing the short-range eigenchannels and the fragmentation channels for the spectrum of interest. Since noble gases have a $p^6\,{}^1S$ ground state, photoabsorption excites their atoms to $J = 1$, odd parity channels with a $p^5\,{}^2P^0$ core whose

† Alternatively, the S matrix may be constructed directly from (7.16) by using $J^{\pm}_{i\alpha}$, but then the subsequent diagonalization of T_d should be performed with the subsidiary condition (8.33) on the $J_{i\alpha}$, thus reducing the size of the matrix from $M + N$ to N.

doublet interval $E_{1/2} - E_{3/2}$ increases from 0.1 to 1.2 eV along the sequence. The fragmentation channels have $l_i = 0$ or 2 and $j_i = \frac{1}{2}, \frac{3}{2}$, and $\frac{5}{2}$. The three values of j_i, combined with the core's $J_i = \frac{1}{2}$ or $\frac{3}{2}$ yield five channels with $J = 1$ and, hence, a fairly complex situation.† However, the analysis is facilitated by a residual degeneracy, since three of the channels have $E_i = E_{3/2}$ and two have $E_i = E_{1/2}$; that is, we deal here with two distinct ionization thresholds only. Spectra with larger numbers of distinct thresholds have been analyzed more recently and with more elaborate computation; we shall refer to them further later.

Here we review the analysis for Ar, whose main objective is to determine five eigenchannel parameters $\delta_\alpha^{(s)}$ and the 5×5 matrix $U_{i\alpha}$ by fitting over 50 known levels of the discrete spectrum with $J = 1$ and odd parity. The fragmentation channels are identified in jj coupling by the pairs of quantum numbers J_i of the core and $l_i j_i$ of the excited electron, as follows:

$$\begin{array}{cccccc} i = & 1 & 2 & 3 & 4 & 5 \\ (J_i, l_i j_i) = & (\frac{3}{2}, 0\,\frac{1}{2}) & (\frac{3}{2}, 2\,\frac{3}{2}) & (\frac{3}{2}, 2\,\frac{5}{2}) & (\frac{1}{2}, 0\,\frac{1}{2}) & (\frac{1}{2}, 2\,\frac{3}{2}) \end{array}. \tag{8.40}$$

The two channels with $i = 4, 5$ have the higher threshold $E_{1/2}$ and would be relabeled as $j = 1, 2$ in the notation of Section 8.4. The short-range eigenchannels α are expected to be identified approximately by LS quantum numbers together with the l of the excited electron, but less acurately so than in the Ne example of Section 7.6. For labeling purposes we set

$$\begin{array}{cccccc} \alpha = & 1 & 2 & 3 & 4 & 5 \\ (l\,L, S) = & (2\,2, 1) & (2\,1, 0) & (2\,1, 1) & (0\,1, 1) & (0\,1, 0) \end{array}. \tag{8.41}$$

8.5.2 Levels and Phase shifts

We take the point of view introduced in Section 8.3.3 of assigning each discrete level below the lowest threshold $E_{3/2}$ to one of the three mock eigenchannels $\rho = 1, 2, 3$ that are identified *above* this threshold by the eigenvalues and eigenvectors of (8.38),

$$F(\Delta, \pi\nu_{1/2}) = \mathrm{Det}\begin{vmatrix} U_{j\alpha} \sin(\pi\nu_{1/2} + \delta_\alpha^{(s)}) \\ U_{i\alpha} \sin(-\Delta + \delta_\alpha^{(s)}) \end{vmatrix} = 0. \tag{8.42}$$

Note that we have entered here the value $\pi\nu_{1/2}$ for the parameter β_j which is appropriate to the long-range Coulomb field external to the core, according to Table 5.1:

$$\beta_j = \pi\nu_{1/2} = \pi/\kappa_{1/2} = \pi[-2(E - E_{1/2})]^{-1/2}. \tag{8.43}$$

Equations (8.42) and (8.43) define Δ as an implicit function of E, $\Delta(E - E_{1/2})$.

† In this section we will drop the bar symbol on J_i and E_i that was used in Chapter 7 for such quantities that pertain to the Ne^+ core.

The levels of the discrete spectrum must satisfy the analogs of (8.29) and the equivalent (8.30), namely,

$$\beta_{i=1,2,3}(E - E_{3/2}) + \Delta(E - E_{1/2}) = \pi\nu_{3/2} + \Delta(E - E_{1/2}) = n\pi, \quad (8.44)$$

$$E_n - E_{3/2} = -\frac{1}{2}\frac{1}{[n - \Delta(E_n - E_{1/2})/\pi]^2}. \quad (8.45)$$

The goal is to determine the parameters $\{U_{i\alpha}, U_{j\alpha}\}$ and $\delta_\alpha^{(s)}$ of (8.42) by requiring Δ to satisfy (8.42) and (8.45) for all discrete levels E_n of the photoabsorption spectrum.

8.5.3 Lu–Fano Plots

This task is approached, as in Section 7.6 for Ne, by representing each observed level E_n as a point on a graph with the coordinates

$$\{\nu_{1/2}, \nu_{3/2}\}_n \equiv \{[-2(E - E_{1/2})]^{-1/2}, [-2(E - E_{3/2})]^{-1/2}\}_{E=E_n}. \quad (8.46)$$

These points must satisfy (8.42), that is, they must lie on three curves—corresponding to the three roots Δ_ρ of (8.42)—that represent this equation. Figure 8.7a shows the three curves, with ordinates $\Delta/\pi = -\nu_{3/2}$ plotted mod 1 and abscissas $\nu_{1/2}$. The dot-dashed lines represent the relationship between $\nu_{3/2}$ and $\nu_{1/2}$ defined by (8.46). The points on the curves represent the experimental values of E_n. These points are so sparse in the lower range of $\nu_{1/2}$ that they would not suffice to define the curves in this range. The function $F(\Delta, \nu_{1/2})$ was actually constructed from the graph of Fig. 8.7b with $\nu_{1/2}$ plotted mod 1; the experimental points have been fitted by allowing the curve $F(\Delta, \nu_{1/2}) = 0$ to shift progressively as a function of the total energy E through a linear dependence of the $\delta_\alpha^{(s)}$ and $\{U_{i\alpha}, U_{j\alpha}\}$ on E. Note that the three curves in Fig. 8.7b may be viewed as forming a single continuous curve on folding back each intersection with the right edge of the figure into the intersection with the left edge and the same ordinate and similarly treating the upper and lower edges. The single curve thus obtained has three points of intersection with any vertical straight line and two with any horizontal one.

8.5.4 Short-Range Eigenphases

The determination of the parameters $\delta_\alpha^{(s)}$ is straightforward because (8.38) requires Δ to coincide with one of the $\delta_\alpha^{(s)}$ at each intersection of the curves $F(\Delta, \pi\nu_{1/2}) = 0$ with the diagonal $\nu_{3/2} = \nu_{1/2}$. [The wave function (8.34) reduces then to a single eigenfunction F_α of the short-range interaction.] The

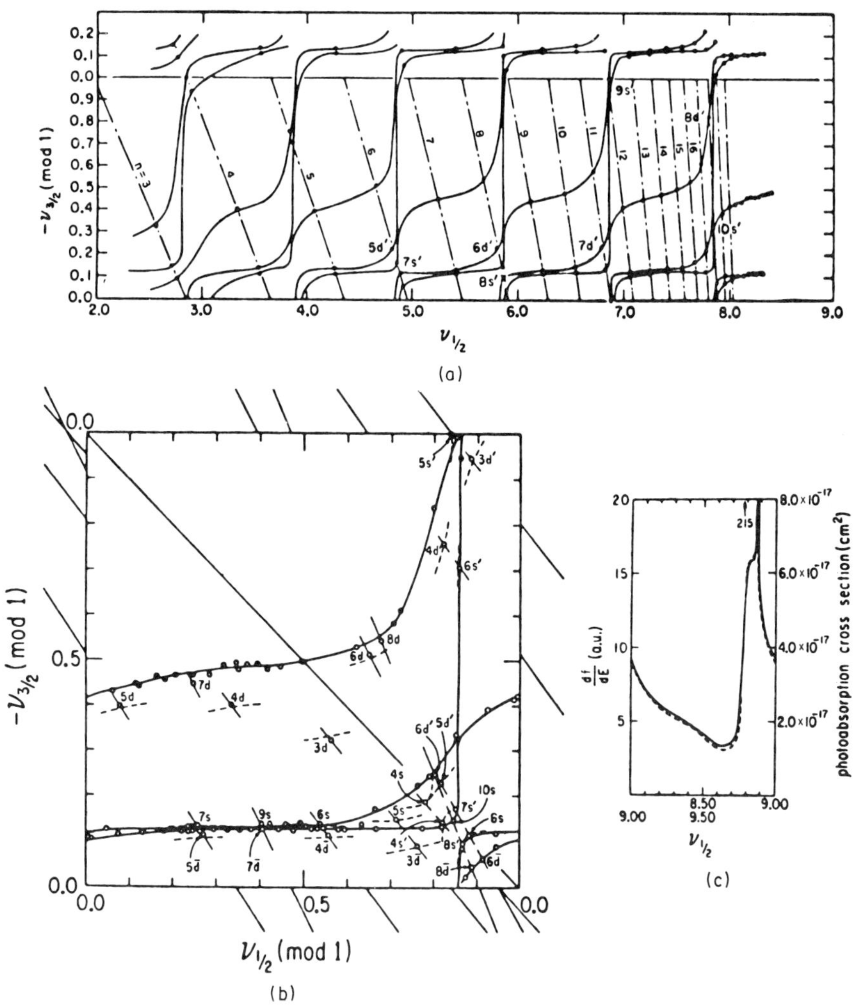

Fig. 8.7. Photoabsorption of Ar. (a) Extended plot of quantum defects (8.46); ——, $F(-\pi\nu_{3/2}, \pi\nu_{1/2}) = 0$ [Eq. (8.42)]; - · -, Eq. (8.45). (b) Reduced (Lu–Fano) plot of (a) modulo 1. Open circles represent observed levels. (c) Autoionization profile for $8.64 \leq \nu_{1/2} \leq 9.64$; — —, experimental data; ——, theoretical fit. (From Lee and Lu [13].)

dependence of each $\delta_\alpha^{(s)}$ on the energy E is determined by requiring the relevant branch of the $F(\Delta, \pi\nu_{1/2})$ curve to traverse the experimental points with lower values of n. For example, the intersection at $\Delta \sim 0.5$ in the center of Fig. 8.7b corresponds to $\alpha = 3$, that is, to nd ^{3}P levels according to (8.41); the increasingly lower positions of the points marked 6d, 5d, 4d, 3d requires the intersection of the upper branch of the curve with the diagonal to shift downward and thus determines the variation of $\delta_3^{(s)}$ as a function of E. The results are represented by

$$\begin{array}{lllll} \alpha = & 1 & 2 & 3 & 4 \\ \delta_\alpha^{(s)}/\pi = & 0.24 + 1.22\epsilon & 0.070 + 0.66\epsilon & 0.500 + 1.84\epsilon & 0.154 - 0.28\epsilon \end{array}$$

$$\begin{array}{c} 5 \\ 0.109 - 0.40\epsilon \end{array}, \qquad (8.47)$$

where $\epsilon = E - E_{3/2}$ a.u.

8.5.5 The Transformation Matrix

Having thus determined the $\delta_\alpha^{(s)}$, it remains to be seen what values must be given to the elements of the orthogonal matrix $U_{i\alpha}$ for the curve $F(\Delta, \pi\nu_{1/2})$ to traverse all the points of the Lu–Fano plot. Since the α channels are presumably close to the LS coupled channels indicated by (8.41), we approximate the U matrix initially by the matrix of angular momentum recoupling coefficients that transforms (jj) into (LS) states. Indicating again $j = 1, 2$ by $i = 4, 5$ as in (8.40), we set initially

$$\{U_{i\alpha}\} \sim [(\bar{L}_i = 1, \bar{S}_i = \tfrac{1}{2})J_i, (l_i\tfrac{1}{2})j_i, 1|(\bar{L}_i, l_i)L_\alpha, (\bar{S}_i, \tfrac{1}{2})S_\alpha, 1), \qquad (8.48)$$

as the recoupling coefficient for passing from (jj) to (LS) coupling (Section 7.4.1). Subsequent adjustments improve the fitting of (8.42) to the experimental points. In the case of argon the necessary adjustments were only in the second or third digit of each element, larger adjustments were needed for Kr and Xe. Moreover, no significant dependence of this matrix on energy was detected. Note that the simple transformation of (jj) to (LS) disregards any coupling of $l = 0$ and 2 in the matrix (8.42). In the absence of this coupling (8.42) would factor into two separate equations for two s and three d channels. The five-branched plot of $F = 0$ in Fig. 8.7b would then separate into two separate plots that intercross freely; the two-branched plot for s channels would be the analog of the plot of Fig. 7.8a for Ne. Very narrowly avoided crossings of the curves are indeed evident in Fig. 8.7b, indicating a weak coupling between s and d channels.

8.5.6 Intensity Distribution

There remains to be considered the determination of the five independent dipole matrix elements $(F_\alpha|x|0)$ through the analysis of the spectral distribution of photoabsorption intensities. Notice at the outset that the three matrix elements with $\alpha = 1, 3, 4$ correspond, at least approximately, to excitation of triplet states from a ground-state singlet and should accordingly be very small. For Ar they were found to be indeed smaller than the others by two or more orders of magnitude.

The fitting to experimental data was based primarily on the intensity profiles of resonances in the range $E_{3/2} < E < E_{1/2}$. The first resonance is shown in Fig. 8.7c; those in succeeding unit intervals of $\nu_{1/2}$ are equal. Two separate resonances are seen, attributable to the s and d closed channels [$i = 4, 5$ in (8.40)] which are weakly coupled to one another. The sharpest resonance belongs to the s channel, which is also weakly coupled to the open channel $i = 1$ because of the small difference of the s eigenphases $\delta_4^{(s)}$ and $\delta_5^{(s)}$. The fitting was then adjusted to optimize agreement with the entire intensity distribution in the discrete and in the low-energy continuum. The resulting values (in atomic units) are

$$\begin{array}{lccccc} \alpha = & 1 & 2 & 3 & 4 & 5 \\ (F_\alpha|x|0) = & 3.0 \times 10^{-3} & 2.1 & -0.020 & 1.0 \times 10^{-4} & 1.4 \end{array} \qquad (8.49)$$

Two aspects of the intensity distribution analysis are noted:

(1) Photoionization to levels $E > E_{1/2}$ yields two groups of photoelectrons, with energies $E - E_{3/2}$ and $E - E_{1/2}$, respectively. The intensity ratio of these groups, $I_{3/2}/I_{1/2}$, can be measured fairly accurately; $I_{3/2}$ equals the sum of transition probabilities to the channels $i = 1, 2, 3$ and $I_{1/2}$ is the sum of the other two. If the matrix $\{U_{j\alpha}, U_{i\alpha}\}$ corresponded exactly to the (LS)-to-(jj) transformation (8.48) this ratio would have the "statistical weight" value of 2. The fitting described previously for Ar involved initially an observed value of 1.98 but was revised later to a newer measurement of 1.87 ± 0.06 near the threshold $E_{3/2}$. (For Xe the corresponding value is $\lesssim 1.6$.)

(2) The analysis of Section 8.4 leads to sorting out the oscillator strength at all $E < E_{1/2}$ into three mock-eigenchannel contributions

$$df^{(\rho)}/dE = [2m(E - E_0)/\hbar^2]\,|\sum_\alpha A_\alpha^{(\rho)}(F_\alpha|x|0)|^2, \qquad \rho = 1, 2, 3, \quad (8.50)$$

where $A_\alpha^{(\rho)}$ $(\alpha = 1, 2, \ldots, 5)$ are the eigenvectors of the matrix in (8.42) corresponding to $\Delta = \Delta_\rho$. The magnitudes of these contributions vary rapidly and periodically as functions of $\nu_{1/2} = [-2(E - E_{1/2})]^{-1/2}$, as shown in Fig. 8.8a for the several ρ values. These contributions are not sorted out experimentally in the continuum at $E > E_{3/2}$, even though we notice that $\rho = 1$ includes the s resonance and indeed is almost coincident with an $l = 0$ classification, while the d resonance belongs to $\rho = 3$. In the discrete spectrum, however, the several ρ values correspond to different branches of the curves in Fig. 8.6 and serve to label different sets of spectral lines as shown in Fig. 8.8b. The points in this panel represent values of $df^{(\rho)}/dE$ derived from the estimated intensity of each line on spectral plates. Note the sharp

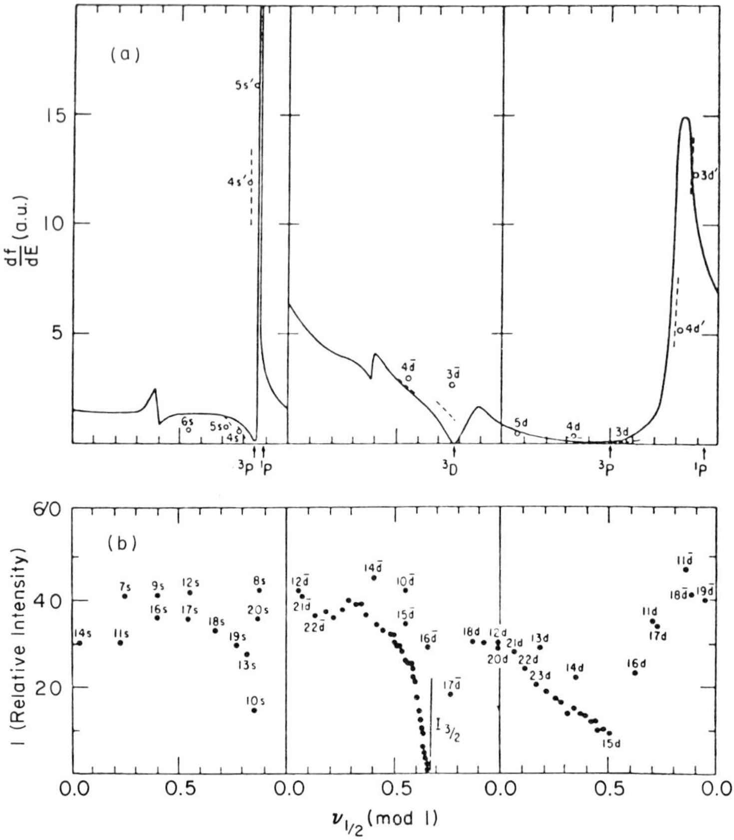

Fig. 8.8. Intensity distribution in Ar for separate ρ channels. (a) Autoionization continuum and (b) discrete spectrum. (From Lee and Lu [13].)

variations along most of the line sequences and particularly the points of zero. As noted in Fig. 8.8a, these zeros correspond to values of $\nu_{1/2}$ that coincide with one of the $\delta_\alpha^{(s)}/\pi$ with a triplet label. This remark accounts for otherwise unexplained anomalies of the spectrum.

In conclusion, the MQDT analysis gives a complete accounting of the spectrum of excitations from the outer p shell of the noble gases. It gives the energy positions and oscillator strengths of discrete levels; the energies, widths, and intensities of autoionizing levels; and the branching ratios for the photoelectron distribution in the continuum region in terms of 22 parameters: the two ionization limits, the ten independent quantities that characterize the $U_{i\alpha}$ matrix, the five short-range eigenphases, and the five dipole matrix elements in (8.49).

8.5.7 Angular Distribution and Spin Orientation of Photoelectrons

Observations of the angular distribution and spin orientation of photoelectrons are increasingly serving to complement the observations of energy levels and photoelectron yield in the process of fitting MQDT parameters. The angular distribution has been discussed in Section 7.5.3 for the photoionization of H_2 in the continuum above the thresholds considered there and again in Section 7.10. In the autoionization region the parameter β is modulated over the resonances much as is the total cross section (Fig. 8.9).

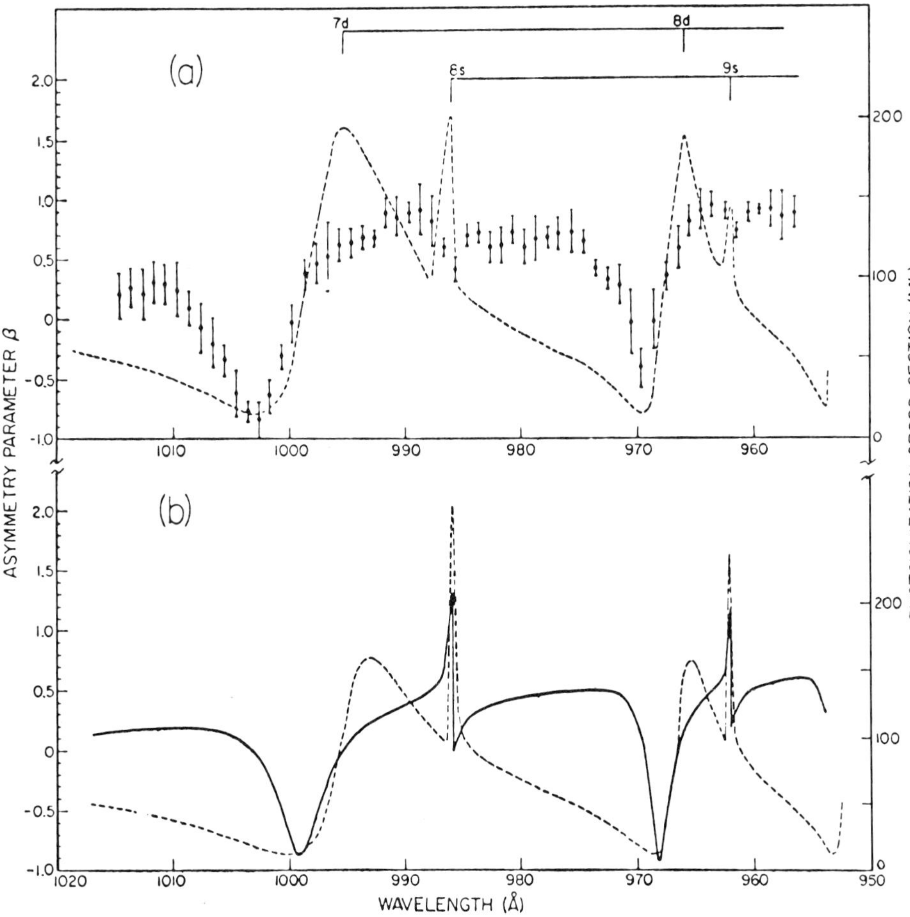

Fig. 8.9. Asymmetry parameter β of photoelectron angular distribution and photoionization cross section for Xe. (a) Experimental β values and photoabsorption resonances (— —); (b) theoretical β values (——) and photoabsorption cross sections (— —). (From Samson and Gardner [14].)

More recently, experimental ability to measure the degree of spin polarization along various axes has developed. Three orthogonal components, generally labeled ξ, η, and ζ, have been measured as functions of energy, particularly for the noble gases. The sharp spectral dependences of these parameters resemble one another, but with appreciable differences [15].

Each of the parameters, ξ, η, and ζ, can be represented analytically as a quadratic function of the amplitudes $A_\alpha^{(\rho)}$ and can be expressed in terms of the short-range parameters $\delta_\alpha^{(s)}$ and $U_{i\alpha}$ and the long-range phases $\delta_i^{(l)}$ and $\beta(\kappa_i)$. Sharp dependences on the energy arise, of course, from these latter quantities characteristic of the long-range field.

8.5.8 Applications to Other Spectra

The analysis described earlier for Ar has been applied to all the noble gas spectra (group 0). The alkalis, and their isoelectronic ions, have single-channel spectra for which the fragmentation and eigenchannels coincide; thereby the $U_{i\alpha}$ matrix reduces to unity, and analysis of the spectral levels provides directly the short-range phases $\delta_\alpha^{(s)}$. The alkaline earths and the elements in the further columns of the periodic system have multichannel photoabsorption and generally more than two ionization thresholds within a few electron volts. The graphical approach to spectral analysis described earlier is then no longer adequate. Nevertheless, it has proved possible to apply equivalent procedures to an increasing number of elements, particularly to the alkaline earths and group IV atoms, essentially by trial-and-error fitting of the parameters of (8.38).

A striking phenomenon identified in these studies is exemplified in Fig. 8.5a. It occurs when two strongly coupled fragmentation channels (i, j) have ionization thresholds separated by $E_j - E_i \sim 2$ eV. The intense absorption due to excitation of one of the lower levels of the j channel ($3s3p^2\ {}^2D$ of Al in the example of Fig. 8.5a) is so broadened by channel coupling as to intermix with the entire Rydberg series absorption of the ith channel ($3s^2nd\ {}^2D$ in our example). The mixing is so strong, over a sizable energy range, that the jth channel line is no longer identifiable except as a modulation of the other spectral series.

In the alkaline earths, the diversity of spectral levels excited by one- and two-photon absorption is very large. Extensive efforts have been devoted to the MQDT analysis of these spectra, with results that remain fragmentary. The splitting of ionic thresholds, now no longer due to spin-orbit coupling but to electron–core interactions, is much larger. Furthermore, additional thresholds corresponding to excitation from lower closed shells lie nearby. In the heavier alkaline earths, for example, the ionic core can be left not only in ns and np states ($n = 6$ for Ba) but also in $(n-1)$d. The normally closed shell $(n-1)$p is also fairly readily excited. The coupling between the various channels converging to these limits is very strong because of close interactions within the valence shell and the modest separation ($\simeq 1$ eV) of the ionization thresholds. MQDT analyses have started with Lu–Fano plots restricted to the two lowest limits but have progressed by computer graphics and other techniques to include several limits [16]. Figure 8.10 illustrates the strong interaction of the two lowest channels. Note the widely avoided

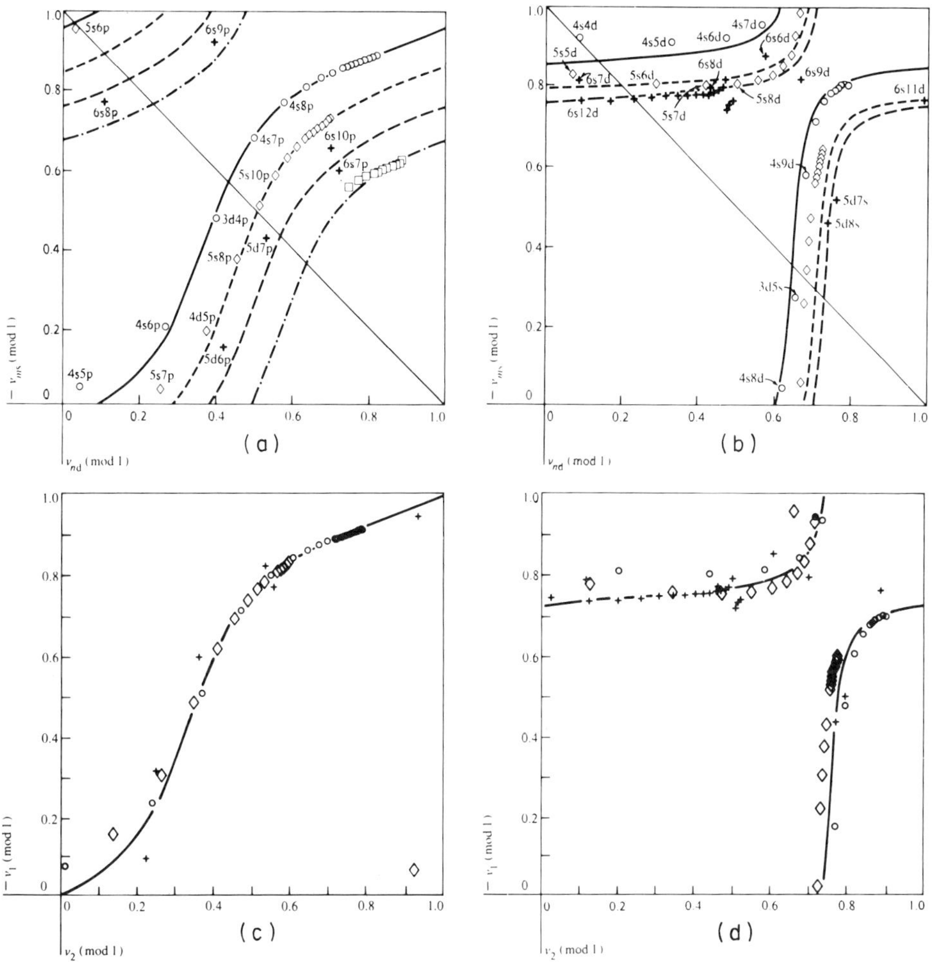

Fig. 8.10. Lu–Fano plots of Rydberg series of the alkaline earths converging to *n*s and $(n-1)$d limits: ○, Ca; ◇, Sr; +, Ba; □, Ra. (a) $J = 1$ odd parity, (b) $J = 2$ even parity, (c) and (d) same as (a) and (b) with scales shifted to display the universal shape. (From Wynne and Armstrong [17]. Copyright 1979 by International Business Machines Corporation, reprinted with permission.)

crossings between channels. A remarkable result for such a case emerged, namely, that the 2×2 matrix $U_{i\alpha}$ is represented by a rotation through $\simeq 45°$. An interpretation of this finding will be given in Section 11.2.3.

The major focus of atomic spectroscopy from 1950 to 1970 has been directed to unraveling the multitude of levels arising from the lowest configurations of transition elements, lanthanides, and actinides, that is, species with several open shells. The MQDT analysis in this section is not relevant to this effort, but applications have begun to the Rydberg series of groups III–VII [18]. We will return to this subject in Section 11.5. Likewise, MQDT studies of the Rydberg series of molecules, both diatomic [19] and polyatomic [20] ones, have begun. Even more complex manifestations of strong channel coupling may emerge from such studies.

8.6 ELIMINATION OF OPEN CHANNELS

The preceding sections have described how the bound states in the closed fragmentation channels of a particle–core complex manifest themselves through resonant anomalies of the open channels. Parametrization of these anomalies, through the effective phase shifts Δ_ρ and the scattering eigenchannel coefficients $T_{i\rho}$, makes it unnecessary to refer to the closed channels explicitly. Similarly, we may wish to eliminate any explicit reference to irrelevant open channels when dealing with elastic scattering within a given channel. The existence of unobserved open channels will generally manifest itself as a diversion of scattering probability away from the process of interest into the unobserved channels, whereas unobserved closed channels manifest themselves as scattering resonances.

The reduction of scattering probability is readily identifiable in the two-channel problem discussed in Section 7.2, where the nonspin-flip process plays the role of elastic scattering. Compare the probability amplitude for nonflip, T^{nf}, shown in (7.7), with the corresponding expression for a single-channel process in (4.6+), $[\exp(2i\delta_l)-1]/2i$. The single-channel phase shift δ_l is first replaced in (7.7) by an average over two channels, $\frac{1}{2}(\delta_{l+} + \delta_{l-})$, which is reasonable. *In addition*, the scattering matrix element $\exp i(\delta_{l+} + \delta_{l-})$ is reduced in magnitude by the factor $\cos(\delta_{l+} - \delta_{l-}) \leq 1$, which represents the diversion of probability amplitude into the spin-flip process. Indeed, the modulus of the spin-flip amplitude in (7.7) is proportional to the complementary factor $\sin(\delta_{l+} - \delta_{l-}) = [1 - \cos^2(\delta_{l+} - \delta_{l-})]^{1/2}$.

The expression of the scattering matrix in terms of Jost matrices [(7.16)] performs the elimination of irrelevant channels automatically for a general problem, especially when the Jost matrices themselves can be factored as in

(8.32). In this case, we have

$$S_{ji} = i^{l_j} e^{i\delta_j^{(1)}} \left[\sum_\alpha U_{j\alpha} e^{2i\delta_\alpha^{(s)}} (U^{-1})_{\alpha i} \right] i^{l_i} e^{i\delta_i^{(1)}}. \tag{8.51}$$

This formula shows how the modulus of S_{ji} would be unity for $i = j$ (and zero otherwise) if the eigenphase shifts $\delta_\alpha^{(s)}$ were all equal; this modulus is certainly smaller than unity otherwise. Two remarks are in order here:

(1) Even though no fragmentation channels appear in (8.51) other than the pair (i, j), the Σ_α extends over the entire set of eigenchannels of the short-range interaction. The same remark applies to the elimination of the closed channels in Sections 8.3 and 8.4. Thus, we have, in fact, eliminated only references to the identity of the irrelevant fragmentation channels, but the dimensionality of the multichannel system has not yet been abridged.

(2) To attain effective reduction of the *number of parameters* required for the evaluation of S_{ji} in (8.51), and in analogous expressions, we must go further by replacing a large part of the Σ_α in these equations by a phenomenological estimate or by a measured value of some relevant parameter. This procedure of representing the aggregate effect of interactions with numerous systems by a few appropriate parameters underlies the definition of familiar macroscopic quantities such as electric resistance, elastic moduli, the relaxation times of spin precessions, and so on. This procedure is a manifestation of the semimacroscopic approach to atomic phenomena emphasized in Chapter 1.

8.6.1 Formulation through Reaction Matrices

We introduce here an alternative formulation of channel elimination that is instructive and in rather wide use. Let us return to the expansion (7.14) of an eigenchannel wave function $F_\alpha(\omega, r)$ into fragmentation channel functions, replacing the $\{f^-, f^+\}$ base pair with the standing wave pair $\{f, g\}$ of Chapter 5. This alternative expansion is

$$F_\alpha(\omega, r) = \sum_i \Phi_i(\omega)[f(\epsilon_i, l_i, r) \cos \delta_\alpha^{(s)} - g(\epsilon_i, l_i, r) \sin \delta_\alpha^{(s)}] U_{i\alpha}, \tag{8.52}$$

as obtained by combining the more explicit form (7.39) with (5.6), (5.7), and (5.14). A superposition of these functions

$$\sum_\alpha F_\alpha(\omega, r) A_\alpha = \sum_i \Phi_i(\omega) \sum_\alpha [f(\epsilon_i, l_i, r) \cos \delta_\alpha^{(s)} - g(\epsilon_i, l_i, r) \sin \delta_\alpha^{(s)}] U_{i\alpha} A_\alpha \tag{8.53}$$

can be expressed in terms of *fragmentation channels only* by means of the alternative set of coefficients

$$C_j = \sum_\alpha U_{j\alpha} \cos \delta_\alpha^{(s)} A_\alpha, \tag{8.54}$$

and a real symmetric reaction matrix, analogous to (4.16c). This matrix, $K_{ij}^{(s)}$, is introduced through the identity

$$U_{i\alpha} \sin \delta_\alpha^{(s)} = \sum_\beta U_{i\beta} \tan \delta_\beta^{(s)} \sum_j (U^{-1})_{\beta j} U_{j\alpha} \cos \delta_\alpha^{(s)} = \sum_j K_{ij}^{(s)} U_{j\alpha} \cos \delta_\alpha^{(s)}. \tag{8.55}$$

The index s reflects the same dependence on short-range interactions as in $\delta_\alpha^{(s)}$. [For convenience, this index will not always be indicated in what follows but throughout the rest of this chapter, K will refer to $K^{(s)}$.] With this matrix, (8.53) is recast in the form

$$\sum_\alpha F_\alpha(\omega, r) A_\alpha = \sum_i \Phi_i(\omega) \sum_j [f(\epsilon_i, l_i, r)\delta_{ij} - g(\epsilon_i, l_i, r_i) K_{ij}^{(s)}] C_j, \tag{8.56}$$

which is familiar in close coupling treatments.

The parametrization of the wave function (8.56) in real form through the reaction matrix $K_{ij}^{(s)}$ is appropriate to our goal because this single matrix incorporates all the $\delta_\alpha^{(s)}$ and $U_{i\alpha}$. Algebraic manipulation to incorporate the appropriate boundary conditions is carried out conveniently on this matrix. Since the channels to be eliminated play a passive role of removing a portion of the scattered wave function, the boundary condition amounts to equating the f_i^- component to zero, which would represent incoming amplitude in those channels. For the purpose of eliminating open channels, we revert to the function $\{f^-, f^+\}$ to the extent of identifying the coefficient of f^-. To this end, we write the expression in brackets in (8.56), using (5.6), (5.7), and (5.14), as

$$\begin{aligned} f_i \delta_{ij} - g_i K_{ij} &= (1/2i)[(if_i - g_i)(\delta_{ij} + iK_{ij}) + (if_i + g_i)(\delta_{ij} - iK_{ij})] \\ &= (2/\pi k_i)^{1/2} (1/2i)[e^{i\delta_i^{(1)}} f_i^+ (\delta_{ij} + iK_{ij}) - e^{-i\delta_i^{(1)}} f_i^- (\delta_{ij} - iK_{ij})]. \end{aligned} \tag{8.57}$$

The condition for channel elimination is then

$$C_i - i \sum_j K_{ij} C_j = (1 - iK_{ii}) C_i - i \sum_{j \neq i} K_{ij} C_j = 0. \tag{8.58}$$

Thereby, the coefficient C_i of the ith channel that is being eliminated is expressed in terms of the remaining C_j.

Here we aim at eliminating a large subset of all channels. In Section 8.4 the closed channels to be eliminated were called j and those to be retained were called i, and this distinction was maintained. An extended capability is provided by a more compact notation of operator algebra, in which the set of

C_i's is regarded as components of a vector **C**.† Two projection operators P and $Q = 1 - P$ select the channels to be retained or eliminated, respectively. The extension of (8.58) to the subset projected by Q is

$$Q(1 - iK)\mathbf{C} = Q[1 - iK(P + Q)]C = (Q - iQKQ)\mathbf{C} - iQKP\mathbf{C} = 0, \quad (8.59)$$

from which the subset to be eliminated is expressed in terms of the one retained as

$$Q\mathbf{C} = i[1 - iQKQ]^{-1}QKP\mathbf{C}. \quad (8.60)$$

Substitution of this formal solution reduces the wave function (8.56) to

$$\begin{aligned} &\sum_{i\epsilon P} \Phi_i(\omega) \sum_{j\epsilon P} \left\{ f(\epsilon_i, l_i, r)\delta_{ij} - g(\epsilon_j, l_j, r)\left[PKP + iPKQ \frac{1}{1 - iQKQ} QKP\right]_{ij} \right\} C_j \\ &\quad + \sum_{i\epsilon Q} \Phi_i(\omega) \left[\left(\frac{2}{\pi k_i} \right)^{1/2} \exp(i\delta_i^{(1)}) f^+(\epsilon_i, l_i, r) \right] \\ &\quad \times \sum_{j\epsilon P} \left(QKQ \frac{1}{1 - iQKQ} QKQ\right)_{ij} C_j. \end{aligned} \quad (8.61)$$

This modified form of the wave function depends only on coefficients C_j of the subset P. The elimination of the remaining C_j, possibly most of the initial set, is reflected here by the occurrence of a *modified non-Hermitian* reaction matrix. Setting in (8.61) $(1 - iQKQ)^{-1} = (1 + iQKQ)/[1 + (QKQ)^2]$ shows that the elimination adds to K one real and one imaginary term.

The occurrence of a non-Hermitian K matrix in the P subspace corresponds to the occurrence of a nonunitary S matrix displayed by (8.51) for an analogous subspace. Both of these effects involve the addition of imaginary terms to real ones. [Thus, the term $\exp[-i(\delta_{l+} + \delta_{l-})] \cos(\delta_{l+} - \delta_{l-})$ in the expression (7.7) of T^{nf} discussed earlier, could be written in terms of an imaginary contribution to the phase shift as $\exp[-i(\delta_{l+} + \delta_{l-} - i\eta_l)]$, where $\eta_l = \ln \cos(\delta_{l+} - \delta_{l-})$]. Imaginary elements occur quite generally in the physicomathematical representation of dissipative phenomena. They arise typically through the imposition of complex boundary conditions as in (8.58), which in turn represent the loss of wave intensity from the subsystem under consideration.

Note in (8.61) that the first line represents the wave function in the channels that are retained, which are typically few in number. The second line displays the outgoing waves whose number and details are of no consequence to observations in the P channels. They have been included here only for the sake of completeness of the wave function. The essential point is that

† Both the substance and notation of this procedure were introduced by Feshbach [21].

experiments on a small subset P bear only on the correspondingly small effective reaction matrix,

$$K_{\mathrm{eff}} \equiv P[K + iKQ(1 - iQKQ)^{-1}QK]P. \tag{8.62}$$

The goal of characterizing experimental data by a few parameters has thus been reached. However, the matrix (8.62) contains in it the much larger matrix QKQ. Therefore, explicit reference to its detailed structure in *ab initio* calculations of (8.62) would largely nullify the purpose of the channel elimination. Here, as in the construction of optical potentials in Section 6.9, the goal of theory should rather be to evaluate a small set of semimacroscopic parameters [namely (8.62)] without explicit reference to its microscopic origins.

In closing, the reaction matrix representation utilized in this section to eliminate open channels also serves for the elimination of closed channels that we had treated previously by Jost matrices. To this end we select out terms of the wave function (8.56) with $\epsilon_i < 0$ by a projection operator Q and replace f and g in these terms by $\{f^-, f^+\}$, now to be drawn from (5.12) and (5.12′). The coefficient of $f^-(\epsilon_i, l_i, r)Q$, which must vanish, is

$$\sum_j Q(\delta_{ij} \sin \beta_i + \cos \beta_i K_{ij})C_j = \cos \beta_i Q(\delta_{ij} \tan \beta_i + K_{ij})C_j = 0. \tag{8.63}$$

The analog of (8.60) is then

$$Q\mathbf{C} = -\frac{1}{Q \tan \beta + QKQ} QKP\mathbf{C}, \tag{8.64}$$

where $\tan \beta$ is a diagonal matrix in the i channels that constitute the Q space. The K matrix in the open channels now becomes

$$PKP - PKQ \frac{1}{\tan \beta + QKQ} QKP, \tag{8.65}$$

where the rapid variations of $\tan \beta_i$ as a function of $\epsilon_i = E - E_i$ produce the same resonance effects as we have described earlier in terms of the effective phase shifts $\Delta_\rho(\epsilon_i)$. Indeed, the eigenvalues of the matrix in (8.65) coincide with $\tan \Delta_\rho$.

More specifically, the matrix (8.65) becomes singular at the roots of the determinant of the matrix in the denominator, that is, when

$$\mathrm{Det}|Q \tan \beta\, Q + QK^{(s)}Q| = 0. \tag{8.66}$$

This is, in fact, the condition of bound-state energy levels in the closed channels as introduced by Seaton [8]. It is equivalent to our corresponding (8.38). The more compact matrix form is made possible by the introduction of the $K^{(s)}$ matrix in (8.55). The equivalence is demonstrated by expanding the sine functions in (8.38′) and factoring out $\cos \beta(\kappa_j)$ from each row and $\cos \delta_\alpha^{(s)}$ from each column; an orthogonal substitution of the columns with the matrix $(U^{-1})_{\alpha j}$ completes the reconstruction of the matrix $K^{(s)}$ for closed channels, denoted by $QK^{(s)}Q$ in (8.66). For a single closed channel, the equation reduces to

$$\tan\beta(\kappa) + \tan\delta^{(s)} = 0, \tag{8.66'}$$

which is equivalent to our initial (5.26).

8.6.2 Decay of Closed-Channel Excitations

The discussion of the dissipation of transition amplitudes from channels of interest into unobserved channels leads us to consider also the dissipation of closed-channel excitations into open channels that need not be observed. Formally, we can use (8.62) again but with P now referring to closed channels and Q to *all* the open channels, that is, we enter the matrix (8.62) in place of $QK^{(s)}Q$ in (8.66). As noted following (8.61), the $(1 - iQKQ)^{-1}$ term introduces an imaginary part in the effective reaction matrix of the closed channels. As a result, the determinantal equation (8.66) will have roots only to the extent that the real and imaginary parts of the sum of the matrices $\tan\beta$ and K_{eff} in (8.62) commute.

In common practice, we regard the imaginary part to be small, being of second order in the coupling PKQ. In this event we find the energies at which a real eigenvalue of the matrix $\tan\beta + \text{Re}\, K_{\text{eff}}$ vanishes and add to them an imaginary part proportional to the mean value of $\text{Im}\, K_{\text{eff}}$ over the eigenvector. In this representation we say that the coupling to the continuum, represented by the second term in K_{eff}, has produced both a shift and a broadening of the energy levels in the closed channel. The determination of complex energy eigenvalues of (8.66) is, however, unrestricted when there is a single closed channel.

We consider here the application to the system of Section 8.3 with a single pair of channels, one open and one closed. Returning to the notation of (8.8), we simply set to zero the coefficient of the ingoing wave $f^-(\epsilon_0, l, r)$ in the open channel $N = 0$:

$$\sum_\Lambda U_{0\Lambda} e^{-i\delta_\Lambda^{(s)}} A_\Lambda = 0, \tag{8.67}$$

which is solved by $A_\Lambda = e^{i\delta_\Lambda^{(s)}}(U^{-1})_{\Lambda 2}$ owing to the orthogonality of $U_{N\Lambda}$. The imaginary element in these A_Λ provides here the dissipative effect in the closed channel of interest ($N = 2$), thus replacing the imaginary $iQKQ$ in the denominator of (8.62). Turning to the channel $N = 2$, we enforce the boundedness condition (8.4) that the coefficient of $f(\epsilon_2, l, r)$ vanish

$$\begin{aligned}\sum_\Lambda U_{2\Lambda}\sin(\beta_2 + \delta_\Lambda^{(s)})A_\Lambda &= \sum_\Lambda U_{2\Lambda}[e^{i(\beta_2+\delta_\Lambda^{(s)})} - e^{-i(\beta_2+\delta_\Lambda^{(s)})}](2i)^{-1}e^{i\delta_\Lambda^{(s)}}(U^{-1})_{\Lambda 2}\\ &= (2i)^{-1}e^{i\beta_2}\left[\sum_\Lambda U_{2\Lambda}e^{2i\delta_\Lambda^{(s)}}(U^{-1})_{\Lambda 2} - e^{-2i\beta_2}\right]\\ &= 0.\end{aligned} \tag{8.68}$$

Note that the combination $\Sigma_\Lambda U_{2\Lambda}\exp(2i\delta_\Lambda^{(s)})(U^{-1})_{\Lambda 2}$ can be written as $\exp[2i(\arctan K)_{22}]$. The long-range phase shift β_2 amounts to $\pi\nu_2$ for the Coulomb field in this example of $e + H_2^+$, whereby the phase factor $\exp(-2\pi i\nu_2)$ is represented by a point of the complex plane that revolves on the unit circle each time $\nu_2 = [2(E_2 - E)]^{-1/2}$ increases by unity. The condition (8.68) would thus be met once for each period of ν_2 if the first term in the brackets had also unit modulus. However, this term constitutes a submatrix of the scattering matrix, equivalent to the factor in the brackets of (8.51), whose modulus is reduced *below* unity by coupling to the unobserved channel. Hence, this very effect makes (8.68) and (8.67) incompatible and thus prevents the occurrence of a stationary bound state.

This situation, familiar in various problems of quantum mechanics, is often dealt with by considering that the condition (8.68) could be fulfilled by attributing to ν_2 a small imaginary part $-i\eta_2$ corresponding to a complex energy eigenvalue and of such magnitude that $\exp(-2\pi\eta_2)$ equals the modulus of the first term in the brackets of (8.68), namely $\exp[-2\,\mathrm{Im}(\arctan K)_{22}]$. Eigenstates of the closed channel corresponding to these complex eigenvalues are, of course, nonstationary and decay in the course of time at a rate proportional to η_2, thus representing the transition of the particle from the bound state into the open channel.

Additional aspects of this decay process are brought out by assuming a bound-state wave function $\psi(\omega, r)$ at an initial time $t = 0$ and representing it as a superposition of the stationary state eigenfunctions constructed in Section 8.3,

$$\psi(\omega, r, t) = \int dE\, c(E) \left[\sum_{\Lambda} F_{\Lambda}(\omega, r) A_{\Lambda} \right]_E e^{-iEt}, \tag{8.69}$$

each of which has components in both channels and satisfies (8.4)—equivalent to (8.68)—but does *not* satisfy (8.67). We know from (8.23) that the bound-state component of each of these eigenfunctions has an amplitude proportional to $(d\Delta/d\beta_2)^{1/2}$, a factor that peaks at the energies for which $\beta_2 = -\pi\mu_2 = -\mathrm{Re}(\arctan \mathrm{K})_{22}$ according to (8.18). The coefficient $c(E)$ will be proportional to the overlap of the eigenfunction (8.23) with the bound-state wave function $\psi(\omega, r, 0)$ and will thus be proportional to $(d\Delta/d\beta_2)^{1/2}$. The closed-channel component of the integrand in (8.69) will then depend on the energy E primarily through *two* identical factors $(d\Delta/d\beta_2)^{1/2}$, one contributed by $c(E)$ and one by the wave function itself. These factors combine to yield the *full Lorentzian factor* (8.18). We can see now under what conditions the closed-channel wave function $\psi(\omega, r, t)$ will decay according to the familiar $\exp(-\frac{1}{2}\Gamma t)$. The Fourier transform of this exponential decaying factor is just the Lorentzian $\frac{1}{2}\Gamma/\pi[E - E_0)^2 + \Gamma^2/4]$. Exponential decay will prevail to the extent that (8.18) can be represented as a Lorentzian function of E over the relevant range of integration in (8.69). The conditions are fulfilled only approximately, at best.

This reinterpretation of closed-channel resonances in terms of an imaginary part of ν_2 introduces a complex quantum defect $\mu_2 + i\eta_2$ in place of μ_2 in (8.19) [8]. Thereby, the expression of $4\hbar/\Gamma_n$ in (8.21), where Γ_n is the line width, and (8.19) for the energy position of the resonance are combined into a single complex energy

$$E = E_2 - \mathrm{Ry}/(n - \mu_2 - i\eta_2)^2 = E_n - \tfrac{1}{2}i\Gamma_n. \tag{8.70}$$

Autoionizing resonances are thus cast in a form analogous to discrete energy levels.

For small but nonzero values of the width parameter ξ^2 in (8.21)—and, correspondingly, for small and positive η_2—comparison of (8.70) and (8.21) shows that

$\xi^2 = \pi\eta_2$. The more general link between these parameters can be seen from the wave function in (8.23). The condition for discrete or quasi-discrete levels is that the coefficient of the exponentially rising function f^- vanish. On admitting complex values of the phase parameter, this condition reduces to $\mathrm{Im}\,\Delta = -\infty$. It then follows from (8.15) that such solutions obtain at $\beta_2 = \pi\nu_2 = n\pi - \pi(\mu_2 + i\eta_2)$ when ξ^2 is identified as

$$\xi^2 = \tanh(\pi\eta_2). \tag{8.71}$$

The expression for the resonance width given earlier by (8.21) holds only for $\xi^2 \ll 1$, but the result in (8.70) and (8.71) for Γ_n as a function of ξ^2 is more generally valid.

8.7 FRAGMENTATION CHANNELS IN EXTERNAL FIELDS

The theory of the Stark and Zeeman effects on spectral lines dates from the early years of quantum physics [22]. These effects could be treated by perturbation methods because the electric and magnetic field strengths available in the laboratory were far weaker than the atomic fields acting on an electron in a valence shell or in a Rydberg level of moderate excitation. We do not concern ourselves here with laser fields or with the high magnetic fields prevailing on neutron stars or in other astrophysical settings, whose strength is indeed comparable to that of atomic fields. We shall deal instead with the effects of laboratory fields on highly excited electrons that attain large radial distances from an atom or molecule, where the atomic fields are themselves weak and quite comparable to the external ones. The treatment of these effects, which have been studied extensively since about 1970, has been largely successful even though it remains incomplete at this time [23].

A key element of this treatment lies in the fact that the external fields need be considered only at large radial distances where an excited electron belongs to a specific fragmentation channel and the field of an atomic or molecular core is purely Coulomb. Accordingly, the first and main task consists of characterizing each fragmentation channel i under the combined influence of the two fields, external and Coulomb. A second task then consists of determining the frame transformation matrix $U_{i\alpha}$ that connects the eigenchannels of short-range interaction to the fragmentation channels modified by the external field. The actions of the external field at short radial distances are generally negligible or amenable to perturbation treatment; the effect of non-Coulomb components of the core's field at moderate ranges will be discussed in Section 9.3.7. By applying the multichannel analysis in a new context, the present discussion will illustrate its flexibility.

This program has been carried out fully for the Stark effect, because the Schrödinger equation for an electron in the combined potential field

$$V_S(\mathbf{r}) = -(e^2/r) + eFz \tag{8.72}$$

(F = external field) is separable in parabolic coordinates [22]. Section 8.7.1 shows how the orbital quantum number l_i of an electron in a central field is replaced by a parameter β_i with discrete values for the potential (8.72) and how this parameter as well as the asymptotic phase shift and a normalization constant depend resonantly on the energy E, the external field strength F, and the magnetic quantum number m of each fragmentation channel i. The frame transformation that connects these channels to the eigenchannels α of the short-range interactions, yielding analogs of the Jost matrices $J^{\pm}_{i\alpha}$ in (8.32), is then described in Section 8.7.2.

The familiar Zeeman effect, linear in the magnetic field $B\hat{z}$, acts on the z component of atomic or molecular angular momenta regardless of the radial distance of an electron. On the other hand, the quadratic—or "diamagnetic"—Zeeman effect is proportional to the area encompassed by the orbital motion, thus increasingly rapidly with radial distance. Combination of this effect with the Coulomb field of an ionic core yields the potential

$$V_Z(\mathbf{r}) = -(e^2/r) + (e^2/8mc^2)B^2r^2 \sin^2\theta. \tag{8.73}$$

Since the Schrödinger equation of an electron in this potential field is not separable, no "good" quantum number replaces the orbital l_i of an electron in central field or the β_i for the Coulomb–Stark potential (8.72). [The magnetic quantum number m remains constant, however, owing to the axial symmetry of (8.73).] Nevertheless, the experimental observation of resonances in photoabsorption and other spectra and the theoretical evidence emerging from various fragmentary treatments strongly suggest the existence of approximate constants of the motion that should characterize fragmentation and resonant channels. Section 8.7.3 describes the main aspects of this problem.

8.7.1 Fragmentation Channels in the Stark Field

The Schrödinger equation for an electron in the potential field (8.72) separates in the parabolic coordinates

$$\xi = r + z = r(1 + \cos\theta), \qquad \eta = r - z = r(1 - \cos\theta), \qquad \phi = \arctan(y/x), \tag{8.74}$$

thus yielding the pair of equations in atomic units [22]

$$\left\{\frac{d^2}{d\xi^2} + \left[\frac{1}{2}\epsilon + \frac{\beta}{\xi} - \frac{1}{4}F\xi - \frac{m^2-1}{4\xi^2}\right]\right\}\xi^{1/2}\Xi_\beta(\xi) = 0, \tag{8.75a}$$

$$\left\{\frac{d^2}{d\eta^2} + \left[\frac{1}{2}\epsilon + \frac{1-\beta}{\eta} + \frac{1}{4}F\eta - \frac{m^2-1}{4\eta^2}\right]\right\}\eta^{1/2}\Upsilon_\beta(\eta) = 0. \tag{8.75b}$$

A complete energy-normalized eigenfunction regular at $r = 0$ will be indicated by

$$\Psi_{\beta,m}(\xi, \eta, \phi) = (2\pi)^{-1/2}e^{im\phi}\Xi_\beta(\xi)\Upsilon_\beta(\eta). \tag{8.76}$$

Both equations (8.75) are structurally analogous to the basic radial equation (5.1), with the centrifugal term and the renormalization coefficient $\xi^{1/2}$ (or $\eta^{1/2}$) appropriate to the axial rather than central symmetry of the potential (8.72). The Coulomb field is seen to be partitioned between the two equations in the ratio $\beta/(1-\beta)$. On the other hand, the *sign difference* of the Stark field potential terms of the two equations (8.75) causes them to play substantially different roles. This difference is reflected in the profiles of the respective potential wells shown in Fig. 8.11. The well in the ξ

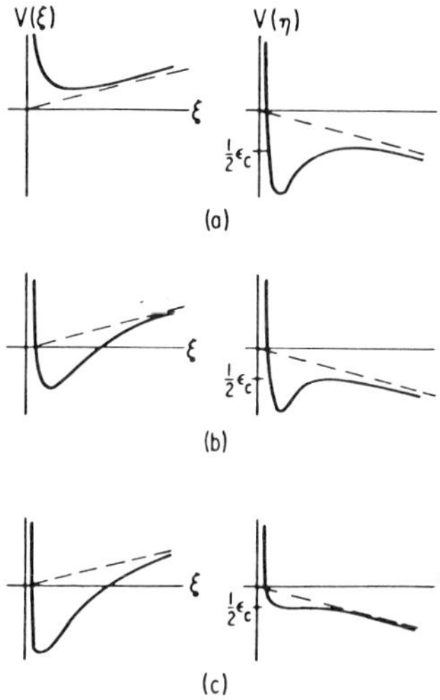

Fig. 8.11. Qualitative plots of the potentials $V(\xi)$ and $V(\eta)$ in Eqs. (8.75) for $m = 1$ and for sample values of β: (a) $\beta = 0.1$, (b) $\beta = 0.4$ and (c) $\beta = 0.9$; — —, pure Stark potentials $\frac{1}{4}F\xi$ and $-\frac{1}{4}F\eta$. (From Harmin [24].)

coordinate confines the oscillations of the wave function $\Xi_\beta(\xi)$ to a finite range of ξ, whereas $\Upsilon_\beta(\eta)$ can oscillate without restriction at sufficiently large values of η regardless of the energy eigenvalue ϵ that ranges continuously from $-\infty$ to ∞. For any given value of ϵ the boundedness of the electron motion along ξ restricts $\Xi_\beta(\xi)$ to a *discrete spectrum* of the charge separation parameter β_i, whose index $i = 0, 1, 2, \ldots$ represents the number of nodes of $\Xi_{\beta i}(\xi)$. The eigenvalues β_i are readily found for given values of ϵ and F, by numerical integration of (8.75a) or by the WKB approximation [24]. The boundedness of motion along ξ in parabolic coordinates, for $F > 0$, parallels the boundedness of the motion along θ for central symmetry and causes the parameter β_i to play the same role that the orbital quantum number l_i plays in characterizing fragmentation channels at $F = 0$.

Equation (8.75b) has a single solution regular at $\eta = 0$ for any given values of (ϵ, β_i, F). At large values of η, where the terms in η^{-1} and η^{-2} are negligible, $\eta^{1/2}\Upsilon_{\beta_i}(\eta)$ is a superposition of the regular and irregular Airy functions of the variable $(\frac{1}{2}F)^{1/3}(\eta + 2\epsilon/F)$ with coefficients $\cos\delta_i$ and $-\sin\delta_i$. The phase shift δ_i is a second channel parameter, companion to β_i.

A third channel parameter, analogous to the parameter B of Table 5.1, relates the amplitude of the eigenfunction (8.76) near the origin to its amplitude at $\eta \to \infty$, which is standardized by energy normalization. The leading term in the expansion of (8.76) has the form

$$\Psi_{\beta_i m_i}(\xi \sim 0, \eta \sim 0, \phi) = N_i(2\pi)^{-1/2} e^{im_i\phi}(\xi\eta)^{|m_i|/2}[1 + O(r)], \tag{8.77}$$

where the factor $(\xi\eta)^{|m_i|/2} = (x^2 + y^2)^{|m_i|/2}$ reflects penetration in the centrifugal barrier about the z axis, and the squared parameter N_i^2 represents the density of states in the ith channel at $\xi \sim \eta \sim 0$. Tunneling through the potential barrier that encloses the well in Fig. 8.11b causes the value of N_i to vary sharply as a function of energy in the range of moderately negative values of ϵ. Peaks of N_i occur for combinations of

$\{\epsilon, F, \beta_i, m_i\}$ that allow an integer number of half-wavelengths of Υ_{β_i} to fit within the well; exponentially small values of N_i prevail otherwise. WKB procedures have been found quite adequate in representing the eigenfunction (8.76) over most of the relevant range of $\{\epsilon, F, m\}$, expressing its parameter $\{\beta_i, \delta_i, N_i\}$ in terms of elliptic integrals [24].

The development of the channel wave functions in the presence of a Stark field is brought to full analogy with the treatment of Chapter 5 by introducing standardized irregular solutions of (8.75) analogous to the g functions of Section 5.4. This final step implies the introduction of a fourth channel parameter analogous to the $\mathscr{G}$ of Table 5.1 and (5.17). However, whereas the parameter $\mathscr{G}$ is generally small and slowly dependent on energy in Chapter 5, the corresponding parameter and its influence on normalization in a Stark field are greatly magnified by tunneling through the potential barrier in Fig. 8.11b. We summarize here the treatment developed by Harmin [24].

Recall from Chapter 5 that radial functions irregular at the origin are introduced to represent the influence of the short-range interactions on the amplitude and phase of the wave function in a fragmentation channel through an appropriate superposition of regular and irregular solutions of the wave equation for that channel. To this end an irregular solution is conveniently chosen whose phase lags that of the regular solution by 90° in the range of radial coordinates close to but outside the core. The parabolic coordinate η replaces for this purpose the radial r in the Stark field problem, the relevant range of η being much smaller than the range covered by the potential barrier in Fig. 8.11b. Accordingly, the irregular solution $\bar{\Upsilon}_\beta(\eta)$ introduced by Harmin [24] as a companion to $\Upsilon_\beta(\eta)$ lags Υ_β in phase by 90° *within* the well. Tunneling through the barrier alters the relative amplitude *and phases* of Υ_β and $\bar{\Upsilon}_\beta$ drastically in the energy range $\epsilon < 0$. Both of these effects are expressed conveniently in terms of a supplementary phase shift γ_i, the fourth parameter, but depend actually on two separate WKB parameters, namely the phase integral over the classically accessible range of η *within* the well and the tunneling integral τ over the inaccessible range through the barrier. This treatment extends to energies above the barrier to take into account the residual wave optics effect of the barrier.

8.7.2 Frame Transformation Matrix for the Stark Field

The connection between the eigenchannels of short-range interaction and the fragmentation channels in the Stark field is established by disregarding the influence of the Stark field on the electron at distances from the core of the order of a few atomic units. In (8.75) a field strength F of 5000 V/cm amounts to $\sim 10^{-6}$ a.u. Accordingly, the terms linear in F are negligible as compared to the Coulomb terms, with $\{\beta, 1-\beta\} \sim 1$, as long as $\{\xi, \eta\} \ll 10^3$ a.u. Thereby (8.75) in parabolic coordinates is equivalent to the wave equation for an electron in a Coulomb field in polar coordinates over a broad range of coordinates, and their respective sets of solutions can be expanded into one another. The coefficients of this expansion will then provide the desired transformation matrix $U_{i\alpha}$.

More specifically, note that (8.75) has the same structure as the radial equation of the H atom, once the terms linear in F are disregarded, the expression in the brackets consisting of an energy term, a Coulomb term, and a centrifugal term. The wave function $\Xi_\beta(\xi)$ and $\Upsilon_\beta(\eta)$ are accordingly represented in terms of confluent hypergeometric functions $F(-n+l+1, 2l+2, r/n)$ in the range $(\xi, \eta) \ll 10^3$ a.u. The integer quantum number n is, however, replaced by the noninteger $\beta/\sqrt{-\epsilon}$ in Ξ_β and by $(1-\beta)/\sqrt{-\epsilon}$ in Υ_β, while l is replaced by $\frac{1}{2}m$ and r by ξ or η.

The wave function (8.76) thus transformed may now be expanded into polar coordinate hydrogenic eigenfunctions setting

$$\Psi_{\beta_i m_i}(\xi, \eta, \phi) = N_i \sum_\alpha U_{i\alpha} f^0(\epsilon, l_\alpha, r) Y_{l_\alpha m_i}(\theta, \phi), \qquad (\xi, \eta) \ll F^{-1/2}, \tag{8.78}$$

with the same coefficients N_i as in (8.77). The parabolic coordinates (ξ, η) are now recast in terms of (r, θ) by (8.74), after which (8.78) is multiplied by $Y_{l_\alpha m_i}(\theta, \phi)$ and integrated over (θ, ϕ). Both sides of the resulting equation are finally represented as power series in r and the coefficients $U_{i\alpha}$ determined in algebraic form by matching the coefficients of equal powers of r on both sides of the equation. Details are described in Harmin [24].

On this basis the photoabsorption of Na vapor in a Stark field has been calculated by Harmin [24], by using a variant of the procedures of Chapters 7 and 8, and has been compared with experimental data as shown in Fig. 8.12. The resonance peaks

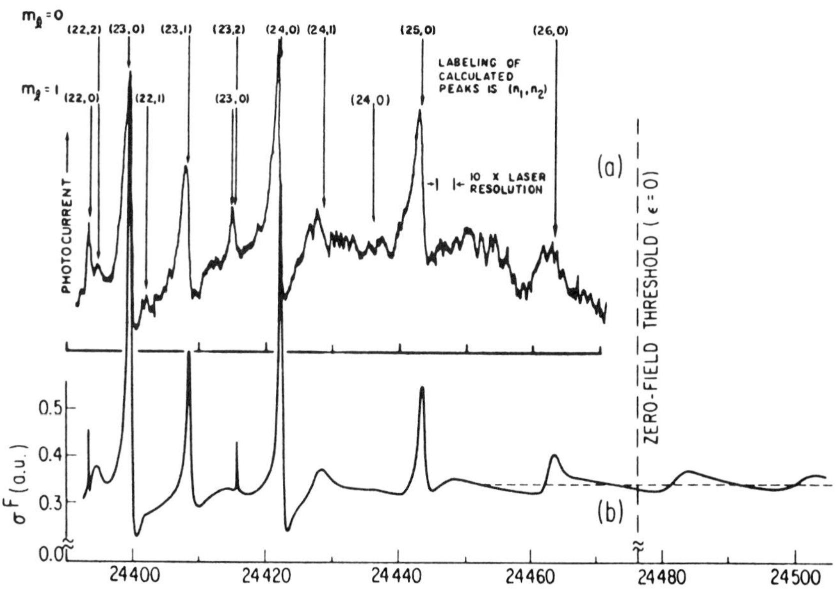

Fig. 8.12. (a) Experimental photoionization spectrum of Na $3^2P_{3/2}$ in a field $F = 3.59$ kV/cm versus photon energy $\hbar\omega$ just below threshold. (b) Theoretical cross section σ^F; — —, $F = 0$. (From Harmin [24].)

correspond to maxima of the amplitude N_i of several fragmentation channels, that is, to density peaks of the final state wave function within the core where the final and ground states overlap. Amplitudes of different fragmentation channels contribute to photoionization coherently, the more so in the presence of an ionic core and thus interfere. The narrow spectral range covered in Fig. 8.12 lies just below the ionization threshold, where the Coulomb field of the ion core and the external field act with comparable strength and combine to yield the most elaborate intensity pattern. The resonance peaks flatten out above the ionization threshold; perturbation treatments also prove adequate to predict the Stark effect at lower excitation energies.

Two remarks of some general significance complement the preceding outline of the connection $U_{i\alpha}$ and its application. Whereas the frame transformation matrices considered in Chapter 7 and elsewhere in the present chapter relate two sets of excitation channels without restriction, the transformation considered in this section has only local significance, within the range of (ξ, η) and r where the field strength F in (8.75) is effectively negligible. Accordingly, the matrix $U_{i\alpha}$ is not shown to be orthogonal, in that $\Sigma_i \tilde{U}_{\alpha i} U_{i\alpha'} \neq \delta_{\alpha\alpha'}$. A second remark relates to the special circumstance of the Na example illustrated in Fig. 8.12, namely, that the spherical symmetry of the Na^+ core permits each wave function of the short-range eigenchannels α to be continued outside the core without being resolved into a superposition of fragmentation channels of the Na atom in zero field. (In other words, the fragmentation channels of Na coincide with its short-range eigenchannels.) Under more general circumstances it might well be appropriate to connect the Stark field fragmentation channels first to the zero-field fragmentation channels and then to connect the latter channels to the short-range eigenchannels through the usual zero-field matrix $U_{i\alpha}$.

8.7.3 Fragmentation Channels and Rydberg Levels in a Diamagnetic Field

The symmetry of the potential (8.73) under rotations about the z axis permits fragmentation channels of a Rydberg electron to be classified by a constant magnetic quantum number m_i. The potential (8.73) is also invariant under the reflection $\theta \to \pi - \theta$ through the xy plane. Parity under this reflection is accordingly a constant of the motion. Odd-parity states, with a node at $\theta = \frac{1}{2}\pi$, where the diamagnetic term of (8.73) peaks, respond to this term in less dramatic fashion and accordingly will be peripheral to our analysis.

Fragmentation is confined to the $\pm\hat{z}$ directions ($\theta = 0$ or π) by the steady rise of the diamagnetic potential in other directions. In the limit $|z| \to \infty$ the Coulomb term of (8.73) reduces to $1/z$, whereby the Schrödinger equation separates in cylindrical coordinates (z, ρ, ϕ). The equation in ρ has the oscillator potential proportional to ρ^2 and hence the eigenvalues $(n_\rho + 1)e\hbar B/2mc$. The fragmentation channels are classified accordingly by the quantum numbers n_ρ and m and have wave functions

$$\Psi_{n_\rho m}(z, \rho, \phi) \underset{|z|\to\infty}{\to} \sin(k_{n_\rho} z + \zeta \ln z + \delta_{n_\rho}) f_{n_\rho m}(\rho) e^{im\phi}$$

$$k_{n_\rho}^2 = (2m/\hbar^2)[E - (n_\rho + 1)e\hbar B/2mc], \tag{8.79}$$

where the phase shift δ_{n_ρ} includes long- and short-range components and the parameter ζ is the same as that in (5.2) and (5.6b). Continuation of these wave functions to finite values of z is inhibited by the nonseparability of the potential (8.73) in polar or cylindrical coordinates. Also inhibited is the construction of the transformation matrix $U^{(m)}_{n_\rho l_\alpha}$ that would connect the fragmentation channels to short-range eigenchannels, classified by the orbital quantum number appropriate to the region where the diamagnetic potential is negligible.

Nevertheless, the absorption spectrum shown in Fig. 8.13a shows a grouping of unsharp lines into families. Their classification suggests the introduction of appropriate quantum numbers, one of which might represent the number of nodes in the wave function plot of Fig. 8.13c. Such quantum numbers need not represent exact constants of the motion but only approximate ones insofar as the observed levels are not quite sharp or their grouping rigorous. (A further occurrence of approximate quantum numbers will emerge in Chapter 10.)

Clues to the interpretation of the grouping of levels emerge from the perturbation analysis applicable at lower excitations where the expectation value of the diamagnetic term of (8.73) is lower than the separation of successive Rydberg levels in the Coulomb field, namely, $1/n^3$ a.u. Qualitatively, the diamagnetic term of (8.73), however weak it may be, breaks the degeneracy of hydrogenic levels of equal n and different l; it also removes the selection rule that restricts photoexcitation to p levels in the example of Fig. 8.13. Indeed, the diamagnetic field manifests itself first through the appearance of satellites to each main absorption line as n increases in Fig. 8.13a.

Application of a degenerate perturbation treatment to the manifold of Rydberg levels with the same quantum number n yields a set of nondegenerate levels arranged according to the expectation value $\langle r^2 \sin^2 \theta \rangle$, that is, mainly according to $\langle \sin^2 \theta \rangle$ since r^2 does not vary much within the manifold. A semiclassical analysis has shown the eigenfunctions of these levels to represent alternative excitations of the libration of an elliptical orbit. The higher energy levels correspond to librations about the xy plane, the lower ones about the z axis. These two classes of motion centered in different regions of space are seen to be separated by an effective potential barrier [23], a phenomenon akin to the separation of eigenfunctions of the asymmetric top with equal angular momentum into two classes centered about the largest and smallest axes of inertia.

Remarkably, this classification holds well beyond the normal limit of the perturbation treatment from which it was derived. The data in Fig. 8.13b and c, obtained by numerical solution of the Schrödinger equation with the potential (8.73), show the groups of levels arising from the field's action on the Rydberg manifolds with different n to fan out in energy, intersecting each other but preserving their identity. In fact, the highest and most intensely excited level of each manifold agrees rather well with experimental data and has long been interpreted semiclassically as corresponding to the nth level of radial oscillation within and about the xy plane.

Note also that the sequence of most intense and highest levels of each manifold extends in Fig. 8.13a beyond the ionization threshold, much as the sequence of resonances with successive values of β_i do in Fig. 8.12. This analogy implies a correspondence between an electron's motion within and about the plane xy, confined to a finite range by the rising potential (8.73) proportional to $x^2 + y^2$, and its motion along the parabolic coordinate ξ in the Stark field. In both instances the electron

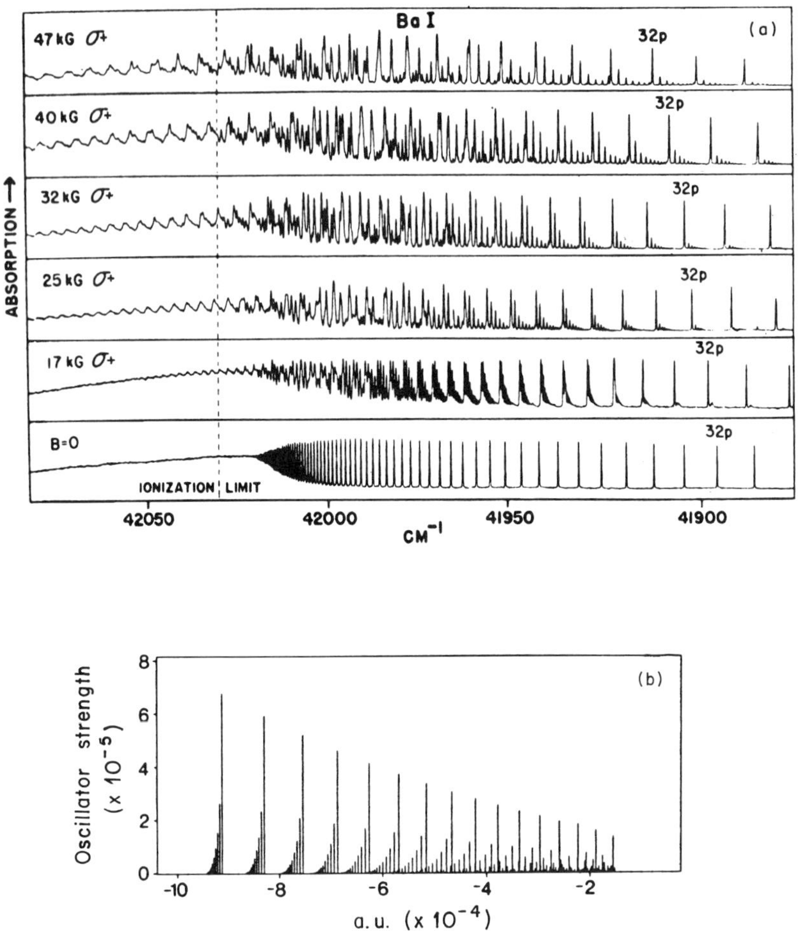

Fig. 8.13. (a) Microdensitometer recording of the plate transmission of the Ba principal series with magnetic field strengths $B = 47$, 40, 32, 27, 17 and 0 kG and with σ^+ polarization. Abscissa is the level energy above the ground state. (b) Calculated absorption spectrum of atomic hydrogen with σ polarization; $B = 47$ kG. Vertical lines represent oscillator strength of discrete levels. Note that the energy scale runs opposite to that in (a). (c) Wave functions corresponding to successive lines of a group in (b) plotted as a function of angle θ for a radial distance related to θ by $-1/r + (8 \times 137^2)^{-1}B^2r^2 \sin^2 \theta = \epsilon$ (energy level in atomic units.). (From Clark *et al.* [23].)

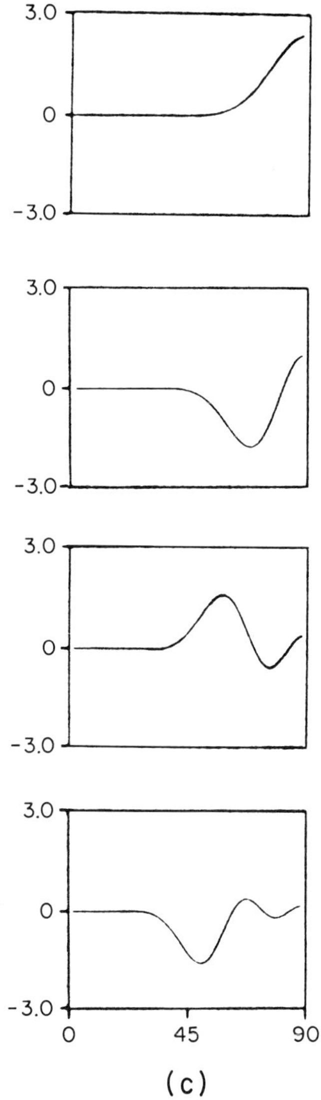

Fig. 8.13. (*Continued*)

eventually escapes this confinement through its motion along the z or η coordinate, respectively.

Qualitatively, the fragmentary interpretations outlined earlier provide a consistent picture of the observed effects of a diamagnetic field. What is missing is a unified quantum-mechanical representation over the whole range of radial distances, analogous to that provided for the Stark field in Section 8.7.1. The channel sets for the two fields should nevertheless differ in that the Stark field channels are coupled only by short-range interactions with the ionic core, represented by the matrix $U_{i\alpha}$, whereas the diamagnetic channels are presumably coupled by the long-range potential (8.73) itself; the mode, location, and strength of this coupling remain obscure. The symmetry of the potential (8.73) about this plane $\theta = \pi/2$, where it peaks as a function of θ, has presumably a key role in the classification of channels; an analogous role of such a symmetry element will emerge in Section 10.6.

Problems

8.1 Equations (8.40) and (8.41) list the i and α channels with $J = 1$ for noble gas atoms with the structure $p^5 l (l = 0, 2)$.

(a) Give the corresponding list for channels with $J = 2$.

(b) Which alternative channels with other J values exist for the same $p^5 l$ structures?

8.2 Find the q value for the resonance in Fig. 8.1b by fitting its profile to (8.2).

8.3 Verify (8.22) by entering in (8.6) the amplitudes A_Λ defined by (8.4), (8.9), and (8.10). The resulting amplitude of the closed channel term in the wave function (8.6) should equal $(d\Delta/d\beta_2)^{1/2}$.

8.4 The Ne spectrum of photoexcitation from $2p^6 \to 2p^5 ns$ resembles those of other rare gases discussed in Section 8.5 except for the mixing of s and d channels, which remains negligible in Ne. The 5×5 matrix $U_{i\alpha}$ reduces accordingly to two blocks, a 2×2 block for s channels and a 3×3 for d. Consider only the s channels.

(a) Verify that the $U_{i\alpha}$ block for s channels coincides with the matrix (7.22) for F^- in the approximation of (8.48).

(b) Given the $\delta_\alpha^{(s)}$ phase shifts $\delta_4^{(s)} = 0.15\pi$, $\delta_5^{(s)} = 0.10\pi$, construct the Lu–Fano plot and compare it with Fig. 7.8a.

(c) The dipole matrix elements are approximately $(F_4|x|0) = 0$, $(F_5|x|0) = 1.4$ a.u. Plot the oscillator strength obtained from (8.50) and contrast with Fig. 8.7c.

8.5 The $6sns\ ^1S_0$ series of Ba is strongly perturbed by a level of a different configuration in the range $14 \leq n \leq 22$. Using the observed line positions (in cm^{-1}): 40,892, 41,093, 41,245, 41,362, 41,441, 41,468, 41,535, 41,596, 41,646, 41,689, and the series limit 42,035 cm^{-1} determine the quantum defects of the lines and plot them against the wave number of the line. Determine the position of the perturber in the spectrum as the abscissa of maximum slope in the plot.

8.6 Using the Jost matrices (8.32a) for a two-channel problem with the $U_{i\alpha}$ in (8.13), construct the S matrix. Show that the squared matrix elements $|S_{ij}|^2$ are functions of the single combination $4\xi^2/(1 + \xi^2)^2$, that is, that the various parameters enter into the cross-section formulas only through the combination given in (8.17e).

References

1. C. E. Kuyatt, J. A. Simpson, and S. R. Mielczarek, *Phys. Rev.* **138**, A385 (1965).
2. G. J. Schulz, *Phys. Rev. Lett.* **10**, 104 (1963).
3. A. Giusti-Suzor and U. Fano, *J. Phys. B* **17**, 215 (1984).
4. U. Fano and J. W. Cooper, *Rev. Mod. Phys.* **40**, 411 (1968).
5. Ch. Jungen and D. Dill, *J. Chem. Phys.* **73**, 3338 (1980).
6. P. M. Dehmer and W. H. Chupka, *J. Chem. Phys.* **65**, 2243 (1976).
7. C. D. Lin, *Astrophys. J.* **187**, 385 (1974); P. F. O'Mahony, *Phys. Rev. A* **32**, 908 (1985).
8. M. J. Seaton, *Rep. Prog. Phys.* **46**, 167 (1983).
9. Z. W. Wang and K. T. Lu, *Phys. Rev. A* **31**, 1515 (1985).
10. P. G. Burke, J. W. Cooper, and S. Ormonde, *Phys. Rev.* **183**, 245 (1969), see especially p. 261; J. Macek, *Phys. Rev. A* **2**, 1101 (1970); Z. W. Wang and K. T. Lu, *Phys. Rev. A* **31**, 1515, 1521 (1985).
11. K. T. Lu, *Phys. Rev. A* **4**, 579 (1971); C. M. Lee and K. T. Lu, *Phys. Rev. A* **8**, 1241 (1973).
12. W. R. Johnson, K. T. Cheng, K. N. Huang, and M. LeDourneuf, *Phys. Rev. A* **22**, 989 (1980).
13. C. M. Lee and K. T. Lu, *Phys. Rev. A* **8**, 1241 (1973).
14. J. A. R. Samson and J. L. Gardner, *Phys. Rev. Lett.* **31**, 1327 (1973).
15. J. Kessler, "Polarized Electrons." Springer-Verlag, Berlin and New York, 1976; C. Heckenkamp, F. Schäfers, G. Schönhense, and U. Heinzmann, *Phys. Rev. Lett.* **52**, 421 (1984).
16. J. A. Armstrong, P. Esherick, and J. J. Wynne, *Phys. Rev. A* **15**, 180 (1977); J. P. Connerade, M. W. D. Mansfield, G. H. Newsom, D. H. Tracy, M. A. Baig, and K. Thimm, *Phil. Trans. R. Soc. London* **290**, 327 (1977); M. Aymar, *Phys. Rep.* **110**, 163 (1984).
17. J. J. Wynne and J. A. Armstrong, *IBM J. Res. Dev.* **23**, 490 (1979).
18. M. L. Ginter, D. S. Ginter, and C. M. Brown, *Appl. Opt.* **19**, 4015 (1980).
19. A Giusti-Suzor and Ch. Jungen, *J. Chem. Phys.* **80**, 986 (1984).
20. J. A. Dagata, G. L. Findley, S. P. McGlynn, J. P. Connerade, and M. A. Baig, *Phys. Rev. A* **24**, 2485 (1981).
21. H. Feshbach, *Ann. Phys.* **19**, 287 (1962).
22. H. A. Bethe and E. E. Salpeter, "Quantum Mechanics of One- and Two-Electron Atoms," Chapter III. Springer-Verlag, Berlin and New York, 1957; L. D. Landau and E. M. Lifshitz, "Quantum Mechanics," 3rd ed., Sections 76, 77, and 112. Pergamon, New York, 1977.
23. C. W. Clark, K. T. Lu, and A. F. Starace *in* "Progress in Atomic Spectroscopy," (H. J. Beyer and H. Kleinpoppen, eds.), p. 247. Plenum, New York, 1984.
24. D. A. Harmin, *Phys. Rev. A* **24**, 2491 (1981); *Phys. Rev. A* **26**, 2656 (1982); *Phys. Rev. Lett.* **49**, 128 (1982).

9 CALCULATION OF INTERACTION PARAMETERS

9.1 INTRODUCTION

The analysis of phenomena in Chapters 7 and 8 has shown how the information contained in a large amount and variety of experimental data can be condensed into a small set of parameters: short-range eigenphases $\delta_\alpha^{(s)}$, transformation matrices $U_{i\alpha}$, and dipole matrix elements $(F_\alpha|x|0)$. The long-range contributions to phase shifts, $\delta_i^{(l)}$ or β_i, are known analytically for standard fields (Table 5.1); their modifications by departures from standard fields can be evaluated numerically with the aid of the last column of that table.

The short-range parameters constitute a *suitable interface* for comparison of experiment and theory, in the sense that it appears particularly meaningful to calculate their values from first principles and compare them with corresponding values culled out of experiments. These values not only constitute a smaller set, but are more nearly *independent of each other* and weakly dependent on energy. The calculation of parameters that depend on short-range interactions, at $r < r_0$, deals directly with the substantive problem of treating the multiparticle core without dispersing one's effort in dealing simultaneously with the region outside the core, where phenomena are simpler. This calculation constitutes the main subject of the present chapter.

A difficulty remains in the outer region, at $r > r_0$, which was mentioned in Section 7.5.1 but soft-pedaled in the following. The factorization of Jost matrices introduced in (7.40), $J_{i\alpha}^{\pm} \propto e^{\pm i\delta_i^{(l)}} U_{i\alpha} e^{\pm i\delta_\alpha^{(s)}}$, may not be fully applicable to multichannel processes, owing to noncentral long-range interactions. The treatment of this problem is discussed in Section 9.3.7. The further limitation to the low-energy range, where only one particle can leave the core, will be lifted in Part D, at least to some extent.

The initial section of this chapter deals with general characteristics of alternative approaches to the calculation of parameters; the following sections deal with specific methods of calculation and examples of results that have been obtained.

9.2 ALTERNATIVE PARAMETRIZATIONS

The effective volume of the core region is defined in the present context as a sphere of radius r_0 centered at the nucleus, whose surface at $r = r_0$ may be reached by no more than one electron because all core electrons are so tightly bound that their probability of reaching r_0 is negligible. This definition aims at ensuring that the exchange interaction of the outermost electron with the core electrons is negligible at $r \geq r_0$. The choice of a spherical volume is appropriate to our dealing explicitly with an atom; its shape can, of course, be modified when dealing with molecules or other aggregates.

The apparent arbitrariness in the choice of the core radius r_0 has been previously noted and discussed at the end of Section 6.2. We pointed out on that occasion that the exact value of r_0 is immaterial, in the sense that a shift of r_0, say by 10%, should have no effect on the determination of the observable short-range parameters. The choice of r_0 is indeed determined by the operational requirement of determining the *limiting value* of each parameter $\{\delta_\alpha^{(s)}, U_{i\alpha}, \text{etc.}\}$ as $r_0 \to \infty$. To this end r_0 should not be very large, which would nullify the intent of our approach, but just large enough to permit a sufficiently accurate extrapolation of parameters to their limiting value. Attention to correcting the extrapolation is preferable, for example, to lowering the threshold value below which the exchange interaction will be disregarded. The same consideration recommends that the effects of long-range anisotropic components of the core's field be treated separately as corrections to the long-range central potential (to be described in Section 9.3.7) instead of being included in the short-range parameters.

In the region of configuration space where the radial coordinate of the outermost electron approaches r_0, a wave function $F_\alpha(\omega, r)$ of the core + electron complex can be represented as a superposition of the N fragmentation channel wave functions $\Phi_i(\omega)$ considered in Chapters 7 and 8. We initially parametrize $F_\alpha(\omega, r)$ over a very short interval about the core boundary by its expansions

$$F_\alpha(\omega, r_0) = \sin \phi_\alpha(r_0) \sum_{i=1}^{N} \Phi_i(\omega) a_{i\alpha},$$

$$\left.\frac{r\partial F(\omega, r)}{\partial r}\right|_{r=r_0} = r_0 \phi'_\alpha(r_0) \cos \phi_\alpha(r_0) \sum_{i=1}^{N} \Phi_i(\omega) a_{i\alpha}. \tag{9.1}$$

The set of mutually orthogonal channel functions Φ_i must be complete in the sense of representing by (9.1) any eigenfunction of the Hamiltonian of core + electron within the energy range at which no two electrons reach the core boundary. (At higher energies, to be considered in Chapter 10, the set of channel functions Φ_i is to be extended to represent the escape of two, or more, electrons. For this purpose, as well as for molecular applications, the set of variables indicated by ω must be extended appropriately.)

Solving the Schrödinger equation within the sphere $r \leq r_0$ is required to determine the coefficients $\{\phi_\alpha(r_0), \phi'_\alpha(r_0), a_{i\alpha}\}$ of (9.1) for a complete set of degenerate eigenfunctions $F_\alpha(\omega, r)$ of the Hamiltonian with convenient specifications. Alternative specifications for this set lead to equivalent sets of boundary parameters to be obtained by different methods, as was discussed in Chapter 6 for the case of a single channel. The problem to be studied here is more complicated than that of Chapter 6 but structurally analogous to it.

As in Chapter 6, the problem formulated for an arbitrary energy E in a certain range contrasts with the familiar eigenvalue problems of wave mechanics. We are prescribing the energy and seeking eigenvalues of boundary parameters such as the $a_{i\alpha}$ or equivalent ones. The problem may also be formulated as an inhomogeneous equation analogous to the determination of the reaction matrix $K^{(s)}$ by (6.18).

9.2.1 Direct Variational Approach

A first procedure to specify a set of degenerate eigenfunctions $F_\alpha(\omega, r)$ is to require them to satisfy a standard boundary condition analogous to (6.26),

$$r \left. \frac{\partial F_\alpha(\omega, r)}{\partial r} \right|_{r=r_0} + \sigma_\alpha F_\alpha(\omega, r_0) = 0, \tag{9.2}$$

holding for *all* ω at $r = r_0$. In view of the expansion (9.1) this condition implies the relationship

$$\sigma_\alpha = - \left. \frac{r\partial F_\alpha(\omega, r)/\partial r}{F_\alpha(\omega, r)} \right|_{r=r_0} = -r_0 \phi'_\alpha(r_0) \cot \phi_\alpha(r_0). \tag{9.3}$$

Alternative values of the parameters σ_α, equal in number to the fragmentation channels $\Phi_i(\omega)$, result as the eigenvalues of a variational procedure. The corresponding eigenvectors $a_{i\alpha}$ are equivalent to the transformation matrix $U_{i\alpha}$ of Chapters 7 and 8.

The variational procedure may consist of selecting a trial wave function

$$\Psi_t(\omega, r) = \sum_n c_n f_n(\omega, r) \tag{9.4}$$

in which each base function $f_n(\omega, r)$ either vanishes on the boundary $r = r_0$ together with $\partial f_n(\omega, r)/\partial r$ or can be represented in the form of (9.1) at $r = r_0$.

With this functional form we evaluate the mean values of the Hamiltonian and the normal logarithmic derivative

$$H_t = \frac{(\Psi_t|H|\Psi_t)}{(\Psi_t|\Psi_t)}, \qquad \sigma_t = -\left.\frac{\partial \ln \Psi_t(\omega, r)}{\partial r}\right|_{r=r_0}, \tag{9.5}$$

and vary the coefficients c_n to find extrema of σ_t subject to the subsidiary condition $H_t = E$, the preselected energy of the system core + electron. Here, as in Section 6.5, the variational condition leads to a linear homogeneous system in c_n, with eigenvalues σ_α.

The implementation of this procedure is discussed in Section 9.3, but two of its aspects are stressed here. First the eigenvalues σ_α, which depend on the choice of the functions $f_n(\omega, r)$, are *underestimates* of the exact values, for the reason discussed in Section 6.5. Second, these eigenvalues depend only on properties of the electron + core system *within the sphere* $r \le r_0$, being wholly unrelated to the extension of $\Psi_\alpha(\omega, r)$ at $r \ge r_0$.

To illustrate the variation of each σ_α as a function of increasing energy, note that σ_α rises with E, diverging periodically when the function $\phi_\alpha(r)$ in (9.1) and (9.3) vanishes at $r = r_0$. Each node of a radial function $\phi_\alpha(r)$ would shift smoothly toward lower r as the energy E increases in a one-dimensional setting according to the familiar theorem of Sturm. Our problem is, in fact, multidimensional, but the correlation of angular and radial nodes of $F_\alpha(\omega, r)$ is represented only by the (generally small) set of eigenvector components $\{a_{i\alpha}\}$ in (9.1). For purposes of orientation we disregard these correlations here and view the nodes of $\phi_\alpha(r)$ as traversing the core boundary at $r = r_0$ periodically as E increases, whereby σ_α diverges periodically as a tangent function.

Indeed, the representation (9.1) of F_α through the phase function $\phi_\alpha(r)$ yields

$$\sigma_\alpha = -r_0 \phi'_\alpha(r_0) \cot \phi_\alpha(r_0) = \left.\frac{d\phi_\alpha}{d \ln r}\right|_{r=r_0} \tan[\phi_\alpha(r_0) - \tfrac{1}{2}\pi]. \tag{9.6}$$

The phase derivative $\phi'_\alpha(r_0)$ is nonnegative by definition and $\phi_\alpha(r_0)$ increases monotonically with both r_0 and E *without any singularity*, thus serving as a more convenient boundary parameter than σ_α itself.

9.2.2 Determination of $\delta_\alpha^{(s)}$ and $U_{i\alpha}$

The definition of the MQDT parameters that represent the effect of short-range interactions rests on the expansion of $F_\alpha(\omega, r)$ [(8.53)] in terms of fragmentation channels throughout the range $r \ge r_0$, namely,

$$F_\alpha(\omega, r) = \sum_{i=1}^{N} \Phi_i(\omega)[f(\epsilon_i, l_i, r) \cos \delta_\alpha^{(s)} - g(\epsilon_i, l_i, r) \sin \delta_\alpha^{(s)}]U_{i\alpha}, \qquad r \ge r_0. \tag{9.7}$$

The pair of base functions $\{f, g\}$ should, of course, be appropriate to the field prevailing at large r, in accordance with Chapter 5, and may be replaced with one of the alternative pairs, $\{f^-, f^+\}$ or $\{f^\circ, g^\circ\}$, with the required changes in the structure of (9.7). Equation (9.7) implies replacing the single r-dependent factor $\sin \phi_\alpha(r)$ in (9.1) with base functions $\{f, g\}$ that evolve with r at slightly different rates in different fragmentation channels, in accordance with the respective values of ϵ_i. The alternative formulation of F_α in (9.7) relates the determination of the short-range parameters $\{\delta_\alpha^{(s)}, U_{i\alpha}\}$ to their intended application more directly; by the same token it intermixes the core interaction problem with values of the functions $\{f, g\}$ pertaining to the long-range field. More specifically, it replaces the single-phase function $\phi_\alpha(r_0)$ in (9.6) with the phase *shift* $\delta_\alpha^{(s)}$, namely, with the *difference* of the phase of each term of (9.7) and the phase of the base radial function $f(\epsilon_i, l_i, r_0)$.

The variational calculation of parameters outlined earlier for σ_α and $a_{i\alpha}$ is readily adjusted to determining the eigenvalues $\delta_\alpha^{(s)}$ directly instead of σ_α. Note only that the $\delta_\alpha^{(s)}$ are eigenvalues of a different operator than the one corresponding to σ_α and that the new operator includes values of the external field functions $\{f, g\}$. The monotonic relation between the energy E and the σ_α, which underlies the variational property of σ_α, nevertheless extends to $\delta_\alpha^{(s)}$ because the connection of σ_α and $\delta_\alpha^{(s)}$ is itself monotonic, as verified by the positive character of $d\sigma_\alpha/d\delta_\alpha^{(s)}$ derived from (9.2) and (9.7),

$$\frac{d\sigma_\alpha}{d\delta_\alpha^{(s)}} = W(f, g)[f(\epsilon_i, l_i, r_0)\cos\delta_\alpha^{(s)} - g(\epsilon_i, l_i, r_0)\sin\delta_\alpha^{(s)}]^{-2} > 0, \quad \text{for each } i. \tag{9.8}$$

Here $W(f, g)$ indicates the Wronskian of f and g, which equals $2/\pi$ according to the normalization equation [(5.16)].

The influence of the long-range field introduced here by the functions $\{f, g\}$ includes their singularities at the ionization threshold. These singularities are removed by replacing the pair $\{f, g\}$ in each channel with the pair $\{f^\circ, g^\circ\}$ normalized at the origin as discussed in Section 5.4; the replacement formulas are given in Section 5.5.2. The short-range parameters $\delta_\alpha^{(s)}$ and $U_{i\alpha}$ are replaced accordingly by corresponding parameters $\delta_\alpha^{(so)}$ and $U_{i\alpha}^\circ$, which are free from threshold singularities. This replacement, which complicates the analytical work, is usually dispensed with in the important case of a long-range Coulomb field where the threshold singularities affect only higher derivatives with respect to energy, as noted in Section 7.8.

9.2.3 Orthogonality and Normalization

Each of the alternative sets of eigenchannel functions $F_\alpha(\omega, r)$, selected by alternative formulations of boundary parameters as in Sections 9.2.1 and

9.2.2, can be cast in orthonormalized form. This result is achieved by extending the procedure of Section 5.8 from a single channel to a multichannel context in which surface integration plays a key role. An important aspect will emerge, namely, a connection between normalization within and without the finite volume of the core region $r \le r_0$.

We apply (5.33) to a pair of eigenfunctions $\{F_\beta(E', \omega, r), F_\alpha(E, \omega, r)\}$, for initially different energies $\{E', E\}$ and with additional integration over all the ω variables on both sides of the equation as understood from the derivation of (5.33). Multiplying again initially by $E' - E$ and recalling the boundary value equations [(9.2) and (9.3)], we have the alternative expressions

$$\begin{aligned}
(E' - E) &\int_0^{r_0} dr \int d\omega\, F_\beta^*(E', \omega, r) F_\alpha(E, \omega, r) \\
&= \tfrac{1}{2} \int d\omega\, W[F_\beta^*(E', \omega, r), F_\alpha(E, \omega, r)]\Big|_{r=r_0} \\
&= \frac{1}{2r_0} [\sigma_\beta(E') - \sigma_\alpha(E)] \int d\omega\, F_\beta^*(E', \omega, r_0) F_\alpha(E, \omega, r_0) \\
&= \tfrac{1}{2} W[\sin\phi_\beta(E', r), \sin\phi_\alpha(E, r)]\Big|_{r=r_0} \sum_i a_{i\beta}(E') a_{i\alpha}(E) \\
&= \frac{1}{4}\Bigg\{\left[\frac{\partial\phi_\alpha(E, r)}{\partial r} + \frac{\partial\phi_\beta(E', r)}{\partial r}\right] \sin[\phi_\beta(E', r) - \phi_\alpha(E, r)] \\
&\quad + \left[\frac{\partial\phi_\alpha(E, r)}{\partial r} - \frac{\partial\phi_\beta(E', r)}{\partial r}\right] \sin[\phi_\beta(E', r) + \phi_\alpha(E, r)]\Bigg\}_{r=r_0} \sum_i a_{i\beta}(E') a_{i\alpha}(E),
\end{aligned} \tag{9.9}$$

where the Wronskian involves radial derivatives and the integral over ω reduces to the orthonormality of the functions Φ_i for different fragmentation channels. Setting $E' = E$ shows that eigenfunction pairs $\{F_\beta, F_\alpha\}$ at equal energy but with $\sigma_\beta \ne \sigma_\alpha$ are *orthogonal over the boundary*. For $\beta \equiv \alpha$, L'Hospital's theorem applies in the limit $E' \to E$ and relates the *volume integral* to the surface integral through $d\sigma_\beta/dE$.

When the same argument is applied to F functions expanded in the form (9.7), Eq. (9.9) takes the form

$$\begin{aligned}
(E' - E) &\int_0^{r_0} dr \int d\omega\, F_\beta(E', \omega, r) F_\alpha(E, \omega, r) \\
&= \frac{1}{2} \sum_i U_{i\beta} U_{i\alpha} W[f(\epsilon_i', l_i, r_0)\cos\delta_\beta^{(s)} - g(\epsilon_i', l_i, r_0)\sin\delta_\beta^{(s)}, \\
&\quad f(\epsilon_i, l_i, r_0)\cos\delta_\alpha^{(s)} - g(\epsilon_i, l_i, r_0)\sin\delta_\alpha^{(s)}],
\end{aligned} \tag{9.10}$$

where the complex conjugate notation has been suppressed, since we deal with real functions. In this expression each of the Wronskians $W(f, g)$ converges to its standard value $2/\pi$ as $\epsilon_i' \to \epsilon_i$ and $r_0 \to \infty$ while $W(f, f)$ and $W(g, g)$ vanish in this double limit. Accordingly, (9.10) reduces to

$$(E' - E) \int_0^{r_0} dr \int d\omega\, F_\beta(E', \omega, r) F_\alpha(E, \omega, r) \xrightarrow[E' \to E]{} \frac{1}{\pi} \sin[\delta_\beta^{(s)}(E') - \delta_\alpha^{(s)}(E)] \times \sum_i U_{i\alpha} U_{i\beta} + O\left(\frac{1}{kr_0}\right), \quad (9.10')$$

where the symbol $O(1/kr_0)$ represents the contribution from Wronskians that vanish only in the limit $kr_0 \to \infty$.

For $\beta \neq \alpha$ both expressions (9.9) and (9.10′) yield the orthogonality relations

$$\sum_i a_{i\beta}(E) a_{i\alpha}(E) = 0, \qquad \sum_i U_{i\beta} U_{i\alpha} = 0, \qquad \beta \neq \alpha. \quad (9.11)$$

For $\beta = \alpha$ we obtain instead conceptually different expressions from (9.9) and (9.10′) in the limit $E' \to E$, namely,

$$\int_0^{r_0} dr \int d\omega\, F_\alpha(E, \omega, r)^2 = \frac{1}{2} \left\{ \frac{\partial \phi_\alpha(E, r)}{\partial r} \frac{\partial \phi_\alpha(E, r)}{\partial E} + \frac{\partial^2 \phi_\alpha(E, r)}{\partial E\, \partial r} \frac{1}{2} \sin[2\phi_\alpha(E, r)] \right\}_{r = r_0} \sum_i a_{i\alpha}^2(E) \quad (9.12a)$$

and

$$\int_0^{r_0} dr \int d\omega\, F_\alpha(E, \omega, r)^2 = \frac{1}{\pi} \frac{d\delta_\alpha^{(s)}}{dE} + O\left(\frac{1}{kr_0}\right). \quad (9.12b)$$

In the first expression, (9.12a), the $\Sigma_i\, a_{i\alpha}^2$ that results from integration over ω has the same dimensions as $\int d\omega\, F_\alpha^2$. The remaining factors represent the effective radius of the core, a measure of its ability to hold the electron. (Note that in atomic units $2E$ is a squared wave number and hence $\partial^2 \phi_\alpha / \partial E\, \partial r$ is a length.) At a resonance of electron + core complex, $\sigma_\alpha\, \partial \phi_\alpha / \partial E$ rises rapidly with energy and the effective radius of the core reaches a maximum.

In the second expression, (9.12b), F_α is normalized per unit energy in the continuum according to (9.7) and the volume integral has the dimension of an inverse energy, as does $d\delta_\alpha^{(s)}/dE$. The integral's value, $\pi^{-1}\, d\delta_\alpha^{(s)}/dE$, represents a time delay divided by h according to Section 4.4. More specifically, it represents the time spent by the electron within the complex in excess of its propagation under the influence of the long-range field alone. The integral of $\pi^{-1}\, d\delta_\alpha^{(s)}/dE$ over a resonance equals unity, a result familiar in solid-state theory under the name of Friedel's sum rule for a core consisting of an electron acceptor. More generally, $\pi^{-1}\, d\delta_\alpha^{(s)}/dE$ is the spectral density of states of the complex in the eigenchannel α.

9.2.4 Reaction Matrix

The effects of the electron–core interaction can be represented by a short-range reaction matrix $K^{(s)}$ without any need for identifying interaction eigenchannels such as those represented by $F_\alpha(\omega, r)$ in Sections 9.2.1 and

9.2.2. The reaction matrix, with elements $K_{ij}^{(s)}$, is a multichannel generalization of the matrix introduced in Section 6.3.1 for a single channel. Only matrix elements diagonal in energy, analogous to the single element $K^{(s)}(E)$ in (6.26), are required to parametrize the effects of interaction, though off-diagonal elements indicated by $(\epsilon_i|K^{(s)}|\epsilon_j)$ in the notation of Section 6.3 are involved in their calculation.

The construction of a reaction matrix $K^{(s)}$ will amount to solving a linear *inhomogeneous* system of equations, that is, inverting a matrix in contrast to the eigenvalue problems considered in Sections 9.2.1 and 9.2.2, which involve matrix diagonalizations. The determination of eigenvalues and eigenchannels provides a greater amount of explicit information on the system under study, but the solution of an inhomogeneous problem is generally more expeditious. In this section we generalize the procedure of Section 6.3 not only by extending it to multichannel processes but also by utilizing a more flexible base set of wave functions. This set serves also for other purposes, such as the close coupling procedure outlined in Section 9.3.4. On the other hand, the following procedure is unlikely to find direct application.

The determination of $K^{(s)}$ often utilizes a double set of base functions $\{\phi_n\}$. One subset consists of functions $\{\phi_i(\bar{E}_i, \epsilon_i)\}$ of the core and electron variables that take the form $\Phi_i(\omega)f(\epsilon_i, l_i, r)$ in the outer region $r \gtrsim r_0$, where f is the wave function of the long-range potential and is regular at $r = 0$. The second subset consists of functions $\{\phi_v\}$ that vanish outside the core. The complete wave function is represented by the superposition

$$\Psi = \sum_n a_n \phi_n, \tag{9.13}$$

analogous to (6.9), where the sum includes integration over the electron energies ϵ_i of the first subset. The Schrödinger equation analogous to (6.11) is

$$(E - \langle n|H|n\rangle)a_n = \sum_{n'} \langle n|H|n'\rangle a_{n'}. \tag{9.14}$$

We utilize here the substitution

$$a_n = -\left[\frac{1}{\pi}\frac{P}{E - \langle n|H|n\rangle}\langle n|K^{(s)}|Ei\rangle - \delta(E - \langle n|H|n\rangle)\right]N_{Ei}, \tag{9.15}$$

equivalent to (6.12) and (6.17), where $|Ei\rangle$ indicates a wave function Φ_i with total energy $E = \bar{E}_i + \epsilon_i$ and N_{Ei} is a normalization constant. This reduces (9.14) to the inhomogeneous equation for the reaction matrix

$$\langle n|K^{(s)}|Ei\rangle = \sum_{n'} \langle n|H|n'\rangle \frac{P}{E - \langle n'|H|n'\rangle}\langle n'|K^{(s)}|Ei\rangle - \pi\langle n|H|Ei\rangle, \tag{9.16}$$

analogous to (6.18).

The parameters that represent the effect of short-range interactions on the electron outside the core emerge in the wave function Ψ from the portion of the Σ_n over the subset of ϕ_n with $n \equiv |Ei\rangle$, which is nonzero at $r > r_0$. The contributions of this portion, from *both* terms of the coefficient (9.15), are represented by wave functions "on the energy shell," that is, by states ϕ_i with energy E. In particular, the principal part integration conducted as in Section 6.3.6 over the first term of (9.15), changes the factor $f(\epsilon_i, l_i, r)$ of a wave function ϕ_i into the companion function g evaluated at the energy $\epsilon_i = E - \bar{E}_i$, $\bar{E}_i$ being the core energy in the fragmentation channel Φ_i.

The wave function Ψ at $r \geq r_0$ thus involves only matrix elements of $K^{(s)}$ diagonal in the total energy E. This energy may be indicated as a parameter of $K^{(s)}$ or even omitted altogether, but channel indexes must be retained as the matrix generally couples channels with $j \neq i$. Thus, we write (9.13) in the form

$$\Psi_{Ei} = -\sum_j \Phi_j(\omega)[g(E - \bar{E}_j, l_j, r)K_{ji}^{(s)}(E) - f(E - \bar{E}_j, l_j, r)\delta_{ji}]N_{Ei}, \qquad r \geq r_0, \quad (9.17)$$

equivalent to (8.56), which has been derived in Section 8.6 from assumed prior knowledge of the parameters $\delta_\alpha^{(s)}$ and $U_{i\alpha}$. The wave function (9.17) has been labeled with the index i of the base function $\Phi_i(\bar{E}_i, \epsilon_i)$ introduced in the structure of the coefficients (9.15). Note that Ψ_i contains the regular function f in the single channel i and the irregular companion function g in all channels.

To determine the normalization coefficients N_{Ei} let us calculate the overlap integrals of pairs of Ψ_i using the orthonormality of Φ_j and assuming that all the j channels are open,

$$\begin{aligned}(\Psi_{Ek}|\Psi_{Ei}) &= N_{E'k}\sum_j \int_0^\infty dr[K_{kj}^{(s)}(E')g(E' - \bar{E}_j, l_j, r) - \delta_{kj}f(E' - \bar{E}_j, l_j, r)] \\ &\quad \times [g(E - \bar{E}_j, l_j, r)K_{ji}^{(s)}(E) - f(E - \bar{E}_j, l_j, r)\delta_{ji}]N_{Ei}\delta(E' - E) \\ &= N_{E'k}\left[\sum_j K_{kj}^{(s)}(E')K_{ji}^{(s)}(E) + \delta_{ki}\right]N_{E'i}\delta(E' - E) \\ &= N_{E'k}\{[K^{(s)}(E)^2]_{ki} + \delta_{ki}\}N_{Ei}\delta(E' - E). \end{aligned} \quad (9.18)$$

[Any closed channels should be eliminated by the procedure of Section 8.4 before constructing (9.18).] The set of stationary states Ψ_{Ei} is thus seen *not* to be orthogonal. Any single state can be normalized simply by setting

$$N_{Ei} = \{1 + [K^{(s)}(E)^2]_{ii}\}^{-1/2}. \quad (9.19)$$

as is apparent from (9.18). Construction of an orthonormal set, on the other hand, requires the superposition of Ψ_{Ei} with the transformation matrix $[1 + K^{(s)}(E)^2]^{-1/2}$, which yields the new set

$$\begin{aligned}\bar{\Psi}_{Ei} &= -\sum_j \Phi_j(\omega)\sum_k [g(E - \bar{E}_j, l_j, r)K_{jk}^{(s)}(E) - f(E - \bar{E}_j, l_j, r)\delta_{jk}] \\ &\quad \times [1 + K^{(s)}(E)^2]_{ki}^{-1/2}. \end{aligned} \quad (9.20)$$

9.2.5 Dipole Matrix Elements: Many-Body Theories

Discussion of light emission and absorption processes has centered in previous chapters on the dipole matrix elements $(f|x|i)$ for the transition of an atom or molecule from an initial state $|i)$ to a final state $(f|$. We have been concerned with factoring out of such matrix elements their dependence on the level of excitation in any one fragmentation channel or on the channel index itself. At the outset of this chapter we have in fact listed among the main parameters of interest the dipole matrix elements $(F_\alpha|x|0)$ from the ground state $|0)$ of an atom to eigenchannel wave function $F_\alpha(\omega, r)$ of the short-range interaction, normalized per unit energy and shorn of any reference to details of the final state.

The calculation of these matrix elements would meet no obstacle for transitions of an electron in a central field, but problems do arise from the presence of many electrons in atomic and molecular cores and especially from their exchange interaction. These problems emerge particularly on testing the accuracy of the matrix elements for dipole transitions between stationary states $|i)$ and $(f|$ that depend on long-range as well as short-range interactions. These matrix elements can be expressed in alternative forms whose values must coincide when $(f|$ and $|i)$ are eigenfunctions of the same Hamiltonian with eigenvalue difference ϵ, thus obeying the equation

$$(f_\alpha|x|i) = -i(f_\alpha|dx/dt|i)\hbar/\epsilon = -(f_\alpha|d^2x/dt^2|i)(\hbar/\epsilon)^2. \qquad (9.21)$$

The results of calculations conducted with apparent accuracy are frequently found to violate (9.21) by factors of the order of 2.

A major source of these departures has been traced to the use of independent-electron approximations. Equation (9.21) holds for transitions of a single electron in the field of a local potential, whereas common schematizations of the exchange interactions—represented, for example, by (6.6b)—amount to the use of a nonlocal potential. Equation (9.21) rests on the commutation properties of the kinetic and potential energies, which are altered drastically when independent electrons are subject to nonlocal potentials. Accordingly, the values of dipole matrix elements have proved unduly sensitive to the treatment of exchange interactions. More generally, they have proved sensitive to inconsistencies in approximate calculations of the initial and final state wave functions.

Removal of these inconsistencies is ensured by procedures that are designated collectively as "many-body theories" and originate from the treatment of electrons in metals or nucleons in nuclear matter [1], whose complete wave functions are not readily formulated. (Atomic applications are more recent [2].) The intent of these procedures is to concentrate their formulation on the variables that affect a *transition* of interest directly, thus omitting any explicit reference to the coordinates of particles that remain unaffected. Accordingly we omit the explicit construction of separate wave functions for the initial and final states of a transition, $|i)$ and $(f|$, considering instead transition operators indicated by $\Gamma_{if} \equiv |i)(f|$, integrated as feasible over less relevant variables. [Note that the set of two-index operators $\Gamma_{\rm if}$ is far more numerous than the sets of $|i)$ or $(f|$.)]

As a starting point for the many-body procedures we consider the Schrödinger equation for Γ_{if}, which combines those for $(f|$ and $|i)$,

$$H|i)(f| - |i)(f|H = (E_i - E_f)|i)(f|, \tag{9.22}$$

$$|i)(f| \equiv (\mathbf{r}_1, \mathbf{r}_2, \ldots, \mathbf{r}_N|\Gamma_{\text{if}}|\mathbf{r}'_1, \mathbf{r}'_2, \ldots, \mathbf{r}'_N), \tag{9.22$'$}$$

observing that dipole matrix elements depend only on integrals of Γ_{if} over all but one electron, represented by the "one-particle transition matrix"

$$(\mathbf{r}_1|\Gamma_{\text{if}}|\mathbf{r}'_1) = \prod_{q=2}^{N} \iint \cdots \int d\mathbf{r}_q\, d\mathbf{r}'_q\, \delta(\mathbf{r}_q - \mathbf{r}'_q) \times (\mathbf{r}_1, \mathbf{r}_2, \ldots, \mathbf{r}_N|\Gamma_{\text{if}}|\mathbf{r}'_1, \mathbf{r}'_2, \ldots, \mathbf{r}'_N), \tag{9.23}$$

$$\left(f\left|\sum_{q=1}^{N} x_q\right|i\right) = N \iint d\mathbf{x}_1\, d\mathbf{x}'_1\, \delta(\mathbf{r}_1 - \mathbf{r}'_1)(\mathbf{r}'_1|\Gamma_{\text{if}}|\mathbf{r}'_1)x_1. \tag{9.24}$$

With the Hamiltonian separated into one- and two-particle terms,

$$H = \sum_i H_0(r_i) + \frac{1}{2}\sum_{ij} V(\mathbf{r}_i - \mathbf{r}_j), \tag{9.25}$$

and with assumed symmetry under particle permutations, the Schrödinger equation [(9.22)] is directly integrated over the position of all but one particle, thus yielding

$$H_0(\mathbf{r}_1)(\mathbf{r}_1|\Gamma_{\text{if}}|\mathbf{r}'_1) - (\mathbf{r}_1|\Gamma_{\text{if}}|\mathbf{r}'_1)H_0(\mathbf{r}'_1) + \iint d\mathbf{r}_2\, d\mathbf{r}'_2[V(\mathbf{r}_1 - \mathbf{r}_2) - V(\mathbf{r}'_1 - \mathbf{r}'_2)](\mathbf{r}_1\mathbf{r}_2|\Gamma_{\text{if}}|\mathbf{r}'_1\mathbf{r}'_2) = (E_i - E_f)(\mathbf{r}_1|\Gamma_{\text{if}}|\mathbf{r}'_1). \tag{9.26}$$

This two-particle equation eliminates complications due to additional particles only apparently, because the two-particle transition density $(\mathbf{r}_1\mathbf{r}_2|\Gamma_{\text{if}}|\mathbf{r}'_1\mathbf{r}'_2)$ in turn obeys another equation, which results by integrating (9.22) over all but two particles and which relates $(\mathbf{r}_1\mathbf{r}_2|\Gamma_{\text{if}}|\mathbf{r}'_1\mathbf{r}'_2)$ to $(\mathbf{r}_1\mathbf{r}_2\mathbf{r}_3|\Gamma_{\text{if}}|\mathbf{r}'_1\mathbf{r}'_2\mathbf{r}'_3)$, and so on.

This procedure thus actually reduces (9.22) to a hierarchy of equations involving increasing numbers of particles, but it stands to reason that the hierarchy can be approximately truncated without undue errors and *preserving consistency* in the treatment of $|i)$ and $(f|$. Two main approaches have been utilized in truncating the hierarchy of equations that begins with (9.26). One of these regards the interaction term $V(\mathbf{r}_1 - \mathbf{r}_2)$ as a small perturbation, truncating the hierarchy by disregarding terms of higher-than-desired order in powers of V. This procedure, to be described in Section 9.3.8, is accordingly called the many-body perturbation theory.

The second approach truncates the hierarchy of equations in a manner to be described in Section 9.3.9, with manifest consistency that ensures compatibility with (9.21) and the sum rules related to it. Both the original and the recent applications of this approach disregard interactions among excited electrons, a step that was initially justified on the ground that effects of such interactions would be minimized by destructive interference among different terms. This argument, superseded currently by other considerations, led nevertheless to calling the approach the random phase approximation.

The many-body theories have been introduced here in the context of dipole matrix elements because, in atomic physics, they have been applied primarily to studies of photoionization. They are equally relevant, of course, to the calculation of the form factors discussed in Section 3.5 and any other off-diagonal matrix element, whose accuracy hinges on consistent treatment of a pair of relevant wave functions.

9.3 METHODS OF CALCULATION

The previous section dealt with general aspects of the determination of electron–core interaction parameters. Here we deal with more specific aspects of alternative methods of calculation, of which there has been a large proliferation as theorists strived to expand their capabilities. The diversity of approaches has been boosted by the historical view that calculating solutions of the well-established Schrödinger equation is an art to be pursued by whichever device may appear convenient from any particular point of view.

The following outline of several approaches has twin purposes. On the one hand, it should provide guidance to the very extensive literature and its interpretation and utilization. Illustrative examples of current results will be presented separately in Section 9.4. On the other hand, it should guide the design of future calculations in the light of the previous chapters and sections. In this light, future calculations should probably rely mainly on R-matrix methods, of which alternative variants will be outlined, but other methods remain necessary or at least relevant to the further end of constructing Jost matrices and thence the various observable parameters. Principal among such tools are those that take into account the long-range anisotropic component of the core's field (Section 9.3.7) and the development of the many-body theories introduced in Section 9.2.5 to ensure consistency in the calculation of transition matrix elements.

9.3.1 *R*-Matrix: The Traditional Approach (Wigner–Eisenbud)

The formulation and calculation of the electron–core interaction effect outlined in Section 6.4 for a single excitation channel extends readily to a multichannel setting. The formulation is generally cast in terms of fragmentation channels by representing energy eigenfunctions near the core's edge in the form $F(\omega, r) = \Sigma_i \Phi_i(\omega) F_i(r)$. The logarithmic derivatives of the radial functions $F_i(r)$ at $r = r_0$ can then be expressed in equivalent terms of reaction matrices $K_{ij}^{(s)}$, as in (9.17), or $K_{ij}^{(so)}$, or rather in terms of an R matrix $R_{ij}(E)$.

As in Section 6.4 we consider here a complete set of eigenfunctions Ψ_ν of the Hamiltonian for $N + 1$ electrons with boundary conditions at $r = r_0$ characterized by a parameter σ_0 as in (6.28) and with eigenvalues E_ν. Each Ψ_ν is represented near the boundary as

$$\Psi_\nu = \sum_i \Phi_i(\omega)\phi_{\nu i}(r), \qquad \left[r\frac{d\phi_{\nu i}(r)}{dr} + \sigma_0 \phi_{\nu i}(r)\right]_{r=r_0} = 0. \tag{9.27}$$

(Different values of σ_0 might be used for different channels, but a single σ_0 is considered here.)

As in Section 6.4, we seek next the boundary parameters for eigenfunctions $\Psi = \Sigma_\nu c_\nu \Psi_\nu$ with a fixed energy eigenvalue E, in contrast to the Ψ_ν whose boundary conditions are fixed but whose energies E_ν are found by diagonalizing the Hamiltonian matrix. These functions have alternative channel expansions within and without the boundary

$$\Psi(\omega, r) = \sum_\nu c_\nu \Psi_\nu(\omega, r) \xrightarrow[r \to r_0]{} \sum_{\nu i} c_\nu \Phi_i(\omega)\phi_{\nu i}(r), \qquad r \leq r_0, \tag{9.28a}$$

$$= \sum_i \Phi_i(\omega) F_i(r), \qquad r \geq r_0. \tag{9.28b}$$

With this notation and owing to the orthonormality of the $\Phi_i(\omega)$, the extension of (6.32) takes the form

$$\begin{aligned} c_\nu &= \int_0^{r_0} dr \int d\omega\, \Psi_\nu(\omega, r)\Psi(\omega, r) \\ &= \sum_i \phi_{\nu i}(r_0)\left[r\frac{dF_i}{dr} + \sigma_0 F_i(r)\right]_{r=r_0} \Big/ 2r_0(E - E_\nu). \end{aligned} \tag{9.29}$$

Proceeding further to extract the essential information from this last equation, we multiply it by $\phi_{\nu i}(r_0)$ and sum it over ν. This procedure introduces the multichannel R matrix

$$R_{ij}(E) = \sum_\nu \frac{\phi_{\nu i}(r_0)\phi_{\nu j}(r_0)}{2r_0(E - E_\nu)} \tag{9.30}$$

and yields the linear relation between the $F_i(r)$ and their derivatives on the boundary

$$F_i(r_0) = \sum_j R_{ij}(E)\left[r\frac{dF_j}{dr} + \sigma_0 F_j(r)\right]_{r=r_0}. \tag{9.31}$$

This relation is once again made explicit by multiplying it by $1/R(E)$,

$$r_0\left(\frac{dF_i}{dr}\right)_{r=r_0} + \sum_j \sigma_{ij}(E)F_j(r_0) = 0, \tag{9.32}$$

with

$$\sigma_{ij}(E) = \sigma_0\delta_{ij} - \left[\frac{1}{R(E)}\right]_{ij}. \tag{9.33}$$

The matrix $\sigma_{ij}(E)$ thus defined and the R matrix itself are equivalent to the matrix $K^{(s)}(E)$ discussed in Section 9.2.4 or its singularity-free form $K^{(so)}(E)$. This connection is made explicit in convenient matrix notation by indicating the set of radial functions $\{F_i(r_0)\}$ as a diagonal matrix F_0 and the set of their derivatives $\{(r\,dF_i/dr)_{r=r_0}\}$ by F'_0. Similarly, the base sets of wave functions $\{f, g\}$ or $\{f^o, g^o\}$ of the long-range field, introduced in Chapter 5, will be denoted by diagonal matrices together with their derivatives. Comparison of the coefficients of Φ_i in (9.28b) and (9.17) yields the connection

$$F_0 = f_0 - g_0K^{(s)}, \qquad F'_0 = f'_0 - g'_0K^{(s)}. \tag{9.34}$$

These equations combine with (9.32), which now reads $F'_0 + \sigma F_0 = 0$, to yield

$$K_{ij}^{(s)} = (g'_0 + \sigma g_0)_i^{-1}(f_0 + \sigma f_0)_j. \tag{9.35}$$

The matrix $K^{(so)}$ is obtained by replacing (f_0, g_0, f'_0, g'_0) by $(f_0^o, g_0^o, f_0^{o\prime}, g_0^{o\prime})$; the result constitutes the extension of (6.37). Note that the radial derivative $r_0(dF_i/dr)_{r=r_0}$ is not simply proportional to the corresponding wave function $F_i(r_0)$ in a multichannel context; direct proportionality is restored by diagonalizing $\sigma_{ij}(E)$ and $R_{ij}(E)$, that is, by introducing their eigenchannels analogous to, but different from, the eigenchannels α of $K^{(s)}$.

The remarks of Section 6.4 on the dependence of the R matrix on E and the difficulty in converging (9.28) near the boundary apply equally in the multichannel context. Additional remarks are in order here, as we deal with realistic calculations that should utilize maximum flexibility in contrast to the idealized case of a single channel:

(1) The form of the eigenfunctions Ψ_ν on the boundary, [(9.27)], may be modified freely, provided only that its connection with the representation of Ψ outside the boundary—(9.28b) or an analog—remains identifiable. Furthermore, the finite volume and its boundary need not be restricted to a spherical shape but may be chosen flexibly.

(2) The representation of Ψ_ν at $r < r_0$ in terms of Slater determinants, considered in Section 6.4, has valuable aspects but is overly restrictive. The separation of factors Φ_i and $\phi_{\nu i}$ at $r \sim r_0$ is essential, of course, but the form of Ψ_ν is fully flexible otherwise, at $r < r_0$. In particular, the restriction to a "frozen core" representation of the Φ_i—epitomized by the factorization $a_i = Q_\alpha q_k$ of the coefficients in (6.10)—is unnecessary.

(3) A general procedure to provide flexibility of the wave functions Ψ_ν consists of

including in their analytical expansion two kinds of terms

$$\Psi_\nu(\omega, r) = \sum_n a_{\nu n} \Psi_{0n}(\omega, r) + \sum_m b_{\nu m} \Psi_{0m}(\omega, r) \tag{9.36}$$

with different specifications. Both of them represent the state of $N + 1$ electrons; the Ψ_{0n} take the form (9.27) when the radial distance of the outermost electron approaches the boundary $r = r_0$, whereas the Ψ_{0m} become negligible when any radial variable does so. The Ψ_{0m} are generally constructed with suitable natural orbitals (also called pseudo-orbitals), properly antisymmetrized. They serve to represent elaborate electron correlations through virtual excitations. The coefficients $\{a_{\nu n}, b_{\nu m}\}$ are often the only variational parameters in (9.36), whereby the diagonalization of the Hamiltonian matrix reduces to a linear algebraic system in these parameters. The flexibility in the choice and number of the $\{\Psi_{0n}, \Psi_{0m}\}$ allows us to incorporate in them whatever information is available on the dynamics of the $N + 1$ electrons in a desired range of excitation energies.

(4) The presence of highly excited orbitals in the wave function components Ψ_{0m} may give rise to spurious resonances in the energy dependence of the R matrix. The spurious effects may then be smoothed out, but extension of the calculation to high energies that allow the escape of two or more electrons remains tentative, pending fuller development of the analysis to be introduced in the later chapters.

(5) This chapter has focused thus far on the calculation of parameters of the short-range electron–core interaction, to serve as input for treatment of electron motion in the external field according to Chapters 7 and 8. When long-range anisotropic interactions make that treatment inadequate, it may be desirable to treat these interactions by a close-coupling procedure (Sections 9.3.4 and 9.3.7) that integrates the set of radial functions $\{F_i(r)\}$ from $r = \infty$ toward lower r. In this event the matrix $R_{ij}(E)$ constructed according to (9.30) need not be studied *per se*. It serves as input to (9.33) and (9.32), the latter of which represents boundary conditions at $r = r_0$ to be satisfied by the $\{F_i(r)\}$ obtained by numerical integration over the range $r_0 \leq r \leq \infty$. These conditions merely restrict the choice of initial values of the $\{F_i(r), dF_i/dr\}$ at $r = \infty$. Results of R-matrix calculations to be described in Section 9.4 have been obtained in this manner.

9.3.2 *R* matrix: The Eigenchannel Approach

The eigenchannel parameters $\delta_\alpha^{(s)}$ [or $\delta_\alpha^{(so)}$] and $U_{i\alpha}$ may be derived from the R matrix (9.30) through the transformation (9.35) followed by diagonalization of $K^{(s)}$. Analogous parameters σ_β or R_β and matrices $U_{i\beta}$ are obtained by diagonalizing σ_{ij} or R_{ij}. It is also interesting to view the latter parameters as resulting from iterative adjustment of the calculation described in Sections 6.4 and 9.3.1.

The procedure leading to the construction of the R matrix (9.30) rests on the consideration that the energy eigenvalues E_ν of the Hamiltonian within a finite core volume and with the boundary condition (9.27) generally do not coincide with the energy E of interest, as was stressed in Section 6.4. Indeed, the R matrix (9.30) diverges whenever one of the eigenvalues E_ν approaches

E. In this limit ($E_\nu \to E$) the matrix $\sigma_{ij}(E)$ approaches the diagonal form $\sigma_0 \delta_{ij}$ according to (9.33), since the determinant of the reciprocal matrix $1/R$ vanishes.

Interest in the diagonal form of the matrix $\sigma_{ij}(E)$ and the related reaction matrices suggests that the eigenstates of the Hamiltonian be studied in the limit $E_\nu \to E$. Locations of this limit can be identified by diagonalizing the Hamiltonian for alternative values of the boundary parameter σ_0, thus determining numerically the functional dependence $E_\nu(\sigma_0)$ and the eigenvalues $\sigma_0 = \sigma_\beta$ at which

$$E_\nu(\sigma_\beta) = E. \tag{9.37}$$

This procedure previously outlined in Section 6.5 for the case of a single channel converges rapidly in practice, as will be illustrated in Section 9.4.1. The eigenvectors $\phi_{\nu i}(r_0)$ corresponding to an energy eigenvalue (9.37) coincide, if properly normalized, with the boundary parameters $U_{i\beta}$, elements of the matrix that diagonalizes the $\sigma_{ij}(E)$ calculated for an arbitrary σ_0. Viewing the functional dependence $E_\nu(\sigma_0)$ is itself of interest as it illustrates the relationship between eigenvalues and boundary parameters described in Section 6.5.

Once a value $\sigma_0 = \sigma_\beta$ is identified, the sum over ν in (9.28a) reduces to the single term for which (9.37) holds, because this term suffices to match (9.28b) at $r \geq r_0$. The problem raised by slow convergence of the expansion over ν is thus bypassed. The identification of σ_β replaces explicit use of the R matrix.

The number of values $\sigma_0 = \sigma_\beta$ that satisfy (9.37) generally equals the number n of fragmentation channels in (9.27) and (9.28). This amounts to saying that plots of the functions $E_\nu(\sigma_0)$ over the range $-\infty \leq \sigma_0 \leq \infty$ consist of n branches that rise monotonically with increasing energy, each of them crossing any preestablished ordinate E in the energy range for which (9.27) is designed.

9.3.3 *R* Matrix: Direct Variational Approach

The study of the functional dependence of eigenvalues on the boundary parameter σ can itself be bypassed by direct variational determination of the roots σ_β of (9.37). To this end we transform the usual variational expression, whose stationary values are energy eigenvalues for a given boundary parameter σ, into a related expression whose stationary values are eigenvalues of σ for a given energy E.

The kinetic energy term of the Hamiltonian should be separated here explicitly from the potential energy, by writing

$$H = -\frac{1}{2}\sum_q \nabla_q^2 + V \tag{9.38}$$

in atomic units, where $\mathbf{V}_q$ indicates the gradient operator of the qth electron ($q = 1, \ldots, N + 1$). The structure of the potential energy V need not be described explicitly. The real trial wave function

$$\Psi(\omega, r) = \sum_k c_k w_k(\omega, r) \underset{r \sim r_0}{\longrightarrow} \sum_k c_k \sum_i \Phi_i(\omega)\phi_{ki}(r), \tag{9.39}$$

analogous to (6.38), is represented here at $r \sim r_0$ in a notation similar to (9.27) with r representing the distance of the farthest electron. (This notation and the specification of the boundary could be modified for molecules or systems other than single atoms.) The set of functions $\{w_k(\omega, r)\}$ need not be orthonormal; on the contrary, linear dependence of its elements proves convenient. The set should be selected so that its superposition can describe all the relevant features of the atom under consideration. In particular, the parameters σ_{ki} in the boundary equation $[r\ d\phi_{ki}/dr + \sigma_{ki}\phi_{ki}(r)]_{r=r_0} = 0$ should span a wide range to permit optimization of the boundary parameter σ to be imposed on $\Psi(\omega, r)$,

$$r\frac{\partial\Psi(\omega, r)}{\partial r} + \sigma\Psi(\omega, r)\bigg|_{r=r_0} = 0. \tag{9.40}$$

The procedure for constructing energy eigenfunctions with a given boundary parameter σ utilizes the variational expression

$$E = \int \Psi[-\tfrac{1}{2}\mathbf{V}^2 + V]\Psi \prod_{j=1}^{N+1} d\mathbf{r}_j \bigg/ \int \Psi^2 \prod_{j=1}^{N+1} d\mathbf{r}_j, \tag{9.41}$$

whose integrals extend over all electron positions with $r_j \le r_0$. The $\mathbf{V}^2$ operator stands for the sum of all the $\mathbf{V}_j^2$ terms for the $N + 1$ electrons. Solution of the system of equations $\partial E/\partial c_k = 0$ would determine the eigenvalues E_ν. An alternative variational expression for the boundary parameter σ in (6.38) is obtained by integrating by parts the $\mathbf{V}^2$ terms in the numerator of (9.41) and bringing the resulting surface integral into the denominator [3],

$$\sigma = \int \{-|\mathbf{V}\Psi|^2 + 2\Psi(E - V)\Psi\} \prod_{j=1}^{N+1} d\mathbf{r}_j \bigg/ r_0 \int \Psi(\omega, r_0)^2\, d\omega. \tag{9.42}$$

Substitution of the expansion (9.39) of Ψ reduces (9.42) to the algebraic form

$$\sigma = \sum_k \sum_l c_k \Gamma_{kl} c_l \bigg/ \sum_k \sum_l c_k \Lambda_{kl} c_l, \tag{9.43}$$

where

$$\Gamma_{kl} = \int [-\mathbf{V}w_k \cdot \mathbf{V}w_l + 2w_k(E - V)w_l] \prod_{j=1}^{N+1} d\mathbf{r}_j, \tag{9.44}$$

$$\Lambda_{kl} = r_0 \sum_i \phi_{ki}(r_0)\phi_{li}(r_0). \tag{9.45}$$

[Equation (9.45) reflects the orthonormality of the set $\{\Phi_i(\omega)\}$ in (9.39).]

The extremum condition $\partial\sigma/\partial c_k = 0$ leads to the generalized eigenvalue problem

$$\sum_l \Gamma_{kl} c_l = \sigma \sum_l \Lambda_{kl} c_l, \tag{9.46}$$

with eigenvalues σ_β and eigenfunctions

$$\Psi_\beta = \sum_k c_k^{(\beta)} w_k(\omega, r). \tag{9.47}$$

Note that the eigenvalue problem (9.46) does not coincide with that of diagonalizing the reaction matrix $K^{(s)}$ at the core boundary, even though the two problems are substantially equivalent; accordingly, the index β has been used here for σ_β and $\{c_k^{(\beta)}\}$. The eigenvector components at r_0, indicated by $\Sigma_k\, c_k^{(\beta)} \phi_{ki}(r_0)$ according to (9.39), represent transformation coefficients $U_{i\beta}$ to within normalization. The eigenchannel parameters $\delta_\alpha^{(s)}$ [or $\delta_\alpha^{(so)}$] and $U_{i\alpha}$ are then obtained from the σ_β and $U_{i\beta}$ through the transformation (9.35). The number of eigenvalues σ_β and eigenfunctions Ψ_β equals the number n of fragmentation channels $\Phi_i(\omega)$, which is the rank of the matrix Λ_{kl}.

9.3.4 Close Coupling

The procedure outlined in Section 6.6 to calculate the radial wave function of an electron interacting with alternative stationary states of a H atom extends readily to multichannel problems. Its extensions, in various forms, encompass probably the majority of past calculations of electron–core interactions.

The essence of this method may be described succinctly as utilizing the wave function expansion (7.14) in terms of fragmentation channels, i.e.,

$$\Psi(\omega, r) = \mathscr{A} \sum_i \Phi_i(\omega) F_i(r), \tag{9.48}$$

within as well as outside the core, that is, at all radial distances from $r = 0$ to $r = \infty$. Use of this expansion in the wave equation $(E - H)\Psi = 0$, followed by multiplication by $\Phi_i(\omega)$ and integration over ω, reduces it to the system of coupled radial equations (in atomic units)

$$\left[\frac{1}{2}\frac{d^2}{dr^2} - \frac{1}{2}\frac{l_i(l_i+1)}{r^2} + E - \bar{E}_i - V_{ii}(r)\right] F_i(r) - \sum_{j \neq 1} V_{ij}(r) F_j(r) = 0. \tag{9.49}$$

Here l_i is the orbital momentum of the electron in the ith channel; the matrix element $(\Phi_i|H|\Phi_i)$ is represented by $\bar{E}_i + V_{ii}(r)$, that is, by the core energy in the ith channel plus the potential exerted by its core on the $(N + 1)$th electron; and the $V_{ij}(r)$ represent the off-diagonal elements $(\Phi_i|H|\Phi_j)$. The potentials include here nonlocal exchange components. Equation (9.49) reduces to (6.46) in the simple case of Section 6.6. The antisymmetrization symbol $\mathscr{A}$ in (9.48) has been indicated explicitly for emphasis, even though it had been dropped for simplicity in most previous equations.

Numerical solution of (9.49), starting from $r = 0$ with $F_j(r) \propto \delta_{ij} r^{l_i+1}$, is straightforward unless the number of channels is too large. The development and applications of this approach predate the emphasis of Chapters 7 and 8 on separate study of phenomena within and outside the core. Applications, to be illustrated in Section 9.4, have dealt mostly with a modest number of channels ($n \lesssim 5$) emphasizing precise evaluation of the influence of anisotropies in the long-range channel interaction potentials $V_{ij}(r)$, to be discussed in Section 9.3.7. Indeed, (9.49) affords the most obvious approach to dealing with long-range anisotropy.

As anticipated at the end of Section 9.3.1, numerical integration of the close-coupled equations (9.49) at large ranges starts conveniently from $r = \infty$, where the off-diagonal $V_{ij}(r)$ vanish as $1/r^2$ or faster. Independent boundary conditions are then imposed on the $F_i(r)$, [typically, exponential decay $\exp(-\kappa_i r)$ in the closed channels with $E - \bar{E}_i = -\frac{1}{2}\kappa_i^2 < 0$, and either $\sin k_i r$ or $\cos k_i r$ oscillations in the open channels with $E - \bar{E}_i = \frac{1}{2}k_i^2 > 0$.] The values of all the $F_i(r)$ and dF_i/dr calculated by alternative integration from $r = 0$ and $r = \infty$ are then matched at an appropriate intermediate $r = r_m$, by an algebraic procedure described by Smith [3a] that restricts the number of free integration parameters in the solutions started from $r = 0$ and $r = \infty$. Only for eigenvalues E of a discrete spectrum can the matching be completed in the energy range below all $\bar{E}_i$.

In the absence of long-range anisotropy, solutions of (9.49) initiated at $r \sim 0$ can be cast in the form of (8.56) at $r > r_0$. In this event the close coupling method can be viewed as an alternative to R matrix procedures. From this point of view, the structure of the wave function (9.48) appears unrealistically restrictive, as the use of a small number of channel functions allows little flexibility for the representation of correlations between the N core electrons and the $(N + 1)$th. This inadequacy has been generally remedied, since the late 1960s, by the procedure of (9.36), which complements (9.48) by a sum of terms $\Sigma_m b_m \Psi_m(\omega, r)$ that become negligible near the core's edge at $r \sim r_0$. The coefficients b_m are optimized by variational procedures. The comments indicated in paragraph (4) at the end of Section 9.3.1 are then relevant to close coupling as well. The addition of such terms to (9.48) is equivalent to the addition of a number of closed channels or the use of optical potential terms introduced in Section 6.6.1. Other comments in Sections 6.6 and 6.6.1, on the effects of truncation in (9.48) and the diminishing return in adding Rydberg states of the core to that expansion, are equally applicable to our present multichannel context.

Reliance on the close-coupling equation (9.49) as a springboard for extensive accurate calculations has been boosted further by a recent development that reduces its solution to a procedure of linear algebra well suited to modern computing [4]. A first preliminary step removes any exchange (i.e., nonlocal) contribution to the terms $\Sigma_j V_{ij}(r) F_j(r)$ of (9.49) by the procedure of Sections 6.3 and 6.7, thus transforming (9.49) into a larger system—akin to

(6.53) and (6.54)—with local potentials only. A more decisive step consists of representing the jth solution of (9.49) in terms of the Green's function of the operator on its left-hand side, that is, in the form

$$F_{ij}(r) = f_i(r)\delta_{ij} + \int dr'\, G_i(r, r') \sum_k V_{ik}(r')F_{kj}(r') \underset{r\to\infty}{\longrightarrow} f_i(r)\delta_{ij} + g_i(r)K_{ij}, \quad (9.49')$$

whose asymptotic terms, cast in the notation of (8.57), are restricted to open channels.

The recasting of (9.49) into its integral form (9.49′) is designed to permit quadrature on a mesh of the integrals on the right-hand side as well as any appropriate formulation or approximation of the Green's function. The equation has the matrix form $M\mathbf{F} = \mathbf{F}$, whose solution is achieved by current vector computers with great numerical stability and for very large matrix dimensions. Integration in (9.49′) may be restricted to a finite range $r \leq r_0$ by imposing an R-matrix boundary condition on F_{ij} at r_0 as in (9.31).

Initial applications to electron–atom and electron–molecule scattering have used the Coulomb Green's function for G_i, but optical potentials can be readily included by supplementing the expansion (9.48) with $(N + 1)$ electron functions as in (9.36). The calculations for electron collisions with hydrogen-like and heliumlike atoms and molecular targets provide probably the most efficient and accurate results to date.

The application of Schwinger's variational principle to solve the matrix equation (9.49′), by varying the Green's function itself, is also under active development [5]†. The illustrative examples in Section 9.4.1 show how the solution of close-coupled equations has proved extremely valuable in the limited context of low excitation energies.

9.3.5 Multiconfiguration Hartree–Fock‡

The elementary concepts and formulas of configuration mixing have been introduced in Section 6.3 in the limited context of constructing the reaction matrix for a single-channel electron–core interaction. Equation (6.9), $\Psi = \Sigma_i a_i \Psi_{0i}$, represents a multielectron wavefunction as a superposition of antisymmetric Slater determinants Ψ_{0i} constructed with products of independent-electron Hartree–Fock wave functions that belong generally to different configurations, that is, to different distributions of electrons among (nl) orbitals.

The superpositions of Ψ_{0i}, with different (nl) distributions, to form the complete wave function Ψ is referred to as the "mixing" of configurations.

† The Schwinger variational method, a variant of those discussed in Section 6.9, is described in Nesbet [6].

‡ See Fischer [7].

The coefficients a_i of the superposition in an eigenstate of energy E are determined as eigenvectors of (6.10)

$$E_{0i}a_i + \sum_{i'} (i|V|i')a_{i'} = Ea_i.$$

These elements underlie the whole treatment of Sections 9.2.4 and 9.3.1–9.3.3. The present section comments on the relations between the role of multiconfiguration Hartree–Fock (MCHF) calculations in the current theory of atomic spectra and in the broader context of spectral and collision phenomena considered in this book and particularly in this chapter.

Multiconfiguration Hartree–Fock is currently the main tool for the precise theoretical determination of the energy of discrete spectral levels. Its extension to heavy atoms, where relativistic effects are important, replaces the Schrödinger Hamiltonian by Dirac's and is accordingly called the Multiconfiguration Dirac–Fock (MCDF) procedure. Multiconfiguration Hartree–Fock represents the manifold correlations of electrons in the corresponding eigenstates through the superpositions $\Sigma_i \, a_i \Psi_{0i}$, which are not always easily interpreted, especially when the mixing is extensive. Calculations of levels and their eigenfunctions by MCHF are well adapted to the characteristics of valence shell levels and also low Rydberg levels. Their extension to higher Rydberg levels is feasible but not as appropriate. Carrying out a separate calculation for each level obscures major regularities, such as existence of quantum defects and other quantitative properties common to whole series.

This shortcoming becomes increasingly apparent in the presence of extensive series perturbations, which often extend into the continuum. Attempts to calculate these effects by ordinary MCHF procedures have failed at times. In certain cases—notably for the levels of double excitations to be treated in Chapter 10—calculations with extensive configuration mixing have yielded, in fact, correct levels and eigenfunctions without providing any obvious cue to the systematics of their results. Lack of clear guidance to the interpretation of extensive mixing thus confines the range of successful applications of MCHF. A more direct restriction lies in the rapidly increasing amount of mixing required to represent higher excitations near to or within the continuum.

The treatment of channel mixing in Chapters 7 and 8, as contrasted to configuration mixing in the MCHF, was in fact motivated by the study of high and perturbed Rydberg levels and by its extension to the adjacent continua. From this point of view MCHF procedures are properly incorporated with other techniques for the construction of reaction and R matrices. The relevant developments of Sections 6.3 and 9.2.4 indicate the general character of the required adaptation of the standard MCHF, namely, the replacement of bound-state radial wave functions of one (or more) highly excited electrons by continuum wave functions or wave functions truncated at the boundary of a core region.

9.3.6 Matrix Variational Methods[†]

The Kohn variational principle, introduced in Section 6.8 for a single channel, has been used extensively in multichannel electron–atom scattering. This method parallels the direct variational R-matrix method of Section 9.3.3 in recasting the familiar variational expression for the energy into the expression for a boundary parameter of the wave function. Whereas the R-matrix procedure confines its attention to a finite volume and considers the logarithmic derivative of the wave functions at the boundary as the relevant parameter, the Kohn and related variational methods (such as the Schwinger principle which is receiving increasing attention) consider the wave function over all space. The tangent of the phase shift at infinity (or its multichannel version, the K matrix) is then the quantity to be determined variationally.

As in (9.36), the full $(N + 1)$-electron wave function is expanded as

$$\Psi_j(\omega, r) = \mathscr{A} \sum_i \Phi_i(\omega) F_{ij}(r) + \sum_m b_{mj} \Psi_m(\omega, r) \tag{9.50}$$

without dividing space into inner and outer regions. All the radial functions in the second term of (9.50) vanish at infinity. As in the close-coupling and R-matrix methods, this term is designed to take into account short-range correlations. A second index j has been added in (9.50) to indicate the single channel whose radial function includes the term f_j regular at $r = 0$, as was done in (9.17). The radial functions $F_{ij}(r)$ satisfy scattering boundary conditions at infinity that can be written in the matrix notation of (9.34) as

$$F_{ij}(r) \underset{r\to\infty}{\longrightarrow} \begin{cases} f_i(r)\delta_{ij} - g_i(r)K_{ij}^{(s)}, & i \text{ open}, \quad (9.51a) \\ 0, & i \text{ closed}. \quad (9.51b) \end{cases}$$

The expansion (9.50) thereby extends to scattering the MCHF of bound-state spectra by letting the radial functions of open channels extend to asymptotic distances.

A trial wave function Ψ_{jt} is constructed in the form of (9.50) with known functions $F_{ij,t}(r)$ that have the asymptotic structure in (9.51) with a trial K matrix $K_{ij,t}^{(s)}$. The procedure of Section 6.8 leads then to the variational approximation

$$K_{ij,v}^{(s)} = K_{ij,t}^{(s)} - \pi \iint \Psi_{it}(\omega, r)[H - E]\Psi_{jt}(\omega, r)\, d\omega\, dr, \tag{9.52}$$

which is analogous to (6.64) but integrated over the complete space (ω, r) of $N + 1$ electrons. Then $K_{ij,v}^{(s)}$ is the desired variational estimate, an improvement over the

[†] See Nesbet [6].

initial $K^{(s)}_{ij,t}$. Typically, the radial functions $F_{ij,t}(r)$ contain open parameters, which are varied along with the coefficients b_m in (9.50) to determine the best stationary estimate. Since the radial functions $F_{ij,t}(r)$ for the open channels are defined over the entire range of r but include a function $g(r)$ with irregular behavior at $r = 0$, this irregularity is eliminated by insertion of a factor analogous to the $(1 - e^{-r})$ in (6.61), which vanishes at $r = 0$.

Evaluation of the integral in (9.52) involves the calculation of radial matrix elements between diverse combinations of bound-state and continuum functions; continuum–continuum elements prove most tedious. Once all the matrix elements have been obtained, (9.52) becomes a function of the variational parameters in $F_{ij,t}$ and b_m. Varying it with respect to the parameters and equating to zero gives a set of linear equations to be solved for their optimal values. The matrix inversion at this stage of the calculation may introduce anomalous singularities, whose proper handling has required much effort over the years, particularly by Nesbet and collaborators [6].

9.3.7 Treatment of Long-Range Anisotropies

The limiting radius r_0 of the short-range interactions has been fixed in this chapter so as to exclude any significant overlap of the wave function of the core electrons with that of one (or more) "outer" electrons. The potential field exerted by an atomic (or molecular) core at $r > r_0$ is thus local and has often approximately spherical symmetry as assumed in Chapters 7 and 8. Departures from spherical symmetry are, however, often far from negligible, as noted in the introduction to this chapter.

A main source of asymmetry lies in the quadrupole moment of atomic cores with nonzero orbital momentum L_i in the ith channel; most molecular cores also possess such a moment. The field of quadrupole moments is represented in (9.49) by components of the off-diagonal potentials $V_{ij}(r)$, which decrease as $1/r^3$ with increasing r. Even more important are the contributions of dipole moments of the core, which decrease only as $1/r^2$ and occur in atomic potential terms $V_{ij}(r)$, when $|\mathbf{L}_i - \mathbf{L}_j| = 1$, as well as in the fields of the cores of polar molecules. These fields cause an outer electron to pass from one fragmentation channel to another at $r \geq r_0$, transitions that have been disregarded in Chapters 7 and 8. Field components of electric 2^k-pole moments, with $k > 2$, decrease more rapidly with increasing r, in fact as $r^{-(k+1)}$, whereby their effects can be estimated more readily. The fields of magnetic dipoles or multipoles have weaker effects, disregarded in the following.

The potential matrix $V_{ij}(r)$ is thus resolved conveniently into multipole components that converge to zero with successive powers of r^{-k},

$$V_{ij}(r) = \sum_{k=0}^{\infty} \frac{V^{(k)}_{ij}}{r^{k+1}}. \tag{9.53}$$

The monopole term ($k = 0$) represents, of course, the Coulomb field of the core which is isotropic and hence diagonal in (i, j),

$$V_{ij}^{(0)}(r) = \delta_{ij} Z_{\text{eff}}/r. \tag{9.53'}$$

This component has been fully taken into account in Chapters 7 and 8. The analytic treatment of motion in a dipole field $V_{ij}^{(1)}(r)$, according to Section 5.5 and Table 5.1, applies when the core energies $(\bar{E}_i, \bar{E}_j)$ are degenerate (at least approximately), as will become apparent.

Our description of the asymmetric long-range field components has been cast in the form of the close-coupling equation [(9.49)], because that equation affords a straightforward approach to calculation of their effects, as noted in Section 9.3.4. Extensive calculations conducted in this manner turned out to require generally a computing effort disproportionately larger than that required to work out the short-range effects, which are inherently more complicated. (These aspects of the problem have been reviewed comprehensively [8].) We turn accordingly to an analysis of the effect of channel coupling in the outer field.

The off-diagonal portion of (9.49) in the space of fragmentation channels consists of the matrix

$$\tfrac{1}{2}k_{ij}^2(r) = (E - \bar{E}_i)\delta_{ij} - V_{ij}(r) - \tfrac{1}{2}l_i(l_i + 1)\delta_{ij}/r^2, \tag{9.54}$$

whose last term is often unimportant at large r. The strength of anisotropy is indicated by the ratios

$$V_{ij}(r)/[\bar{E}_i - \bar{E}_j - V_{ii}(r) + V_{jj}(r) - \tfrac{1}{2}[l_i(l_i + 1) - l_j(l_j + 1)]/r^2], \tag{9.55}$$

whose rate of convergence to zero as $r \to \infty$ depends strongly on the separation of core levels $\bar{E}_i - \bar{E}_j$. Each off-diagonal term of (9.49) may be treated as a small perturbation as soon as the corresponding ratio (9.55) has dropped well below unity. In fact, the reduction of electron–dipole interactions to an isotropic polarization field [(6.68)], which converges as $r^{-2(k+1)}$, was achieved by such a perturbation treatment. Switching the integration of (9.49) to a perturbation treatment of each term as (9.54) drops below a suitable threshold is, however, not readily compatible with computing routines and does not appear to have been utilized in such a direct form. Accordingly, we consider here alternative procedures of integration at $r > r_0$.

The goal of the integration is viewed in Section 7.3 as the calculation of the Jost matrices $J_{i\alpha}^{\pm}$. These matrices would reduce to $U_{i\alpha} e^{\pm i\delta_\alpha^{(s)}}$ in the absence of any long-range force. The effect of diagonal potentials $V_{ii}(r)$ has been represented in Chapters 7 and 8 by the diagonal phase factor $e^{\pm i\delta_i^{(l)}}$ contributed by the long-range potential $V_{ii}(r)$ in each fragmentation channel. The presence of off-diagonal potentials $V_{ij}(r)$ may then be viewed as contributing successive increments to $J_{i\alpha}^{\pm}$, each of them proportional to the square root of the matrix (9.54).

This approach may be formulated in terms of an r-dependent variant of the Jost matrices, $\bar{J}^{\pm}_{i\alpha}(r)$, which coincides with $J^{\pm}_{i\alpha}$ at $r \to \infty$, with $U_{i\alpha}e^{\pm i\delta^{(s)}_{\alpha}}$ at $r = r_0$ and obeys the equation

$$\frac{d\bar{J}^{\pm}_{i\alpha}}{dr} = \pm i \sum_j k_{ij}(r)\bar{J}^{\pm}_{j\alpha}(r) \tag{9.56}$$

with k_{ij} as defined in (9.54). The formal solution of (9.56),

$$\bar{J}^{\pm}_{i\alpha} = \sum_j [\exp \pm i \int_{r_0}^{r} k(r)\, dr]_{ij} U_{j\alpha} e^{\pm i\delta^{(s)}_{\alpha}}, \tag{9.57}$$

becomes more transparent utilizing a diagonalized form of the matrix $k_{ij}(r)$, namely,

$$\bar{k}_{\mu}(r)\delta_{\mu\mu'} = \sum_{ij} \tilde{\bar{U}}_{\mu i}(r)k_{ij}(r)\bar{U}_{j\mu'}(r). \tag{9.58}$$

This notation shows the long-range phase shift $\delta^{(l)}_i$ to be replaced by the aggregate of differential contributions $\bar{k}_{\mu}(r)\, dr$, which operate in a sequence of r-dependent channels μ. The matrix $\tilde{\bar{U}}_{\mu i}(r)$ becomes diagonal, as $k_{ij}(r)$ does, in the limit $r \to \infty$, where the ratio (9.55) vanishes, thus establishing a one-to-one correspondence between the μ and i channels.

A numerical application would construct the matrix $\exp(i \int_{r_0}^{\infty} k\, dr)$ by disregarding the variations of $\bar{U}_{\mu i}$ within each suitably small interval δr. This procedure falls within the scope of the phase–amplitude methods outlined in Section 5.10. It converges faster than $k^2_{ij}(r)$ converges to diagonal form, by a factor $1/r$.

An equivalent application of phase amplitude methods has been developed into a computer routine by the following procedure. Assume that the diagonal potential $V_{ii}(r)$ of (9.49) has been incorporated into standard pairs of base functions $\{f, g\}$ for each fragmentation channel and express the solutions $F_i(r)$ of (9.49) in terms of other $\{f, g\}$ and an additional pair of more slowly variable functions $\{u_i(r), v_i(r)\}$ by the pair of relations analogous to (5.58) and (5.60),

$$F_i(r) = u_i(r) f(\epsilon_i, l_i, r) - v_i(r) g(\epsilon_i, l_i, r) \tag{9.59a}$$

$$\frac{dF_i(r)}{dr} = u_i(r)\frac{df(\epsilon_i, l_i, r)}{dr} - v_i(r)\frac{dg(\epsilon_i, l_i, r)}{dr}. \tag{9.59b}$$

The new functions $\{u_i, v_i\}$ are not independent, being related to a single common factor $\xi_i(r)$ by

$$du_i/dr = g(\epsilon_i, l_i, r)\xi_i(r), \qquad dv_i/dr = f(\epsilon_i, l_i, r)\xi_i(r). \tag{9.60}$$

These definitions reduce (9.49) to

$$-\frac{1}{2} W(f, g)\xi_i(r) = \sum_j V_{ij}(r)F_j(r), \tag{9.61}$$

with the Wronskian normalized at $2/\pi$ as in Chapter 5. The integral of (9.60) is then

$$u_i(r) = u_i(r_0) - \pi \int_{r_0}^{r} dr' g(\epsilon_i, l_i, r') \sum_j V_{ij}(r')F_j(r'), \tag{9.62}$$

with F_j from (9.59a), and a companion expression for v_i. These Volterra equations are amenable to numerical solution.[†]

The limiting values of $\{u_i(r), v_i(r)\}$ at $r \to \infty$ now yield the Jost functions if the initial values are set at

$$u_i(r_0) = U_{i\alpha} \cos \delta_\alpha^{(s)}, \qquad v_i(r_0) = U_{i\alpha} \sin \delta_\alpha^{(s)}. \tag{9.63}$$

Comparing (9.59a) with (9.7) and (7.14) then yields

$$\begin{aligned} J_{i\alpha}^{\pm} &= e^{\pm i\delta_i^{(1)}}[u_i(\infty) \pm iv_i(\infty)] \\ &= \left\{ e^{\pm i\delta_i^{(1)}} U_{i\alpha} e^{\pm i\delta_\alpha^{(s)}} - \left(\frac{2\pi}{k_i}\right)^{1/2} \int_{r_0}^{\infty} dr'\, e^{\pm ik_i r'} \sum_j V_{ij}(r') F_j(r') \right\}, \end{aligned} \tag{9.64}$$

with $F_j(r')$ drawn from the solution of (9.59) at $r' \geq r_0$.

9.3.8 Many-Body Perturbation Theory (MBPT)

The MBPT approach, introduced in Section 9.2.5, utilizes a convenient system of diagrams analogous to the Feynman diagrams of quantum electrodynamics and illustrated in Fig. 9.1. Each diagram represents a specific transformation of atomic structure that may have been generated by an external field, an interaction among electrons, or a combination of these. Listing the diagrams whose contributions to a particular phenomenon are taken into account describes a calculation completely, at least in principle. A procedure is specified to convert each diagram into an analytical expression to be evaluated numerically. The conversion of diagrams into matrix elements proceeds typically through a zero-order basis drawn from the independent-electron model.

Figure 9.1a concerns the elementary process of photoabsorption—represented by a dashed line connecting "electron lines" to a ball—which excites a ground-state electron to a vacant excited orbital k (ascending line) leaving a vacancy p (descending line) in the ground-state structure. Figure 9.1b represents a variant of Fig. 9.1a in which the initial photoabsorption is followed by an electron–electron interaction (dashed line joining two dots), which causes the electron raised initially to k' to fill its vacancy q while a second electron rises to k, leaving the vacancy p. Figure 9.1c represents a further variant in which the electron–electron interaction acts initially on the ground state, raising two electrons to excited orbitals (k, k') and leaving the vacancies (p, q). A subsequent photoabsorption shifts an electron from the excited orbital k' back into the vacancy q, achieving the same final state as Fig. 9.1a or b. Including diagrams such as Fig. 9.1c, with electron–electron interactions acting on the ground state "before" photoabsorption, is the main

[†] A computer program is available from M. LeDourneuf, Observatoire de Paris, 92195 Meudon, France.

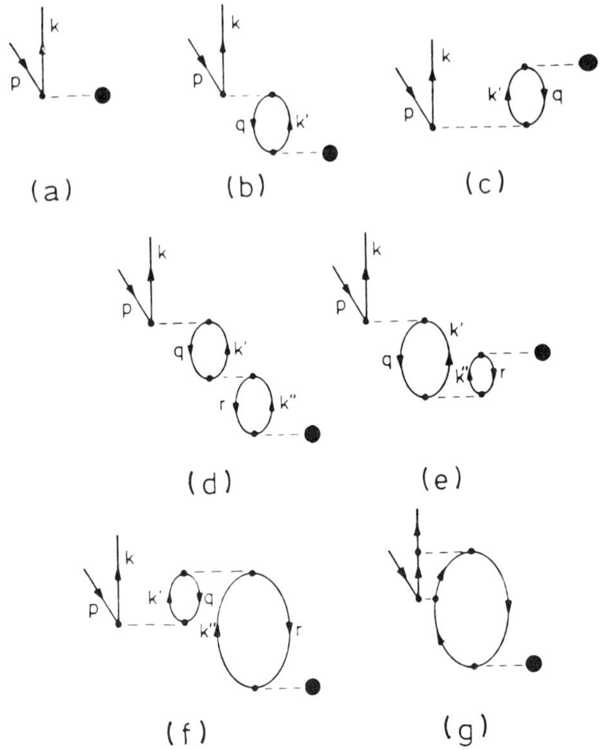

Fig. 9.1. Diagrams contributing to photoionization matrix elements. Dashed lines ending in a filled dot indicate matrix elements of the dipole operator, and dashed lines represent the electron–electron interaction. Corresponding "exchange" diagrams are not shown but are included in the atomic calculations. See text for discussion of individual diagrams. (From Starace [9].)

feature that distinguishes the "many-body" treatments from other approaches, as stressed in Section 9.2.5. Note that Figs. 9.1b and c are topologically equal, though differing in the arrangement of their component parts; they are accordingly treated on an equal footing. Figure 9.1d–g represent second-order perturbations through the occurrence of two-electron interaction lines.

The presence of an atomic ground state in a specified configuration is thus implied by each diagram but is not represented explicitly, being regarded as a background "vacuum" of filled (occupied) subshells, analogous to the actual vacuum of quantum electrodynamics which consists of filled negative energy states. The photoabsorption line of Fig. 9.1a translates for analytic calculation into a dipole matrix element $(k|x|p)$, while the electron interaction line of Fig. 9.1b translates into a matrix element $(i|V|i')$ of (6.10), with $i \equiv (pk)$ and

$i' \equiv (qk')$, divided by the "resonance denominator" energy difference $(\epsilon_k - \epsilon_p) - (\epsilon_{k'} - \epsilon_q)$. Each diagram implies multiplication of all analytical factors and summation or integration over the energies of all orbitals between two nodes (dots). The sums over (p, q) indexes range over the orbitals occupied in the ground state, those over k indexes over all bound and continuum orbitals to infinite energy values ϵ_k.

The whole set of orbitals considered in this treatment should be calculated for a single self-consistent field. Orthodox Hartree–Fock theory leads, however, to a grossly unrealistic difference between the effective potentials acting on orbitals occupied or unoccupied in the ground state, for the following reason. The repulsion among the ground-state electrons includes for each of them a self-energy term $(pp|V|pp)$, which vanishes—as it should—owing to cancellation between its direct and exchange components. Accordingly, each ground-state electron is, in fact, repelled only by the remaining $N - 1$ electrons of the ground-state configuration and is acted on by a residual nuclear attraction at any radial distance. Formal consistency leads the field V^N acting on unoccupied orbitals to include repulsion by all N electrons and, hence, vanish at large distances from a neutral atom, even though this analysis becomes inappropriate when one of the N atomic electrons has been excited to an unoccupied orbital.

The MBPT remedies this apparent shortcoming by complementing the V^N field acting on the orbitals k of Fig. 9.1, with the effect incorporated in the sequence of diagrams b, d, ... summed to infinite order. The resulting orbital k is then the same as that generated in the field V^{N-1} of the atom *with the vacancy p*. This field V^{N-1} is, in fact, generally utilized for the calculation of excited or ionized orbitals, its use being justified formally by the argument outlined here.

Orbitals calculated for excited electrons with full consideration of associated vacancies are often indicated by double lines as in Fig. 9.2a. Note at this point that the Hartree–Fock procedure affords two alternative variants for the effective potential V^{N-1} acting on an excited electron in the presence of a vacancy p. The "unrestricted" variant regards the vacancy as distributed uniformly over its subshell with quantum

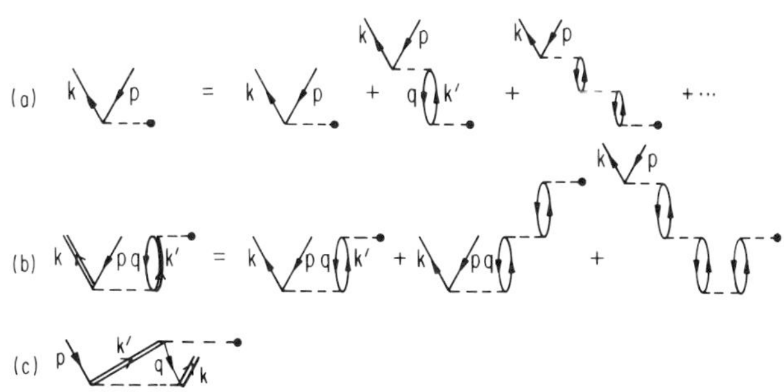

Fig. 9.2. Double-line diagrams defined as sums of chains of "bubble" diagrams from Fig. 9.1. The sum in (a) is equivalent to solving an integrodifferential equation and amounts to the use of a $V^{N-1}(LS)$ field in place of V^N for the Hartree–Fock orbital k. The sum of diagrams (a), (b), and (c), the exchange counterpart of (b), constitutes the RPAE.

numbers *nl*. A "restricted" variant generally treats the ground-state vacancy of the core as coupled to the excited orbital into a state with total spin and orbital numbers (SL); the potential thus obtained, indicated by $V^{N-1}(SL)$, may depend sensitively on the values of S or L, on the order of 10 eV.

Apart from the infinite sequence of diagrams incorporated in the field $V^{N-1}(SL)$, MBPT applications have obtained remarkable success by considering only diagrams with one or two electron interaction lines. Figure 9.3 shows the agreement with experiment attained by the MBPT diagrams in Fig. 9.2, contrasted with results obtained by separate calculation of the excited and ground-state orbitals using the first (L) and second (V) expressions in (9.21). Additional illustrations of MBPT results, most of which originate from H. P. Kelly's school, are shown in Section 9.4.2.

The data in Fig. 9.3 pertain to photoabsorption by Ar. Most calculations by this method indeed concern photoabsorption by rare gases or alkaline earths and Zn-group elements whose ground-state configuration consists of filled subshells. The substance of MBPT is equally relevant to open shell configurations, but technical factors have hindered applications to photoabsorption by ground states with nonzero angular momentum.

Consistency in the treatment of the initial and final states of a transition, evidenced by verification of (9.21), at least to within the order of approximation, is ensured by taking into account all diagrams relevant to a given level of approximation. Typically, this effect is attained by including Fig. 9.1c together with Fig. 9.1b in the first order. Progress of calculations beyond the second order has, however, made it impractical to include all the diagrams required to ensure manifest consistency. Selection of higher-

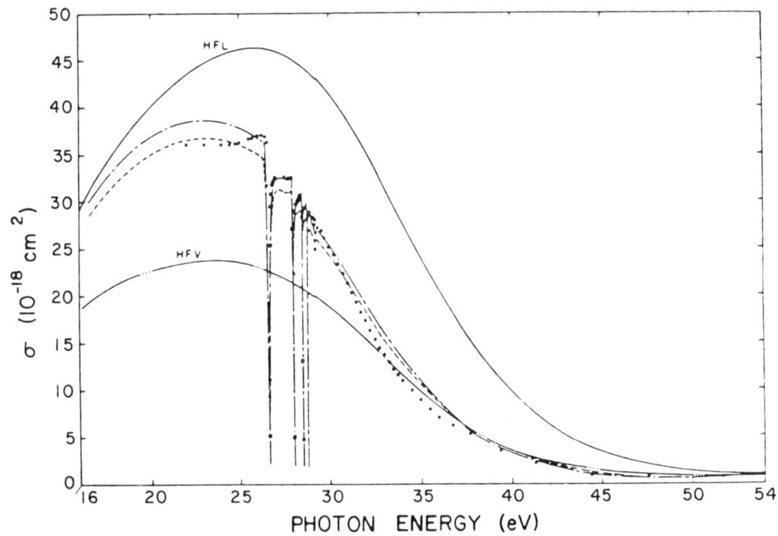

Fig. 9.3. Photoabsorption spectrum of the $n = 3$ shell of argon. Dots are experimental points. ·–·–·, MBPT length; - - - MBPT velocity; HFL, Hartree–Fock length; HFV, Hartree–Fock velocity. (From Starace [9].)

order diagrams required to account for specific phenomena must then be accompanied by consistency checks. Examples presented in Section 9.4.2 will illustrate how inconsistencies emerging in the course of a calculation would lead to order-of-magnitude errors unless properly corrected. Additional comments on the significance of MBPT appear at the end of the next section.

9.3.9 Random-Phase Approximation

The random-phase approximation (RPA) method, introduced briefly in Sec. 9.2.5, amounts to taking into account the infinite subset of MBPT diagrams from Fig. 9.1 which appear to have a dominant effect on photoabsorption spectra. The introduction of this concept into atomic physics by Amusia [2] about 1970, with quantitative evaluation of its implications, accounted for the main unexplained features of photoabsorption, to be illustrated in Section 9.4.

The relevant MBPT diagrams are those indicated in Fig. 9.2 using the condensed notation of double lines to indicate orbitals obeying an appropriate integrodifferential equation. (Each double line is equivalent to an infinite sum over a subset of Fig. 9.1 diagrams, as noted in Fig. 9.2.) The occurrence of two such orbitals in diagram 9.2b implies that they satisfy a system of coupled integrodifferential equations [9]. The equations result from a reformulation and truncation of (9.26) patterned after the familiar statistical analysis of multivariable correlations.

Statistics defines the correlation of two variables (x, y) in terms of the mean value of their product $\langle xy\rangle$, more precisely in terms of the excess (or defect) of $\langle xy\rangle$ with respect to the product of mean values $\langle x\rangle\langle y\rangle$, that is, in terms of the correlation proper $\langle xy\rangle - \langle x\rangle\langle y\rangle$. Similarly, the two-electron transition matrix in (9.26) is resolved into the product of two single-electron matrices $\langle \mathbf{r}_1|\Gamma_{\text{if}}|\mathbf{r}'_1\rangle$ and $\langle \mathbf{r}_2|\Gamma_{\text{r}}|\mathbf{r}'_2\rangle$—pertaining to the excited electron and to the core residue, respectively—and of a correlation proper, by writing

$$(\mathbf{r}_1\mathbf{r}_2|\Gamma_{\text{if}}|\mathbf{r}'_1\mathbf{r}'_2) = \langle \mathbf{r}_1|\Gamma_{\text{if}}|\mathbf{r}'_1\rangle\langle \mathbf{r}_2|\Gamma_{\text{r}}|\mathbf{r}'_2\rangle + \langle \mathbf{r}_1\mathbf{r}_2|\Gamma^{(\text{c})}_{if}|\mathbf{r}'_1\mathbf{r}'_2\rangle. \tag{9.65}$$

The correlation term $\Gamma^{(\text{c})}_{if}$ is then regarded as a second-order quantity, to be truncated off (9.26) and transferred to an analogous equation for the two-particle $\langle\Gamma^{(\text{c})}_{if}\rangle$. Equation (9.26) reduces to a linear equation in $(\mathbf{r}_1|\Gamma_{\text{if}}|\mathbf{r}'_1)$ since Γ_{r} depends only on the ground-state wave function. This truncation procedure could be (but has not been) extended to take into account higher-order correlations.

The selection of the first-order diagrams in Fig. 9.2 as responsible for major effects stems from the role of corresponding diagrams in singling out electron density waves in plasmas from the manifold of alternative electron excitations. This selection is justified by its success in accounting for photoabsorption spectra rather than by obvious relevance of the plasma wave model to atomic mechanics. [No spectral feature of atoms has emerged with the unique aspect of plasma waves, which hinges on the introduction of collective coordinates, but the introduction of Fig. 9.2b modifies the contribution of the elementary diagram (Fig. 9.2a) very substantially.]

Exchange contributions to electron interaction matrix elements are generally understood to be included in the MBPT diagrams for atomic applications, in contrast to being often disregarded in plasma or nuclear contexts; inclusion of exchange is often stressed by labeling the method RPAE instead of RPA. Early atomic calculations with RPA stressed the explicit sum over the contributions of infinite sequences of basic diagrams such as those shown in Fig. 9.1, but later works by Amusia condense these contributions into matrices Γ governed by equations analogous to the reaction matrix equation [(9.16)]. This form of RPA equations affords a proof that the sum rules [(2.24) and (2.25)] implied by (9.21) are satisfied by the RPA results, in contrast to their gross violation when Fig. 9.2b is disregarded. The progress thus attained reflects the basic consistency of the formulation through (9.26).

Both the RPA and the MBPT generally represent major features of photoabsorption through extensive superposition of highly excited virtual states whose radial functions extend to large distances. This description tends to obscure the fact that the important effects thus calculated reside mostly within the radius of the ground-state configuration. The location of the effects *within the core region* becomes directly apparent when solving the integrodifferential equations for the double-line orbitals of Fig. 9.2. Specifically, the procedure of Section 5.7.3 leads us to evaluate the matrix elements in the coefficients of the equation with reference to short-range interactions only, thus introducing the effects of longer-range interactions separately through normalization coefficients as in (5.40).

The framework of (9.26) also casts a different perspective on the *collective* aspects stressed by the original application of RPA to plasma waves and the multitude of simultaneous virtual excitations implied by the diagrams of Fig. 9.1. Fig. 9.2b involves no more than two noninteracting excited orbitals; the number N of electrons of p, d, or f shells nevertheless contributes to photoabsorption and to the relevant electron interactions through a factor $\sqrt{N}$ in the matrix elements of (9.21) and (9.26).

Variants of the RPA treatment have arisen independently of it in order to extend its range of applications. The time-dependent Hartree–Fock method is equivalent to the RPA method described earlier, as is seen by Fourier transformation of the time variable. Spin–orbit interactions and relativistic effects are included in the RRPA treatment [10] which utilizes Dirac instead of Schrödinger Hamiltonians. This more recent treatment has been adapted to yield parameters equivalent to the $\delta_\alpha^{(s)}$ and $U_{i\alpha}$, which represent the effect of short-range interaction on an excited electron. Adaptations of RPA to atoms with open-shell configurations are also in progress [9].

9.4 ILLUSTRATIVE EXAMPLES

Sets of quantitative data on atomic properties are presented in this section that were observed or calculated since the mid-1960s by the various procedures outlined in Section 9.3. Only some of these procedures were conducted within the framework of concepts and goals described at the beginning of this chapter, but their goals and actual results converge toward equivalent endpoints. Recalling that our framework stems from developments in earlier chapters, which were themselves rooted in the analysis of quantitative data, the examples to be presented here may be viewed as a counterpoint and complement to those reported in Chapters 7 and 8. Further examples of calculations by the methods of Section 9.3 are shown in Chapter 11.

The first set of illustrations deals with early results, obtained with the more modest computing facilities available about 1970 and mostly by the close-coupling method discussed in Section 9.3.4. A second group in Section 9.4.2 deals with more recent results obtained by current techniques. These latter illustrations deal explicitly with photoionization but contain much implicit information on collision processes, since the subdivision of photoionization into different channels bears on the magnitude of the corresponding elements of Jost matrices. Indeed, Chapter 7 constructed photoionization and scattering cross sections from the same Jost matrices.

The material in Section 9.4.2 includes an extension of MBPT calculations to the simultaneous detachment of two electrons from an atomic core, a topic that exceeds the main scope of the present chapter, thus affording an introduction to the following chapters. Complements to the material presented in this section and in the preceding one are found in Burke *et al.* [8], particularly its earlier chapters. The same chapters also provide guidance to standardized and available computer programs. Broad and systematic reviews of photoionization theory and experiment are also available [9, 11].

An aspect of atomic phenomena will emerge from the illustrations of this section that we have not mentioned earlier, namely, the considerable influence of correlations between electrons of different shells. These correlations, foreign to the independent electron approximations, fit naturally within the frame of many-body theories. They can also fit within other procedures of Section 9.3, provided suitable allowance is made for them.

9.4.1 Early Results

a Collisions $e + H$ *and* $e + He^+$ Collisions of an electron with a one-electron atom have provided early illustrations of the capabilities and limitations of the close-coupling procedure (Section 9.3.4) at various levels of approximation. Particular attention was given to collisions at energies just below and just above the thresholds for reaching the lowest excited levels, to test the theory's ability to reproduce resonances and threshold effects in cross sections. The wave functions representing such collision processes also represent final states of photoabsorption by the ground states of the two-electron systems H^- and He.

Figure 9.4 shows cross sections for collisions $e + \mathrm{H}(n = 1) \rightarrow e + \mathrm{H}(2\mathrm{s}$ or $2\mathrm{p})$ just above their thresholds. Note the resonance just above the threshold, also displayed in the experimental data of Fig. 1.5 and to be discussed further

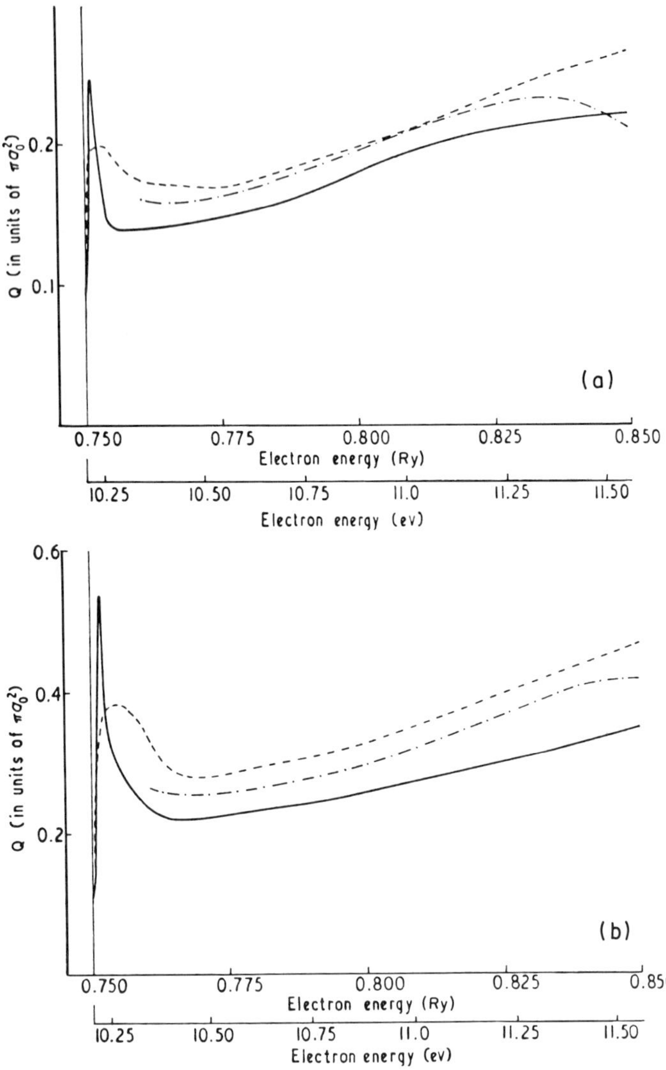

Fig. 9.4. Inelastic cross sections from close-coupling calculations for the excitation of (a) H(2s) and (b) H(2p) by electron impact on ground-state hydrogen: ----, 3 state; ·-·-·, 6 state; —, 3 state + correlation terms in (9.36). (From Taylor and Burke [12].)

at the end of Section 10.5.4. Note also the evolution of results obtained from three progressively better representations of $e + \mathrm{H}$, namely, (9.48) with functions Φ_i corresponding to the three lowest states of H (1s, 2s, and 2p), the six states (1s, 2s, 2p, 3s, 3p, 3d), and with three-state Φ_i complemented by supernumerary terms Ψ_{0m} as in (9.36).†

† For more recent results on e–H, see Callaway [13].

Only elastic cross sections are observed below the lowest excitation threshold, displaying a rich resonance structure at the approach to the threshold. They are conveniently represented in terms of phase shifts according to Section 8.2, regardless of whether these phase shifts are resolved into long- and short-range contributions as was done in Section 8.3. Figure 9.5, calculated without any such analysis and with a base of only three H states, shows a striking sequence of sharp rises of $\delta_{^1P}$ for $e + He^+$ collisions below the lowest excitation threshold. Note the varying sharpness of successive resonances, to be discussed and compared with experimental data in Section 10.2ff. The resonance positions and the corresponding features of the wave functions calculated by close coupling were reproduced independently by MCHF calculations of the discrete states of the complex $e + He^+(n = 2)$ decoupled from the continuum channel $e + He^+(1s)$; representative data thus obtained are shown in Table 9.1.

b Electron–Alkali Collisions Excitation of the alkali atoms from their *n*s ground state to *n*p has much lower energy thresholds than excitation of H or He and correspondingly higher cross sections, features to be discussed further in Chapter 11. Detailed calculations of electron collision effects by close coupling were thus undertaken early, including analysis of spin polarization akin to that of Section 7.2. Samples of the data thus obtained are shown in Table 9.2 and Fig. 9.6.

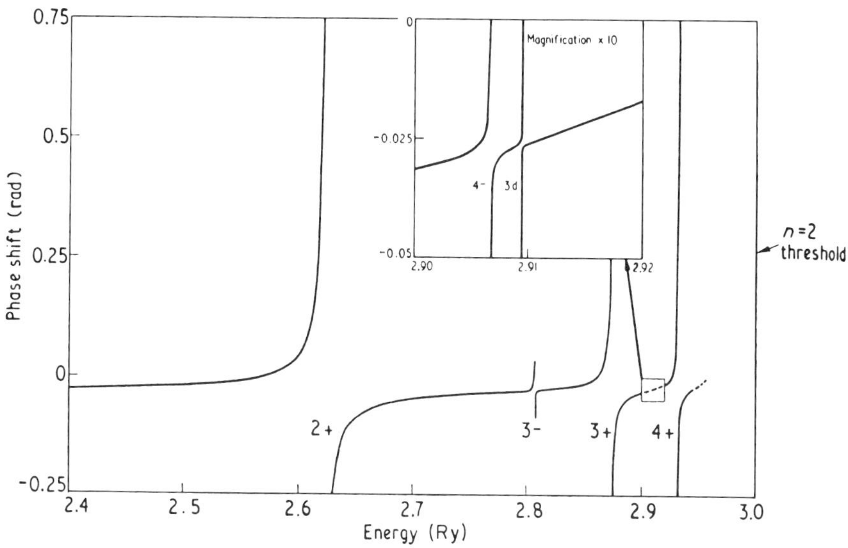

Fig. 9.5. Phase shift for elastic scattering e–He^+ $^1P^o$. The inset is magnified by ten in both dimensions. (From Burke and McVicar [14].)

Table 9.1

Eigenvectors of the First Six $^1P^o$ Resonances in He below $He^+(n = 2)$ from MCHF Calculations: Coefficients of Individual Configurations and Energy Values[a]

Hydrogenic component	Coefficients for each component					
	1	2	3	4	5	6
2s2p	0.87487	−0.00844	−0.18948	−0.03071	−0.02738	−0.10716
23sp+	−0.34637	0.01498	−0.53503	0.20583	0.13623	−0.29119
24sp+	−0.05940	−0.01491	0.68484	−0.30884	−0.18517	−0.15064
25sp+	−0.03123	−0.00351	−0.06351	0.11904	0.05477	0.80937
26sp+	−0.02027	−0.00194	0.01063	−0.00443	−0.00390	−0.29210
23sp−	0.08304	0.89870	0.05733	−0.03358	0.26053	0.00969
24sp−	0.02031	−0.34768	−0.13965	−0.55157	0.45318	−0.00467
25sp−	0.01264	−0.05433	0.01736	0.40143	−0.47345	0.16485
2p3d	−0.30058	0.21881	−0.15096	−0.41102	−0.21052	−0.04977
2p4d	−0.07400	−0.12935	0.39458	0.45132	0.54538	−0.04670
2p5d	−0.04715	−0.02722	−0.02349	−0.07079	−0.31947	0.31385
Energy (eV)	60.27	62.76	63.67	64.11	64.12	64.47

[a] Symbols $2nsp\pm$ refer to linear combinations $(2snp \pm 2pns)/\sqrt{2}$. (From Lipsky and Russek, [15]. More extensive and recent data are in Lipsky *et al.* [16].)

Table 9.2

Phase Shifts (in radians) for *e*–K Scattering[a]

Energy (eV)	^1S	^3S	^1P	^3P	^1D	^3D	^1F	^3F
0.1	2.400	2.953	0.264	2.305				
0.4	1.641	2.349	0.387	1.984	0.259	0.240		
0.8	1.157	1.894	0.403	1.701	0.595	0.388		
1.2	0.866	1.580	0.531	1.491	0.968	0.501	0.230	0.238
1.6	0.673	1.338	1.047	1.325	1.449	0.587	0.303	0.340

[a] From Karule [17].

c Electron Collisions with He Atoms The occurrence of low excited states in the alkali atoms results from the presence of a closed-shell ionic core that removes the degeneracy of s and p states in their valence shell. This removal of degeneracy also enriches the spectrum of excitations, but much more enrichment results when the ionic core has an open-shell structure whose nonzero angular momentum couples with an excited electron into different excitation channels, as discussed in Chapter 7. A fairly simple yet

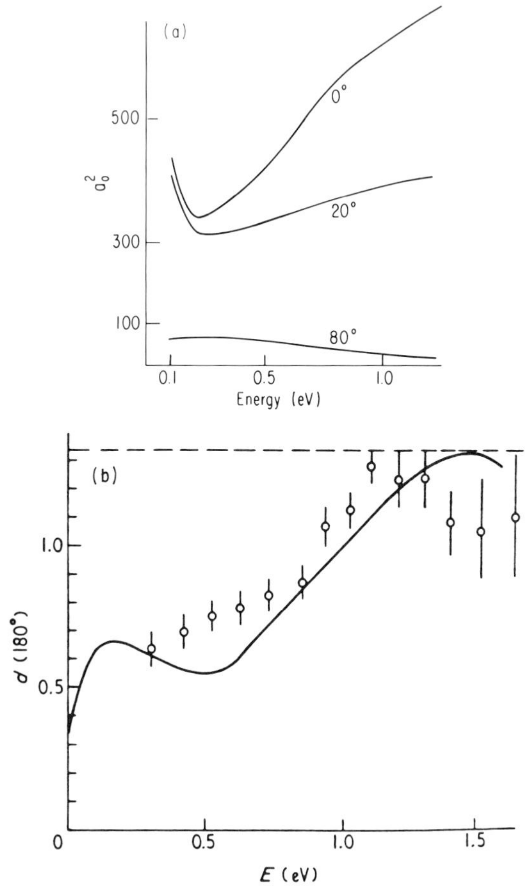

Fig. 9.6. Elastic scattering of electrons from potassium. (a) Differential scattering cross section for sample scattering angles. (b) Ratio of exchange scattering to the total differential cross section at 180° as compared with experiment. (From Karule [17].)

striking manifestation of this enrichment is displayed by the collisional excitation of He to its $n = 2$ levels with threshold near 20 eV.

Six closely spaced levels of excitation occur in this spectrum, labeled as 2^3S_1, 2^1S_1, $2^3P_{0,1,2}$, and 2^1P_1, coupled with the incident electron into numerous channels, as shown in Fig. 9.7. The figure shows the phase shifts as functions of energy for channels of only four (LS) classifications of the He^- complex, but the richness of the excitation spectrum manifests itself in the rapid increase of the number of eigenchannels of the complex within each class as the energy traverses each threshold. The example displayed here was the first to unravel this degree of complexity. Note how the phase shift of each

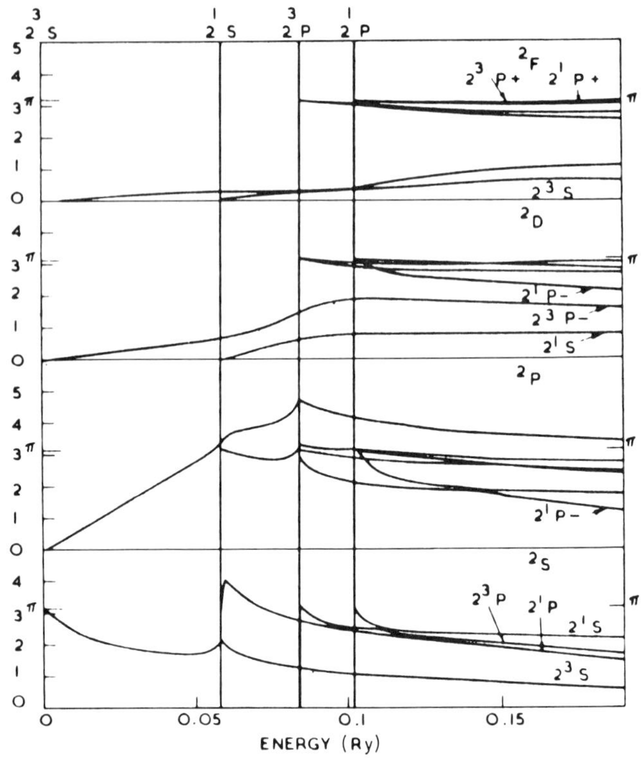

Fig. 9.7. Phase shifts of ^{2}S, ^{2}P, ^{2}D, and ^{2}F eigenchannels of e–He scattering. Fragmentation channel labels are attached to curves where one channel is dominant. (From Burke *et al.* [18].)

newly opened channel starts at a multiple of π, in accordance with the threshold laws of Chapter 4, and the discontinuity of the derivatives $d\delta/dE$ at each threshold.

d Electron Collisions with N and O Atoms The incomplete valence shells of N and O atoms affords an opportunity for the incident electron to be incorporated into a negatively charged collision complex, as it does in collisions with alkali atoms. Figure 9.8 shows results of measurements and an early close-coupling calcuation for e + O. Note that inelastic processes with excitation of metastable ground-state terms (^{1}D and ^{1}S) seem to contribute little to the total calculated cross section.

Figure 9.9 shows the phase shift in the ^{3}P channel of e + N collisions at low energies, calculated by the R-matrix procedure of Section 9.3.1. This channel of the N$^-$ complex is isoelectronic to the ground state of the O atom; the sharp rise of its phase shift corresponds to the resonant formation of a

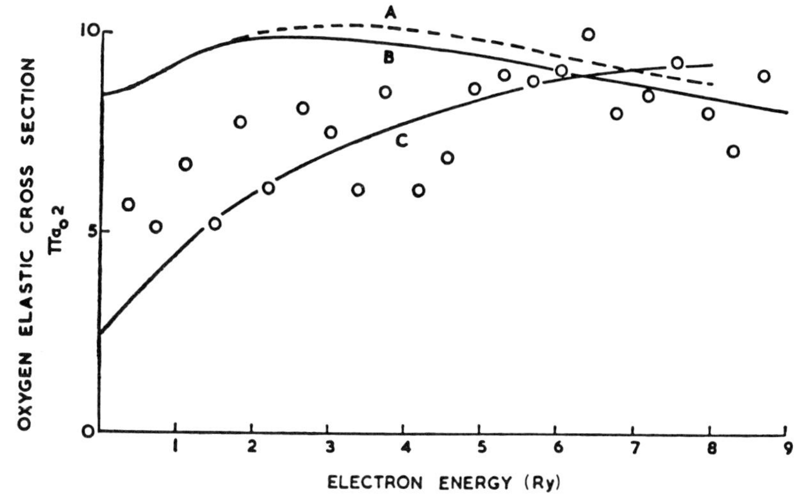

Fig. 9.8. Elastic $^3P \to {}^3P$ (curve *B*) and total $^3P \to {}^3P, {}^1S, {}^1D$ (curve *A*) cross sections in *e*–O scattering compared with earlier results (curve *C*) and experimental data (○) on the total cross section. (From Smith *et al.* [19].)

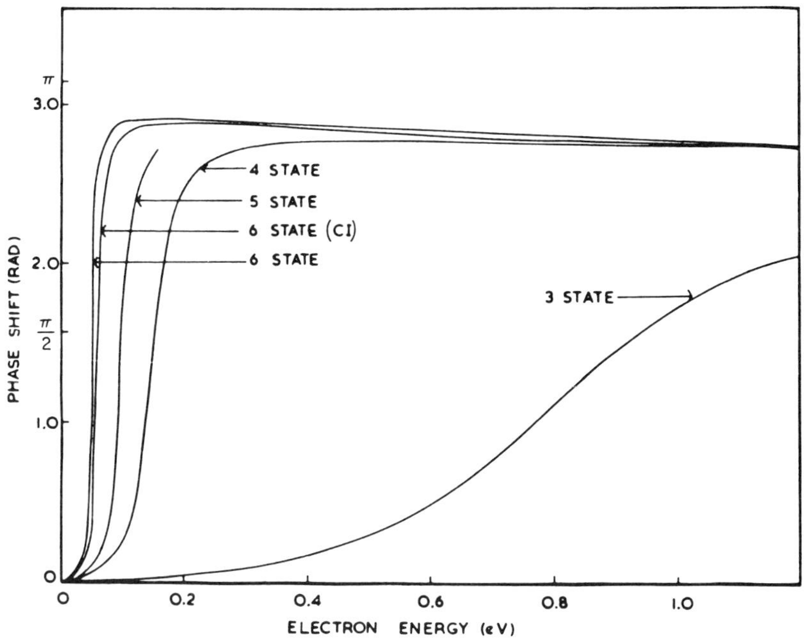

Fig. 9.9. ^3P phase shifts in *e*–N scattering, indicating the convergence of the *R*-matrix calculation. (From Burke and Robb [20].)

quasi-bound state of N^- analogous to the O ground state. A main objective of the calculation, with alternative wave functions [(9.36)] of increasing complexity, was to determine the sensitivity of the position and width of the resonant state to the wave function's detail. The actual position and width are currently known experimentally to be 0.07 and 0.016 eV, respectively.

e MQDT Parameters of the $e + Ar^+$ *Complex* The parameters $\{\delta_\alpha^{(s)}, U_{i\alpha}, (F_\alpha|x|0)\}$ of the Ar spectrum determined semiempirically in Section 8.5 will be compared here with values calculated independently by an expeditious application of the R-matrix method of Section 9.3.2. Input data of the calculation were the wave functions of the $N = 17$ electrons of Ar^+, drawn from an HF calculation of the Ar ground state and entered in the channel functions $\Phi_i(\omega)$ of (9.7), and the experimental value of the doublet interval of ground state Ar^+. The wave function of the eighteenth electron was cast in the form of (9.4) [with $f_n(\omega, r) = r^{p_n}\exp(-q_n r)Y_{lm}(\hat{r})$, ($l = 0$ or 2)] and fitted to (9.7), rather than (9.28), on the core boundary $r = r_0$. Trial values of $\mu = \delta_\alpha^{(s)}/\pi$ and of $U_{i\alpha}$ were entered in the boundary condition (9.7) and the remaining free parameters in the wave functions were varied to determine the energy eigenvalues $E_\nu(\mu)$ plotted in Fig. 9.10. These plots represent the monotonic relation between energy eigenvalues and boundary parameters described in Sections 6.5 and 9.2; note the avoided crossings among the five branches of the plot, corresponding to near degeneracies of the eigenchannels of the electron–core interaction. [The classification of eigenchannels and fragmentation channels for the relevant spectrum of Ar was given in (8.40) and (8.41).]

The values of the eigenphase shifts $\delta_\alpha^{(s)} = \pi\mu_\alpha$ for a desired energy E are read off the abscissas of the intersection of the plots in Fig. 9.10 with the

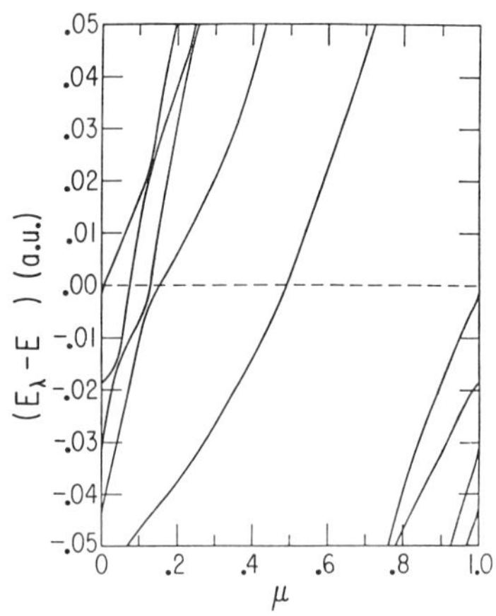

Fig. 9.10. Dependence of energy eigenvalue E_λ on the boundary parameter μ in an R-matrix calculation of the $e + Ar^+$ system. (From Lee [21].)

ordinate E, a procedure equivalent to the solution of (9.37). The small departures of the transformation parameters $U_{i\alpha}$ from their theoretically estimated values [(8.48)] were readily obtained by iterative fit of the whole procedure. The dipole matrix elements $(F_\alpha|x|0)$ were obtained by quadrature from the wave functions of ground state Ar and those of (9.4) with the variational eigenvector parameters c_n. Table 9.3 compares the results thus obtained with the semiempirical data from Section 8.5. Figure 9.11 compares the theoretical profile of an autoionization line determined from the calculated parameters with the observed profile.

The single major discrepancy between calculated and observed parameters in Table 9.3 concerns the value of μ_α for the $\alpha = 2$ channel, designated as $d^1\text{P}$ in (8.41). The calculated value of 0.01 underestimates the actual value 0.07 by a large amount; this discrepancy reflects the sensitivity of the ^1P channel to the exchange interaction, which is not treated adequately in independent-electron calculations, as noted in Section 9.2.5 and documented in Fig. 9.3. Note also that the calculated values of $U_{i\alpha}$ imply an appreciable superposition of s and d orbitals in the ^3P eigenchannels ($\alpha = 3$ and 4), which is not reflected in the semiempirical data; the experimental data may be insensitive to this mixing, owing to the weak excitation of triplet channels.

Table 9.3

Frame Transformation Matrix $U_{i\alpha}$, Eigenphases μ_α, and Dipole Matrix Elements $D_\alpha = (F_\alpha|x|0)$ at the First Ionization Threshold from an R-Matrix Eigenchannel Calculation for Ar[a]

α	1	2	3	4	5
$\mu_{\alpha,0}$	0.20 (0.22)	0.01 (0.07)	0.51 (0.50)	0.12 (0.15)	0.07 (0.11)
$\mu'_{\alpha,0}(\text{eV}^{-1})$	0.08 (0.05)	0.00 (0.02)	0.04 (0.07)	−0.01 (−0.01)	−0.01 (−0.02)
$\mu''_{\alpha,0}(\text{eV}^{-2})$	−0.01 (0.00)	0.00 (0.00)	−0.01 (0.00)	0.00 (0.00)	0.00 (0.00)
$D_{\alpha,0}$(a.u.)	0.03 (0.03)	−2.10 (−2.12)	−0.01 (0.12)	0.03 (0.00)	−1.00 (−1.40)
$D'_{\alpha,0}$(a.u./eV)	−0.02 (0.00)	−0.13 (0.00)	−0.01 (0.00)	0.00 (0.00)	0.18 (0.00)
$D''_{\alpha,0}$(a.u./eV^2)	0.00 (0.00)	0.06 (0.00)	0.00 (0.00)	0.00 (0.00)	0.00 (0.00)

Transformation matrix $U_{i\alpha}$

Calculated					Semiempirical				
0.70	0.59	0.39	0.08	0.01	0.70	0.60	0.38	0.00	0.01
−0.34	0.77	−0.54	−0.10	0.01	−0.32	0.75	−0.58	0.00	−0.02
−0.63	0.25	0.72	0.13	0.01	−0.64	0.28	0.72	0.00	0.02
0.00	0.00	−0.15	0.81	−0.56	0.00	0.00	0.02	0.82	−0.58
0.00	−0.01	−0.10	0.55	0.83	0.00	0.00	−0.02	0.58	0.82

[a] Semiempirical values from an MQDT fit are in parentheses. Primes represent first and second derivatives of μ_α and D_α with respect to energy. (From Lee [21].)

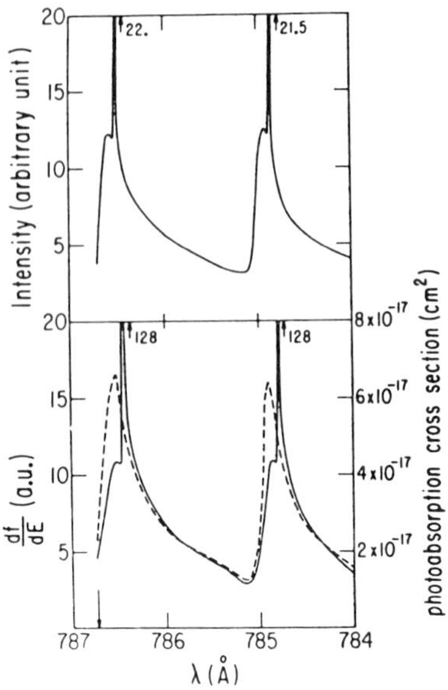

Fig. 9.11. Theoretical (—) and experimental (- - -) profiles of autoionization levels in the photoabsorption spectrum of argon. (From Lee [21].)

f Photoabsorption by Inner-Shell Electrons As noted at the end of Chapter 2, the spectral distribution of this process peaks close to the energy threshold for detaching an electron from each particular subshell, but the peak is shifted to somewhat higher energies by any centrifugal barrier that hinders the escape of electrons with higher orbital momentum. The peak's position and sharpness thus depend on opposing influences and are accordingly sensitive to the theoretical model. Figure 9.12 displays the agreement achieved in 1971 for the photoabsorption by the 4d electrons of Xe, a major success of the RPA method (Section 9.3.9). Calculations without reference to ground-state interactions had yielded an excessively sharp peak. The same or equivalent procedures had been required to account for the outer shell photoabsorption of Ar illustrated in Figs. 2.5 and 9.3.

9.4.2 Recent Results on Photoionization Spectra

The study of partial cross sections for photoionization into different channels, introduced in Section 2.6, has achieved substantial results by calculation as well as by experiment. The alternative channels may be distinguished by the direction and energy of the photoelectrons, through the emission of secondary radiation or electrons by the core residue, or through the core's charge–mass spectrum. Diverse aspects of these phenomena will be illustrated here by the results of recent studies.

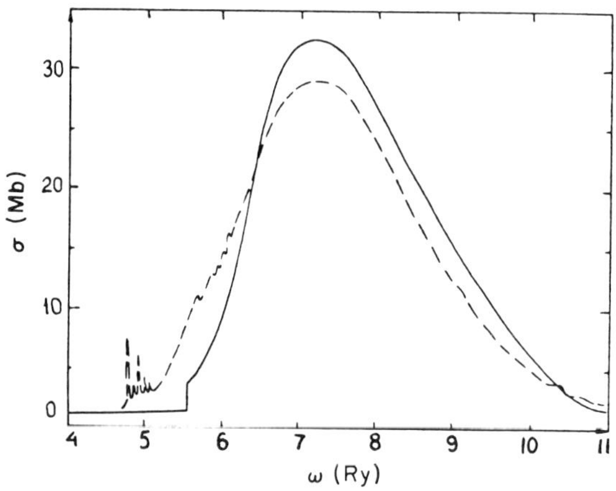

Fig. 9.12. Photoionization cross section for the 4d shell of xenon; —, RPAE; - - -, experimental. (From Amusia [2].)

a Channels of He *Ionization* Most of He photoionization leaves the He^+ residue in its ground state but the fraction leading to excited or even doubly ionized residues is nonnegligible. Among the excited residues the $He^+(ns)$ residues are metastable and hence readily isolated experimentally; their degeneracy with other levels (to within a minute fine structure) presents, however, a problem to theory because electron–dipole interaction affects the orbital motion of the residue up to very large distances of the escaping photoelectron. (This last aspect will be taken up in Section 10.5.5.) Figure 9.13 presents data on the total and partial cross sections, showing very good agreement between recent R-matrix calculations and experiment, whereas earlier calculations by MBPT—with lesser attention to long-range interactions—had yielded incorrect values of the $He^+(2s)/He^+(2p)$ ratio.

The MBPT, on the other hand, extends readily to the treatment of double photoionization, represented to lowest order by the diagram shown in the inset of Fig. 9.14 analogous to those of Fig. 9.1. Good agreement with experiment is displayed in the figure. Note the low values of the cross sections thus determined, whose observation is facilitated by the high sensitivity of ion charge spectrometry.

b Angular Distribution of Photoelectrons This distribution, expressed in terms of a single parameter β according to Sections 7.5.3 and 7.10, generally depends smoothly on the energy of photoelectrons. (See also Section 8.5.7.) Sharp variations do occur nevertheless, especially where the cross section itself is low for the channel under consideration, owing to near cancellation of different contributions to photoemission in the particular channel. Parameters such as β are obviously sensitive to these cancellations. An example of

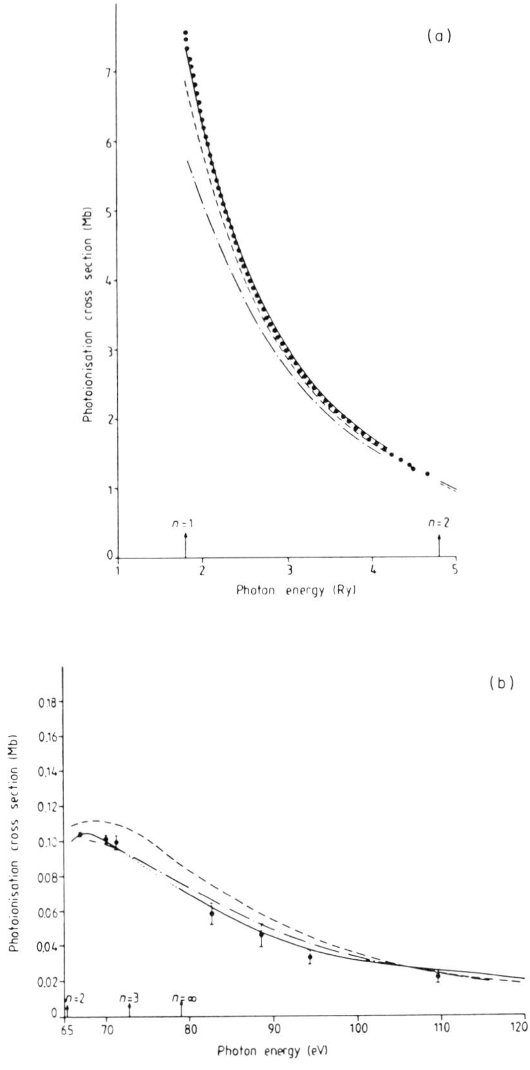

Fig. 9.13. Photoionization of the helium atom. (a) Different theoretical calculations for $h\nu + \mathrm{He} \rightarrow e + \mathrm{He}^{+}(1s)$ compared with experimental points. (b) Cross section for $h\nu + \mathrm{He} \rightarrow e + \mathrm{He}^{+}(n = 2)$: theory and experiment. (c) Ratio of 2p/2s yield from different theories and experimental measurements. (From Berrington *et al.* [22].)

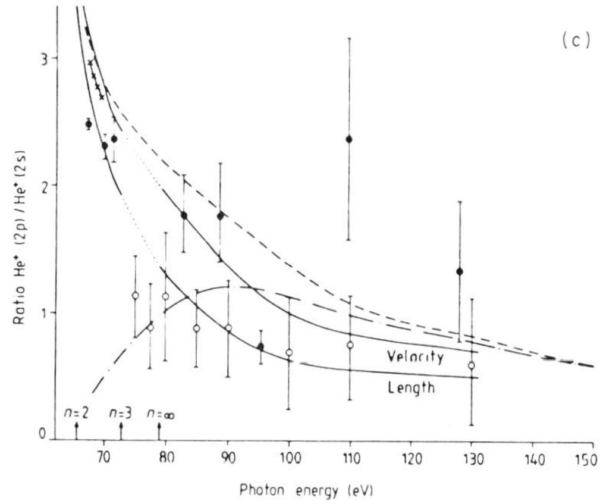

Fig. 9.13. *(Continued)*

the variations of β across the width of autoionization lines has been shown in Fig. 8.4. Figure 9.15 shows instead an example of sharp variations of β over an extended energy range, which illustrates several aspects of photoionization.

The near-zero minimum of β nearly coincides with a Cooper minimum of the total cross section, at which photoabsorption changes character, as noted in Section 2.7. At higher photon energies the value of β approaches 2 once again, but the cross section remains far smaller than it had been at lower energies. (Figure 2.4 shows the analogous behavior of the Ar cross section.)

A further feature shown in Fig. 9.15 is the strong influence of the 4d inner shell on the distribution of photoelectrons from the 5p outer shell. Note that the 4d electrons cannot absorb photons of energy lower than their threshold marked on the figure, yet their coupling with the valence shell affects the latter strongly even at lower energies. Indeed, the intershell effect shifts the position of the Cooper minimum toward lower energies and raises the cross section of valence shell photoabsorption below as well as above the 4d threshold. This effect and analogous ones are amply documented by Starace [9] and elsewhere.

c Branching Ratios of Related Channels The spin–orbit interaction often has a strong influence on photoabsorption, in spite of its weak strength, because photoionization channels classified by different fine-structure levels of the ionic residue are coupled to different subsets of other channels. Nonstatistical values of the branching ratio of fine structure channels, discussed in Section 8.5.6 with reference to the MQDT analysis of phenomena near threshold, may persist over large energy ranges, much as the

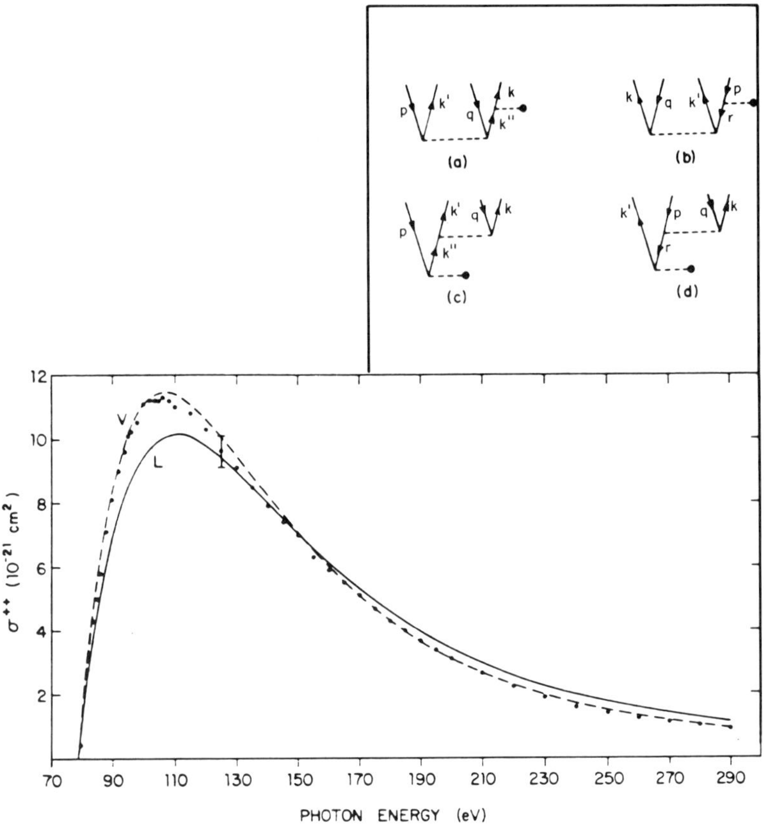

Fig. 9.14. Photo double-ionization of He from MBPT and experiment [*L* and *V* refer to length and velocity forms of (9.21).] Inset shows the lowest-order diagrams contributing to this two-electron excitation. (From Kelly [23].)

energy dependence of angular distribution does, and are similarly affected by cross section minima and intershell coupling. Figure 9.16 shows substantial variations of this ratio for the valence shell photoionization of Xe. Note the relation between the spectra shown here and in Fig. 9.15.

d Relaxation of the Ionic Residue Photoionization implies the generation of a vacancy in the shell structure of the residual ion and thereby a relaxation of this whole structure under the influence of nuclear attraction. This relaxation, which accompanies the photoelectron's escape, influences the final energy of the photoelectron as well as the branching ratio of escape in different channels. These effects often remain small and are accordingly disregarded but are occasionally quite conspicuous.

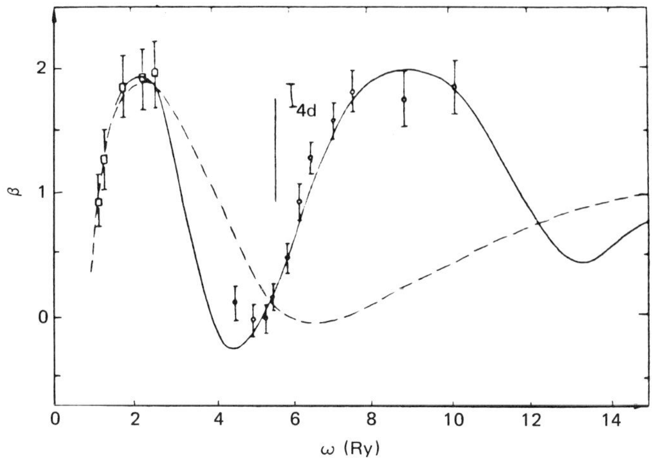

Fig. 9.15. Angular distribution parameter β of photoelectrons from the 5p shell of xenon. Theoretical curves are compared with experimental data points; ——, RPAE; - - -, calculation without the action of $4d^{10}$. (From Amusia [2].)

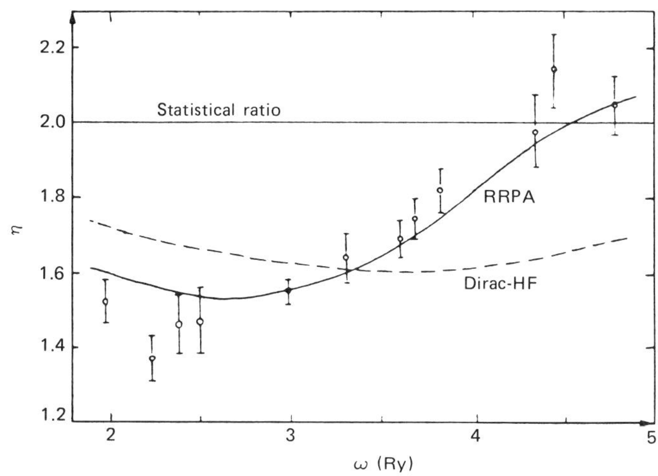

Fig. 9.16. Branching ratio $^2P_{3/2}/^2P_{1/2}$ of photoelectrons from the 5p shell of xenon. (From Amusia [2].)

Figure 9.17 compares the results of alternative calculations of the cross section for photoionization of the 4d shell of Ba with one another and with experiment. The relaxation effect may be boosted in this case by the "softness" of the Ba valence shell, which contracts greatly upon ionization of an inner shell. Note the proximity of the peak vaues of this cross section and of the corresponding cross section of Xe in Fig. 9.12, even though the peak

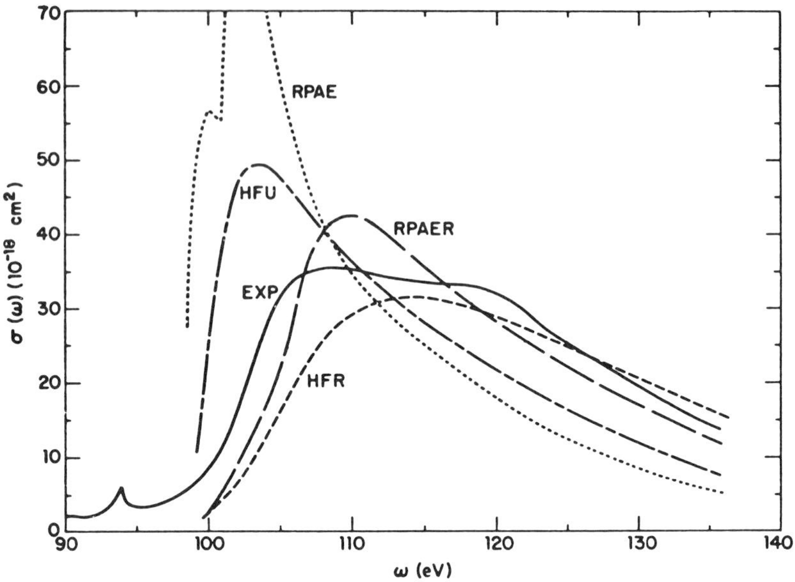

Fig. 9.17. Photoionization cross section for the 4d shell of barium from different many-body theories and experiment. (From Kelly [23].)

positions differ by $\sim 20\%$ for an increase of atomic number from $Z = 54$ to $Z = 56$.

e Effects of Inner Subshell Interaction The structure of transition elements is characterized by the presence of d electrons in an incomplete subshell (n, $l = 2$) whose wave functions extend to radial distances shorter than those of valence electrons in an outer shell. The np subshell of these

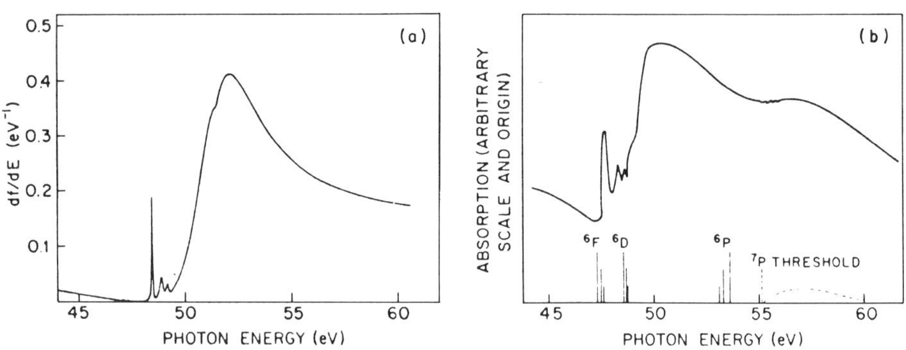

Fig. 9.18. Photoabsorption by the 3p shell of manganese. (a) Experiment and MCHF calculations. (b) Calculations including spin–orbit effects and super-Coster–Kronig transitions. (From Davis and Feldkamp [24].)

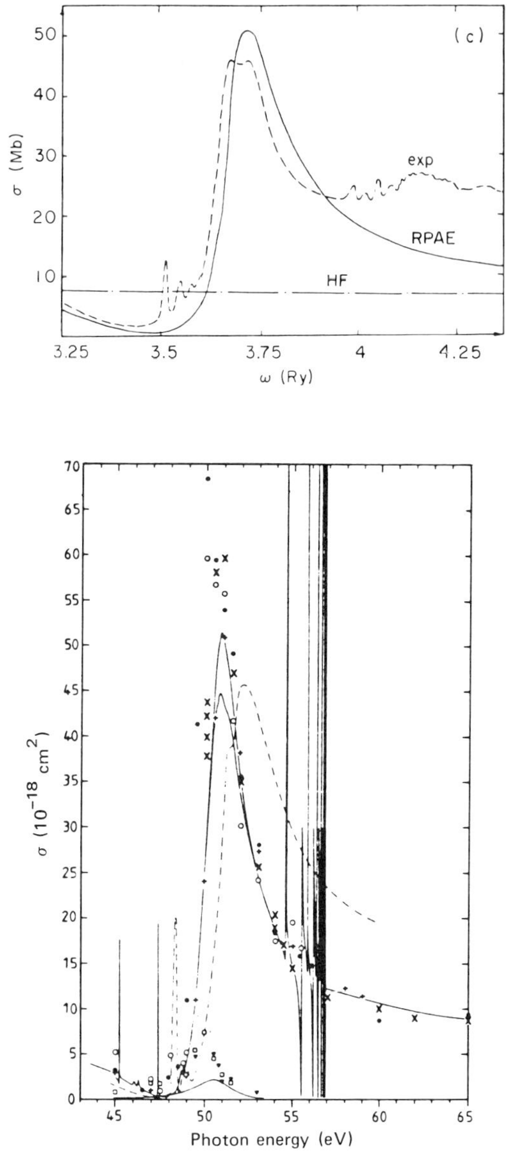

Fig. 9.18. Photoabsorption by the 3p shell of manganese. (c) RPA calculations compared with experiments. (From Amusia [2].) (d) Composite including more recent MBPT calculations and experiments. The structures between 54 and 57 eV represent $3p^6d^5 \rightarrow 3p^5d^5(^7P)nd$, ns resonances. (From Garvin *et al.* [25].)

elements is filled, but its electrons are readily excited to *n*d levels by photons in the range of 30–60 eV. The absorption spectra of these excitations form sharp peaks with a varying structure, which also appear in the spectra of the solid phases with only minor changes. Figure 9.18 shows observed and calculated spectra for the case of Mn. The width of the peak represents, in essence, the rate of the Auger transition in which the vacancy in the *n*p shell is filled by an *n*d electron and a second electron is ejected. (Auger transitions between subshells of the same shell are particularly fast and are called super-Coster–Kronig transitions.)

Calculations of this spectrum have been conducted by three different procedures, namely, an RPA calculation that stresses its collective aspects, an MBPT calculation, and a calculation by spectroscopic techniques within the scope of the MCHF method of Section 9.3.5. The latter method views the absorption peak as consisting of an aggregate of lines (3p → 3d) with different (LS) quantum numbers, each of them superposed to direct photoionization and broadened by Auger decay and coupling to the continuum. The intensity and broadening of different lines varies greatly, depending on (LS) values, as evidenced by the occurrence of smaller and sharper peaks. Analogous and even more conspicuous phenomena occur in the 4d → 4f excitations of rare earths.

f Photoionization of the Ca *Valence Shell* Photoionization of the valence shells of alkaline earths at low energies is dominated by resonances due to autoionization of doubly excited levels—$3dnl\,{}^1P^0$ and $4pn'l'\,{}^1P^0$ in Ca—because direct ionization to $ns\epsilon p\,{}^1P^0$ is nearly suppressed by Cooper minima (Section 2.7). The task of calculating the positions and profiles of numerous autoionization resonances accurately poses a serious challenge to theory. This challenge has been met recently for Ca by independent calculations using the R-matrix and MBPT methods, with the results shown in Fig. 9.19.

Note particularly that the "relaxation diagrams," shown in Fig. 9.19a and analogous to those of Fig. 9.14, have a major role in the MBPT calculation, whereas their effect is taken into account implicitly in the variational part of the R-matrix procedure. Relaxation diagrams are responsible in the MBPT approach for the strong correlation energy of doubly excited independent-electron orbitals.

A substantial effort was required to identify and evaluate the diverse effects relevant to these calculations, some of the evaluations proceeding more readily by one of the methods and some by the other one. Both methods utilized independent-electron orbitals, whereby the strong correlation of doubly excited orbitals had to be approximated by extensive configuration mixing. The MBPT calculation, which included excitations of the $3p^6$ shell besides those of $4s^2$, utilized 36 continuum channels and only 10 discrete orbitals, the effect of other discrete orbitals being evaluated by interpolations

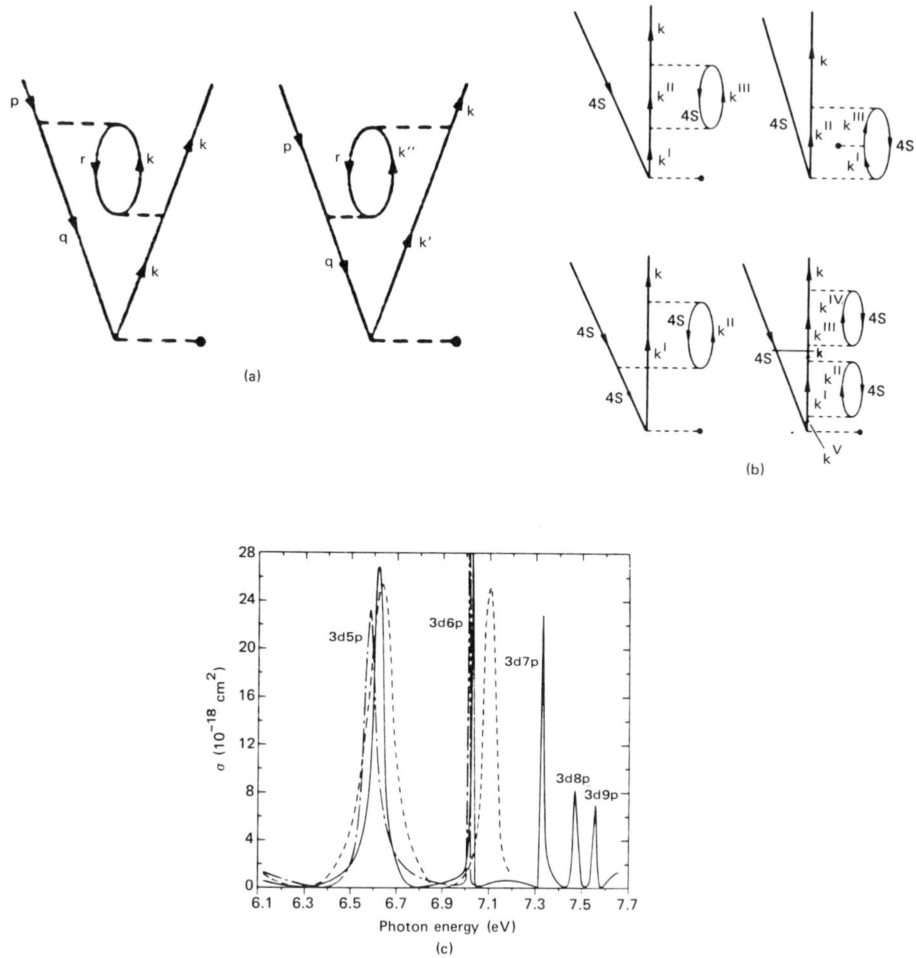

Fig. 9.19. Photoionization of calcium between the 4s ^{2}S and 3d ^{2}D states of the Ca$^+$ ion. (a) MBPT relaxation diagrams. (b) MBPT double excitation diagrams. (c) Results showing resonance dominated cross section; ——, MBPT; -·-, experiment; ---, MQDT fit to experiment. (d) Angular distribution parameter β from calculations and experiment. (From Kelly [23].)

akin to the procedure of Section 5.7.8. Correlation effects were expressed by MBPT largely in terms of the relaxation diagrams in Fig. 9.19a. (Disregarding their influence raised cross sections by an order of magnitude.) The *R*-matrix calculation utilized a set of base orbitals for all subshells of Ca {1s, 2s, 2p, 3s, 3p, 3d, 4s, 4p}, as well as $4\bar{d}$ and $4\bar{f}$ pseudo-orbitals to represent correlations between M- and N-shell electrons. Optimal mixing of the resulting configurations was then achieved by the variational procedure, whereas double excitation diagrams $(k', k'') \equiv (3dnl)$ or $(4pn'l')$ are taken into account individually by MBPT. Resonance profiles, such as those shown in Fig. 9.19c, were found to vary dramatically along a Rydberg series, owing to the occurrence of a Cooper minimum in their spectral range, which changed the sign of channel interferences.

References

1. J. M. Irvine, "Nuclear Structure Theory," Sections 15.2 and 18.3. Pergamon, Oxford, 1972; P. Ring and P. Schuck, "The nuclear Many Body Problem," Sections 8.4 and 8.5. Springer-Verlag, Berlin and New York, 1980; D. Pines, "Elementary Excitations in Solids," Sections 3.3 and 3.5. Benjamin, New York, 1963; G. D. Mahan, "Many Particle Physics," Chapter 5. Plenum, New York, 1981.
2. H. P. Kelly, *Adv. Th. Phys.* **2**, 75 (1968), *Adv. Chem. Phys.* **14**, 129 (1969); M. Ya. Amusia, *Adv. At. Mol. Phys.* **17**, 1 (1981).
3. C. H. Greene, *Phys. Rev. A* **28**, 2209 (1983); *Phys. Rev. A* **32**, 1880 (1985); H. LeRouzo and G. Raseev, *Phys. Rev. A* **29**, 1214 (1984).

3a. K. Smith, "The Calculation of Atomic Collision Processes." Wiley, New York, 1971.

4. L. A. Collins and B. I. Schneider, *Phys. Rev. A* **24**, 2387 (1981), *Phys. Rev. A* **27**, 101 (1983), *Phys. Rev. A* **29**, 1695 (1984).
5. K. Takatsuka and V. McKoy, *Phys. Rev. A* **30**, 1734 (1984).
6. R. K. Nesbet, "Variational Methods in Electron–Atom Scattering Theory." Plenum, New York, 1980.
7. C. F. Fischer, "The Hartree–Fock Method for Atoms." Wiley, New York, 1977.
8. P. G. Burke, W. B. Eissner, D. G. Hummer, and I. C. Percival, eds., "Atoms in Astrophysics," Chapter 6. Plenum, New York, 1983.
9. A. F. Starace *in* "Handbuch der Physik," Vol. 31, (S. Flügge and W. Mehlhorn, eds.), Section 18. Springer-Verlag, Berlin and New York, 1982.
10. W. R. Johnson *in* "Atomic Physics 8," (I. Lindgren, A. Rosen, and S. Svanberg, eds.), p. 149. Plenum, New York, 1983.
11. J. A. R. Samson *in* "Handbuch der Physik," Vol. 31, (S. Flügge and W. Mehlhorn, eds.), p. 123. Springer-Verlag, Berlin and New York, 1982.
12. A. J. Taylor and P. G. Burke, *Proc. Phys. Soc.* **92**, 336 (1967).
13. J. Callaway, *Phys. Rep.* **45**, 89 (1978).
14. P. G. Burke and D. D. McVicar, *Proc. Phys. Soc.* **86**, 989 (1965).
15. L. Lipsky and A. Russek, *Phys Rev.* **142**, 59 (1966).
16. L. Lipsky, R. Anania, and M. J. Conneeley, *At. Data Nucl. Data Tab.* **20**, 127 (1977).
17. E. Karule, *J. Phys. B* **5**, 2051 (1972).
18. P. G. Burke, J. W. Cooper, and S. Ormonde, *Phys. Rev.* **183**, 245 (1969).
19. K. Smith, R. J. W. Henry, and P. G. Burke, *Phys. Rev.* **157**, 51 (1967).
20. P. G. Burke and D. Robb, *Adv. At. Mol. Phys.* **11**, 143 (1975).
21. C. M. Lee, *Phys., Rev. A* **10**, 584 (1974).
22. K. A. Berrington, P. G. Burke, W. C. Fon, and K. T. Taylor, *J. Phys. B* **15**, L603 (1980).
23. H. P. Kelly *in* "Atomic Physics 8," (I. Lindgren, A. Rosen, and S. Svanberg, eds.), p. 305. Plenum, New York, 1983.
24. L. C. Davis and L. A. Feldkamp, *Phys. Rev. A* **17**, 2012 (1978).
25. L. J. Garvin, E. R. Brown, S. L. Carter, and H. P. Kelly, *J. Phys. B* **16**, L269 (1983).

Part D

MULTIELECTRON EXCITATIONS

Chapters 5, 7, and 8 have dealt with the motion of a single electron in the field around an atomic or molecular core, thus treating in detail the multichannel excitation or ionization of that electron. On the other hand, higher energies of excitation provide for two or more electrons to exit from the core into the surrounding field. The interactions among these external electrons are then comparable to their interactions with the core, thereby causing their motion to be strongly correlated. Moreover, diverse correlations are accessible in the large volume external to the core, in contrast to sharp restrictions generated within the core by the exclusion principle.

The effect of such strong correlations can be evaluated in principle by diagonalizing numerically the Hamiltonian matrix in a large base of independent-electron single-configuration states, in accordance with Section 9.3.6. This procedure has indeed yielded very useful results without any ready interpretation, as we shall see; the same holds for other general procedures described in Chapter 9. Intepretations emerge instead from the dependence of eigenfunctions on "collective variables" that represent suitable combinations of electron coordinates. This dependence is studied by constructing approximate solutions of the full Schrödinger equation for all the excited electrons, an equation with numerous nonseparable variables but with internal symmetries that facilitate the task and lead to novel classifications of energy levels.

The direct study of many-particle Schrödinger equations has broad relevance to atomic and molecular processes but has been developed thus far primarily for a pair of excited electrons. Within this limited scope, the simplest two-electron systems, He and H^-, serve as prototypes, lending the language and the techniques for describing doubly excited states, just as the H atom has served for states of single excitation. Also, as in the analogy with H, the prototype examples of doubly excited states of He and H^- exhibit a higher symmetry because they involve only two electrons and a bare nucleus; these examples are treated in Chapter 10, while Chapter 11 will introduce more extended applications. Frequent references for details and bibliography will be made to a review quoted as RPP.†

Note at the outset that a most important source of doubly excited states lies in the inelastic collision of a slow electron with an atom or ion. As soon as one of the atomic electrons has been excited, its kinetic energy becomes comparable to that of the incident electron. The incident electron and the target atom should then be viewed as a doubly excited complex, as anticipated in Section 1.3.

The study of a number of particles with comparable interaction is relevant to the study of the collisions of atoms and molecules and to the whole field of chemical transformations. This broader relevance has been a motivating factor of the development to be presented here, even though it is not considered explicitly. Initial applications to molecular processes have, however, been made (Section 10.7.6; RPP, Section 2.2.5).

The concepts and techniques to be described for the treatment of two or more electrons have not yet been fully developed. They may be viewed as the initial phase of a major complement and extension of the previous chapters, which dealt essentially with a single electron and an atomic core.

† U. Fano, *Rep. Prog. Phys.* **46**, 97 (1983).

10 DOUBLY EXCITED He AND H^-

10.1 INTRODUCTION

The two-electron atoms He and H^- have served as prototypes for the spectroscopy of multiple excitations, much as the H atom has served historically with respect to the familiar spectra of single excitations. General principles suffice to establish the gross structure of the spectral levels of these atoms as an introduction to the description and analysis of more significant features.

In Fig. 10.1 the top strip represents the ordinary channels of single excitation of He considered in Section 7.2.3, in which the atom consists of a He^+ ion in its ground state and an excited electron. Each of these channels, with alternative sets of quantum numbers (SLJM), includes discrete levels below the ionization threshold at 24.58 eV and a continuum above the threshold. Each of the next three strips in Fig. 10.1 represents channels of double excitation in which the atom is viewed as consisting of a He^+ ion excited to its level with quantum number $N = 2$, 3, or 4, respectively, and a second electron with equal or higher excitation. Each strip is characterized by its relevant ionization threshold, namely,

$$24.58 + 4 \times 13.605(1 - 1/N^2)\,\text{eV}. \tag{10.1}$$

Strips with $4 < N \leq \infty$ are omitted in the diagram. Finally, the lowest strip represents all the channels of double ionization with alternative sharing of the escape energy

$$E - 24.58 - 4 \times 13.605\,\text{eV} \tag{10.2}$$

between the two electrons as well as alternative distribution of orbital momenta.

The quantum numbers (SLJM) are insufficient to classify the channels of

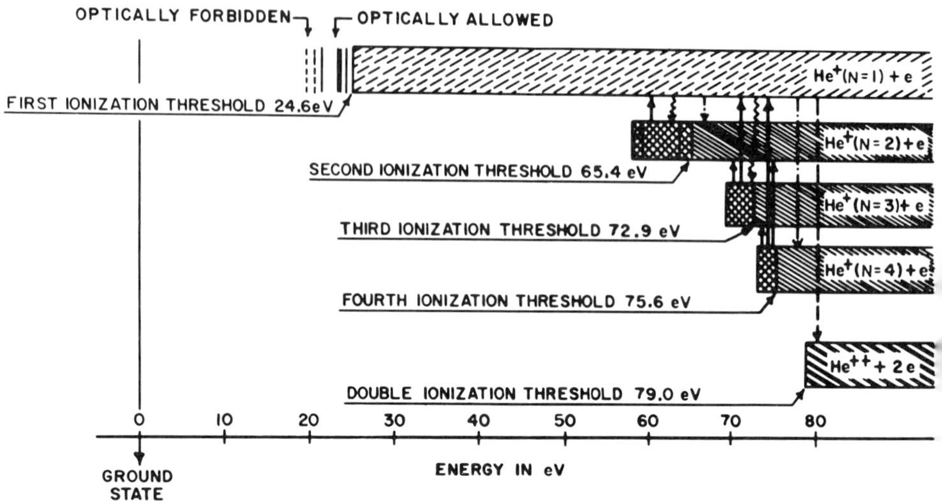

Fig. 10.1. Structure of the helium spectrum along with autoionization and other interconnections between states of different configuration.

any strip with $N > 1$ because the corresponding levels of He^+ are degenerate. This degeneracy makes it impractical to characterize the channels by the orbital angular momenta (l_1, l_2) of the two electrons, as was done previously in Part C (for instance, in Section 7.4.1). The degenerate values (l_1, l_2) are strongly mixed together by the electron–electron interaction. Therefore, collective parameters (quantum numbers of the pair) will provide the requisite classification of channels rather than quantum numbers of the individual electrons. This chapter deals with current progress in developing such a description of the pair of electrons.

Figure 10.1 also indicates various transitions processes between channels belonging to different strips. Levels belonging to various strips may be reached from the ground state of He by photoabsorption, as described in Chapter 2, or by collision of a fast electron, as described in Chapter 3. Transitions also occur between levels of different strips, as we have seen in Chapters 7 and 8, without the intervention of external agents and, hence, between levels of equal energy; energy matching requires, then, at least one of the levels to belong to a continuum $He^+ + e$. Note the reciprocal processes of autoionization and electron capture ("dielectronic recombination"). Note particularly that the infinite array of strips and channels shown or implied in the diagram represents an array of accessible states that diverges rapidly with increasing excitation energy.

The spectrum of the negative ion H^- has the same gross structure as that shown in Fig. 10.1 for He, except that the separation of its discrete and continuum levels is called a "detachment threshold" and occurs at different energies. These energies are obtained from (10.1) and (10.2) by replacing the 24.58-eV value of He by 0.75 eV and deleting the factor 4, which multiplies the Rydberg energy (13.605 eV). Moreover, whereas the discrete levels of He are expected to form a Rydberg series below each ionization threshold, those of H^- are bound only by a dipole-type potential due to the polarization of the H residue and should form a series according to (5.29).

10.2 EARLY OBSERVATIONS: EXPERIMENTAL AND THEORETICAL

Evidence of a primary role of correlations in shaping the states of an electron pair emerged suddenly in 1963 from the first observation of a full series of doubly excited $^1P_1^o$ helium levels. Alternative configurations—2s*n*p, 2p*n*s, and possibly 2p*n*d, belonging to the second strip of Fig. 10.1—would have been expected to contribute with comparable intensity to He photoabsorption in the 60–65-eV range. A single intense series was observed instead, intercalated by a faint one (Fig. 10.2). This finding was viewed as a novel regularity of the joint motion of an electron pair free from the confinement of a filled shell. The width of the spectral lines, proportional to the rate of autoionization of the excited levels, was found to parallel their rate of photoexcitation.

An early interpretation of these findings regarded the new doubly excited states as superpositions of 2s*n*p and 2p*n*s configurations in nearly equal parts and thus quite different from either configuration. It is, in fact, generally observed that the superposition of two transitions with equal amplitudes yields constructive interference, while cancellation results if the amplitudes are equal and opposite. The observation of doubly excited series by photoabsorption was soon followed by the observation of other doubly excited series by inelastic collision of fast electrons or ions showing analogous results (Fig. 10.3).

Soon thereafter calculations of the entire spectrum of doubly excited series of He in the 57–65 eV range were carried out by numerical solution of the two-electron Schrödinger equation by the approximation method of Sections 9.3.4 and 9.3.5. Their results agreed with the initial and later experimental findings and extended them greatly by including spectra less accessible to direct experimentation. They revealed the systematic occurrence of different series of levels with the same (SLJ) labels but with rates of excitation and decay differing by orders of magnitude (Fig. 10.4, see also Section 9.4.1). In other words, transitions to and from different series seemed subject to selection rules of unknown origin. Moreover, the very occurrence of different

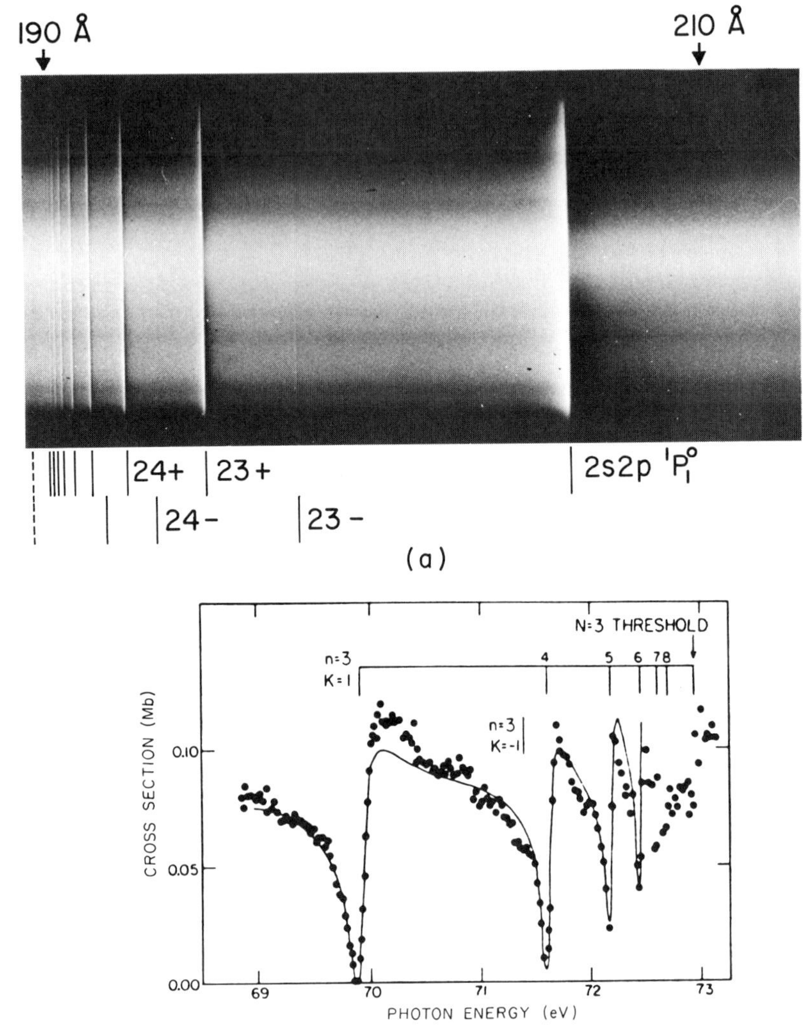

Fig. 10.2. (a) Photoabsorption spectrum of helium between 19 and 21 nm (59–65-eV photons) showing two series of doubly excited state resonances, one intense, the other faint, below the $He^+(N = 2)$ limit. (From Madden and Codling [1]. Courtesy of the authors and the Astrophysical Journal. Copyright 1965 by the American Astronomical Society.) (b) and (c) Further data on He** below the higher $He^+(N = 3)$ limit to illustrate the richness and generality of the phenomenon first seen in (a). (From Woodruff and Samson [2].)

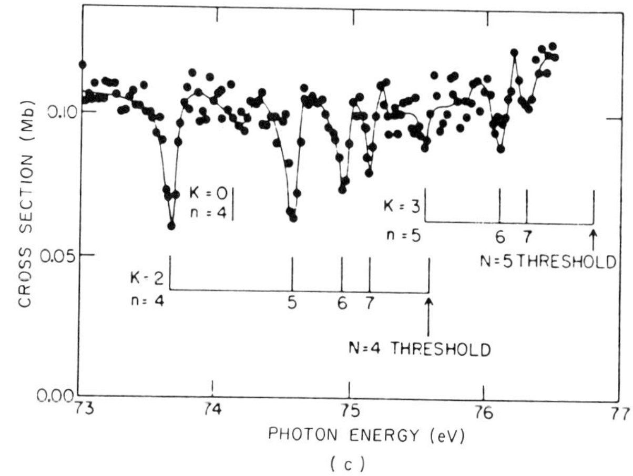

Fig. 10.2. (*Continued*)

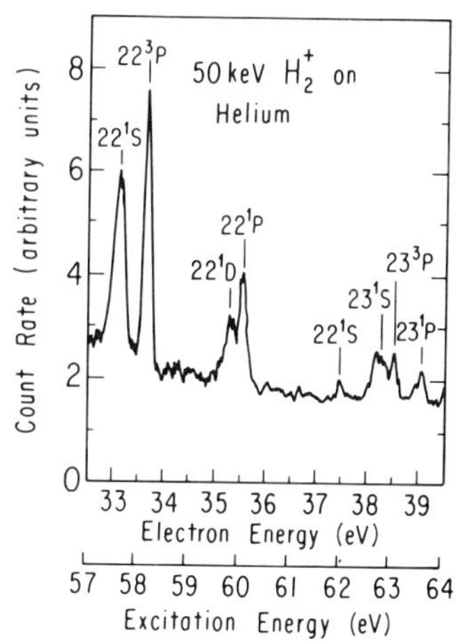

Fig. 10.3. Spectrum of electrons ejected from He by impact of H_2^+ ions. Peaks result from the formation of doubly excited autoionizing states He**. Triplet states of He** arise by electron exchange between H_2^+ and He. (Courtesy of M. E. Rudd).

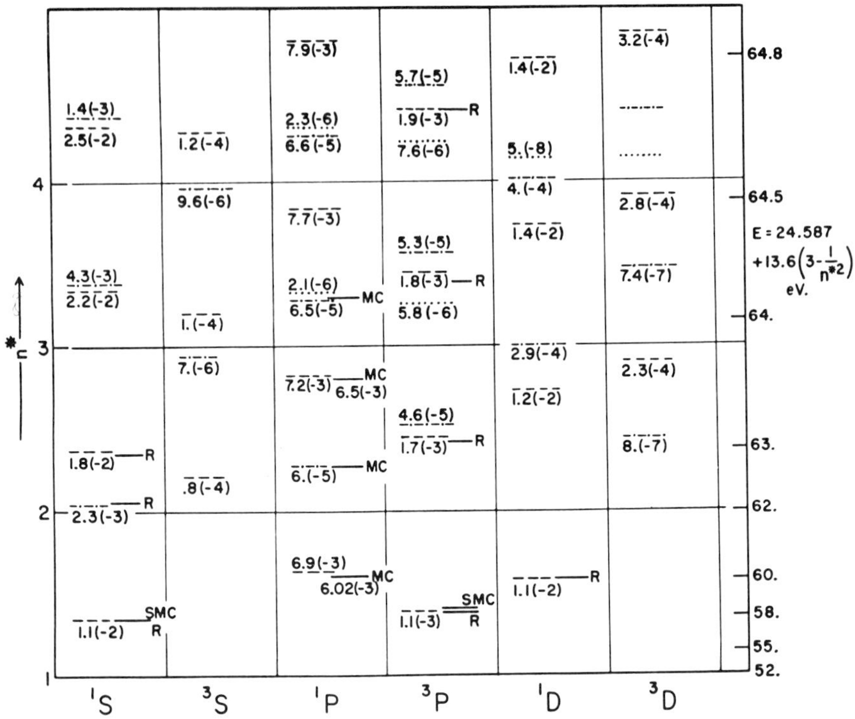

Fig. 10.4. Diagram of doubly excited states of He below 65 eV. Experimental results are shown by solid lines, and results of calculations are shown by dots or dashes. State labels represent reduced widths $\frac{1}{2}\Gamma(\text{a.u.})n^{*3}$ in exponential notation (From Fano [3].)

series remained obscure as the properties of each level emerged from computer outputs without any indication of their relation to those of other levels. This situation provided impetus to the developing of alternative theoretical approaches.

10.3 ANGULAR AND RADIAL CORRELATIONS

Figure 10.5 represents the positions of two electrons in the field of a He^{2+} nucleus or any other positive ion. The complete wave function of the electron pair depends on the size and shape of the V-shaped diagram as well as its orientation in space about the center of mass of the system. The electron correlations, however, depend only on the *shape* of the V, that is, on the relative length of its arms and on the angle between them.

The *angular correlation* of the pair is represented by the wave function's

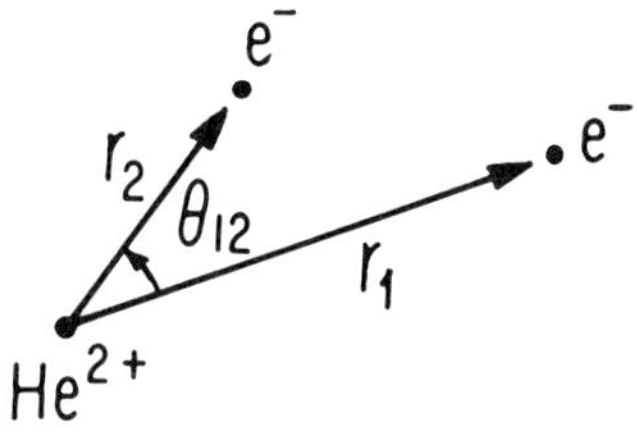

Fig. 10.5. Diagram of two electron configuration.

dependence on the angle θ_{12} between the arms of the V. A strong correlation implies a sharp dependence on θ_{12} and, therefore, the presence of high-order harmonics in its expansion. Accordingly, the analysis of angular correlations involves only familiar concepts, at least in principle.

For a given value of θ_{12} the shape of Fig. 10.5 depends on the ratio of the lengths of its arms, r_2/r_1. The wave function's dependence on this parameter, or an equivalent one, characterizes the *radial correlation* of the pair. For this purpose we usually consider the variable

$$\alpha = \arctan(r_2/r_1), \tag{10.3}$$

whose values range from 0° to 90°, as shown in a plot of radial coordinates (Fig. 10.6). This parameter represents an angle in Fig. 10.6 rather than in physical space. It does represent an angle in the six-dimensional configuration space of the two electrons and may be referred to as a "mock angle." Here again a strong radial correlation implies a sharp dependence of the wave function on α and, therefore, the presence of high-order harmonics in its expansion. Figure 10.7 illustrates the meaning of radial correlations by comparing the amplitude distributions of two alternative approximate wave functions of He in its ground state. Note how the distribution of the correlated wave function stretches out along the two coordinate axes, meaning that the probability of finding one electron far from the nucleus is larger when the other electron lies closer to the nucleus. Concentration of the

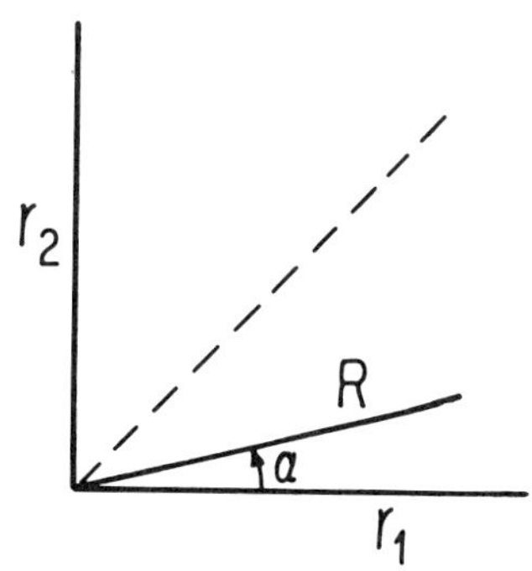

Fig. 10.6. Cartesian and hyperspherical coordinates.

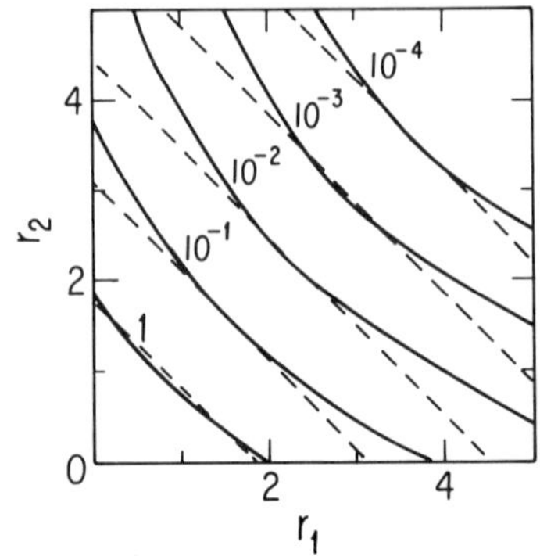

Fig. 10.7. Contour plots of approximate wave functions for the ground state. Solid lines represent correlated Eckart functions

$$N[\exp(-\beta r_1 - \gamma r_2) + \exp(-\beta r_2 - \gamma r_1)],$$

$N = 8.91$, $\beta = 2.18$, $\gamma = 1.19$; dashed lines represent uncorrelated Hartree–Fock function $N_1 \exp[-\mu(r_1 + r_2)]$, $N_1 = 19.22$, $\mu = \frac{27}{16}$. (Courtesy of P. F. O'Mahony).

wave function along the cartesian axes (r_1, r_2) marks the onset of separation of an electron from the rest of the system, since ionization implies $r_{1,2} \to \infty$, that is, $\alpha \to 0$ or $\alpha \to \frac{1}{2}\pi$.

A cautionary note is appropriate here regarding the use of the word *correlation*, which differs in mathematical statistics and in its common, imprecise usage in atomic physics. The introduction of the terms *angular* and especially *radial correlation* in this section should accordingly be viewed as tentative and subject to elaboration in the rest of this chapter.

10.3.1 Alternative Parametrizations

The dependence of the wave function $\Psi(\mathbf{r}_1, \mathbf{r}_2)$ on the angular coordinates θ_{12} and α represents angular and radial correlations directly and separately, but its dependence on other parameters also proves useful. As noted earlier, the strength of correlations is represented by the order of harmonics in the expansion of $\Psi(\mathbf{r}_1, \mathbf{r}_2)$, that is, by quantum numbers complementary to the space coordinates θ_{12} and α. Again, the electron correlations are defined independently of the orientation of the diagram of Fig. 10.5, which is itself represented by Euler angles (Φ, Θ, Ψ); nevertheless, the total orbital momentum $\mathbf{L}$ conjugate to these angles affects the correlations through centrifugal forces. We should then be prepared to use alternative sets of coordinates and quantum numbers in the study of correlations.

For certain applications it will prove appropriate to expand $\Psi(\mathbf{r}_1, \mathbf{r}_2)$ in spherical harmonics of the single electrons with orbital quanta (l_1, l_2). Since the total orbital quantum numbers of the pair (L, M) are constants of the motion, this expansion proceeds stepwise, in terms of the two-electron harmonics defined by

$$\mathscr{Y}_{l_1 l_2 LM}(\hat{r}_1, \hat{r}_2) = \sum_{m_1 m_2} (l_1 l_2 LM | l_1 m_1, l_2 m_2) Y_{l_1 m_1}(\hat{r}_1) Y_{l_2 m_2}(\hat{r}_2), \tag{10.4}$$

with Wigner coefficients. The transformation of these harmonics into the alternative set, expressed in terms of eigenfunctions of the symmetric top and of θ_{12}, i.e.,

$$\sum_{\mu} D^{L}_{m\mu}(\Psi, \Theta, \Phi)e^{i\mu\theta_{12}} \tag{10.5}$$

is available but nontrivial. Only for $L = 0$, however, can the coordinate θ_{12} be disentangled completely from the Euler angles.

Coordinates that do not resolve angular and radial correlations completely nevertheless may prove useful. One such set reflects the ratios among the axes of inertia of the electron pair in Fig. 10.5. Taking a z axis orthogonal to the diagram and the x axis along the diagram's major axis of inertia, we represent the cartesian coordinates of $(\mathbf{r}_1, \mathbf{r}_2)$ as

$$x_{1,2} = R\cos\chi\cos\tfrac{1}{2}(\phi \mp \tfrac{1}{2}\pi), \qquad y_{1,2} = R\cos\chi\sin\tfrac{1}{2}(\phi \mp \tfrac{1}{2}\pi), \qquad z_{1,2} = 0, \tag{10.6}$$

where

$$R = (r_1^2 + r_2^2)^{1/2} \tag{10.7}$$

represents the length of the vector (r_1, r_2) in Fig. 10.6. [This length is called a hyperradius because it represents the radial variable in polar coordinates of the six dimensional space $(\mathbf{r}_1, \mathbf{r}_2)$.] In (10.6) χ runs from 0 to $\pi/4$ and $\tan^2\chi$ equals the ratio of the two lowest among the principal moments of inertia of the pair $(mR^2\cos^2\chi, mR^2\sin^2\chi, mR^2)$, while ϕ runs from 0 to 2π and distinguishes positions $(\mathbf{r}_1, \mathbf{r}_2)$ with the same moments of inertia but with different configurations. The set of six coordinates $(R, \chi, \phi, \Phi, \Theta, \Psi)$ is equivalent to $(\mathbf{r}_1, \mathbf{r}_2)$. A wave function with quantum numbers (L, M) can then be expanded in the form

$$\Psi_{LM}(\mathbf{r}_1, \mathbf{r}_2) = \sum_{\mu} D^{L}_{M\mu}(\Phi, \Theta, \Psi) f_{\mu}(R, \chi, \phi). \tag{10.8}$$

The Euler angles (Φ, Θ, Ψ) relate, in this case, laboratory coordinates to the principal axes of inertia of the electron pair, which constitute an intrinsic "body frame" (RPP, Section 2.2.3).

Yet another parametrization emerges from the expansion of a wave function with given (L, M) into symmetrized products of hydrogenic eigenfunctions of two independent electrons with principal quantum numbers (N, n),

$$\Psi_{LM}(\mathbf{r}_1, \mathbf{r}_2) = \sum C_{LMNn}\, g_{Nn}(\mathbf{r}_1, \mathbf{r}_2). \tag{10.9}$$

Orbital momenta $(\mathbf{l}_1, \mathbf{l}_2)$ of the separate hydrogenic orbitals are not constants of the motion in the presence of electron interaction. Also not constant are the alternative Lenz vectors $(\mathbf{b}_1, \mathbf{b}_2)$, which represent the major axes of elliptical orbits in the classical

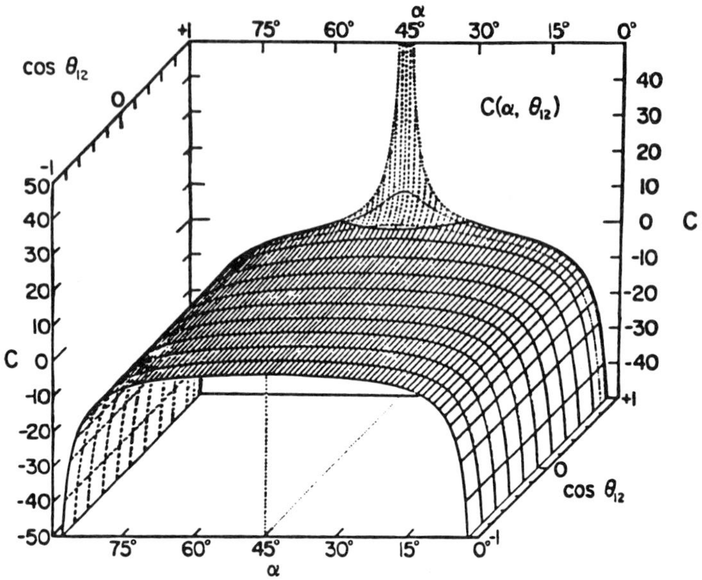

Fig. 10.8. Potential function $C(\alpha, \theta_{12})$ for H^-. (From Lin [4].)

limit. Two combinations of these vectors have been found, however, to be approximate constants, as we shall see, together with L and M; these are

$$|(\mathbf{l}_1 \pm \mathbf{b}_1) + (\mathbf{l}_2 \mp \mathbf{b}_2)|^2 = J^{\pm}(J^{\pm} + 1), \tag{10.10}$$

with $J^{\pm}$ integer or half-integer. Integer quantum numbers (K, T) are defined by

$$\tfrac{1}{2}(N - 1 + K \pm T) = J^{\pm} \tag{10.11}$$

with

$$N \leq n, \qquad 0 \leq T \leq \min(L, N - 1), \qquad \pm K = N - T - 1, N - T - 3, \ldots, 1 \text{ (or 0)}. \tag{10.11$'$}$$

The interpretation of K and T in terms of angular correlations will be described later (RPP, Sections 2.2.4 and 3.4.2).

10.4 ANALYSIS OF THE SCHRÖDINGER EQUATION

An early achievement of wave mechanics was to account for the observed radii of atoms as resulting from the stable equilibrium of opposite tendencies. The Coulomb attraction by the positive charge of the nucleus strives to condense the electrons into a small volume; hence expenditure of work in inverse ratio to the atomic radius would be required to boost this radius. On the other hand, reducing this radius would raise the electron kinetic energy in inverse ratio to the *square* of the radius.

This balance of kinetic and Coulomb energies is displayed most clearly by evaluating their mean values for a wave function of the H atom with a free parameter and observing their dependence on the parameter as it changes the mean radius. The demonstration of the same stability principle for atoms with many electrons is less familiar but can be achieved by proper analysis of the potential and kinetic energy terms in the Schrödinger equation. This analysis will provide a foundation for the study of doubly (or multiply) excited states of atoms.

10.4.1 Potential Energy

The potential energy of a pair of electrons in the Coulomb field of a nucleus depends on both the size and shape of the diagram in Fig. 10.5. The main point here is that size and shape act on the potential energy separately. A change in size that, for example, doubles all interparticle distances reduces *each* Coulomb interaction by a factor of two and, hence, reduces the entire potential energy of the atom by this factor. Analytically, the root-mean-square distance of the electrons from the nucleus, that is, $\frac{1}{2}R$ according to (10.7), may serve as a scale parameter of distances in the expression of the potential energy

$$V(\mathbf{r}_1, \mathbf{r}_2) = -\frac{Ze^2}{r_1} - \frac{Ze^2}{r_2} + \frac{e^2}{|\mathbf{r}_1 - \mathbf{r}_2|} = \frac{e^2}{R} C(\alpha, \theta_{12}). \qquad (10.12)$$

Here we have indicated by

$$C(\alpha, \theta_{12}) = -\frac{Z}{\cos\alpha} - \frac{Z}{\sin\alpha} + \frac{1}{(1 - \sin 2\alpha \cos\theta_{12})^{1/2}} \qquad (10.12')$$

a number that represents the sum of effective charge products of all particle pairs in the atom. This sum depends only on the *shape* of the diagram in Fig. 10.5 through the angles α and θ_{12}, whereas the total potential energy (10.12) *scales* in inverse ratio to the size parameter R.

The index Z in (10.12) represents the nuclear charge, as usual. The construction of (10.12′) utilizes geometrical relations from Figs. 10.5 and 10.6, such as $r_1 = R\cos\alpha$. The alternative parametrizations indicated in Section 10.3.1, of course, could have been used here to express the effective charge C in terms of coordinates other than (α, θ_{12}). However, the separation of the potential energy into its factors C and $1/R$ is central to our present purposes.

Since the potential function (10.12) has three independent variables, it can be represented graphically by alternative mappings in which one or another of these variables is kept fixed. Figure 10.8 shows a relief of the function

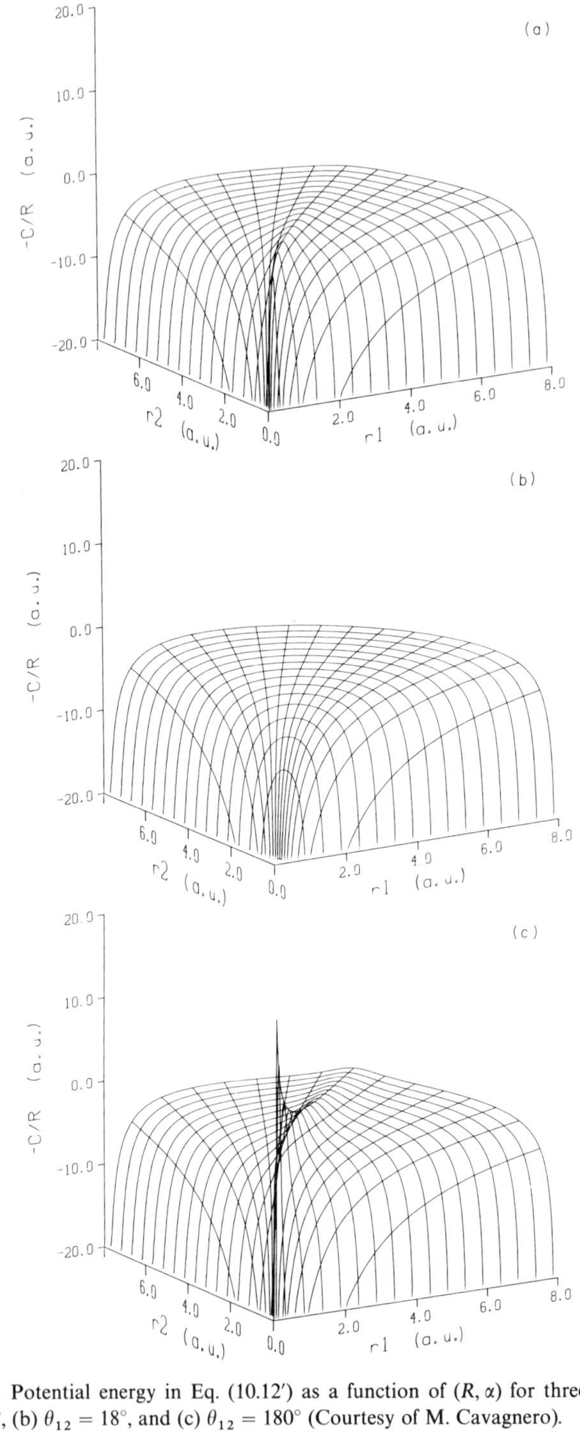

Fig. 10.9. Potential energy in Eq. (10.12′) as a function of (R, α) for three values of θ_{12}: (a) $\theta_{12} = 10°$, (b) $\theta_{12} = 18°$, and (c) $\theta_{12} = 180°$ (Courtesy of M. Cavagnero).

$C(\alpha, \theta_{12})$, that is, the potential V at constant R. Note the flat saddle at ($\alpha = 45°$, $\theta_{12} = 180°$), the valleys at $\alpha = 0°$ and $90°$, where either electron approaches the nucleus regardless of θ_{12}, and the spike where the electrons approach each other. This structure of the surface will be seen to have an essential influence on electron correlations. Most important is the saddle point, which constitutes a position of *unstable equilibrium* at each value of R. This instability results from the increase of mutual screening between the electrons, which occurs whenever their radial distances become unequal at constant R, a phenomenon that escapes attention unless the potential is mapped in configuration space. Figure 10.9 shows complementary reliefs of $V(\mathbf{r}_1, \mathbf{r}_2)$ as a function of $(r_1, r_2) \equiv (R, \alpha)$ for three fixed values of θ_{12}. The ridge in Fig. 10.9 and the saddle point of the potential surface in Fig. 10.8 will play a major role in the structure of doubly excited states and in the pathways to their excitation. The spike of Fig. 10.8 appears in Fig. 10.9 as a sharpening of the ridge at $r_1 = r_2$ for low values of θ_{12}.

10.4.2 Kinetic Energy

The Laplacian operator, which represents the kinetic energy for a single electron, $-(\hbar^2/2m)\nabla^2$, separates in polar coordinates into radial and centrifugal parts $(\hbar^2/2m)[-r^{-1}(d/dr)^2 r + \mathbf{l}^2/r^2]$, both of which are inversely proportional to the squared radial coordinate r^2. An analogous separation is obtained for the kinetic energy of a pair of particles. This separation utilizes the hyperradius R [(10.7)] in place of r and a "grand angular momentum" $\mathbf{\Lambda}$ in place of the orbital momentum $\mathbf{l}$,

$$-(\hbar^2/2m)(\nabla_1^2 + \nabla_2^2) = (\hbar^2/2m)[-R^{-5/2}(\partial^2/\partial R^2)R^{5/2} + (\mathbf{\Lambda}^2 + \tfrac{15}{4})/R^2]. \tag{10.13}$$

The operator $\mathbf{\Lambda}^2$ has the following structure in terms of the orbital momenta $(\mathbf{l}_1, \mathbf{l}_2)$ of the two separate electrons and the radial correlation angle α [(10.3)],

$$\mathbf{\Lambda}^2 \equiv \frac{\mathbf{l}_1^2}{\cos^2\alpha} + \frac{\mathbf{l}_2^2}{\sin^2\alpha} - \frac{1}{\sin\alpha\cos\alpha}\left(\frac{\partial^2}{\partial\alpha^2}\right)\sin\alpha\cos\alpha - 4, \tag{10.13'}$$

and takes alternative forms when the five coordinates $(\theta_1, \phi_1, \theta_2, \phi_2, \alpha)$ are replaced by others in accordance with Section 10.3.1.

The expression (10.13) of the two-particle kinetic energy displays the inverse proportionality to the squared radial coordinate R^2 in full analogy to the single-particle expression. Important novel elements appear, however, in (10.13) and (10.13′). First, we see that the gradient $\partial/\partial\alpha$ with respect to the mock angle α contributes to the grand angular momentum $\mathbf{\Lambda}$ much as the gradients with respect to the space angles $(\theta_1, \phi_1, \theta_2, \phi_2)$ do within the orbital momenta $(\mathbf{l}_1, \mathbf{l}_2)$. Recall that the angle α has been introduced as a radial correlation parameter, thus relating space coordinates *indirectly* in contrast to the other angular coordinates. That α can thus be incorporated into angular coordinates for purposes of physicomathematical analysis has important consequences because α has a finite range, $0 \leq \alpha \leq \frac{1}{2}\pi$. A second novelty lies in the occurrence of the positive coefficient $\frac{15}{4}$, which adds to the nonnegative $\mathbf{\Lambda}^2$ in (10.13); this additional term implies that the centrifugal contribution to the two particle kinetic energy *never vanishes*, in contrast to the single-particle case.†

The eigenvalues and eigenfunctions of the squared grand angular momentum $\mathbf{\Lambda}^2$ are known analytically as they are for the orbital momentum $\mathbf{l}^2$. The eigenvalues are represented by

$$\mathbf{\Lambda}^2 = \lambda(\lambda + 4), \qquad \lambda = 0, 1, \ldots, \infty, \tag{10.14}$$

and are degenerate for $\lambda \neq 0$. The degenerate eigenfunctions *may* be classified by the five quantum numbers (l_1, l_2, n_{rc}, L, M) pertaining to the representation (10.13′) of $\mathbf{\Lambda}^2$; l_1 and l_2 are here the orbital quantum numbers of the two separate electrons, n_{rc} is the analogous number of nodes in α, and L and M are the total quantum numbers as in (10.4). In this representation λ is partitioned in the form

$$\lambda = l_1 + l_2 + 2n_{rc}, \tag{10.14′}$$

where the coefficient 2 reflects that n_{rc} is the number of nodes in a range $\frac{1}{2}\pi$ of α, whereas $l_1 = (l_1 - m_1) + m_1$ is the number of nodes in ranges π of θ_1 and ϕ_1. The eigenfunction of $\mathbf{\Lambda}^2$ with these quantum numbers and total orbital numbers (L, M) is

$$\begin{aligned}\phi_{n_{rc}l_1l_2LM}&(\alpha, \hat{r}_1, \hat{r}_2)\\ &= \cos^{l_1+1}\alpha \sin^{l_2+1}\alpha \; {}_2F_1(-n_{rc}, n_{rc} + l_1 + l_2 + 2, l_2 + \tfrac{3}{2}; \sin^2\alpha)\mathcal{Y}_{l_1l_2LM}(\hat{r}_1, \hat{r}_2),\end{aligned} \tag{10.15}$$

where $\mathcal{Y}$ is given by (10.4) and the hypergeometric function F reduces to a Jacobi polynomial [5]. Alternative sets of angular coordinates lead, of course, to alternative sets of eigenfunctions labeled by different quantum numbers. An angular correlation quantum number n_{ac} can also be defined as

$$n_{ac} = \tfrac{1}{2}(l_1 + l_2 - L). \tag{10.15′}$$

Like n_{rc} in (10.14′) it is a quantum number characteristic of the pair of electrons. Together they serve as indexes of radial and angular correlation.

† Equation (10.13) gives the polar coordinate representation of the Laplacian in six dimensions. *Only* for three dimensions does this representation fail to include a nonzero constant in addition to the analog of $\mathbf{\Lambda}^2$.

10.4.3 Correlations at Extreme Ranges of R

The expressions of the kinetic and potential energies, (10.12) and (10.13), combine to yield the Schrödinger equation in hyperspherical coordinates,

$$\left\{\left(\frac{\hbar^2}{2m}\right)\left[-R^{-5/2}\frac{\partial^2}{\partial R^2}R^{5/2}+\frac{\Lambda^2+15/4}{R^2}\right]+\frac{e^2}{R}C(\alpha,\theta_{12})\right\}\Psi(\mathbf{r}_1,\mathbf{r}_2)=E\Psi. \tag{10.16}$$

The dominance of the centrifugal term—the second term in (10.16)—over the potential term ensures that the radial and angular coordinates can be separated approximately for small values of R. More specifically, a complete set of eigenfunctions of (10.16) may be constructed as a series in powers of R and ln R; the first term of each function consists of a product

$$R^{\lambda}\phi_{n_rcl_1l_2LM}(\alpha,\hat{r}_1,\hat{r}_2), \tag{10.17}$$

the exponent λ being the positive root of the relevant indicial equation.

Complete solutions of (10.16) must, of course, take into account that Λ^2 and $C(\alpha, \theta_{12})$ do not commute, as $\partial/\partial R$ does not commute with $1/R$ and $1/R^2$. The potential $V(\mathbf{r}_1, \mathbf{r}_2)$ may, however, be taken into account at small R by a degenerate perturbation procedure that diagonalizes the matrix of $C(\alpha, \theta_{12})$ in the finite base of harmonics (10.15) with the same value of λ. Full knowledge of the particle correlations at small R is thus acquired for a complete set of approximate solutions in this range $R \sim 0$.

The electron correlations are also readily identified in the opposite limit of $R \to \infty$, in which one electron lies at a very large distance while the other remains generally bound to the nucleus in a hydrogenic orbital. [Detachment of both electrons from the nucleus may give rise to more elaborate correlations to be described in Section 10.6.] An angular correlation is generally established in this event between the two electrons, because the excited hydrogenic orbitals are readily polarized and the polarized orbital exerts a potential r^{-2} on the detached electron (Section 5.6). The correlation is then readily evaluated, again by a degenerate perturbation treatment of the electron interaction.

More specifically, the wave function factors approximately as $R \to \infty$ much as it does for $R \to 0$ in (10.17). If we indicate the position of the detached electron by $\mathbf{r}_1$, with $r_1 = R\cos\alpha \gg r_2 = R\sin\alpha$, whence $\alpha \sim 0$ and $R\cos\alpha \sim R$, the wave function takes the form

$$f(R)g_{LM}(R\alpha,\hat{r}_1,\hat{r}_2), \qquad R\to\infty \tag{10.18}$$

with

$$g_{LM}=\sum_{l_1m_1l_2m_2}c^{LM}_{l_1m_1l_2m_2}Y_{l_1m_1}(\hat{r}_1)h_{l_2m_2}(R\alpha,\hat{r}_2). \tag{10.18'}$$

Here $h_{l_2m_2}(\mathbf{r}_2)$ indicates one of the degenerate hydrogenic orbitals of the bound electron, whose correlation with $\hat{r}_1$ is characterized by the coefficients $c^{LM}_{l_1m_1l_2m_2}$.

[The formulation (10.18) of an eigenfunction at $R \to \infty$ has been oversimplified by identifying the position of the detached electron as $\mathbf{r}_1$, thus disregarding the identity of the two electrons. The correct form of (10.18) includes its antisymmetrization as well as the necessary labeling of the spin orientations. The same remark applies to the expression (10.15) of the eigenfunctions at $R \to 0$.]

The two limiting forms of the wave function, (10.17) and (10.18), contrast sharply in their dependence on α. The $R \sim 0$ expression (10.17) displays standing-wave behavior over the whole range of α, from 0 to $\frac{1}{2}\pi$, while the hydrogenic factor $h_{l_2m_2}(\mathbf{r}_2) \equiv h_{l_2m_2}(R\alpha, \hat{r}_2)$ in (10.18′) is confined at large R to $\alpha \ll 1$ by its exponential decay, $\exp(-R\alpha/n_2)$. This contrasting behavior results from the potential barrier in Figs. 10.8 and 10.9 that rises with increasing R, where it confines the wave function to the well at $r_2 \ll r_1$ (or, symmetrically, at $r_1 \ll r_2$) but allows free propagation at small R over the whole range of α from the r_1 coordinate axis ($\alpha = 0$) to the r_2 axis ($\alpha = \frac{1}{2}\pi$). That is, the potential barrier prevents the radial interchange of the two electrons at large (but *constant*) values of R but allows them to interchange freely at low R where their kinetic energy prevails. (These contrasting aspects are restricted to negative values of the energy, whereas both electrons may escape from the atom at $E > 0$ as represented by a wave function that propagates above the barrier at all R. The remaining effects of the barrier at $E > 0$ will be touched on in Section 10.6.)

The wave function at large R may be said to display a strong radial correlation in the sense of Section 10.3 insofar as its significant values are restricted to very small ranges of α. In fact, this distribution of the two electrons means that one of them is much closer to the nucleus than the other; in this sense the radial motions are hardly correlated at all, that is, are practically independent of each other. Indeed, the motion of the electron pair at large R amounts to that of an $He^+ + e$ pair as treated in Chapters 7 and 8.

This initial assessment of correlations in the opposite ranges of $R \sim 0$ and $R \to \infty$ presents a picture analogous to that encountered in an initial study of chemical bonding. Bonding of two atoms in the limiting cases of very small or very large internuclear distances is characterized by the alternative labels of "united atom" and "separate atom." The transition between these limits is generally described by viewing the state of the bonding electrons as a function of the internuclear distance and by following the evolution of this state from the united to the separate atom limit in a "Born–Oppenheimer approximation." We shall see next how an analogous procedure affords a description of the evolution of electron correlations from $R \sim 0$ to $R \to \infty$, returning again to these "condensation" and "fragmentation" limits in Sections 10.5.3 and 10.5.5.

10.4.4 Evolution of Correlations in Adiabatic Channels

We have seen in the previous section how the correlation of an electron pair is represented by the eigenfunction of the operator $\mathbf{\Lambda}^2$ in the limit $R \to 0$ because the potential term of (10.16) is negligible in this limit. It was also noted that the potential's influence can be taken into account to lowest order by considering the particular set of eigenfunctions of $\mathbf{\Lambda}^2$, $\{\phi_{\lambda\mu}\}$, for which the matrix of the effective charge $C(\alpha, \theta_{12})$ remains diagonal within each degenerate manifold,

$$(\lambda\mu|C(\alpha, \theta_{12})|\lambda\mu') = (\lambda\mu|C(\alpha, \theta_{12})|\lambda\mu)\delta_{\mu\mu'}. \tag{10.19}$$

The wave functions $\{\phi_{\lambda\mu}\}$ represent the correlations correctly as long as matrix elements of the potential energy off-diagonal in λ, $(e^2/R)(\lambda'\mu'|C|\lambda\mu)$, remain negligible as compared to the difference in eigenvalues of the centrifugal energy $(\hbar^2/2m)(\mathbf{\Lambda}^2 + \frac{15}{4})/R^2$ for λ and λ'.

The influence of the Coulomb potential energy on electron correlations increases rapidly at larger values of R, where the centrifugal term of the Hamiltonian no longer dominates. The joint influence of the centrifugal and Coulomb terms is then reflected more adequately in the eigenfunctions of the combined operator

$$H_R = (\hbar^2/2m)(\mathbf{\Lambda}^2 + \tfrac{15}{4})/R^2 + (e^2/R)C(\alpha, \theta_{12}), \tag{10.20}$$

which depend on R parametrically as indicated by their defining equation

$$H_R\Phi_\mu(R; \alpha, \hat{r}_1, \hat{r}_2) = U_\mu(R)\Phi_\mu. \tag{10.21}$$

These wave functions represent a particular class of correlations each of which is represented by one of the eigenfunctions $\phi_{\lambda\mu}$ of $\mathbf{\Lambda}^2$ at $R \to 0$ and one of the functions (10.18) at $R \to \infty$.

Before discussing the evolution of the $\{\Phi_\mu\}$ from $R = 0$ to $R = \infty$, we indicate their relevance to the solution of the complete Schrödinger equation (10.16), whose Hamiltonian may be cast as

$$H = -\frac{\hbar^2}{2m} R^{-5/2} \frac{\partial^2}{\partial R^2} R^{5/2} + H_R. \tag{10.22}$$

Note at the outset that the $\{\Phi_\mu\}$ form a complete set over $(\alpha, \hat{r}_1, \hat{r}_2)$ at each value of R and that they may be labeled by quantum numbers LM because H_R is invariant under rotations of physical space. Accordingly, an eigenfunction $\Psi_{ELM}(R, \alpha, \hat{r}_1, \hat{r}_2)$ of the Hamiltonian (10.22) may be expanded into $\{\Phi_\mu\}$ at each value of R,

$$\Psi_{ELM}(R, \alpha, \hat{r}_1, \hat{r}_2) = \sum_\mu F_\mu^{ELM}(R)\Phi_{\mu LM}(R; \alpha, \hat{r}_1, \hat{r}_2). \tag{10.23}$$

The full electron correlation in the eigenstate Ψ_{ELM} is thus represented as a superposition of the separate $\Phi_{\mu LM}$ with the R-dependent amplitudes $F_\mu^{ELM}(R)$. The representation of Ψ_{ELM} in (10.23) may be viewed as analogous to the representation of multichannel functions in Chapters 7–9, the more so as the form (10.18) of a component of Φ at large R represents a particular i channel of separation of an electron from the atomic residue. Accordingly, we shall refer to the Φ_μ as channel functions and to the several terms of (10.23) as representing component channels.

Entering the expansion (10.23) in the Schrödinger equation (10.16) and projecting the result in turn onto the several $\{\Phi_\mu\}$ yields a set of ordinary differential equations in the amplitudes $F_\mu^{ELM}(R)$. These equations would be uncoupled were it not that the derivative $\partial^2/\partial R^2$ also operates on the parameter R of the base set $\{\Phi_\mu\}$. The expansion (10.23) is called adiabatic because it originates from approximation procedures in which each base function analogous to Φ_μ was understood to vary slowly, that is, adiabatically, as a function of the parameter corresponding to R. Diatomic molecules provide a prototype example in which the electron wave function varies slowly with the internuclear distance, which serves as the adiabatic parameter. In this example the success of the approximation is usually attributed to the nuclei moving far more slowly than the electrons, owing to their much higher mass. Generally, however, different circumstances may cause the Φ_μ to vary slowly with R. The validity of adiabatic approximations should accordingly be assessed from the magnitude of the coupling terms in the equations for $F_\mu(R)$.

These coupling terms consist of projections of derivative $\partial\Phi_\nu/\partial R$ onto the base set Φ_μ,

$$P_{\mu\nu}(R) = \int_0^{\pi/2} d\alpha \int_{4\pi} d\hat{r}_1 \int_{4\pi} d\hat{r}_2 \Phi_\mu \frac{\partial\Phi_\nu}{\partial R} = -P_{\nu\mu}. \tag{10.24}$$

Indicating the matrix $P_{\mu\nu}$ by $\mathbf{P}$, the matrix $U_\mu(R)\delta_{\mu\nu}$ by $\mathbf{U}$, and the set of $\{F_\mu^{ELM}(R)\}$ by $\mathbf{F}(R)$, (10.23) and (10.16) yield the linear system

$$\left[-\frac{\hbar^2}{2m} R^{-5/2}\left(\mathbf{I}\frac{\partial}{\partial R} + \mathbf{P}(R)\right)\left(\mathbf{I}\frac{\partial}{\partial R} + \mathbf{P}(R)\right)R^{5/2} + \mathbf{U}(R) - \mathbf{I}E\right]\mathbf{F}(R) = 0. \tag{10.25}$$

The coupling effect of $\mathbf{P}$ depends in part on the magnitude of $\partial\mathbf{F}/\partial R$ and the differences $|U_\mu(R) - U_\nu(R)|$. The notation of (10.25) affords the option of utilizing a base set $\{\Phi_\mu(R)\}$ that does not diagonalize the operator H_R fully, in which case the matrix $\mathbf{U}(R)$ is no longer quite diagonal. An alternative set of "diabatic" functions may instead minimize $\mathbf{P}$. More generally, the ability to utilize alternative sets $\{\Phi_\mu\}$ affords flexibility in the formulation and analysis of correlations.

Our view of the expansion (10.23) of the eigenfunction Ψ is thus the following. Each Φ_μ at a given R represents a particular correlation pattern, radial and angular, through its dependence on α and $(\hat{r}_1, \hat{r}_2)$ or on alternative parameters. The dependence of Φ_μ on R describes instead the evolution of this pattern from $R \sim 0$ to $R \to \infty$. Amplitude variations of the radial functions $F_\mu(R)$ reflect the changing contribution of each Φ_μ to the full eigenfunction Ψ_{ELM}. The expansion serves thus to articulate an analysis of correlations, particularly when it receives substantial contributions from only a few terms. This approach will prove effective for examples presented in following sections. However, the examples will deal mainly with channels of low energy $U_\mu(R)$, which remain more nearly independent owing to their larger values of the energy intervals $U_\mu(R) - U_\nu(R)$.

We anticipate here qualitative aspects of the results of detailed calculations, to be discussed in Section 10.5:

(1) Each of the coupling matrix elements $P_{\mu\nu}(R)$ [(10.24)] remains very small over broad ranges of R, thus allowing the radial integrations of $F_\mu(R)$ and $F_\nu(R)$ in (10.25) to proceed independently over those ranges. This result reflects a smooth dependence on R of the channel functions Φ_μ and Φ_ν and rests presumably on the smooth variation of the relative magnitude of the terms in (10.16) proportional to $1/R$ and $1/R^2$. It implies that the observable manifestations of channel coupling stem from identifiable circumstances prevailing in specific limited ranges of R, as will be detailed especially in Sections 10.5.3–10.5.5.

(2) The range in which a coupling element $P_{\mu\nu}$ is significant is generally identifiable on plots of potential curves $U_\mu(R)$, such as those shown in Fig. 10.10, as the position of an "avoided crossing" of U_μ and U_ν, that is, as the position where the curves approach and then diverge again. This phenomenon parallels the one observed in diatomic molecules, whose electronic transitions are liable to occur at avoided crossings of potential curves of the nuclear motion, with a resulting energy transfer between electrons and nuclei.

(3) Single avoided crossings generally arise from the occurrence of small interaction terms among the variables $(\alpha, \hat{r}_1, \hat{r}_2)$ of Φ_μ and Φ_ν. Disregarding such a term in the

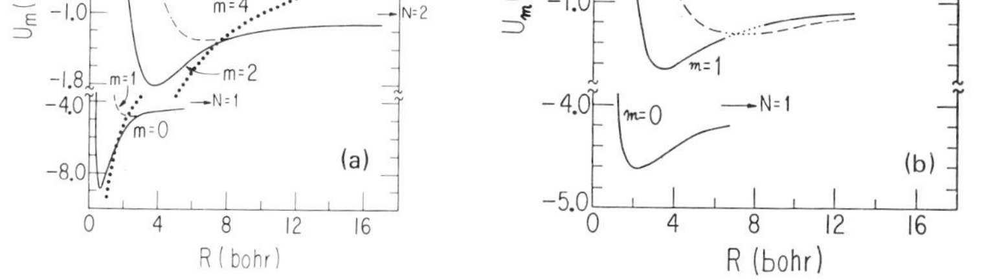

Fig. 10.10. Adiabatic potential curves $U_m(R)$ for He, labeled by index m, which corresponds to μ in the text. (a) ——, ^{1}S; – – –, ^{3}S. Dotted curve representing $C(\alpha = 45°, \theta_{12} = 180°)/R$ gives the potential at the saddle point of the potential surface in Fig. 10.8. (b) ^{1}P^o. (Courtesy of C. D. Lin.)

operator H_R of (10.21) would yield a local separability of these variables and cause the pair of potential curves $\{U_\mu(R), U_\nu(R)\}$ actually to cross; it would also smooth out the local variation of eigenfunctions $\{\Phi_\mu, \Phi_\nu\}$, thus changing them from adiabatic to diabatic.

(4) A most important array of avoided crossings is observed in Fig. 10.10 and generally in analogous plots. Its locus is marked by the dotted curve in the figure, whose ordinate equals $(e^2/R)C(\alpha = 45°, \theta_{12} = 180°)$; that is, it represents the potential energy of the electron pair on the saddle point of Fig. 10.8 and along the ridge line of Fig. 10.9a. This line represents the locus of positions at which successive eigenfunctions Φ_μ of (10.21) evolve rapidly from their low-R oscillating ("standing-wave") behavior to their confinement in potential valleys at large R.

This identification of the mode and place of a rapid transition from the radial correlations prevailing at $R \sim 0$ to that prevailing at $R \to \infty$ is a major contribution of the analysis in terms of hyperspherical coordinates and the adiabatic expansion (10.23). This transition takes place, for each channel, near the critical radius $R_{c\mu}$ at which the channel energy $U_\mu(R)$ dips below the ridge of the potential in Fig. 10.9a. The equation that determines $R_{c\mu}$ is, then, in view of (10.12′)

$$U_\mu(R_{c\mu}) = \frac{e^2}{R_{c\mu}} C(\alpha = 45°, \theta_{12} = 180°) = -\frac{e^2}{R_{c\mu}} 2\sqrt{2}(Z - \tfrac{1}{4}). \quad (10.26)$$

In the last expression, $Z - \frac{1}{4}$ represents the nuclear charge acting on each electron, the combined factor $\sqrt{2}/R_{c\mu}$ represents the reciprocal root-mean-square distance of each electron from the nucleus [in view of the definition of R in (10.7)], and the factor 2 allows for the potential energy of *two* electrons.

Note how each curve $U_\mu(R)$ in Fig. 10.10 reaches a minimum at a value of R a little lower than $R_{c\mu}$; it would then rise to intersect the next curve of its family (full or dashed) were it not kept away from it in a pattern of avoided crossing. Figure 10.11 shows how matrix elements $P_{\mu\nu}(R)$ that couple

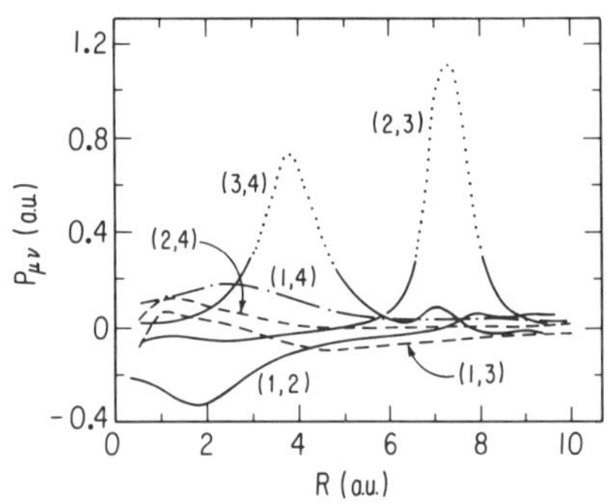

Fig. 10.11. Coupling matrix elements $P_{\mu\nu}$ for $^1P^o$ channels of He. The $\mu = 1$ channel converges to $He^+(N = 1)$, and the $\mu = 2$–4 channels converge to $He^+(N = 2)$. (Courtesy of C. D. Lin).

successive channels of the same family peak sharply in the range of R between the $R_{c\mu}$ values of the two channels. This peaking causes a major departure from adiabatic behavior of the coupled system (10.25) in the region of Fig. 10.10a along the dotted line, which will prove the main vehicle for transfer of an electron pair to higher and higher channels of excitation (Section 10.5.4).

10.5 LOW DOUBLY EXCITED STATES

Figure 10.2 presented the evidence for doubly excited states in He below the He^+ ($N = 2$–5) thresholds. The analysis of the two-electron Schrödinger equation in hyperspherical coordinates and the study of radial and angular correlations in such states has been developed mostly for the states below the thresholds of $He^+(N = 2)$ and $He^+(N = 3)$ and their counterparts in H^-.

This task will develop further the analogy to the study of diatomic molecules. The adiabatic treatment that separates the motion in R from that in the angular coordinates leads to potential wells as shown in Fig. 10.10, which hold series of doubly excited states, much as molecular potential curves in the internuclear distance R contain series of vibrational states of the molecule. Generally, the minima of potential curves represent an equilibrium between the kinetic energy, which increases when the system condenses at small R, and the potential energy, which rises at increasing R. Significant differences between our problem and that of diatomic molecules will emerge as well.

10.5.1 Adiabatic Channels and Potential Wells

Adiabatic channels are obtained by solving (10.21), the Hamiltonian H_R in angular coordinates alone with R regarded as a parameter. Among alternative schemes for solving such a partial differential equation in several variables, one that has proved convenient is to expand Φ_μ into the two-electron spherical harmonics $\mathscr{Y}_{l_1 l_2 LM}(\hat{r}_1, \hat{r}_2)$ defined in (10.4). The coefficients $g_\mu^{l_1 l_2}(R; \alpha)$ of this expansion are functions of α and depend on R parametrically. Retention of a few values of (l_1, l_2) that are compatible with the given total angular momentum L and parity Π leads to a small set of coupled equations in α that can be solved numerically. The smallness of this set reflects the weak dependence of the angular correlations on R for the low-lying adiabatic channels.

Figure 10.12 provides an example of the eigenvalues $U_\mu(R)$ for $^1P^o$ states that converge to the He^+ ($N = 2$) threshold as $R \to \infty$. Three potential curves $U_\mu(R)$ converge to this threshold. This set of three replaces as many series of independent electron configurations of $^1P^o$ symmetry, namely 2s*n*p,

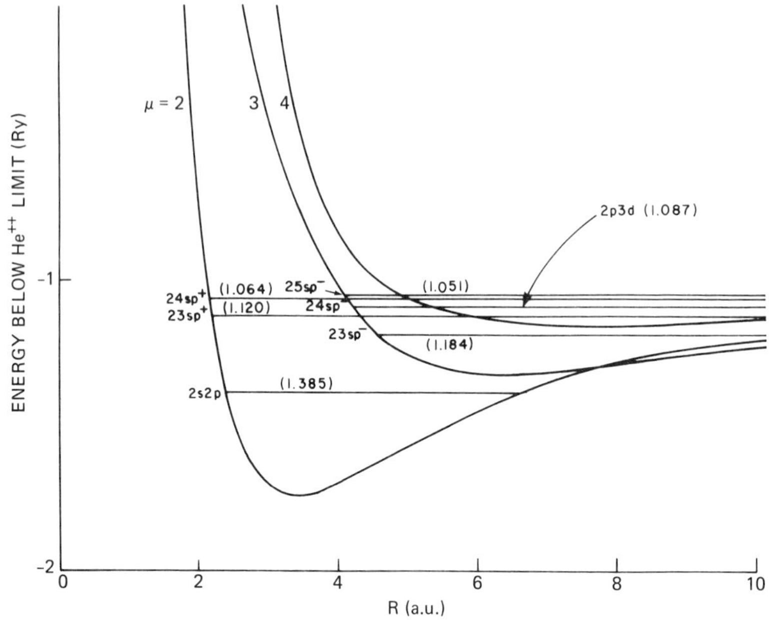

Fig. 10.12. Adiabatic potentials $U_\mu(R)$ and energy levels of $^1P^o$ symmetry below He$^+$($N = 2$). (From Macek [6].)

2pns, and 2pnd. Qualitative examination of the eigenfunction Φ_μ for the lowest curve ($\mu = 2$) shows it to consist primarily of a superposition of the two (l_1, l_2) pairs (0, 1) and (1, 0), nearly symmetric under the radial interchange $r_1 \leftrightarrow r_2$, that is, $\alpha \to \frac{1}{2}\pi - \alpha$. The curve $U_2(R)$ may accordingly be labeled sp+. Similarly, the $\mu = 3$ curve, which shares the same (l_1, l_2) structure but with wave functions nearly antisymmetric under $\alpha \to \frac{1}{2}\pi - \alpha$, may be labeled sp−. The $\mu = 4$ curve is found to have primarily (1, 2) and (2, 1) character, without any analogous symmetry, as will be discussed later. [A lower curve ($\mu = 1$) is not shown because it converges to the He$^+$ ($N = 1$) limit.]

Numerical solution of the coupled differential equations in α presents problems, particularly as an increasing number of (l_1, l_2) pairs is included, leading to many coupled equations. An alternative procedure for obtaining the adiabatic potential curves expands Φ_μ into eigenfunctions of Λ^2 in (10.15) with coefficients obtained by matrix diagonalization; however, this expansion converges poorly as R increases. Currently, a variational procedure is applied, with base functions consisting of harmonics of $(\hat{r}_1, \hat{r}_2)$ and analytical expressions in α (RPP, Section 3.1.2).

Calculations have pointed to two simplifying elements: first, that matrix elements of $C(\alpha, \theta_{12})$ in (10.12′) off-diagonal in (l_1, l_2) are generally smaller than the diagonal elements by more than half an order of magnitude and, second, that truncation of Φ_μ

to a single value of the quantum number n_{rc} in (10.14′) serves as an initial approximation. Restriction to a single n_{rc} means that successive values of μ correspond to increasing values of n_{rc} for a given (l_1, l_2). The resulting truncated eigenvectors Φ_μ are represented in a form that explicitly incorporates the Pauli principle, that is, symmetry (antisymmetry) for $S = 0(1)$ under the interchange $\alpha \to \frac{1}{2}\pi - \alpha$, $\hat{r}_1 \leftrightarrow \hat{r}_2$,

$$\Phi_\mu(R; \alpha, \hat{r}_1, \hat{r}_2) = g_{n_{rc}}^{l_1 l_2}(R; \alpha)\mathscr{Y}_{l_1 l_2 LM}(\hat{r}_1, \hat{r}_2) + (-)^S g_{n_{rc}}^{l_1 l_2}(R; \tfrac{1}{2}\pi - \alpha)\mathscr{Y}_{l_1 l_2 LM}(\hat{r}_2, \hat{r}_1). \tag{10.27}$$

Figure 10.10, which shows the potential curves $U_\mu(R)$ for $^{1,3}S^e$ states was, in fact, constructed in this manner with the further restriction that $n_{ac} = \frac{1}{2}(l_1 + l_2) = 0$. For these $L = 0$ states, the two $\mathscr{Y}$ functions in (10.27) are identical and the wave function's symmetry concerns only its α dependence; singlets (triplets) have an even (odd) number of nodes in α. This factorization of the dependences on α and $(\hat{r}_1, \hat{r}_2)$ does not hold for general values of $\{L, S, \Pi\}$. Note in Fig. 10.10 that singlet and triplet curves merge in pairs at large R, indicating that the addition of one node in α has hardly any effect on $U_\mu(R)$ in the range of R where the wave function tunnels through the potential barrier at $\alpha = 45°$ in Fig. 10.9. At low and intermediate R, the absence of a node at $\alpha = 45°$ acts instead to lower the singlet curves relative to the triplets.

Several truncated eigenvectors (10.27) are generally included in a calculation of the adiabatic potential curves $U_\mu(R)$ and the corresponding eigenvectors. For $^1P^o$ symmetry, the $U_\mu(R)$ so generated coincide with the solutions of coupled differential equations shown in Fig. 10.12. It suffices for qualitative analysis to display the principal truncated component for each channel. Figure 10.13 thus shows the functions $g_{n_{rc}}^{l_1 l_2}(R; \alpha)$ for the lowest curves ($\mu = 2$ and $\mu = 3$) plotted against α for different values of R. The $\mu = 3$ curves show a node near $\alpha = 45°$. On the other hand, for $\mu = 2$, the functions exhibit an antinode near $\alpha = 45°$ at small and intermediate values of R. With increasing R, the amplitude at $\alpha = 45°$ falls off past the critical radius $R_{c\mu}$, which marks the onset of tunneling under the barrier in the potential. The transition of the wave functions $g_{n_{rc}}^{l_1 l_2}$ in Fig. 10.13a and b to tunneling for $R > R_{c\mu}$ resembles the transition of molecular wave functions, bonding or antibonding, in the separated atom limit of dissociation. The correspondence of potential curves $U_\mu(R)$ in Figs. 10.10 and 10.12 to molecular potentials in the Born–Oppenheimer approximation has further aspects. The steep rise at small R reflects a generalized angular momentum barrier, $\lambda(\lambda + 4)/R^2$ from (10.14), analogous to the resistance of atoms to interpenetration. The softer rise toward large R reflects in both cases the binding by Coulomb interactions.

Curves converging to higher N limits in Fig. 10.10 show stronger repulsion at small R and are displaced toward larger R relative to the lowest curve that converges to $N = 1$. Since the ground and singly excited states of the atom are held in the lowest potential well for each L value while the doubly excited states reside in the curves with larger N, an important characteristic is immediately evident: The overlap between the wave functions of these two classes of states is generally small. The sharpness of the doubly excited features in a spectrum such as the one shown in Fig. 10.2 and the weakness of their excitation reflect this small overlap. In the total oscillator strength of

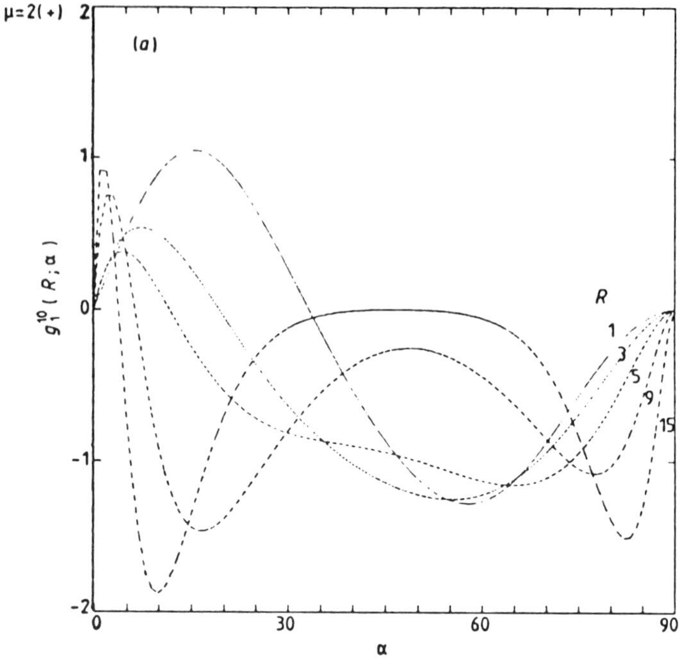

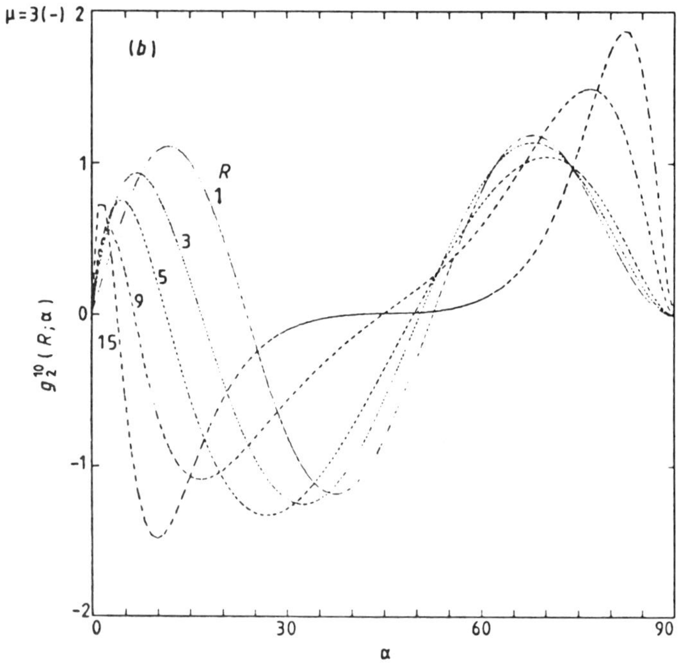

Fig. 10.13. Channel functions $g^{10}_{n_{rc}}(R;\alpha)$ in Eq. (10.27) for ${}^1P^o$ channels of He at different R values: (a) $n_{rc} = 1$ and (b) $n_{rc} = 2$. (Courtesy of C. D. Lin).

the ground state of He, double excitations amount, therefore, to no more than a small fraction ($\simeq 10\%$), although they stand out in the narrow energy ranges in which they are concentrated. In other atoms to be discussed in Chapter 11, the higher potential curves in a figure such as Fig. 10.10 will show increasing overlap with the lowest one, so that double excitations are of greater importance even in the aggregate.

10.5.2 Channel Classification

Increasing numbers of potential curves occur when our discussion is extended to states converging to He^+ limits with higher N. Enumeration of the alternative independent particle configurations of ${}^1P^o$ symmetry (3snp, 3pns, 3pnp, and 3dnf) points to five potential curves converging to He^+ $(N = 3)$ in place of the 3 in Fig. 10.12. The independent particle labels are again inadequate because of the strong correlations. The simple numerical ordering in μ, which sufficed in Fig. 10.12, no longer provides adquate discrimination.

A classification scheme in terms of quantum numbers appropriate to the pair is accordingly introduced. The pair of quantum numbers (K, T) introduced in (10.11′) as replacements of the pair (l_1, l_2) of separate orbital quantum numbers incorporates much more adequately the influence of electron–electron interaction on angular correlations. This is evidenced by verifying numerically that the matrix of $1/r_{12}$ is nearly diagonal in the base of the quantum numbers (K, T) [7]. The qualitative interpretation of (K, T) proves simplest in the region of large R where one electron lies far from the residual He^+ or H and the interaction polarizes the residue strongly (linear Stark effect). That is, the states of the residue degenerate in l are mixed by the electric field of the distant electron into dipole eigenstates, labeled by (K, T). The quantum number K represents the projection of the dipole moment of the residue onto the direction of the distant electron. This projection is indicated by the expectation value $\langle -r_< \cos\theta_{12}\rangle = \langle -\mathbf{r}_< \cdot \hat{r}_>\rangle$ where $r_<(r_>)$ refer to the smaller (larger) of (r_1, r_2). This expectation value is in turn proportional to $\langle -\hat{b}_> \cdot \mathbf{b}_<\rangle$ within a fixed (N, n) subspace, as anticipated in Section 10.3.1.

Thus, the interaction energy increases primarily as K decreases. The orientation of the orbital angular momentum of the atomic residue with respect to the direction of the distant electron is also relevant and is conveniently expressed by the quantum number $T = \langle(\mathbf{L}\cdot\hat{r}_>)^2\rangle$. These quantum numbers take the values given in (10.11′),

$$0 \le T \le \min(L, N-1), \qquad \pm K = N - T - 1, N - T - 3, \ldots, 1 \text{ (or 0)}. \tag{10.28}$$

The eigenvalues of the potential energy thus increase with decreasing K and with decreasing T at equal K. Further aspects of this interaction at large R will emerge in Section 10.5.5.

In the region where one electron is farther than the other, the polarization of the atomic residue implies much of the radial correlation as well because the number of radial nodes in the wave function of the residue equals $N - l_< - 1$ for each of its orbital components $l_<$. Thereby, most of the information in the radial correlation quantum number n_{rc} is already implied by the value of K. The residual information concerns the behavior of the wave function under the operation of radial interchange, $\alpha \to \frac{1}{2}\pi - \alpha$. Section 10.5.1 has pointed to an important element of this behavior, namely, the distinction between states that have an antinode and those that have a node near $\alpha = \pi/4$, the ridge of the potential surface. This symmetry was indicated in Section 10.5.1 by a $\pm$ index.

An additional type of correlation function Φ_μ was identified in Section 10.5.1 by the index $\mu = 4$, which corresponds to a pd configuration. In this case, because of the centrifugal repulsion $6/r_>^2$ from the orbital quantum number $l_> = 2$, the pair hardly penetrates to the region of the ridge $\alpha = \pi/4$. Corresponding states with this characteristic of being confined to the valleys $\alpha = 0$ and $\pi/2$ occur at higher N, whenever the orbital quantum number $l_>$ exceeds the value $N - 1$. In these states the two electrons hardly come close radially. The same holds for the channel of single excitation converging to He^+ ($N = 1$). This absence of significant radial correlation will be indicated by the zero value of an additional quantum number A that also takes the values $A = \pm 1$ to represent the class of states with strong radial correlation, indicated previously as $\pm$. The three alternative values of A (± 1 and 0) provide a partial characterization of the radial correlation along with the information implied by the value of K.

Together, $(K, T)^A$ serve to label uniquely the potential curves for any $\{L, S, \Pi\}$ and ionization limit N, thus describing completely the angular correlation and the approximate symmetry in the α coordinate [8]. The full set of $U_\mu(R)$ curves for a given N are enumerated by finding all possible (K, T) from (10.28) and ordering them at large R with the maximum K at the bottom and then in decreasing steps upward. The T values are similarly ordered within each K manifold, starting with the largest T at the bottom.

The value of A does not enter the enumeration of the number of $U_\mu(R)$ curves, but once the number of (K, T) pairs has been determined the curves are labeled by A according to the following prescription [8]. The number of $A = +1$ curves equals the number of $\{L, S, \Pi\}$ states compatible with the Pauli principle when both electrons are in the N shell; the number of $A = -1$ curves is the number of analogous states for the alternative value of S for the electron pair. This prescription amounts to setting

$$A = (-)^{S+N-K-1}\Pi \qquad \text{for all} \quad K > L - N.$$

All the remaining (K, T) pairs, that is, all those with $K \le L - N$ are labeled $A = 0$. Note that for any N and $L > 2(N - 1)$, all $\frac{1}{2}N(N + 1)$ curves have $A = 0$. The ordering in energy of the $U_\mu(R)$ curves at low and intermediate R follows the rule that

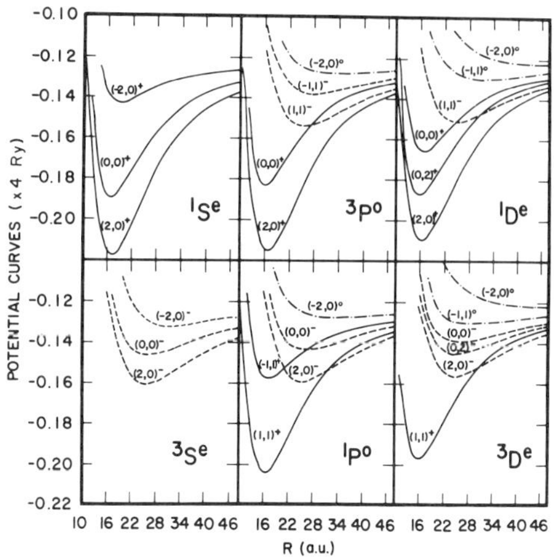

Fig. 10.14. Adiabatic potential curves $U_\mu(R)$ for He below the $He^+(N = 3)$ limit. Labels indicate values of $(K, T)^A$. (From Lin [8].)

the $A = +1$ curves lie lowest in energy, followed by $A = -1$ with the $A = 0$ curves lying highest. Within each of these sets, the ordering of (K, T) is according to the highest K at the bottom, then in decreasng steps upward and, within each K manifold, a similar ordering of the values of T.

Figure 10.14 illustrates this classification for the potential curves that converge to the He^+ $(N = 3)$ limit. The $A = \pm 1$ curves are shown to cross at large R. Crossings and avoided crossings at small, intermediate, and large R will be discussed separately in the next three sections. These regions are the sites of the major mixing of adiabatic channels because coupling between channels has the strongest influence in the vicinity of avoided crossings, as in the molecular analog.

10.5.3 Condensation Region

The radial coordinates of both electrons vanish in the limit $R \to 0$, which represents a compressed system analogous to the united atom limit of a diatomic molecule. In the condensation region of low R, the eigenvalues $U_\mu(R)$ depend mainly on the repulsive kinetic energy, which scales as $1/R^2$. According to Section 10.4.2, the Hamiltonian H_R in (10.20) reduces to its first term in the limit $R \to 0$, and its eigenvalues are represented by $(\hbar^2/2mR^2)(\lambda + 3/2)(\lambda + 5/2)$, which is equivalent to (10.14). The corresponding eigenfunctions ϕ_λ in (10.15), of course, must be antisymmetrized.

The contribution of the potential energy in the Hamiltonian (10.20) becomes significant with increasing R. On factoring $1/R^2$ from this Hamiltonian, we consider the operator

$$\tfrac{1}{2}\Lambda^2 + RC(\alpha, \theta_{12}) \tag{10.20'}$$

in atomic units, with eigenvalues $R^2U_\mu(R)$. This form proves more convenient than H_R in (10.20) because the hyperspherical radius R enters (10.20′) as a factor multiplying $C(\alpha, \theta_{12})$, and the eigenvalues $R^2U_\mu(R)$ are linear in R at small R. Figure 10.15, which replots the data on $U_\mu(R)$ in Fig. 10.10, shows a linear evolution of the $R^2U_\mu(R)$ curves over an extended range of R. From each set of degenerate eigenvalues of $\mathbf{\Lambda}^2$ at $R = 0$, these curves fan out with increasing R, thus showing avoided crossings where curves originating from different eigenvalues of $\mathbf{\Lambda}^2$ would otherwise intersect.

As an example of avoided crossings in the condensation region, consider the one near $R \simeq 4$ a.u. between the $\mu = 3$ and 4 curves of $^1P^o$ symmetry shown in Fig. 10.12 and again in the replotted Fig. 10.15. We noted in Section 10.5.1 that $\mu = 3$ has dominant character $(l_1, l_2) = (0, 1)$ and corresponds to $n_{ac} = 1$, $n_{rc} = 2$, while $\mu = 4$ has (1, 2) with $n_{ac} = 3$, $n_{rc} = 0$. In the limit $R \to 0$, the larger value of n_{rc} for the $\mu = 3$ curve yields a higher value of $\lambda = 5$, whereas $\mu = 4$ converges to $\lambda = 3$. The classification by $\mu = 3, 4$ or by the set of quantum numbers $(K, T)^A = (-1, 0)^{-1}$ and $(0, 1)^0$ has been established in Sections 10.5.1 and 10.5.2 on the basis of adiabatic energy in the range of intermediate R. The adiabatic connection to the condensation range is represented in Fig. 10.15 by the avoided crossing of the corresponding curves R^2U_3 and R^2U_4. Near the avoided crossing, the adiabatic correlation functions Φ_3 and Φ_4 are no longer characterized by constant values of the quantum numbers (n_{ac}, n_{rc}). In fact, a comparison of the correlation plots of Φ_3 and Φ_4 in the coordinate plane (α, θ_{12}) shows that these functions interchange their values of (n_{ac}, n_{rc}) gradually over the interval $1 \leq R \leq 6$ a.u. (Fig. 26 of RPP).

A whole sequence of analogous avoided crossings is expected to occur systematically near the interface of condensation and intermediate ranges. The evolution of the

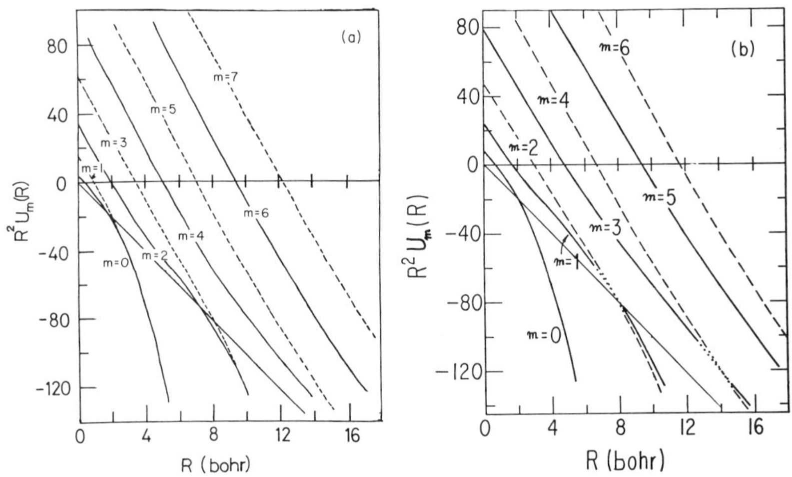

Fig. 10.15. $R^2U_m(R)$ for helium: (a) ^1S (——) and ^3S (---)and (b) $^1P^o+$ (——) and $^1P^o-$ (---). Note avoided crossings for ^1S in (a) and for + channels in (b). The label m corresponds to μ in the text. The straight line through the origin tracks the saddle point according to (10.26). (From Lin [4].)

correlation plots of Φ_μ in (α, θ_{12}) should also parallel that of the prototype discussed earlier and should have appreciable effects on both sides of the interface. We return to this subject in Section 10.5.6. Its general effect is to introduce higher values of l in the lower-energy function Φ_μ in the condensation region, as shown by the introduction of the (pd) configuration into the $\mu = 3$ function of $^1P^o$.

10.5.4 Excitation Ladder on the Potential Ridge

Section 10.4.4 noted a major array of avoided crossings in the ^1S curves of Fig. 10.10 lying along the dotted line that represents the value of the potential energy (10.26) at its saddle point, $-(e^2/R)C(45°, 180°)$. These crossings are associated with the rapid variation of the distribution in α of the factors $g_{n_{rc}}^{l_1 l_2}(R, \alpha)$ in (10.27) as they pass from oscillation to tunneling under the ridge, at the interface between the intermediate and fragmentation ranges of R (Fig. 10.13). This passage occurs just at the combination of energy and radius that constitutes the locus of the avoided crossing.

A main characteristic of these avoided crossings is that they couple amplitude variations of radial functions $F_\mu(R)$ pertaining to potential curves $U_\mu(R)$ that converge to He^+ limits with different N values. This effect derives from selection rules concerning the magnitude of the coupling coefficients $P_{\mu\nu}(R)$. These selection rules favor first the matrix elements $P_{\mu\nu}$ between channels with the same A quantum number, as a consequence of the similar behavior of their wave functions Φ_μ near $\alpha \simeq \pi/4$. Second, they favor $P_{\mu\nu}$ between curves with the highest K values that converge to the different N [9].

The net result of these selection rules is to provide a path for making $F_\mu(R)$ nonzero in channels converging to higher values N of He^+ by a sequential coupling from one N to the next higher one. Note that *direct* excitation to a channel with high N from, say, the ground state by photoabsorption is unlikely because of the large separation in R of the potential wells $U_\mu(R)$ which results in scarce overlap of the corresponding radial functions $F_\mu(R)$. Excitation proceeds thus by sequential transitions from one strip to the other in the diagram of Fig. 10.1. This argument applies equally to the excitation by electron impact of a ground-state atom because the collision begins in the lowest channel for any given $\{L, S, \Pi\}$. These remarks are documented by illustration of the correlation functions in Section 10.5.6 and the detailed treatment of excitation to high N in Section 10.6.

The comparison of the effect of the avoided crossings discussed in this section and in the previous one helps in visualizing their respective features. It was noted at the end of Section 10.5.3 that the pair of correlation functions Φ_μ and Φ_ν interchange their quantum numbers (n_{ac}, n_{rc}) as R traverses the range of avoided crossing. These quantum numbers would instead be preserved if the wave functions were altered to reflect fully diabatic behavior in which the potential curves would actually cross. On the other hand, in the array of avoided crossings considered here, each adiabatic correlation function preserves its quantum numbers. Transition to a diabatic representation would instead preserve the values of the correlation function on the ridge that is not represented by a quantum number. This phenomenon is illustrated further in Section 10.5.6.

10.5.5 Fragmentation Region

We consider here the behavior of potential curves and their correlation functions in the fragmentation region, that is, the region where an independent particle description becomes relevant. Basically, the pattern of correlations should no longer change in this region. Nevertheless, we notice in Figs. 10.12 and 10.14 the occurrence of very narrowly avoided crossings that, in fact, have been treated diabatically in the graphical presentation and in the assignment of μ and (K, T) labels. This third class of crossings involves pairs of potential curves with $A = \pm 1$, a prominent example being the one in Fig. 10.12 near $R \simeq 8$ a.u. At low and intermediate R, the $A = +1$ curve is more attractive than $A = -1$, which has an extra node in α (Fig. 10.13). At large R, however, when tunneling under the potential ridge has set in, both states have little amplitude in the vicinity of $\alpha = \pi/4$ and the presence or absence of a node in that region is of little consequence to the contribution of kinetic energy to $U_\mu(R)$. The dependence of the energy on $(K, T)^A$ in this region is controlled instead by the value of $\langle -\hat{b}_< \cdot \mathbf{b}_> \rangle$ according to Section 10.5.2. The sign of this interaction depends on $(K, T)^A$ and for the $^1P^o$ states of Fig. 10.12 is positive for the $\mu = 2$ curve with $K = A = +1$. Retention of the labels $(K, T)^A$ then implies a diabatic crossing that inverts the sign of $U_2(R) - U_3(R)$. The classification scheme in Section 10.5.2 has been designed to retain these quantum numbers because adherence to the adiabatic representation would cause an extremely sharp switch of the eigenfunction from plus to minus behavior (the minimum separation between the adiabatic curves would be $\sim 10^{-3}$ a.u. for the example in Fig. 10.12).

Since the wave function in the fragmentation region is confined to the valleys at $\alpha = 0$ and $90°$ (Fig. 10.13), it is natural to describe the channels in this region in the independent electron picture rather than in hyperspherical coordinates. We revert then to the procedures of Chapters 5 and 7. According to Section 5.6, the long-range dipole potential of the excited He^+ residue is conveniently combined with the centrifugal potential of the outer electron, which has the same inverse quadratic dependence on the radial distance $r_>$ of the outer electron, i.e.,

$$[(\hbar^2/2m)\mathbf{l}_>^2 \, \delta_{l_<, l'_<} + e^2(nl_< | r_< \cos\theta_{12} | nl'_<)]/r_>^2 . \tag{10.29}$$

(The Coulomb potential of the He^+ residue, $-e^2/r_>$, is common to all channels at large R.) Eigenvalues of (10.29) within the base of the degenerate states of $He^+(N)$ are easily calculated and indicated by $(\hbar^2/2m)\lambda_d(\lambda_d + 1)/r_>^2$ as in Section 5.6 with an effective angular momentum λ_d. (A subscript d has been attached here to avoid confusion with the eigenvalues of the generalized hyperspherical angular momentum.) The value of λ_d is a function of the hyperspherical quantum numbers $(K, T)^A$. For the example of $^1P^o$ and $N = 2$ only the combination $(-1, 0)^{-1}$ yields a negative eigenvalue for (10.29).

The repulsive effect of the potential (10.29) for the other two curves below the $N = 2$ limit becomes dominant in the $^1P^o$ states of H^- where the attractive Coulomb potential $-e^2/r_>$ is absent. This is illustrated in Fig. 10.16, the analog of Fig. 10.12 for the H^- system. Note how the repulsive potential generates a potential barrier for the $A = +1$ curve which actually prevents the occurrence of bound states in this well. The crossing between $A = \pm 1$ curves in this example is also shifted to higher R, $R \simeq 14$ a.u., by the absence of the Coulomb potential. The attractive dipole potential for $A = -1$ suffices instead to bind an electron, as will be illustrated in Section 10.5.7.

The examples presented thus far deal with $^1P^o$. Triplet channels behave differently because the assignment of the quantum number A depends on the spin quantum number S according to Section 10.5.2. In fact, the correspondence between the eigenvalues of (10.29) and the values of A is reversed for the $^3P^o$ channel. Note, for instance, that no crossing of the lowest $A = \pm 1$ curves occurs in the $^3P^o$ curves of Fig. 10.14.

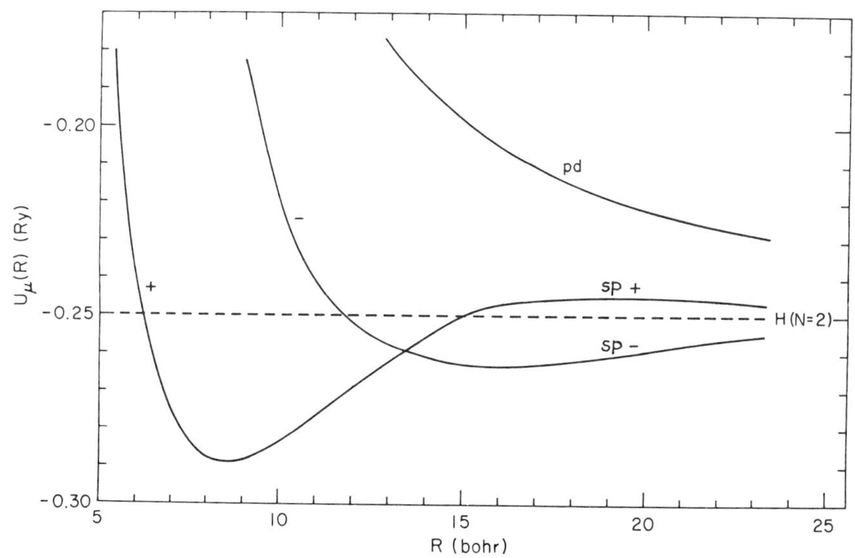

Fig. 10.16. Potential curves $U_\mu(R)$ for $^1P^o$ states of H^- converging to $H(N = 2)$. Compare with similar curves for He $^1P^o$ in Fig. 10.12. (From Lin [10].)

10.5.6 Correlation Patterns

The discussion of the eigenvalues $U_\mu(R)$ of H_R in the previous four sections is complemented by closer examination of the corresponding eigenvectors $\Phi_\mu(R; \alpha, \hat{r}_1, \hat{r}_2)$. Figure 10.13 has provided a preliminary look at the distribution of these functions in α for the $^1P^o$ states. Alternative plots, particularly of $|\Phi_\mu|^2$ at fixed R as functions of the angles α and θ_{12}, provide a more complete picture of the adiabatic eigenvectors (RPP, Section 3.1.2). These plots exhibit at a glance the nature of the radial (α) and angular (θ_{12}) correlation patterns in the adiabatic states.

Figure 10.17 shows $|\Phi_\mu|^2$ for the two ^1S channels below He$^+(N = 2)$ at two values of R. Recall that the two channels here have the quantum numbers $(K, T)^A = (\pm 1, 0)^+$; both of them have + symmetry. Neither channel has a node in α near $\alpha \simeq \pi/4$, the (1, 0) function has no node in θ_{12}, whereas $(-1, 0)$ has one angular node. For either channel, the behavior in θ_{12} remains substantially constant with R. This is true in general for such plots of $|\Phi_\mu|^2$,

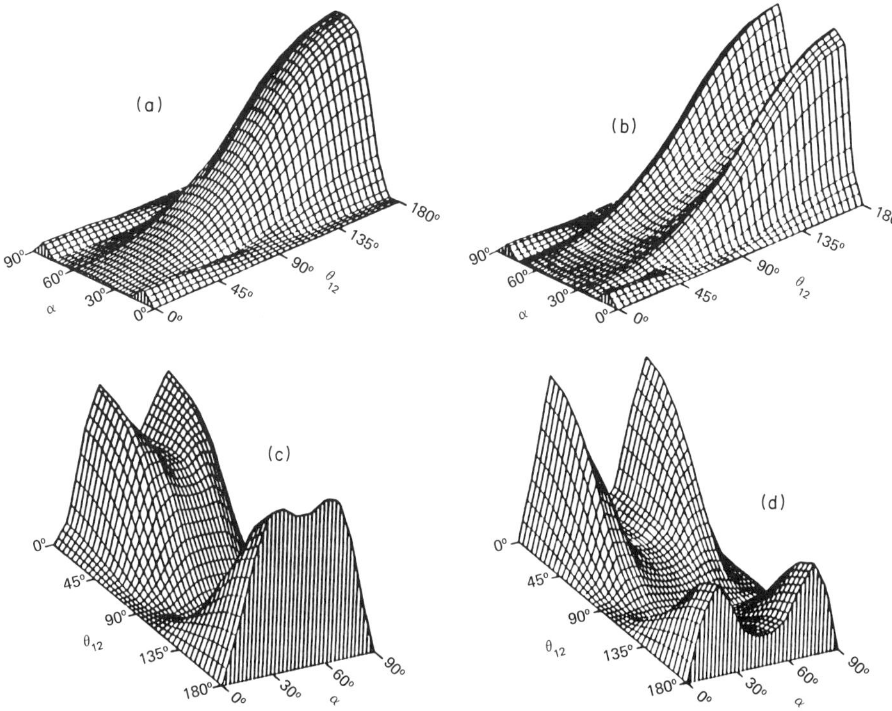

Fig. 10.17. Adiabatic correlation functions $|\Phi_\mu(R; \alpha, \theta_{12})|^2$ for ^1S states below the He$^+(N = 2)$ limit: (a) $K = 1$, $R = 5$ a.u.; (b) $K = 1$, $R = 7.5$ a.u.; (c) $K = -1$, $R = 5$ a.u.; and (d) $K = -1$, $R = 7.5$ a.u. (From Lin [11].)

illustrating the weak dependence of the angular correlation on R [8, 9]. The dependence on α, however, does evolve with R with the major feature already encountered in Fig. 10.13, namely, the onset of tunneling at $R_{c\mu}$ under the potential ridge. At lower values of R, the + states have substantial wave function density at $\alpha = 45°$, but this density drops away rapidly into the valleys as R exceeds $R_{c\mu}$. Of particular significance is the concentration of the wave function density on the saddle ($\alpha = 45°$, $\theta_{12} = 180°$) of the potential, which occurs generally in all + states for $R < R_{c\mu}$ and will prove crucial in Section 10.6.

Similar plots for $^1P^o$ states are shown in Fig. 10.18 for the $A = \pm 1$ and 0 states below the $He^+(N = 2)$ limit. In this case of $L \neq 0$, the dependence on θ_{12} does not decouple from the three Euler angles contained in $(\hat{r}_1, \hat{r}_2)$. Therefore, an angular average of $|\Phi_\mu|^2$ over these Euler angles must be performed to obtain the plots in Fig. 10.18. Other features discussed in the context of ^1S channels remain valid. The dependence on θ_{12} is insensitive to R; the $A = +1$ state has high density on the saddle when R is small but drops off into the valleys at large R. The $A = -1$ and 0 states never have high density at the saddle. These features of the correlations in the adiabatic channels have also been observed for states below higher $He^+(N)$ limits. Such studies have shown the similarity of patterns for the same $(K, T)^A$ but different $\{L, S, \Pi\}$ [8].

The behavior of the correlation patterns $|\Phi_\mu|^2$ for $A = +1$ channels along the array of avoided crossings at intermediate R (Section 10.5.4) is particularly noteworthy. We just noted that the lowest $A = +1$ channel for $N = 2$ has

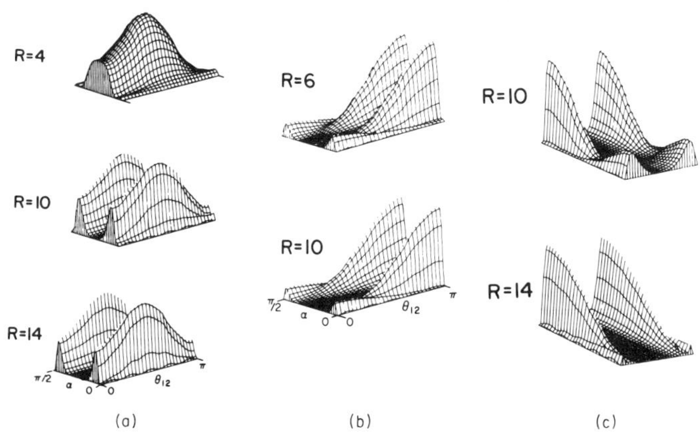

Fig. 10.18. Adiabatic correlation functions $|\Phi_\mu|^2$ for $^1P^o$ states below the $He^+(N = 2)$ limit, corresponding to the potential curves in Fig. 10.12: (a) $A = +1$, (b) $A = -1$, and (c) $A = 0$. (Courtesy of C. D. Lin.)

substantial concentration at the saddle until R exceeds $R_{c,\mu=2}$ beyond which the amplitude falls into the valleys. However, around this value of R, this potential curve has an avoided crossing with the lowest $A = +1$ curve of the next higher N as shown, for instance, in Fig. 10.10. Examination of this channel with higher N shows a substantial concentration of its $|\Phi_\mu|^2$ in the saddle to much higher values of R—in fact, out to the larger critical radius $R_{c,\mu=4}$ for that channel, where it then undergoes an avoided crossing with the $A = +1$ curve of the next N. Figure 10.19 shows plots of $|\Phi_\mu|^2$ for the $A = +1$ curve with largest K for a series of N values, in each case for values of R that bracket the value of $R_{c\mu}$. These plots demonstrate the role of the array of avoided crossings in maintaining wave function density at the saddle. A state that would pass diabatically across this array would have a wave function that maintains substantial amplitude in the saddle. This state is presumably the vehicle for transfer of excitation to higher energy (Section 10.6).

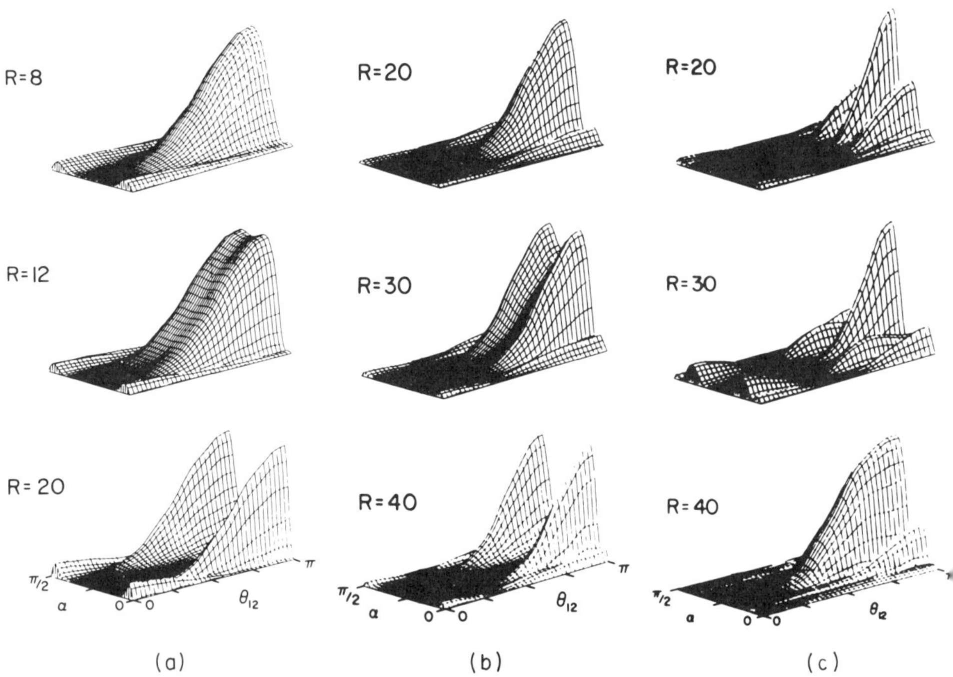

Fig. 10.19. Adiabatic functions $|\Phi_\mu|^2$ for ^1S channels of H^- below various $H(N)$ limits: (a) $N = 2$, (b) $N = 3$, and (c) $N = 4$. Note how the probability amplitude drops off from $(\alpha = \pi/4, \theta_{12} = \pi)$ in each case as R increases but continues to larger R in the next higher N. (Courtesy of C. D. Lin.)

10.5.7 Rydberglike Levels

The adiabatic channels serve as a basis for expanding the full wave function in (10.23). The coefficients of the expansion are the radial wave functions satisfying the coupled linear differential equations (10.25). The diagonal matrix of potentials $U_\mu(R)\delta_{\mu\nu}$ represents a separate potential well in each equation of this linear system and the nonadiabatic matrix elements $P_{\mu\nu}(R)$ act as coupling coefficients.

An initial approximation neglects the effect of coupling entirely, treating each μ equation independently. Justification of this procedure comes from the discussion in previous sections of potential curves U_μ and wave functions Φ_μ in different ranges of R:

(1) In the large R fragmentation region, once the major crossings between $A = \pm 1$ curves are handled diabatically according to Section 10.5.5, the remaining coupling terms between channels are small and can be handled perturbatively. A slow fall-off of these terms with R proves troublesome and is discussed further in the next section.

(2) At intermediate R, the potential curves $U_\mu(R)$ converging to $He^+(N)$ with low N remain fairly far apart and the effects of coupling correspondingly weak.

(3) Avoided crossings in the small R region lie generally at low values of R, where $U_\mu(R)$ exceeds the energies of interest and acts as a centrifugal barrier to penetration of F_μ. Because of the resulting small values of $F_\mu(R)$ in this region, coupling effects remain small.

Together, these arguments suggest the complete neglect of coupling in the system of radial equations (10.25) as a starting point, to be supplemented by perturbative treatment of $P_{\mu\nu}$ in higher approximations. This initial step of calculating eigenvalues in each potential well $U_\mu(R)$ is the analog of the computation of vibrational states of nuclear motion in the Born–Oppenheimer approximation for molecules.

The horizontal lines in Fig. 10.12 represent eigenvalues of $^1P^o$ states calculated in this manner. The energy values of the lowest states and of their analogs with other $\{L, S, \Pi\}$ values agree well with the experimental data in Fig. 10.4 for $\mu = 2$ and 3. No states of the $\mu = 4$ series have yet been observed. The observation in Section 10.5.1 that potential curves of the $-$ states are displaced to higher R relative to the $+$ states is reflected in their radial functions $F(R)$ having smaller amplitudes at low values of R. As a result, these functions hardly overlap those of the ground state of He, which resides in the lowest potential curve in Fig. 10.10a. This accounts in part for the much lower intensities shown in Fig. 10.2 for photoexcitation of the $-$ states. The $A = 0$ states have even weaker excitation. Also, all channels converging to higher N limits, whose $U_\mu(R)$ curves and radial functions $F_\mu(R)$ overlap even less with the ground state, are not accessed appreciably by direct excitation from the ground state. Instead, their excitation involves initial excitation to $N = 2$ followed by passage through the array of avoided crossings to higher N (Section 10.5.4). This is considered in Section 10.6.

Chapter 5 showed that the nature of the bound-state eigenvalue spectrum depends crucially on the long-range potential. In the current context, the dependence $-e^2/R$ of the $U_\mu(R)$ curves of He at large R means that each potential well holds an infinite Rydberg series of bound states. In the large-R region, however, independent-particle coordinates are appropriate so that the potential reduces to the single electron Coulomb potential of $He^+(N)$ seen by the distant electron combined with the dipole potential (Section 10.5.5). The series of Rydberg levels below the $He^+(N)$ limit is given by the Rydberg formula [(5.28)]. All the effects of joint two-electron motion in the hyperspherical potential wells at small and intermediate R are lumped into the quantum defects for the series in the Rydberg formula. Note in Fig. 10.4 that the difference between $A = \pm 1$ states of $^1P^o$ symmetry amounts to a difference of 0.5 in their quantum defects.

In the H^- system, on the other hand, the long-range behavior of $U_\mu(R)$ is quite different from that in He. According to Section 10.5.5, only a long-range dipole potential acts in H^-, which is attractive only for $A = -1$ and repulsive for $A = +1$ and 0. The $-$ curve in Fig. 10.16 does support a bound state of very weak binding observed experimentally (Fig. 1.5b) as a sharp resonance below the $H(N = 2)$ threshold. [Actually, the attractive dipole potential in the $-$ channel is strong enough to hold an infinite number of bound states described by (5.30) but even the lowest of these lies very close to the threshold and is the only one that has been observed. Additional resonances have been detected in the $+$ channels converging to $H(N = 3$, 4 and 6).] The $+$ potential well in Fig. 10.16 proves incapable of holding a bound state. However, integration over the well and the potential barrier in the $A = +1$ $U_\mu(R)$ curve in Fig. 10.16 yields a shape resonance just above the threshold $H(N = 2)$. This resonance is also displayed in Fig. 1.5b, together with its experimental verification (RPP, Section 3.1.3). The rich structure described here for the photodetachment of H^- in the range 10–14 eV contrasts sharply with the simple spectrum in Fig. 7.2 at lower energies.

10.5.8 Channel Coupling

The previous sections have provided an initial description and classification of low-lying doubly excited states. Residual effects of coupling between channels with different potentials $U_\mu(R)$ remain. These effects may be taken into account by the close-coupling procedure of Section 9.3.4 or equivalent methods. The most significant values of the coupling terms $P_{\mu\nu}(R)$ remain generally confined to regions of avoided crossing as shown in Fig. 10.11. Their effect is proportional to the ratio $[P_{\mu\nu}(R)]^2/[U_\mu(R) - U_\nu(R)]$; the magnitude of this ratio as compared with unity indicates the coupling strength even when it is not small enough to allow perturbative treatments. Note how this index peaks in the region of the avoided crossing where $U_\mu(R) - U_\nu(R)$ reaches a minimum.

Extensive calculations of the nonadiabatic effects have been discouraged thus far by the difficulty in obtaining reliable values of the matrix elements $P_{\mu\nu}(R)$. A more basic obstacle stems from the lack of an appropriate treatment of the array of avoided crossings at intermediate R that leads to excitation to higher states according to the description in Section 10.5.4. This problem is discussed further in Section 10.6.

The dominant coupling terms $P_{\mu\nu}(R)$ for adiabatic channels of $^1P^o$ symmetry below $He^+(N = 2)$ are P_{23} and P_{34} in Fig. 10.11. Adoption of a diabatic description for the 2–3 crossing eliminates the peak in P_{23} in Fig. 10.11 as discussed in Section 10.5.6. A similar procedure for all crossings with large peaks in $P_{\mu\nu}$ is one avenue for improving on the adiabatic approximation and has been explored recently [9].

A major problem in these calculations of levels at low N is that the coupling $P_{\mu\nu}(R)$ between adiabatic levels falls off slowly with R at large R. The $1/R^n$ dependence of these terms, with n varying typically from 1 to 3, hinders the convergence and accuracy of the calculations. This difficulty remains to be remedied in the future, probably by avoiding the use of hyperspherical coordinates in the fragmentation region.†

Alternative methods for studying the nonadiabatic couplings in the radial equations (10.25) have seen limited exploration. One method replaces these second-order equations by a double set of first-order differential equations, linear in the coupling matrix $P_{\mu\nu}$ and with a structure similar to that of spin precession equations (RPP, Sections 2.2.1 and 2.2.2). This approach has proved effective for numerical applications. It has also served to resolve the effect of coupling into two steps: the first constitutes a "postadiabatic" approximation that corrects adiabatic channels without causing transitions between them while true nonadiabatic passage between channels occurs in a second step.

A different approach, as yet unexplored, would cast (10.25) in an alternative form through the transformation

$$\mathbf{F}(R) = \exp\left(-\int^R \mathbf{P}(R')\,dR'\right)\tilde{\mathbf{F}}(R). \tag{10.30}$$

The transformed set of functions $\tilde{\mathbf{F}}(R)$ then obey the alternative set of equations

$$\left\{-\left(\frac{\hbar^2}{2m}\right)R^{-5/2}\mathbf{I}\left(\frac{\partial^2}{\partial R^2}\right)R^{5/2} + \left[\exp\left(\int^R \mathbf{P}(R')\,dR'\right)\mathbf{U}(R)\exp\left(-\int^R \mathbf{P}(R')\,dR'\right)\right] - \mathbf{I}E\right\}\tilde{\mathbf{F}}(R) = 0. \tag{10.31}$$

The effect of the $P_{\mu\nu}$ terms is again split into two parts, one that reshuffles $\mathbf{F}(R)$ in (10.30) and a second one in which the term in square brackets in (10.31) receives off-diagonal contributions only from the part of $\mathbf{P}$ that does not commute with $\mathbf{U}$. The

† A recent application along these lines for s-wave e–H scattering phase shifts has, in fact, given excellent results [12].

transformation (10.30) and this particular splitting of the coupling into two parts bear a resemblance to the use of the interaction represèntation of time-dependent perturbation theory. Note, however, that the differential equations now involve derivatives of second order in R. Such a system remains to be studied analytically and numerically.

10.5.9 Summary

The present analysis of two-electron states has provided a view of correlations in successive steps. For states of low N, the behavior of $|\Phi_\mu|^2$ in α and θ_{12} gives essentially a complete picture of the correlations. The adiabatic separation of motion in R from that in (α, θ_{12}) applies generally to low-N channels, and nonadiabatic effects of coupling between different patterns $|\Phi_\mu|^2$ remain small once certain crossings are handled diabatically, as discussed in Section 10.5.8. With increasing excitation, however, the mixing between different channels, particularly those converging to different limits N, become important. Simultaneous treatment of the radial coordinate R and angular coordinates α and θ_{12}, but especially α, is implied by this mixing in N. This study becomes a central element of the structure and excitation of states of high N to be discussed in Section 10.6.

10.6 HIGH EXCITATIONS AND DOUBLE IONIZATION

This section deals with states of a pair of electrons both of which reach simultaneously far from the nucleus, to distances of the order of 10^2–10^3 a.u. and larger. Both electrons may eventually escape if their combined energy exceeds the limit of 79.0 eV above the He ground state indicated in Fig. 10.1 or the corresponding limit of 14.35 eV for H^-. The path of such high excitations runs through the array of avoided crossings along the potential ridge considered in Section 10.5.4, but this array becomes increasingly dense at larger distances owing to the diverging number of adiabatic channels that become accessible in the range of 75–79 eV as indicated by dots in Fig. 10.1.

A wave function that describes the main aspects of propagation of the excited electron pair, all the way from the nucleus to such large distances above the ridge, should differ radically from the format (10.23), which would consist of tens or hundreds of terms. No adequate alternative format has yet been developed but much fragmentary evidence, experimental and theoretical, provides a qualitative but consistent picture of the high double excitations and their electron correlations.

An essential character of these correlations was pointed out long ago by Wannier (RPP, Section 3.2.2) in the context of the ionization of neutral atoms (or positive ions) by collision of an electron with energy barely sufficient to remove a second electron to infinite distance from the residue. (The role of this process has been indicated in Section 1.3). Wannier pointed out first that the motion and correlations of the electron pair at large distance hardly depend on the structure of the ionic residue; accordingly, the prototype collision $e + \mathrm{He}^+$, or $e + \mathrm{H}$, provides general guidance. In fact, since the threshold behavior of any process is insensitive to the specific pathways that connect the initial to the final state (Sections 4.8 and 5.6), the processes $h\nu + \mathrm{He}$ or $h\nu + \mathrm{H}^-$ with photon energies barely sufficient to remove two electrons lead to the same behavior as the corresponding collisions with electrons mentioned earlier.

Wannier identified the path to detachment of both electrons as proceeding very close to equidistance of both electrons from the nucleus, in the region $r_1 \sim r_2$ that we call the ridge of the potential (10.12) and in opposite directions with respect to the nucleus. This configuration, indicated by ($\alpha = 45°$, $\theta_{12} = 180°$), is the saddle point of the potential factor $C(\alpha, \theta_{12})$ shown in Fig. 10.8. The restriction to $r_1 \sim r_2$ is particularly crucial. The instability in the mutual screening of the electrons, identified in Section 10.4.1, would amplify any departures from this configuration so that one electron would then end up trapped in the field of the positive residue even above the threshold for double escape. Recall from Fig. 10.1 that in this energy range the double ionization channel is degenerate with an infinity of single-electron continua.

It was then sufficient for Wannier and his followers to consider the truncated expansion of the potential (10.12) near the saddle,

$$V = \frac{e^2}{R} C(\alpha, \theta_{12}) = \frac{e^2}{R} [-Z_0 - Z_\alpha(\tfrac{1}{4}\pi - \alpha)^2 + Z_\theta(\pi - \theta_{12})^2], \quad (10.32)$$

with

$$Z_0 = 2\sqrt{2}(Z - \tfrac{1}{4}), \quad (10.33a)$$

$$Z_\alpha = 3\sqrt{2}(Z - \tfrac{1}{12}), \quad (10.33b)$$

$$Z_\theta = \sqrt{2}/16. \quad (10.33c)$$

The restricted study of the electrons' motion in this potential—at large R and near the saddle, and hence disconnected from other aspects of the excitation—is described in Section 10.6.2. A framework of key experimental observations and their semiempirical interpretations will be presented first, in order to bridge over from the treatment of Section 10.5 to Wannier's approach.

10.6.1 Evidence from Key Experiments

A first question to be considered is whether higher double excitations are reached with comparable or decreasing probability. A relevant experiment has dealt with electron–helium collisions that form a complex with three electrons rather than two. However, one electron remains bound tightly in the ground state of the He^+ ion and hardly affects the motion of the highly excited pair. (The effect of the unpaired spin of the inner electron on the outer two will be discussed in Section 11.4). To study double excitations the experiment detects very slow electrons emerging from the collision with a small residual energy $\Delta \sim 0.02$ eV. This detection implies that one electron of the pair has retained only sufficient energy to leave the complex. All the remaining energy, fractionally lower than the incident energy E, has then served to excite the He atom.

Figure 10.20 shows the count rate of slow electrons as a function of the incident energy E. We see at a glance that the probability of excitation remains fairly uniform over the whole range of the experiment. The peaks on the left of the graph correspond to excitation of discrete levels of He, broadened by imperfect monochromaticity of the incident electrons; the intensity averaged over these peaks is roughly constant. The smooth portion on the right of the plot corresponds to ionizations of He in which the unobserved electron has retained most of the available energy. (The sharing of energy between two ejected electrons is discussed in Section 10.6.2).

The notch in the middle of the spectrum constitutes an important departure from its general uniformity. Note how precisely the deepest point of this notch corresponds to the ionization threshold of He at 24.58 eV.

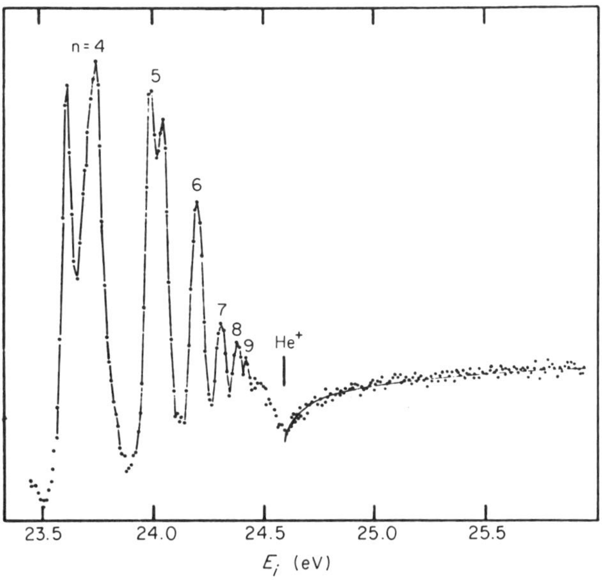

Fig. 10.20. Yield of very low-energy electrons resulting from e–He collisions. Solid line through the data points above the ionization limit at 24.6 eV has the energy dependence $(E_i - 24.6)^{0.127}$. (From Cvejanovic and Read [13].)

(Indeed the primary goal of the experiment, developed in the laboratory of F. H. Read in Manchester, England, was to investigate threshold phenomena.) Since the experiment detects electrons that have retained a bare minimum of energy required to escape, the lower rate of detection near threshold implies that a delicate balance in the energy sharing between electrons is less likely to be maintained at large radial distances (clearly so beyond the radius of the He level $n = 10$). Note also that the depression of intensity near threshold is smoothed out by the limited resolving power of the experiment, much as the discrete lines are broadened. (We shall see later that the notch in the spectrum would probably extend down to zero intensity in a spectrum with perfect resolving power. The notch will also be seen to extend far from the threshold, in principle, but with hardly detectable slope.)

The counting rate shown on the ordinate scale of Fig. 10.20, for a given energy E lower than the ionization threshold, measures the probability of electron ejection from the channel (or channels) μ whose adiabatic potential energy $U_\mu(R \to \infty)$ lies within ~ 0.02 eV below the energy E. This probability is, in turn, proportional to the squared amplitude $|F_\mu(R \to \infty)|^2$ in the wave function analogous to (10.23) for the He^- complex formed in the collision $e + He(1s^2)$ with energy E. Since channel coupling is weak in the fragmentation region at large R, according to Section 10.5.5, the value of $|F_\mu(R \to \infty)|^2$ should be approximately the same as $|F_\mu(R_{c\mu})|^2$, its value at the critical radius at which the correlation function Φ_μ ceases to propagate along the ridge to become confined in the potential valleys.

Even though we are dealing here with a negative ion, $U_\mu(R)$ is mainly dominated by the value of the Coulomb field (10.26) in the region of the interface between intermediate and fragmentation zones. The effects of the dipole field that are large in H^- and lead to the barrier in Fig. 10.16 are less important in He^-, the asymptotic $U_\mu(R \to \infty)$ being no longer degenerate now. Recalling that $U_\mu(R_{c\mu}) = 24.58\text{ eV} - 3e^2/\sqrt{2}R_{c\mu}$ from (10.26) with $Z = 1$ (whose scale is shifted by 24.58 eV with respect to the present one) and noting from Fig. 10.10a that $U_\mu(R_{c\mu})$ is but a little lower than $U_\mu(R \to \infty) \sim E$, the counting rate at energy E reflects the probability of propagation along the ridge up to a radial distance approaching

$$R_{c\mu} \simeq \frac{2e^2(\sqrt{2}/3)}{24.58\text{ eV} - E}. \tag{10.34}$$

For a value of E corresponding to the $n = 10$ peak of Fig. 10.20, the observed counting rate indicates that the wave function of the excited electron pair in He^- remains large on the potential saddle up to a distance $R_{c\mu} \sim 100$ a.u.

This concentration of the wave function on the saddle differs, of course, from its distribution in doubly excited Rydberg states of neutral He, which extends toward large R in the potential valleys with one electron more excited than the other. In the negative ion H^-, the autoionizing ${}^1P^o$ state just below $H(N = 2)$, with the $A = -1$ classification, also extends far into the valleys with a Rydberglike wave function; on the other hand, the shape resonance just above the threshold, with $A = +1$, has its wave function concentrated at the saddle.

Direct evidence of concentration at the saddle for He^- emerges from the observation of series of resonances of the negative collision complex e + He. This observation was achieved by an experiment in which collisions occurred at the crossing of electron and helium beams and their effect was observed by detecting the presence of metastable excited He atoms at some distance from the collision. Sensitive detection in this arrangement, which was also developed in Manchester, afforded a very fine energy tuning of the incident electrons and thereby the observation of long sequences of resonances, even though the resonances become increasingly narrow with increasing excitation of the He^- complex, as shown in Fig. 10.21.

The very ability to detect such high resonances implies their belonging to the most readily excited class labeled by $A = +1$. Semiempirical analysis of the resonance energies demonstrates their connection with the potential (10.32) that prevails on the ridge.

Earlier measurements of the energy levels of single resonances in negative ion complexes had shown each of them to lie generally below an energy level of the corresponding neutral atom, by a fraction of 1 eV which represents the electron affinity of the neutral's level. Series of resonances of the negative complexes were thus loosely associated with Rydberg series of levels of the neutral. Accurate measurements of extended series of resonances, such as the three series shown in Fig. 10.21, afforded a refined fitting to a generalized Rydberg formula,

$$E_{\text{res}} = I - [2(Z - \sigma)^2/(n - \mu)^2]13.6 \qquad \text{eV}, \tag{10.35}$$

where I indicates the ionization threshold of the neutral, 24.58 eV for He. The fitting showed the quantum defect μ to be nearly constant along each series, as it is for the Rydberg spectra of neutrals. More important, the parameter σ, which represents the effective screening of the nuclear charge, lies unmistakably close to the value $\frac{1}{4}$, which corresponds to the potential at the saddle according to (10.26) or (10.32) [15, 16]. In fact, the factor of two on the right of (10.35) expresses the "double Rydberg" nature of the resonances that follows on expressing the potential in the saddle (10.32) in terms of independent electron coordinates:

$$-(e^2/R)2\sqrt{2}(Z - \tfrac{1}{4}) \xrightarrow{\mathbf{r}_1 = -\mathbf{r}_2} -e^2(Z - \tfrac{1}{4})[(1/r_1) + (1/r_2)]. \tag{10.35$'$}$$

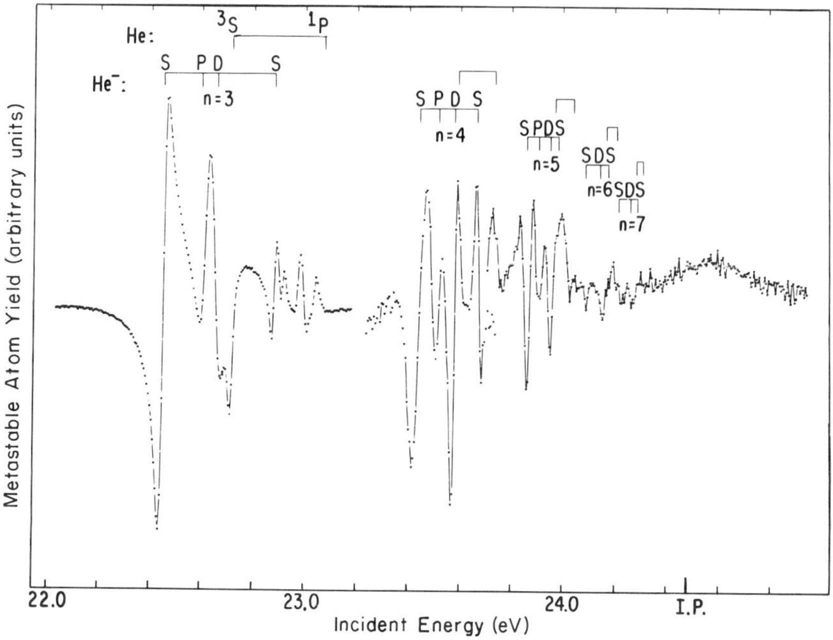

Fig. 10.21. Yield of He metastable excitations showing resonances due to high double excitations of He^-. (From Buckman *et al.* [14].)

Equivalently, (10.35) can be viewed, with a reinterpretation of the parameters σ and n, as a single Rydberg formula appropriate to bound states in the six-dimensional potential (10.32). The pair of electrons is viewed thereby as a single entity with n a principal quantum number of the pair [16]. In this manner, a further step is taken beyond the treatment of Section 10.5.2 in which quantum numbers $(K, T)^A$ were introduced as labels appropriate to the pair. In the present description of doubly excited states at even higher excitation the Rydberg formula, the quantum number n, and the ionization limit are all characteristic of the pair.

A principal characteristic of the wave functions of highly excited electron pairs is thus their apparent propagation along the ridge without major reduction of intensity up to the critical radial distances $R_c = Z_0 e^2/(I - E)$, where I is the ionization threshold, provided R_c does not exceed the order of 100 a.u. Propagation to larger distances appears to involve a loss of intensity according to Fig. 10.21. This loss of intensity at energies close to the ionization threshold, in fact, had been predicted by Wannier, as will be seen in the next section.

10.6.2 The Wannier Theory of Double Escape

The Wannier theory deals with the motion of the electron pair at asymptotically large distances from an ionic core and near the potential saddle, that is, subject only to the potential represented by (10.32). Within this limited scope, the theory is equally applicable to escape from atomic or molecular complexes with an ionic core of charge *Ze*, whether excited by photoabsorption or collision with an electron or other particle.

The theory may be developed semiclassically, since it deals with a smooth potential and very high quantum numbers, or even classically; it takes advantage of the radial scaling emphasized in Section 10.4 and the quadratic dependence on α and θ_{12} in (10.32). Indeed, Wannier represented trajectories of the electron pair in the form $\{\alpha(R), \theta_{12}(R)\}$. Because of the scaling, the functions $\alpha(R)$ and $\theta_{12}(R)$ obey differential equations with constant coefficients and are thus represented in terms of a pair of basic solutions

$$|\tfrac{1}{4}\pi - \alpha_{\pm}(R)| \propto R^{\zeta_{1\pm}}, \qquad |\pi - \theta_{12}(R)| \propto R^{\zeta_{2\pm}}, \tag{10.36}$$

with exponents $\zeta_{i\pm}$ determined as the alternative roots of quadratic algebraic equations

$$\begin{aligned} \zeta_{1\pm} &= -\tfrac{1}{4} \pm \tfrac{1}{4}[(100Z - 9)/(4Z - 1)]^{1/2}, \\ \zeta_{2\pm} &= -\tfrac{1}{4} \pm \tfrac{1}{4}i[(9 - 4Z)/(4Z - 1)]^{1/2}. \end{aligned} \tag{10.37}$$

WKB solutions of the corresponding Schrödinger equation have wave fronts orthogonal to the trajectories (10.36) [17].

The motion of the electron pair along the coordinate θ_{12} is confined by the rising parabolic potential in (10.32). Indeed, the exponent $\zeta_{2\pm}$ is complex (for $Z < \frac{9}{4}$), indicating an oscillatory motion, and the corresponding factor of the wave function is an oscillator function with scale parameter proportional to $R^{-1/4}$. Accordingly, the confinement of θ_{12} near π sharpens with increasing R, just as we would expect, owing to the repulsion between two slow electrons. The motion of the pair along the coordinate α may instead appear wholly unrestricted, since the potential decreases as $(\frac{1}{4}\pi - \alpha)^2$ increases. The trajectory $\alpha(R)$ with the exponent ζ_{1+} does diverge from the ridge, but a nontrivial correlation between α and R causes the exponent ζ_{1-} to be negative, whereby the corresponding trajectory converges to the ridge as $R \to \infty$.

This correlation of R and α, manifested by the converging trajectories and in the corresponding WKB wave functions, proves crucial for the threshold behavior of the cross section, whereas the focusing of θ_{12} about π proves rather incidental. The critical role of the correlation of α and R for the motion on the ridge reflects the drastic change in the α dependence of the adiabatic functions $|\Phi_{\mu}|^2$, displayed in Section 10.5.6, as R traverses the critical range

$R \sim R_{c\mu}$. By contrast, the θ_{12} dependence of $|\Phi_\mu|^2$ hardly changes in this same range.

Wannier's trajectories, and their quantum analogs, pertain to the "Coulomb zone" of radial distances (Section 5.5), where the potential energy far exceeds the eventual kinetic energy, $E - I$, of the escaping electron pair. This zone terminates at the "Wannier radius" R_W determined by the formula, analogous to (10.34),

$$R_W = Z_0 e^2/(E - I). \tag{10.38}$$

Wannier argued that the trajectories no longer diverge from (or converge to) the ridge once the kinetic energy of the pair predominates over the potential at $R > R_W$. This circumstance ensures the escape of both electrons unless the pair has dropped so far off the ridge at R_W that one electron remains irretrievably bound to the ionic core. Wannier's prediction of a threshold law for escape through diverging trajectories rests on this argument.

Consider at this point that the limiting value of tan α attained by a trajectory at $R \gg R_W$ represents a constant ratio of radial distances r_2/r_1, and hence the ratio of escape velocities v_2/v_1 and the corresponding partition of the escape energy $E - I$ between the electrons. The same holds in a wave picture where R becomes constant over the wave fronts and the direction of $\mathbf{R}$ maps the ratio of electron momenta, p_2/p_1. A constant amplitude of the wave function for small depatures of α from the ridge, that is, a uniform population of trajectories, implies then a uniform distribution of the partitions of kinetic energy between the escaping electrons. Uniformity of this distribution was verified over a range $E - I \lesssim 2$ eV by an experiment subsidiary to the measurements displayed in Fig. 10.20 and in several subsequent experiments.†

Escape of two electrons from an excited complex is thus achieved near threshold with uniform probability along the trajectories that traverse all points of a small window of values of α,

$$-\delta \leq \alpha - \tfrac{1}{4}\pi \leq \delta, \tag{10.39}$$

at the outer edge of the Coulomb zone $R = R_W$. Accordingly, the cross section for joint escape is proportional to the half-width δ of this window. However, fewer and fewer diverging trajectories traverse this window as its distance R_W increases with decreasing energy $E - I$. The dependence of the cross section on $E - I$ is then made explicit by expressing the probability of traversing the window δ at the energy-dependent radius R_W in terms of the

† In the opposite limit of very large collision energies, the treatment of Chapter 3 applies, predicting that most of the energy is generally retained by the incident electron. Experiments and theoretical results on the Wannier picture of double escape have been reviewed [18, 19].

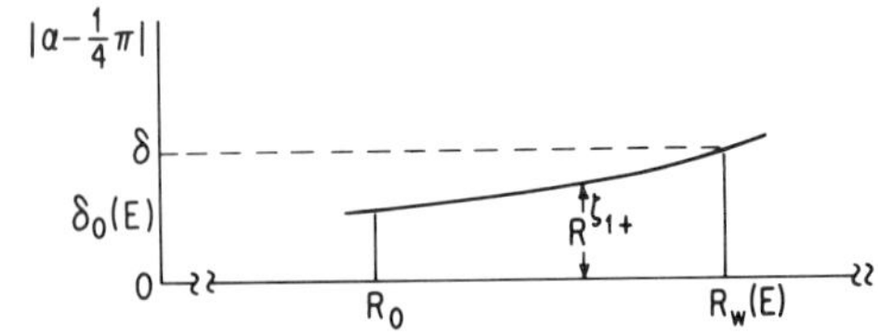

Fig. 10.22. Diagram showing dependence of double escape window on R_W and hence on excess energy above threshold.

probability of traversing an equivalent window $\delta_0(E)$ at a fixed distance R_0. Figure 10.22 shows how $\delta_0(E)$ is related to δ by (10.36) and the energy dependence of R_W in (10.38). We thus obtain the cross section dependence

$$\sigma(E) \propto \delta_0(E) = \delta\left[\frac{R_0}{R_W(E)}\right]^{\zeta_{1+}} \propto (E - I)^{\zeta_{1+}}. \tag{10.40}$$

The numerical value of ζ_{1+}, 1.127 for $Z = 1$, approaches unity as $Z \to \infty$.

In the limit $\zeta_{1+} \to 1$ each trajectory lies straight along a radius vector $\mathbf{R}$ and the cross section is simply proportional to $E - I$, that is, to the number of alternative partitions of the escape energy between the two electrons, a statistical factor equivalent to the accessible volume of phase space.† For finite values of Z, the excess of ζ_{1+} over unity stems from a nontrivial dynamical factor, namely, the instability of motion along the ridge, which depresses the cross section as E approaches the threshold I. This depression is apparent, but not striking, in the plot of the experimental cross section for photo-double-detachment of H^- (Fig. 10.23).

A less direct, but much more critical, comparison of (10.40) with experiment is afforded by the plot in Fig. 10.20 at $E \geq I$, whose ordinates represent a fraction $\Delta/(E - I)$ of the cross section and are thus proportional to

$$\frac{\Delta}{E - I}\sigma(E) \propto (E - I)^{\zeta_{1+} - 1} = (E - I)^{0.127} \tag{10.40'}$$

for ionization of a neutral atom. The data in Fig. 10.20 yield a best fit with the exponent 0.131 ± 0.019, quite consistent with Wannier's 0.127.

† The same result follows from the considerations in Chapter 5 on threshold laws. Recall from Section 5.5 that, when one electron escapes from an ion, the cross section is independent of energy because the QDT normalization parameter B in (5.22′) is independent of energy near threshold. When two electrons escape, similarly no energy dependence arises from $B(k_1)B(k_2)$, where k_1 and k_2 are the wave numbers of the electrons: $E - I = \epsilon_1 + \epsilon_2$, $\epsilon_i = \hbar^2 k_i^2/2m$. The only energy dependence arises from the integration over all possible energy partitions, $\int_0^{(E-I)/2} d\epsilon_1$ and is thus proportional to $E - I$. Likewise, for the escape of n electrons, we would expect $\sigma(E) \sim (E - I)^{n-1}$.

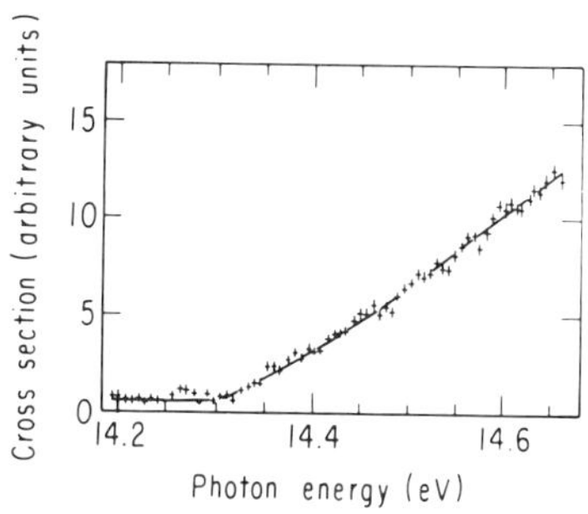

Fig. 10.23. Cross section for photo double detachment of H^-. Threshold energy is 14.35 eV. (From Donahue *et al.* [20].)

This theory predicts, of course, a zero value of the plot in Fig. 10.20 at the abscissa $E = I$. The ordinate should thus approach zero with increasing energy resolution in the experiment, but this is not quite certain in view of the schematizations underlying the theory. The experiment also shows the depression of intensity to extend symmetrically below the threshold as well as above it, as noted in Section 10.6.1. Note that the slope of the function $|E - I|^{0.127}$ decreases rapidly as $|E - I|$ increases, without vanishing altogether.

The experiments that led to Fig. 10.20 not only established the energy dependence in (10.40′) and the uniformity in the partitioning of $E - I$ between the two electrons but a coincidence measurement of the outgoing electrons also verified the dependence of θ_{12} given by (10.36). In particular, the average value of $|\pi - \theta_{12}|$ was seen to be proportional to $(E - I)^{1/4}$, the result expected from (10.36)–(10.38). This energy dependence of the peaking in θ_{12} translates into a superposition of a large number of orbital angular momentum pairs (l_1, l_2) of the electrons up to values $l_i \propto (E - I)^{-1/4}$. Consideration of the angular distribution of the electrons requires, however, the recognition that the final state of double escape that results from electron impact ionization involves a superposition of $\{L, S, \Pi\}$ values of the pair.

The trajectories (10.36) and their corresponding wave functions are symmetric about the ridge, that is, under the operation $\alpha \to \frac{1}{2}\pi - \alpha$. This symmetry is compatible with most values of $\{L, S, \Pi\}$ except for the states $^1P^e$ and $^3S^e$, which necessarily have a node at the ridge (Section 10.5.5). The threshold exponent ζ_{1+} in (10.40) is much larger for these states (RPP, Section 3.2.2). For all other $\{L, S, \Pi\}$ the result in (10.40) applies, but states with odd values of $L + S, \Pi(-1)^S$, or $\Pi(-)^L$ have angular nodes in the saddle and thereby exclude the configuration $\theta_{12} = \pi$. The angular aspects of two-electron escape depend, therefore, on the values of $\{L, S, \Pi\}$ [18].

In the case of photo-double-detachment of H^- leading to a $^1P^o$ final state, the angular distribution of one electron with respect to the linear polarization of the photon is expected to be $\sin^2 \theta$ (asymmetry parameter $\beta = -1$), analogous to the parity unfavored transitions discussed in Section 7.10. The same result is also expected on the other side of the threshold, that is, for single detachment of H^- leaving behind H in an excited state (RPP, Section 3.4.1) [18].

10.6.3 Current Problems

The treatment of two-electron escape outlined earlier will remain isolated until we can show how a wave function propagating astride the potential ridge emerges from a superposition of the wave functions of Section 10.5.

The wave functions considered in this section, for energies near the threshold for double escape, differ from those of (10.23) in that they solve the wave equation only in a limited region of configuration space where the potential is represented by the truncated expansion (10.32) of $C(\alpha, \theta_{12})$ near its saddle point. The contrast of these wave functions illustrates a fundamental complementarity between configuration space and the Hilbert space of states. States adapted to represent features characteristic of a limited region of space are introduced when very large numbers of states of a complete base set are required to represent those same features, whether the base set be adiabatic, as in Section 10.5, or stem from the more conventional independent particle model.

A more varied set of problems stands in the way of describing analytically the phenomena displayed in Figs. 10.20 and 10.21. The dip of the curve in Fig. 10.20 just below the threshold energy $I = 24.58$ eV matches the rise above this threshold, in the sense that both phenomena reflect the low probability of an electron escaping with near-zero energy when the incident energy E is close to threshold. This lower probability is presumably compensated by an increased probability of escape with higher energy leaving the residual atom with excitation appreciably lower than the threshold energy. The small rise in the detection of metastables shown in Fig. 10.21 at $E \sim I$ may reflect this increased probability.

Theories of the yield of He excitations near the threshold and the formation of high-lying resonances in the He^- complex involve a qualitative element foreign to the Wannier theory, namely, the onset of tunneling in the radial propagation at large R on and near the ridge. This tunneling sets in on a "turning surface" $R = R_c(\alpha, \theta_{12})$, where the negative potential (10.32) equals the energy defect to reach double detachment, $E - I < 0$.

Reflection of the wave function at the turning surface $R = R_c$ generates a standing wave. Resonances arise presumably when the phase of this wave matches that of the wave propagating up the ridge from the nucleus. Construction of the standing wave is a preliminary to verification of this surmise. A few qualitative aspects of this phenomenon are listed here:

(1) Conditions for reflection at the turning surface are met by wave fronts orthogonal to Wannier's *converging* trajectories.

(2) Diverging trajectories are instead diverted sharply away at the approach to a turning surface.

(3) This diversion appears to transfer the amplitude of a diverging wave to the converging mode of propagation.

(4) Wave functions propagating along converging trajectories are confined in transverse directions, that is, their amplitude decays away from the ridge as either $|\pi - \theta_{12}|$ or $|\frac{1}{4}\pi - \alpha|$ increases. The wave front may, however, extend with substantial amplitude over a range of α of the order of $\pi/10$ on either side of the ridge.†

10.7 EXTENSION TO MORE THAN TWO ELECTRONS

The previous sections of this chapter have dealt with hyperspherical calculations of two-electron states in He and H^-. Two-electron excitations in other atoms will be studied in Chapter 11. Here we survey initial steps of the extension to multiply excited states. The definition of hyperspherical coordinates for three or more electrons is straightforward. We consider first the case of three electrons and in Section 10.7.5 the extension to a larger number.

In dealing with two electrons the six coordinates $\mathbf{r}_1$ and $\mathbf{r}_2$ were grouped into the directions $\hat{r}_1$ and $\hat{r}_2$ and the two radial distances (r_1, r_2); the latter pair was then represented in circular coordinates (R, α). A similar procedure defines a nine-dimensional hypersphere for three electrons [22]. The nine coordinates $(\mathbf{r}_1, \mathbf{r}_2, \mathbf{r}_3)$ are first rearranged as $(\hat{r}_1, \hat{r}_2, \hat{r}_3)$ and (r_1, r_2, r_3), and the three-dimensional space of the latter set is then expressed in spherical polar coordinates through:

$$R = (r_1^2 + r_2^2 + r_3^2)^{1/2}, \tag{10.41}$$

$$r_1 = R \sin \alpha_2 \cos \alpha_1, \tag{10.42a}$$

$$r_2 = R \sin \alpha_2 \sin \alpha_1, \tag{10.42b}$$

$$r_3 = R \cos \alpha_2. \tag{10.42c}$$

Two hyperspherical angles, with range $0 \le \alpha_1, \alpha_2 \le \frac{1}{2}\pi$, now appear as appropriate to describe radial correlations in the system.

10.7.1 Potential Energy

As in the earlier sections of this chapter, we deal here with a bare nucleus, rather than with a structured ion core. The three electron systems we consider are, therefore,

† Results up to 1982 have been reviewed by U. Fano [21].

H^{--}, He$^-$, and Li and their isoelectronic analogs. The potential energy

$$V(\mathbf{r}_1, \mathbf{r}_2, \mathbf{r}_3) = -Ze^2\left(\frac{1}{r_1} + \frac{1}{r_2} + \frac{1}{r_3}\right) + e^2\left(\frac{1}{r_{12}} + \frac{1}{r_{23}} + \frac{1}{r_{31}}\right) \tag{10.43}$$

is recast as

$$V(R, \alpha_1, \alpha_2, \hat{r}_1, \hat{r}_2, \hat{r}_3) = \frac{e^2}{R} C(\alpha_1, \alpha_2, \theta_{12}, \theta_{23}, \theta_{31}), \tag{10.43'}$$

the analog of (10.12′). Note that C depends on two hyperspherical angles and three angles in physical space. As before, the nine coordinates in the problem have been grouped into one (R) that defines the size of the system, five that define $C(\alpha_1, \alpha_2, \theta_{12}, \theta_{23}, \theta_{31})$ and thereby the shape of the system, and three remaining angles included in $(\hat{r}_1, \hat{r}_2, \hat{r}_3)$ that play no role in the potential but give the orientation of the system with respect to a laboratory frame.

A plot of the function C in (10.43′), analogous to Fig. 10.8, will show deep valleys at $\alpha_i = 0, \frac{1}{2}\pi$, peaks at positions corresponding to the coincidence of two $\mathbf{r}_i$ (e.g., $\alpha_1 = \frac{1}{4}\pi, \theta_{12} = 0$) and saddle point regions (Fig. 10.24). In particular, the electron configuration analogous to the Wannier saddle point in the two-electron potential surface has the electrons at the vertices of an equilateral triangle centered at the nucleus (Fig. 10.25). This configuration is described by

$$\alpha_1 = \tfrac{1}{4}\pi, \qquad \alpha_2 = \cos^{-1}(1/\sqrt{3}), \qquad \theta_{12} = \theta_{23} = \theta_{31} = 2\pi/3; \tag{10.44a}$$

the potential at this point takes the value

$$V = Z_0 e^2/R, \qquad Z_0 = 3(\sqrt{3}Z - 1). \tag{10.44b}$$

This symmetric configuration plays a crucial role for states in the vicinity of the threshold for triple escape, as we will see in Section 10.7.4.

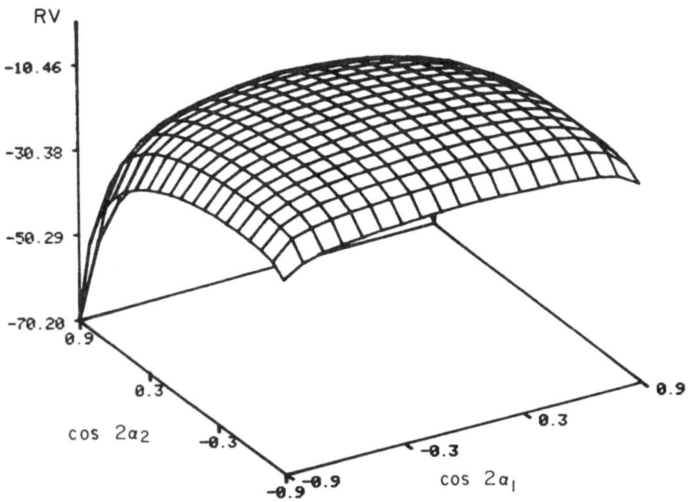

Fig. 10.24. Three-electron potential function in (10.43′) plotted against α_1 and α_2 after averaging over the angles θ_{ij}. (From Clark and Greene [22].)

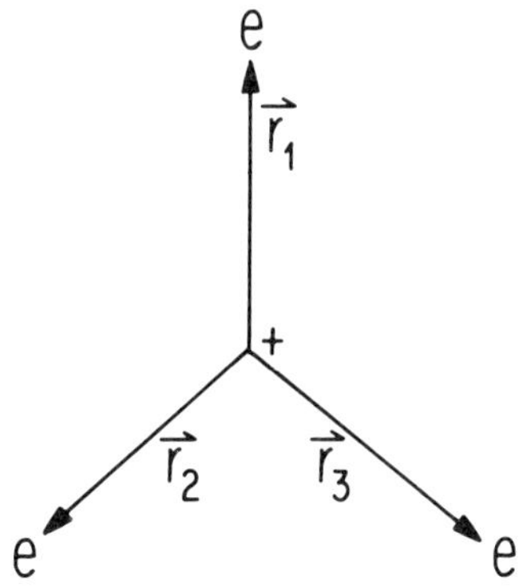

Fig. 10.25. Symmetric configuration for the three-electron system corresponding to a saddle point of the potential surface (10.43′).

10.7.2 Kinetic Energy

The three-electron kinetic energy operator can be written as

$$-(\hbar^2/2m)(\nabla_1^2 + \nabla_2^2 + \nabla_3^2) = (\hbar^2/2m)[-R^{-4}(\partial^2/\partial R^2)R^4 + (\Lambda^2 + 12)/R^2], \qquad (10.45)$$

which is the counterpart of (10.13). The grand angular momentum Λ^2 now has the form

$$\Lambda^2 = \frac{\mathbf{l}_1^2}{\sin^2\alpha_2\cos^2\alpha_1} + \frac{\mathbf{l}_2^2}{\sin^2\alpha_2\sin^2\alpha_1} + \frac{\mathbf{l}_3^2}{\cos^2\alpha_2}$$
$$+ (\sin^2\alpha_2\cos\alpha_2\sin\alpha_1\cos\alpha_1)^{-1}\mathbf{l}^2(\alpha_2,\alpha_1)(\sin^2\alpha_2\cos\alpha_2\sin\alpha_1\cos\alpha_1) - 12, \qquad (10.46)$$

where $\mathbf{l}_i^2$ are the individual angular momentum operators of the ith electron, and

$$\mathbf{l}^2(\alpha_2,\alpha_1) = -(\sin\alpha_2)^{-2}[\{\sin\alpha_2(\partial/\partial\alpha_2)\}^2 + (\partial/\partial\alpha_1)^2] \qquad (10.46')$$

is the angular momentum in the mock angles (α_2, α_1).

As in Section 10.4.2 the eigenvalues and eigenfunctions of the Λ^2 operator have analytical expressions

$$\Lambda^2\phi_{\lambda\nu l_1l_2l_3lLM} = \lambda(\lambda+7)\phi_{\lambda\nu l_1l_2l_3lLM} \qquad (10.47)$$

with

$$\lambda = l_1 + l_2 + l_3 + 2n, \qquad n = 0, 1, 2, \ldots$$
$$\nu - l_1 - l_2 = 0, 2, 4, \ldots \qquad (10.47')$$

and

$$\phi_{\lambda\nu l_1l_2l_3lLM}(\alpha_2,\alpha_1,\hat{r}_1,\hat{r}_2,\hat{r}_3) = N\{[Y_{l_1}(\hat{r}_1)Y_{l_2}(\hat{r}_2)]^l Y_{l_3}(r_3)\}_M^L \sin^{l_2}\alpha_1\cos^{l_1}\alpha_1$$
$$\times \sin^{\nu}\alpha_2\cos^{l_3}\alpha_2\ {}_2F_1(\tfrac{1}{2}(l_1+l_2-\nu),$$
$$\times \tfrac{1}{2}(l_1+l_2+\nu)+2, l_2+\tfrac{3}{2}; \sin^2\alpha_1)$$
$$\times\ {}_2F_1(\tfrac{1}{2}(l_3+\nu-\lambda), \tfrac{1}{2}(l_3+\nu+\lambda)+\tfrac{5}{2}, \nu+3; \sin^2\alpha_2). \qquad (10.47'')$$

The set of quantum numbers consists of the orbitals l_1, l_2, and l_3 of the three electrons, an intermediate value $\bar{l}$ in their coupling ($\mathbf{l}_1 + \mathbf{l}_2 = \mathbf{l}$), and two hyperspherical angular momenta λ and ν. In (10.47″), N is a normalization constant and the ${}_2F_1$ are hypergeometric functions with the first index equal to a nonpositive integer, that is, they are Jacobi polynomials [5].

The enforcement of the Pauli principle, that is, the construction of wave functions with the proper antisymmetrization under $\mathbf{r}_i \leftrightarrow \mathbf{r}_j$ and interchange of spin coordinates, starting from (10.47″), is more complicated than that in (10.27) for two electrons but proceeds according to a standard scheme. The procedure has been developed for N fermion hyperspherical functions in the nuclear physics literature and can be taken over for our purposes [22, 23] (RPP, Sections 2.2.5 and 5).

10.7.3 Adiabatic Treatment

The discussion in Sections 10.4.4 and 10.5 applies quite generally to adiabatic analysis of N-electron excited states, wherein the motion in the angular coordinates is separated from the motion in R. The Schrödinger equation for the former motion is again (10.21) with H_R as in (10.20) except for the modifications in Sections 10.7.1 and 10.7.2. Potential curves $U_\mu(R)$ are again generated by expanding $\Phi_\mu(R; \{\alpha_i\}, \{\hat{r}_i\})$ in terms of the basis in (10.47″) and solving the resulting partial differential equations numerically by some convenient procedure. Calculations to date remain very limited. Figure 10.26 gives early results for ${}^{2,4}S^e$ in Li in which only states with $l_1 = l_2 = l_3 = l = L = 0$ were retained [22]. Compared with the two-electron hyperspherical calculations, the complexity now is much greater both in the computation and in the number of potential curves that converge to each asymptotic limit. The second lowest curve in the figure describes the doubly excited series $1s2s(^3S)ns\,^2S$. The similar quartet series $1s2s(^3S)ns\,^4S$ is given by the lowest curve in Fig. 10.26b. The first triply excited states of Li, described by $2s^2ns$ are contained in the fifth lowest curve for ($R < 3$) in Fig. 10.26a. A set of $U(R)$ curves for ${}^4P^e$ states of He^- is shown in Fig. 10.27 [24]. (See also Section 11.4.) Two distinctive groups are seen. A lower set converges asymptotically to singly excited $1snl\,^3l$ states of He; these hold the doubly excited states of He^-. The second group on top "converges" to doubly excited states of He, forming channels for triply excited states, the lowest of which are loosely described as $2s2p^2\,^4P^e$ [22]. A remarkably large affinity is suggested by the curves for this state, reflecting a very strong correlation. The wave function of the dominant sp^2 contribution has the symmetry of the vector combination $\hat{r}_1 \times \hat{r}_2 + \hat{r}_2 \times \hat{r}_3 + \hat{r}_3 \times \hat{r}_1$ and reaches its minimum potential energy for precisely the symmetric geometry shown in Fig. 10.25. The angular wave function is then fully antisymmetric so that it can combine with a symmetric function in spin and a symmetric radial function, analogous to the + behavior of two-electron states. We will return to this symmetry in Section 10.7.4 when considering three-electron escape.

Once the $U_\mu(R)$ curves have been obtained, the adiabatic calculation of the motion in R proceeds as in Sections 10.4.4 and 10.5.1. The expansion (10.23) gives the coupled equations for $F(R)$ in (10.25). In the calculations to date in Li, He^-, and H^{--}, only the diagonal terms of the coupling elements $P_{\mu\nu}$ between channels have been retained. Only a fully adiabatic analysis has been carried out, in which radial eigenstates are studied in the individual U_μ potential wells.

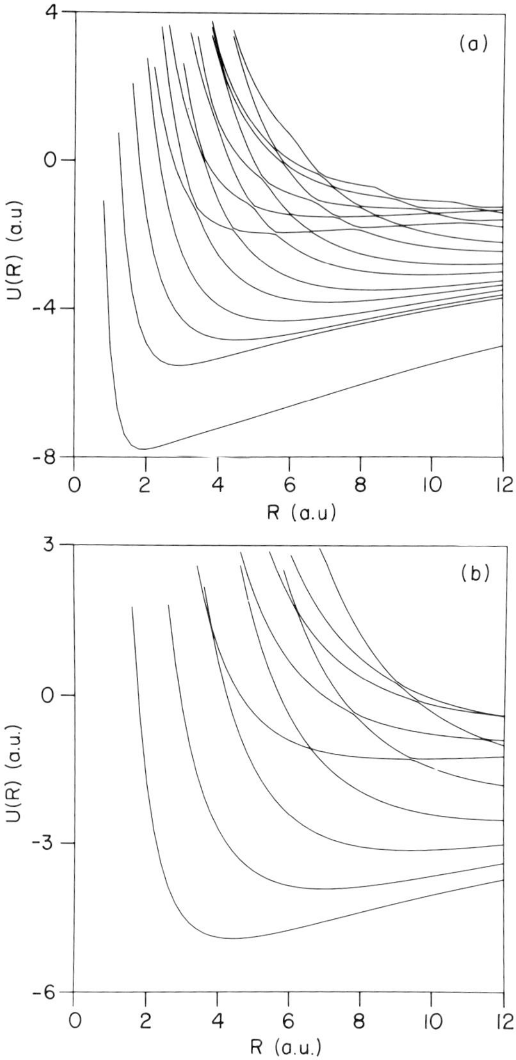

Fig. 10.26. Hyperspherical potential curves for three-electron states ($l_1 = l_2 = l_3 = 0$) in Li: (a) ^2S and (b) ^4S. (From Clark and Greene [22].)

10.7.4 Three-Electron Escape near Threshold

The close similarities between the hyperspherical analyses for two and three electrons points to further features in the analysis of triply excited states. For example, nonadiabatic couplings become important with increasing excitation toward the threshold for three-electron escape.

The role of the symmetric configuration in Fig. 10.25 in this excitation is exemplified by the chain curve in Fig. 10.27, which describes the potential (10.44b) in

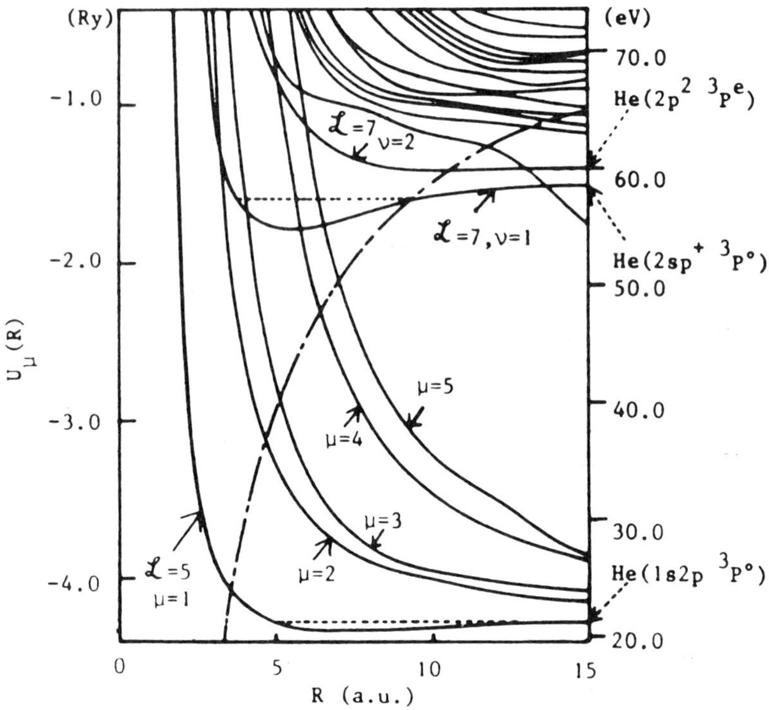

Fig. 10.27. Adiabatic potential curves for $^4P^e$ channels of He^-. Dashed lines give positions of the resonances. Chain curve gives potential at the saddle point of the three-electron potential corresponding to Fig. 10.25. (From Watanabe *et al.* [24].)

the saddle. As in Fig. 10.10 for two-electron states, this line is the locus of avoided crossings between the potential curves. Classical considerations along the lines of Wannier theory (Section 10.6) also point to the importance of maintaining the geometry shown in Fig. 10.25 over the range of the Coulomb zone in order that all three electrons escape. In these aspects, therefore, the three-electron problem poses the same questions as did the two-electron one for the passage from the low-lying states described by an adiabatic scheme to the high-lying ones near threshold where the coupling between R and the angles (particularly, the mock angles α_i) become important.

The three-electron escape near threshold has been subjected to the same Wannier analysis as in Section 10.6.1 by expanding the Schrödinger equation around the saddle and seeking solutions in the Coulomb zone [25]. One unstable motion in terms of departures from $r_1 = r_2 = r_3$ gives rise to the analog of ζ_{1+} and leads to a suppression beyond the initially expected $\sigma(E) \propto E^2$ [18]. The cross section above threshold then has the dependence

$$\sigma(E) \propto E^{2.83}, \tag{10.48}$$

when three electrons escape from the field of a singly charged ion; the exponent 2.83 is reduced to 2.27 when the ion is doubly charged, and it approaches 2 gradually as the charge of the ion increases. There are no experimental results yet on either triple ionization or high-lying triply excited states to test this extension of Wannier theory.

10.7.5 More than Three Electrons

We indicate only briefly the further extension of the hyperspherical description to four or more electrons. There has as yet been no analysis of such states, but the definition of coordinates and the broad lines of the procedure itself seem straightforward. Thus, for N electrons, we define a hyperspherical radius R, $(N - 1)$ hyperspherical angles α_i and $(2N - 3)$ angles to describe the potential;† three Euler angles complete the $3N$ coordinates of the problem. The potential energy has the structure $-(e^2/R)C(\{\alpha_i\}, \{\theta_{ij}\})$, with C a function only of $(3N - 4)$ angular coordinates. The kinetic energy has the form analogous to (10.45),

$$(\hbar^2/2m)[-R^{-(3N-1)/2}(\partial^2/\partial R^2)R^{(3N-1)/2} + \{\mathbf{\Lambda}^2 + \tfrac{1}{4}(3N - 1)(3N - 3)\}/R^2]. \qquad (10.49)$$

The grand angular momentum operator $\mathbf{\Lambda}^2$ has eigenvalues and eigenfunctions as in (10.47). In particular,

$$\mathbf{\Lambda}^2\phi = \lambda(\lambda + 3N - 2)\phi; \qquad (10.50)$$

$(N - 1)$ quantum numbers, in addition to the familiar ones of angular momenta and coupling schemes for an N-electron system, characterize the eigenfunctions.‡

The R dependence of the kinetic and potential energies remaining the same for any N, the analysis of the Schrödinger equation in terms of adiabatic channels and potential curves can be expected to proceed as in Sections 10.5 and 10.7.3. Likewise, we can expect that saddle points of the potential surface, whose number increases with N, will play a major role in controlling nonadiabatic phenomena.

10.7.6 Hyperspherical Treatment of Chemical Reactions

The discussion in this chapter has been couched in language appropriate to single atoms, as it was generally throughout the book. Our approach is nevertheless appropriate to the broader setting of chemical transformations, as noted in Chapter 1. Extensive investigations of reactive collisions that parallel the treatment of two-electron excitations in hyperspherical coordinates have developed independently in recent years [26].

The atom–diatom reaction represented by

$$A + BC \rightarrow AB + C \qquad (10.51)$$

will serve here as a basic example of chemical transformation. Experimental studies of these processes reveal energy transfers between the translational, vibrational, and rotational motions of atoms analogous to those that occur between the radial and

† For $N > 3$, there are $\frac{1}{2}N(N - 1)$ angles θ_{ij}, only $2N - 3$ of which are linearly independent.

‡ See also M. Cavagnero, *Phys. Rev. A*, in press (1986).

rotational motions of two electrons in a doubly excited atom. In the treatment of chemical reactions, the electronic part of the molecular Hamiltonians is generally diagonalized in a preliminary but laborious step that represents its eigenvalues as potential functions of the nuclear motions, $U(\mathbf{r}_A, \mathbf{r}_B, \mathbf{r}_C)$, taking advantage of the large disparity of electron and nuclear masses through the Born-Oppenheimer approximation [27].

The construction of hyperspherical coordinates representing $(\mathbf{r}_A, \mathbf{r}_B, \mathbf{r}_C)$ with reference to their center of mass—that is, a hyperradius R, ratios of radial distances, and space angles—proceeds generally as in earlier sections. It is, however, complicated by the unequal masses of nuclei. The mass ratio may be taken into account by weighting each position vector $\mathbf{r}_i$ by the corresponding mass root $\sqrt{m_i}$ in order to cast the kinetic energy in terms of $\partial/\partial R$ and $\mathbf{\Lambda}$ as in (10.13). Several angular variables are also generally "frozen" or otherwise eliminated to reduce the computational effort in calculating the whole potential $U(\mathbf{r}_A, \mathbf{r}_B, \mathbf{r}_C)$. Most studies of the process (10.51) have thus restricted themselves to a "collinear model" that confines the nuclei to move on a straight line. Key features of the hyperspherical analysis are nevertheless preserved in the analysis of this model.

Figure 10.28a shows the collinear representation of the reaction (10.51) in terms of only two coordinates $\{r_1 = |\mathbf{r}_A - \mathbf{r}_B|, r_2 = |\mathbf{r}_B - \mathbf{r}_C|\}$, the third distance $|\mathbf{r}_A - \mathbf{r}_C|$ being equal to $r_1 + r_2$. The potential function U is accordingly mapped on a plane (r_1, r_2). On the other hand, the kinetic energy expression in these coordinates,

$$T(\dot{r}_1, \dot{r}_2) = \frac{1}{2}\frac{m_A(m_B + m_C)}{m_A + m_B + m_C}\dot{r}_1^2 + \frac{m_A m_C}{m_A + m_B + m_C}\dot{r}_1\dot{r}_2 + \frac{1}{2}\frac{(m_A + m_B)m_C}{m_A + m_B + m_C}\dot{r}_2^2, \quad (10.52)$$

contains a cross term (dots represent time derivatives). This cross term is eliminated by plotting (r_1, r_2) on oblique axes (Fig. 10.28b) while laying on orthogonal axes the alternative pair of coordinates

$$x = [m_A(m_B + m_C)]^{1/2} r_1 + [(m_A + m_B)m_C]^{1/2} r_2 \sin\gamma, \quad (10.53a)$$

$$y = [(m_A + m_B)m_C]^{1/2} r_2 \cos\gamma, \quad (10.53b)$$

with

$$\sin^2\gamma = \frac{m_A m_C}{(m_A + m_B)(m_B + m_C)}. \quad (10.53c)$$

The kinetic energy has now the form appropriate to independent coordinates

$$T(\dot{x}, \dot{y}) = \tfrac{1}{2}(\dot{x}^2 + \dot{y}^2), \quad (10.54)$$

while the motion of a point that represents the structure of the collinear complex $\{\mathbf{r}_A, \mathbf{r}_B, \mathbf{r}_C\}$ *remains confined* to the sector of Fig. 10.28 defined by $(r_1 \geq 0, r_2 \geq 0)$. The potential energy $U(r_1, r_2)$ is similarly mapped on this sector encompassed by OP and Ox as illustrated in Fig. 10.29.

That kinetic and potential energies cannot be mapped on a single orthogonal base is an essential consequence of unequal atomic masses. Alternative plots have been utilized, but orthogonalizing the kinetic energy as in (10.54) affords an easier

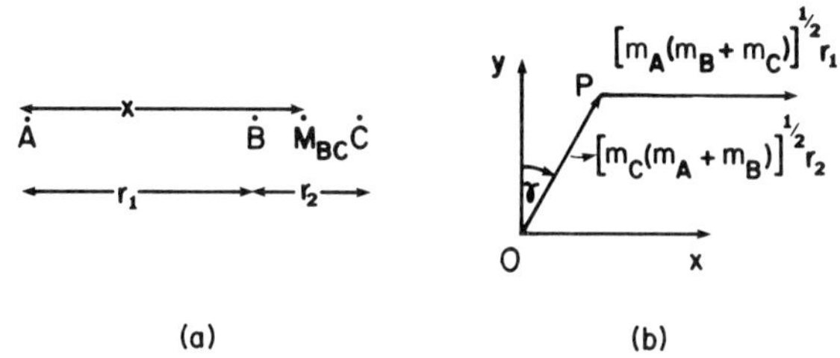

Fig. 10.28. Coordinates for a collinear reaction $A + BC \to AB + C$. (a) Internuclear separations r_1 and r_2 in initial and final states; center of mass of B and C is also shown. (b) Mass-weighted and scaled coordinates x and y along with skew angle γ.

appreciation of inertial effects. The map in Fig. 10.28b involves two numerical parameters, namely, the "skew angle" γ and a "scale factor"

$$[(m_A + m_B)m_C/m_A(m_B + m_C)]^{1/2}.$$

The collision dynamics depend only on the potential surface and on these two parameters. Many studies have dealt with symmetric collisions, with $A \equiv C$ and B a light atom such as H. In this case the skew angle γ ranges from 30° when A also coincides with B to 83° when A and C are iodine atoms and B is hydrogen; γ reaches 90° in the limit $m_A/m_B \to \infty$.

The reaction (10.51), under the influence of the two-valley potential shown in Fig. 10.29, is described in the following terms: The wave function ψ_1 in the initial fragmentation channel represents the translational motion of A through its propagation parallel to the r_1 axis from large r_1 toward the origin and represents the vibration of BC through its variation along lines parallel to the r_2 axis. The wave function ψ_2 in the exit channel propagates instead mainly in the potential valley along the r_2 axis, with the translation of C proceeding to $r_2 \to \infty$ and the vibration of AB parallel to r_1.

A hyperspherical treatment (with $y/x = \tan\alpha$ and $x^2 + y^2 = m_A(m_B + m_C)R^2$) centers instead on adiabatic channel eigenfunctions $F_\mu(R)\phi_\mu(R;\alpha)$ that propagate from the origin to $R \to \infty$ in both potential valleys or prevalently in a single one. In either case the reaction (10.51) is represented by a superposition of channel eigenfunctions with different $\phi_\mu(R;\alpha)$ and different channel energies $U_\mu(R)$. The dynamical role of the skew angle is now apparent in that a smaller skew angle implies a larger spacing of eigenchannel energies $U_\mu(R)$ and hence a smaller probability of the nonadiabatic

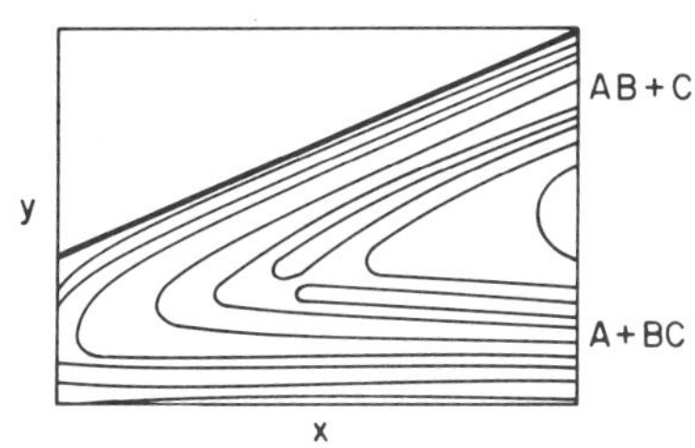

Fig. 10.29. Schematic potential surface for a collinear reaction showing entrance and exit channels and a saddle point.

transitions required for the reaction to proceed. Resonances in chemical reactions, which have been observed experimentally and accounted for classically by temporary localization of trajectories on the potential ridge of Fig. 10.29, also find a ready wave-mechanical interpretation as analogs of the two-electron resonances discussed in Section 10.6.1.

References

1. R. P. Madden and K. Codling, *Astrophys. J.* **141**, 364 (1965).
2. P. R. Woodruff and J. A. R. Samson, *Phys. Rev. A* **25**, 848 (1982).
3. U. Fano *in* "Atomic Physics 1," (B. Bederson, V. W. Cohen, and F. M. J. Pichanik, eds.), p. 209. Plenum, New York, 1969.
4. C. D. Lin, *Phys. Rev A* **10**, 1986 (1974).
5. M. Abramowitz and I. A. Stegun, eds., "Handbook of Mathematical Functions," Applied Mathematics Series 55, Chapters 15.4 and 22. U. S. Govt. Printing Office, 1964.
6. J. H. Macek, *J. Phys. B* **1**, 831 (1968).
7. D. R. Herrick, *Phys. Rev. A* **12**, 413 (1975).
8. C. D. Lin, *Phys. Rev. A* **29**, 1019 (1984).
9. B. L. Christensen-Dalsgaard, *Phys. Rev. A* **29**, 470 (1984).
10. C. D. Lin, *Phys. Rev. Lett.* **35**, 1150 (1975).
11. C. D. Lin, *Phys. Rev. A* **25**, 76 (1982).
12. B. L. Christensen-Dalsgaard, *Phys. Rev. A* **29**, 2242 (1984).
13. S. Cvejanovic and F. H. Read, *J. Phys. B* **7**, 1841 (1974).
14. S. J. Buckman, P. Hammond, F. H. Read, and G. C. King, *J. Phys. B* **16**, 4039 (1983).
15. F. H. Read, *Austr. J. Phys.* **35**, 475 (1982).
16. A. R. P. Rau, *J. Phys. B* **16**, L699 (1983); *in* "Atomic Physics 9," (R. S. Van Dyck and E. N. Fortson, eds.), p. 491. World Scientific, Singapore, 1984.
17. G. Wannier, *Phys. Rev.* **90**, 817 (1953); A. R. P. Rau, *Phys. Rev. A* **4**, 207 (1971); R. Peterkop, *J. Phys. B* **4**, 513 (1971), *J. Phys. B* **16**, L587 (1983).
18. A. R. P. Rau, *Phys. Rep.* **10**, 369 (1984).
19. F. H. Read *in* "Electron Impact Ionization," (G. Dunn and T. Mark, eds.). Springer-Verlag, Berlin and New York, 1984.
20. J. B. Donahue, P. A. M. Gram, M. V. Hynes, R. W. Hamm, C. A. Frost, H. C. Bryant, K. B. Butterfield, D. A. Clark, and W. W. Smith, *Phys. Rev. Lett.* **48**, 1538 (1982).
21. U. Fano *in* "Atomic Physics 8," (I. Lindgren, A. Rosen, and S. Svanberg, eds.), p. 5. Plenum, New York, 1983.
22. C. W. Clark and C. H. Greene, *Phys. Rev. A* **21**, 1786 (1980); C. H. Greene and C. W. Clark, *Phys. Rev. A* **30**, 2161 (1984).
23. M. S. Kil'dyushov and E. L. Surkov, *Sov. J. Nucl. Phys.* **22**, 333 (1975).
24. S. Watanabe, M. LeDourneuf, and L. Pelamourgues, *J. de Physique* Suppl. 11 **43**, C2-223 (1982).
25. H. Klar and W. Schlecht, *J. Phys. B* **9**, 1699 (1976).
26. J. A. Kaye and A. Kuppermann, *Chem. Phys. Lett.* **78**, 846 (1981); V. K. Babamov and R. A. Marcus, *J. Chem. Phys.* **74**, 1790 (1981); E. Pollack, *J. Chem. Phys.* **78**, 1228 (1983); J. M. Launay and M. Le Dourneuf *in* "Electronic and Atomic Collisions," (J. Eichler, I. V. Hertel, and N. Stolterfoht, eds.), p. 635. North-Holland Publ., Amsterdam, 1984.
27. P. J. Kuntz *in* "Dynamics of Molecular Collisions," Part B, (W. H. Miller, ed.), p. 53. Plenum, New York, 1976.

11 EXCITATIONS WITH A STRUCTURED CORE

11.1 INTRODUCTION

The concepts and techniques introduced in the study of doubly excited He and H^- are relevant to analogous phenomena in other atoms. Such phenomena are far more varied in atoms whose excited electrons can exchange energy and angular momentum with ionic cores that include other electrons. We approach this broader problem by presenting a set of qualitative considerations followed by results of initial quantitative studies, which cast new light on the spectra of most atomic species and on electron collisions with them.

The phenomena described in Chapter 10 manifest themselves at energies sufficient to raise two electrons out of their confinement in the 1s shell, namely, ~ 10 eV for H^- and ~ 60 eV for He. Other atoms generally afford sufficient opportunity for electron correlations to develop *within* incomplete valence shells, whereby double excitations already occur at much lower energies. Some of the phenomena have indeed been observed long ago and were partially interpreted within the independent electron model complemented by configuration or channel mixing. The approach of Chapter 10 will complement and enrich previous interpretations. A major part of the material in this chapter will deal with correlations within the valence shell.

The occurrence of strong correlations within open valence shells casts the role of the hyperspherical approach into a new light. The introduction of Part D emphasized the need of a new approach to deal with a pair of electrons emerging from an ionic core at radial distances larger than the core radius r_0. Excitation of a pair of electrons from a partially filled valence shell allows them instead to develop the correlations *within* that shell, that is, before moving beyond its confines. The approach developed in the previous chapter will nevertheless bear on the study of excitations within the valence shell.

Electron vacancies resulting from photoabsorption in certain inner shells have been found to transform into double vacancies in adjoining subshells, through a process called a "giant Coster–Kronig transition," which has been studied and reviewed by Wendin [1]. This production of two strongly correlated vacancies in a shell starting from a single vacancy may be viewed as analogous to the excitation of two correlated electrons by the collision of a single electron with a ground-state atom. This analogy will not be pursued here beyond remarking that it might be profitably explored.

11.2 CORRELATIONS WITHIN VALENCE SHELLS: SURVEY OF PHENOMENA

Electrons in valence shells with principal quantum number $N \geq 2$ have wave functions that penetrate into inner shells. The non-Coulomb field prevailing in inner shells removes the degeneracy of states with equal n and different orbital l; it also removes the degeneracy of spin orientation through the spin-orbit coupling, whose strength increases rapidly with the atomic number (Section 5.9). It is the occurrence of nondegenerate states within the valence shell that affords the production of double excitations *within the same shell* as the ground state, at modest energy cost, by shifting electrons, for example, from s to p subshells. Figure 11.1 compares potential curves of He converging to the threshold He^+ $(N = 2)$ from Fig. 10.12 with the corresponding ones converging to Be^+ $(N = 2)$ described in Section 11.3; note the split of Be^+ thresholds. Excitations of elements of groups IIA, IIIA, and so on, will be discussed here explicitly; those of the B groups and of transition elements should display analogous effects.

The loss of degeneracy resulting from penetration into the inner shells also manifests itself in the multiplicity of ionization or detachment thresholds, which are instead degenerate for each shell of doubly excited He or H^-. This nondegeneracy permits the treatment of higher levels of each series to follow the pattern of Chapters 7 and 8, as we shall see.

11.2.1 Circumscribed Treatment of an Electron Pair

The description of correlations introduced in Chapter 10 through the wave function of an electron pair in hyperspherical coordinates applies generally in any region of space where each electron is subject only to the Coulomb field of an ionic core and the repulsion by its companion electron. Such a region is delineated in the diagram of Fig. 11.2, analogous to Fig. 10.6. The hatched portion of the diagram represents positions of the electron pair with radial

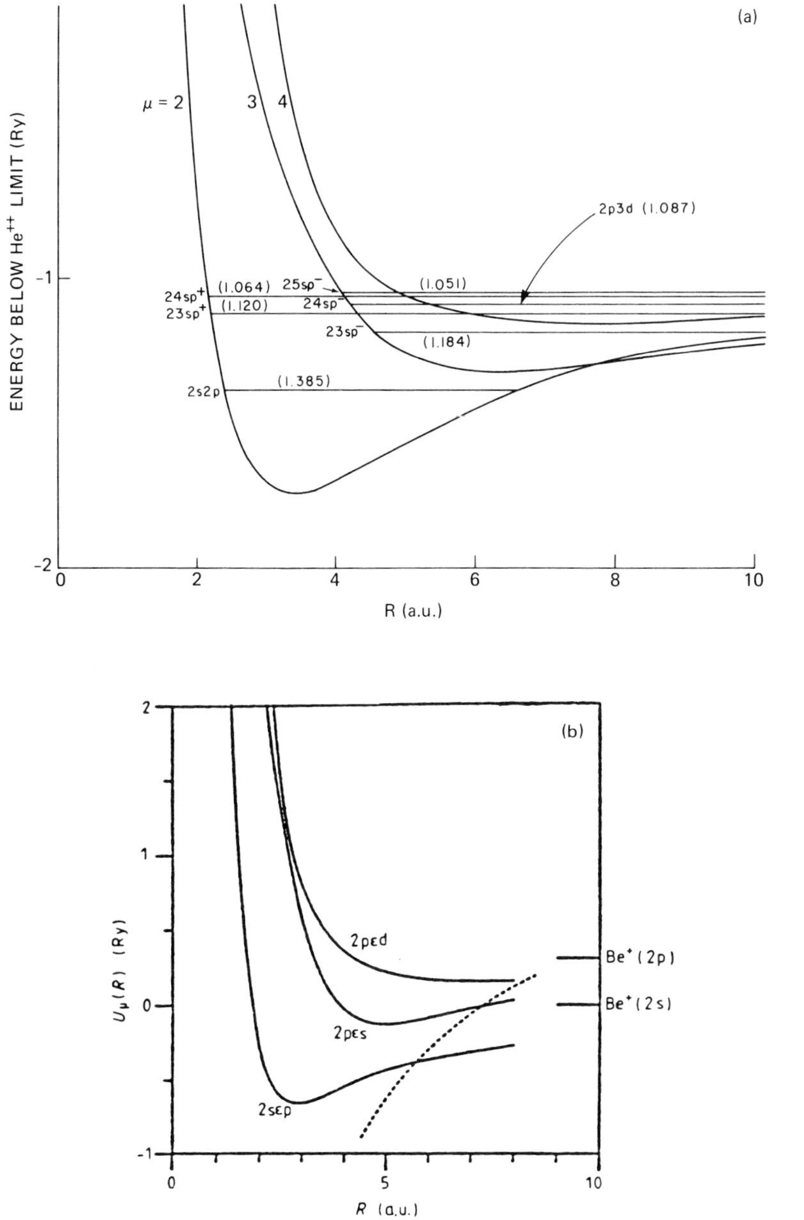

Fig. 11.1. (a) Potential curves for He from Fig. 10.12. (b) Potential curves $U_\mu(R)$ of Be $^1P^o$ converging to Be$^+$ ($N = 2$). Dashed curve represents the potential at the saddle. (From Greene [2].)

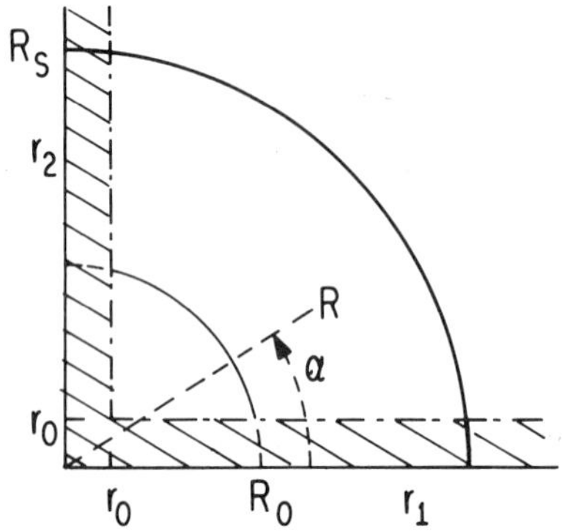

Fig. 11.2. Hyperspherical and Cartesian coordinates analogous to Fig. 10.6 showing corresponding core boundaries R_0 and r_0. Outside the separation radius R_s independent particle coordinates r_1 and r_2 become appropriate.

coordinates $r_1 < r_0$ and/or $r_2 < r_0$, where r_0 indicates the radius of the ionic core. This region is generally excluded from constructing wave functions as in Chapter 10 because the nuclear field is non-Coulomb within the core. A wave function in hyperspherical coordinates can be constructed outside the hatched region with a boundary condition on that region's edge, at $r_1 = r_0$ and $r_2 = r_0$. This condition represents the effects of one electron's penetration into the core semiempirically and is embodied in the value of $d \ln \psi/dr_{1,2}$, the normal logarithmic derivative as determined by an R-matrix treatment of electron + core or by an equivalent phase shift.

This boundary condition, applied at fixed values of the Cartesian coordinates (r_1, r_2), would, however, spoil the approximate separability in the polar coordinates (R, α) which underlies the hyperspherical analysis. This shortcoming can be remedied by restricting the hyperspherical treatment further to the region external to the circle $R = R_0$, with a value of R_0 such that r_2 (or r_1) can be replaced at r_0 by $R_0 \alpha$ (or R_0 $[\frac{1}{2}\pi - \alpha]$).

Both of these restrictions can be bypassed in the simple case of the valence electron pair of alkaline-earth atoms or their isoelectronic ions. The ionic cores of these species have a closed-shell structure with spherical symmetry and high excitation energy. Accordingly, they do not exchange energy or angular momentum with the pair of valence electrons whose kinetic energy is low. Under these circumstances the hyperspherical treatment can be extended to penetration into the core by replacing in the potential $C(\alpha, \theta_{12})/R$ of (10.12) the terms $Z/R \cos \alpha$ and $Z/R \sin \alpha$ by non-Coulomb effective potentials $V(R \cos \alpha)$ and $V(R \sin \alpha)$, which represent the electron–core interaction adequately throughout the diagram of Fig. 11.2. This procedure has been applied without difficulty in the calculations described in Section 11.3 for low excitations of valence electrons, but its validity remains to be tested for higher excitations.

Note finally that the hyperspherical analysis of correlations becomes superfluous, in all atomic species other than He and H^-, outside a circle of sufficiently large radius indicated in Fig. 11.2 by R_s. The value of R_s for a particular channel μ is determined by requiring the correlation function $\Phi_\mu(R; \alpha, \hat{r}_1, \hat{r}_2)$ [(10.21)] to tunnel at $R \geq R_s$ under the potential barrier as displayed by Fig. 10.13 for $R \sim 15$. That is, the potential barrier should prevent any further interchange of the electrons, which thus remain effectively separated at all R exceeding the separation radius R_s. Substantial angular correlations of the two electrons do persist and develop at $R > R_s$ in He and H^-, where the distant electron strongly polarizes the hydrogenic residue whose levels are degenerate. The corresponding effect is, however, negligible in any other atomic species or can be dealt with in accordance with Sections 6.9 and 9.3.7. The calculation of the correlation function Φ_μ by (10.21) can accordingly be confined to the range $R < R_s$ beyond which their eigenvalues $U_\mu(R)$ approach $U_\mu(\infty) - e^2/R$ for neutral atoms other than He. The values of Φ_μ at R_s and those of the companion radial function $F_\mu(R)$ provide the input data μ_α and $U_{i\alpha}$ for extending the calculation to larger radial distances by the procedure of Section 8.5 as is described in Section 11.3 for the example of Be.

In conclusion, the construction of a two-electron wave function should be extended to atoms with more than two electrons in the region of Fig. 11.2 delimited radially by $R_0 < R < R_s$ and outside the hatched region. The aim of this construction may be viewed as extending to $R = R_s$ the results of R-matrix calculations confined to $R < R_0$ and the hatched region. Initial tests of this construction, described in Sections 11.3 and 11.4, have been carried out under special conditions. The general problem is discussed further in Section 11.5.

11.2.2 Angular Coupling with the Ionic Core

The correlation functions Φ_μ considered in Chapter 10 are labeled by the spin quantum number S of the electron pair, corresponding to singlet or triplet states, because the correlation depends substantially on the exchange interaction. The transfers of angular momentum between the ionic core and the correlated pair, contemplated in this chapter, spoil the conservation of the spin for the correlated pair.

Specifically, the boundary conditions to be imposed on Φ_μ at the edges of the hatched zone of Fig. 11.2 are formulated with reference to spin coupling between the core and whichever of the excited electrons comes into contact with it. This is a particular case of the coupling adopted in Chapters 7 and 8 to characterize the i channels of an electron that separates from the core. Accordingly each eigenfunction Φ_μ of (10.21) must be expanded near the boundaries into such spin-coupled components for proper application of boundary conditions at the edge of the hatched strip of Fig. 11.2. The occurrence of alternative couplings complicates the solution of (10.21) to some extent but the required procedure has been carried out successfully in the calculation of He^- to be described in Section 11.4.

Transformation of Φ_μ to this spin coupling is also appropriate in the proximity of the separation radius R_s shown in Fig. 11.2. Indeed, the exchange interaction vanishes as the electrons separate and thus no longer selects out the *LS* coupling of the electron pair.

11.2.3 Critical Evolution of Correlations

The evolution of electron correlations represented by functions Φ_μ from their low-R oscillating behavior to their confinement in potential valleys at large R goes through a rapid and critical stage, as noted at the end of Section 10.4.4 and illustrated in Fig. 10.13. During this stage, the amplitude of Φ_μ at $\alpha \sim 45°$, that is, near the ridge of the potential $C(\alpha, \theta_{12})$, decreases rapidly with increasing R. In atoms with more than two electrons the critical stage presents new aspects that account for an otherwise unexplained feature of optical spectra.

The correlation functions Φ_μ for the two lowest adiabatic channels of an electron pair in a shell with $n \geq 2$ have quite generally + and − character as was described for He in Section 10.5, meaning that Φ_μ has an antinode or a node, respectively, in the vicinity of $\alpha \sim 45°$ as long as it propagates as a standing wave across the ridge. This character persists in He and H^- even when Φ_μ tunnels through the barrier, in the sense that the + correlation function is nodeless under the barrier whereas the − function has a node; the diabatic crossing of the corresponding potentials $U_\mu(R)$, Fig. 10.12, shows the + or − classification to remain a "good" quantum number.

In other atoms, on the other hand, the + or − character of the adiabatic Φ_μ functions no longer obtains when Φ_μ gets confined in a potential valley at $R \gtrsim R_s$, because the potential approaches different ionization thresholds in the two alternative valleys and the electron pair no longer resonates across the barrier. Accordingly, the + wave function, with the lower potential $U_\mu(R)$, gets confined in the valley with lower threshold and the − function in the valley with higher threshold thus losing their + or −character, as documented by calculations reported in Section 11.3.1 and in Fig. 11.3. Their potentials $U_\mu(R)$ do not display diabatic behavior.

Remarkably, however, the + or − character actually manages to persist in the final solutions through *nonadiabatic* behavior that requires the superposition of two adiabatic eigenfunctions Φ_μ to be labeled here as Φ_1 and Φ_2. Upon construction of a nonadiabatic wave function of the type (10.23),

$$\Psi = F_1(R)\Phi_1(R; \alpha, \hat{r}_1, \hat{r}_2) + F_2(R)\Phi_2(R; \alpha, \hat{r}_1, \hat{r}_2), \qquad (11.1)$$

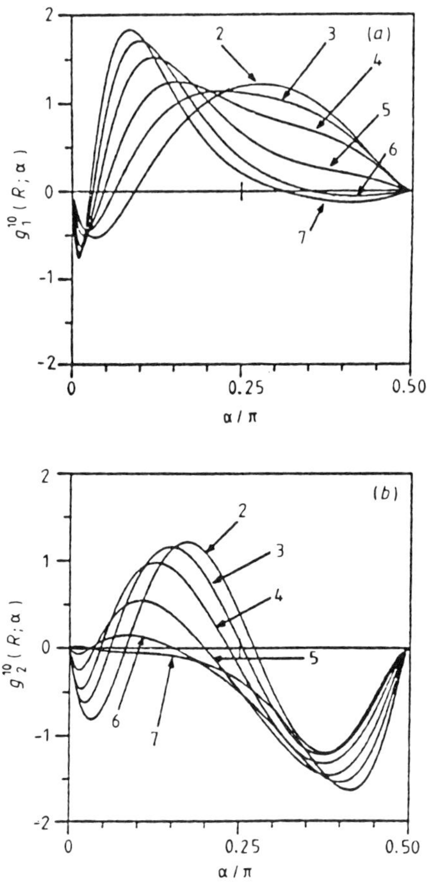

Fig. 11.3. Channel functions $g_{n_{rc}}^{10}(R;\alpha)$ for Be $^1P^o$ at the different values of R indicated on the curves: (a) $n_{rc} = 1$ (2sεp) and (b) $n_{rc} = 2$ (2pεs). (From Greene [2].) Compare with the analogous functions for He in Fig. 10.13.

the following has been observed, as detailed in Section 11.3.2. When the integration in R starts from $R \sim 0$ with $F_2(R \sim 0) = 0$, the coupling of F_2 with F_1 remains negligible up to the critical range $R \sim R_c$, whereby $\Psi \sim F_1\Phi_1$, thus displaying + character in this range. Strongly nonadiabatic behavior emerges instead in the critical range, where the rapid condensation of Φ_1 into one of the valleys induces a rapid rise of the coupling matrix element P_{12} [(10.24)] and a consequent rise of $F_2(R)$. When R has reached the separation radius R_s, the end of the critical range, F_2 has been found to attain a value near that of F_1. Thereby the superposition of Φ_1 and Φ_2 at $R > R_s$, each of them confined in a separate valley, amounts to a correlation function analogous to the Φ_+ of He that tunnels through the barrier with comparable values in both potential valleys. We may visualize the wave function Ψ as being sheared evenly into two separate components $F_1\Phi_1$ and $F_2\Phi_2$ in their respective valleys by the rising potential barrier as it propagates from R_c to R_s.

The wave function Ψ identified as $F_1\Phi_1$ at small R thus evolves at large R into a superposition of components that propagate on in the two valleys with nearly equal amplitudes but with different wavelengths in R corresponding to the adiabatic kinetic energies $E - U_1(R)$ and $E - U_2(R)$. Having noted in Section 11.2.1 that the values of Ψ at and near R_s provide the input data for the construction of MQDT wave functions at $R \geq R_s$, with each valley identified as a fragmentation channel i, we interpret the result as follows: the near equality of $F_1(R_s)$ and $F_2(R_s)$ in (11.1), which represents the lowest excitation channel considered here, leads to the construction of eigenfunctions Ψ_α whose expansion at large R [(8.52)] consists of two terms with nearly equal coefficients $U_{i\alpha}$. This fact had indeed been observed empirically as noted in Section 8.5.8.

11.2.4 Collisions

The strong coupling implied by the observed values of the coefficients $U_{i\alpha} \gtrsim 1/\sqrt{2}$ for the low channels of open valence shells with $n \geq 2$, discussed earlier, has a major influence on collision cross sections. An electron that approaches an open-shell atom or ion with kinetic energy ($\lesssim 10$ eV) and low orbital momentum l, forms with it an excited complex in one of its lower adiabatic channels. According to the general equations (7.14) and (7.40), large values of the coefficients $U_{i\alpha}$ imply that for each eigenstate of total angular momentum:

(1) inelastic cross sections are comparable to the elastic cross section, and

(2) accordingly these cross sections will be generally of the order of the geometric size of the target or the squared electron wavelength, whichever is larger.

These effects may be amplified by the superposition of partial waves with different l. Large cross sections have indeed been measured in extensive experiments, especially by Gallagher and co-workers, on slow electron collisions with alkali atoms and alkaline-earth ions leading to emission of resonant light [3]. Numerous theoretical calculations of the same cross sections have also been conducted by close-coupling procedures (Sections 9.3.4 and 9.4.1).

This behavior differs sharply from that observed or predicted for electron collisions with He, He^+, or atomic H. For these systems with 1s electrons the complex forms with the incident electron in low channels, widely separated in energy from the next channels with equal angular momentum, spin, and parity. The $U_{i\alpha}$ coefficients coupling to channels with atomic excitation are of order 10^{-1} or smaller under these conditions and the inelastic cross sections, which are proportional to $U_{i\alpha}^2$, are even smaller. The same holds for electron collisions with closed-shell structures such as those of rare-gas atoms or their isoelectronic ions.

11.2.5 Hyperspherical Treatment of Ionic Core

The introduction of hyperspherical coordinates in the two-electron Schrödinger equation (Sections 10.3 and 10.4) extends readily to an indefinite number of electrons, at least at a formal level. Its extension to three electrons has been dealt with in Section 10.7. Further extension would include all electrons of the core together with the valence electrons of primary interest. It would thus remove any need for excluding the hatched area of Fig. 11.2 from application of the hyperspherical procedure.

This general approach rests on the fact that the dimensional analysis of the kinetic and potential energy, in Sections 10.4.1 and 10.4.2, does not depend at all on the number of electrons. The potential energy keeps the structure $(e^2/R)C$ [(10.12)], but the dimensionless function C depends now on $3N - 4$ angles and consists of a great many terms. The analytical determination of the eigenvalues and eigenfunctions of the squared grand angular momentum Λ^2 also holds, thus yielding for N electrons the eigenvalues $\Lambda^2 = \lambda(\lambda + 3N - 2)$ [(10.50)], eigenfunctions expressed as products of Jacobi and Legendre polynomials and replacement of the coefficient $\frac{15}{4}$ in (10.13) by $(3N - 1)(3N - 3)/4$ [(10.49)]. The quantum number λ is an integer whose lowest value depends on N as required by the exclusion principle [4].

The adiabatic analysis of correlations through the characterization of the eigenfunctions Φ_μ of (10.21) also remains relevant regardless of the number of electrons. In fact, only the Φ_μ corresponding to a few lowest potential curves $U_\mu(R)$ are most relevant to optical spectra and collisions of slow electrons. The substantive question in the study of many-electron atoms by any of the various available methods lies in the adaptability of each method to select the coordinates relevant to each phenomenon and to determine the dependence of the wave function on those variables only. The application of hyperspherical coordinates for more than two or three electrons remains to be assessed. One possible approach would be to utilize hyperspherical coordinates for all the electrons outside of a closed-shell ionic core, regarding this core as a fixed non-Coulomb potential well as was done in the case of Be. We return to this subject in Section 11.5.

11.3 EXCITATIONS IN THE VALENCE SHELL OF ALKALINE EARTHS

The atoms of the alkaline-earth group have the ground-state configuration $ns^2\ {}^1S$ ($n = 2$ for Be, $n = 3$ for Mg, etc.) and are excited by photoabsorption to $nsn'p\ {}^1P^o$. The open-shell character of these atoms manifests itself in these excitations through heavy ($\sim 40\%$) admixture of doubly excited $npn'l\ {}^1P^o$ and $(n-1)dn'l\ {}^1P^o$ configurations, the latter only for $n \geq 4$. The independent-electron picture combined with configuration mixing has accounted for this large amount of double excitation character in each level without revealing directly how the double and single excitations have comparable roles in photoabsorption and in related processes (Fig. 9.19). This role is apparent in the ground state as well, where $np^2\ {}^1S$ and/or $(n-1)d^2\ {}^1S$ are heavily admixed into ns^2.

An exploratory study of these admixtures in ${}^1P^o$ channels has been conducted in hyperspherical coordinates for Be, with some extension to other species with two valence electrons for the purpose of verifying their similarity [2]. The results of that study are reviewed here.

11.3.1 Adiabatic Potentials and Their Correlation Functions

The three lowest ${}^1P^o$ potential curves $U_\mu(R)$ for the Be atom have been shown in Fig. 11.1 for comparison with the corresponding curves for He. Each curve is labeled with the configuration appropriate to escape of an electron at $R \to \infty$. The three curves converge to two separate limits at $R \to \infty$, corresponding to 2s and 2p levels of Be^+, as anticipated in Section 11.2. Note the avoided crossing of the two lowest curves near their intersection with the dotted curve that represents the height of the saddle point, $C(\alpha = 45°, \theta_{12} = 180°)e^2/R$.

The potentials $U_\mu(R)$ for Be have been calculated as eigenvalues of (10.21) with the potential modified to reflect the stronger than Coulombic attraction within the Be^{2+} core as discussed in Section 11.2.1. The eigenfunctions $\Phi_\mu(R; \omega)$ of (10.21) represent the radial and angular correlations of the two valence electrons. These correlation functions were expanded into spherical harmonics (10.18′), which were properly antisymmetrized with coefficients $g_\mu(R; \alpha)$ as in (10.27). The latter coefficients were calculated by Greene [2] by point-by-point numerical integration of the relevant equation [(10.21)], a slowly converging procedure that can be bypassed by the variational fitting of analytical expressions mentioned in Section 10.5.1.

The radial correlation functions $g_\mu(R; \alpha)$ corresponding to the two lowest $U_\mu(R)$ are displayed in Fig. 11.3, the analog of the He case (Fig. 10.13). At low R the Be plots show + and − character, respectively, like those of He. Again all these plots, for Be as for He, evolve very rapidly in the critical range—at

$R \sim 5$ a.u. for Be—where the potential curves drop below the potential barrier $C(45°, 180°)e^2/R$. However, this evolution differs dramatically in Be and He in that the Be $g_\mu(R; \alpha)$ shown in the two frames of Fig. 11.3 concentrate at large R on opposite sides of the $\alpha = 45°$ abscissa that marks the ridge of the potential barrier. This behavior, anticipated in Section 11.2.3, contrasts with the preservation of the $\pm$ symmetry in He. This behavior of the $g_\mu(R; \alpha)$ functions for Be shows that adiabatic variation of each correlation function leads it to be confined at large R within a single potential valley in which the correlation corresponds to one of the labels 2sϵp or 2pϵs, respectively.

Figure 11.4 shows the lowest $^1P^o$ potential curves $U_\mu(R)$ for the Ca atom, analogous to those of Be shown in Fig. 11.1. Note the insertion of the curve labeled 3dϵp, which converges to the threshold Ca^+(3d), lying below the Ca^+(4p) level. [An additional curve 3dϵf (not shown) would lie mostly higher than the other curves.] Note also the staggering of successive avoided crossings at increasingly large values of R. Figure 11.4 also shows the analogous graph for the ion K^- isoelectronic to Ca. The curve 3dϵp is not shown for K^- because it lies far above the 4pϵs, as the centrifugal potential barrier repels 3d electrons more effectively in K atoms than in Ca^+ ions. The K^- plot includes $^3P^o$ curves relevant to photodetachment from K^- near the doublet threshold whose splitting (not shown) is small but nonnegligible.

11.3.2 Nonadiabatic Coupling

The adiabatic approximation to the Schrödinger equation in hyperspherical coordinates utilizes wave functions with the form of a simple product, $F_\mu(R)\Phi_\mu(R; \alpha, \hat{r}_1, \hat{r}_2)$, where $F_\mu(R)$ represents the radial propagation in the adiabatic channel with potential $U_\mu(R)$ and correlation function Φ_μ. This approximation, generally adequate for the study of the lower channels of He, disregards the channel-coupling matrix P in the complete equation (10.25). On the other hand, it breaks down completely for Be and similar atoms at all or most relevant energies owing to the large values of the coupling matrix element P_{12} shown in Fig. 11.5.

This departure from adiabatic behavior, anticipated in Section 11.2.3, has been treated explicitly [2] by solving numerically the basic equation [(10.25)] truncated to include only the *two* lowest $^1P^o$ channels, that is, disregarding all the coupling matrix elements $P_{1\nu}$ and $P_{2\nu}$ with $\nu > 2$. This truncation proved adequate for producing data to be compared with the main spectroscopic observations. Accuracy of these data has, however, been improved by R-matrix calculations conducted according to Section 9.3.2, as discussed in Section 11.5. The two-term wave function actually used has been represented by (11.1), where the channels 1 and 2 display + or − character, respectively,

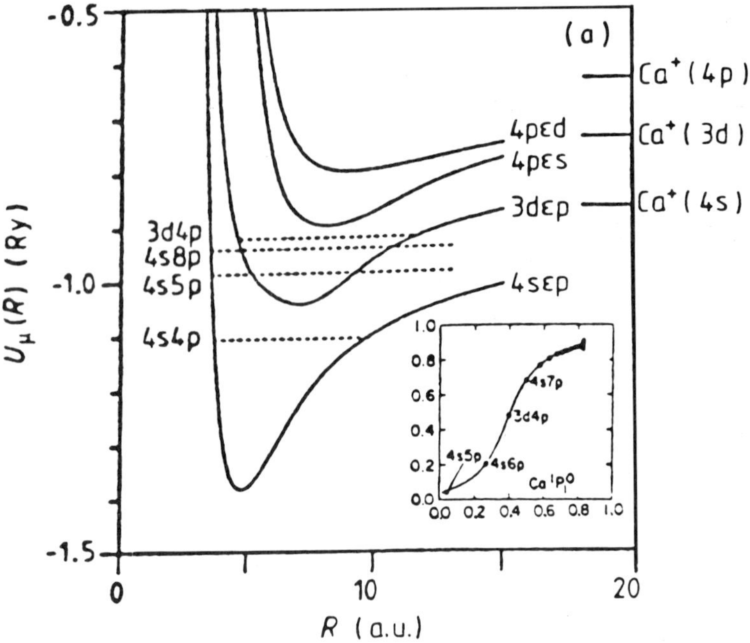

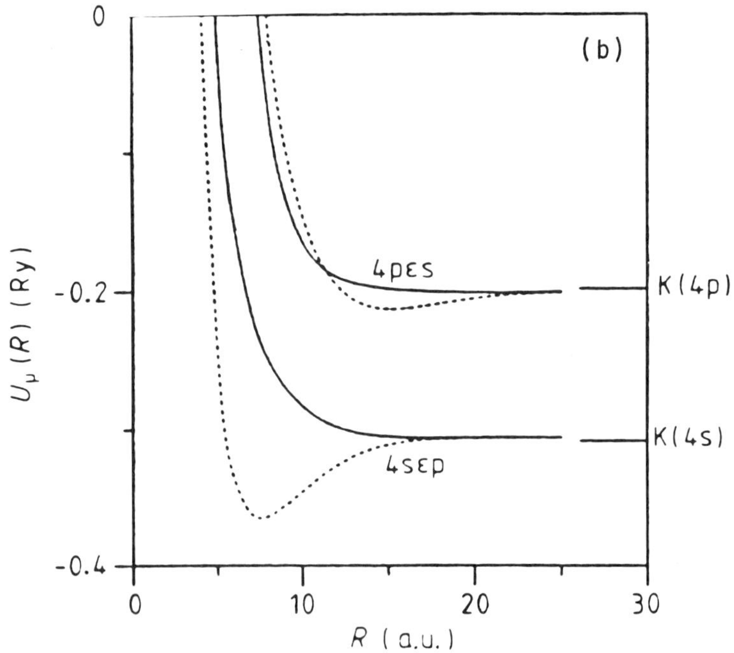

Fig. 11.4. (a) Potential curves $U_\mu(R)$ of Ca $^1P^o$. Dashed lines denote energies of the lowest eigenvalues of the uncoupled radial equations. Inset reproduces MQDT plot of bound levels from Fig. 8.10 and displays the effects of channel coupling. (b) Potential curves of K^- P^o: ——, singlets; ---, triplets. (From Greene [2].)

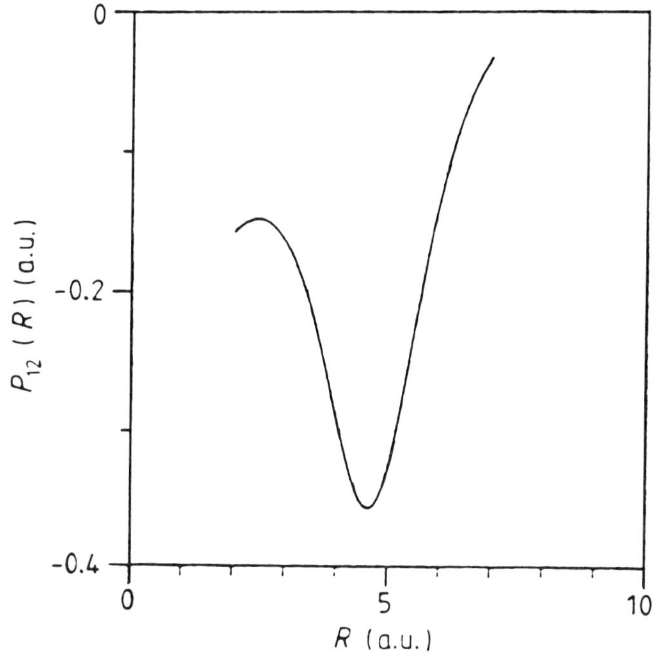

Fig. 11.5. Channel coupling matrix element $P_{12}(R)$ of Be $^1P^o$ (From Greene [2].)

at $R \lesssim 5$ a.u. Note the contrast between the coupling of the two lowest $^1P^o$ channels in He and Be. In He this coupling is negligible and the + and − potential curves appear to cross diabatically while in Be the coupling is strong and the crossing is widely avoided in Fig. 11.1.

As anticipated in Section 11.2.3, the coupling of the + and − channels has a major effect of mixing the correlation functions $\Phi_1(R; \alpha, \hat{r}_1, \hat{r}_2)$ and Φ_2. The mixing attains its full effect, wherein the radial coefficients $F_1(R)$ and $F_2(R)$ oscillate with comparable amplitude, when the excitation energy E reaches or exceeds the threshold for ionization to $Be^+(2s)$. At much lower energies, on the other hand, the coupling is hardly effective and the radial propagation proceeds adiabatically. That is, an integration of the coupled radial equations that starts in channel 1, with $F_1(R) \neq 0$ and $F_2 = 0$ at low R, yields only small values of $F_2(R)$ at larger radial distances on, or beyond, the critical range $R \sim R_c = 5$ a.u.

A quantitative formulation and analysis of the effect of channel coupling is provided by terminating the radial integration of two independent wave functions, $F_1(R)\Phi_1$ and $F_2(R)\Phi_2$, at the separation radius R_s discussed at the end of Section 11.2.1 and representing their values and radial derivatives at R_s in either of the alternative MQDT forms (8.53) or (8.56). In the present context the latter equations involve two fragmentation channels ($i = 1$ or 2) corresponding to the alternative structures $Be^+(2s)\epsilon_1 p\,^1P^o$ and $Be^+(2p)\epsilon_2 s\,^1P^o$. The adiabatic functions Φ_1 and Φ_2 reduce to the respective fragmentation channel functions at $R = R_s$, as is apparent from Fig. 11.3. Matching to (8.56) determines the three independent elements of the 2×2 symmetric reaction matrix $K_{ij}^{(s)}$; matching to (8.53) determines instead the two eigenvalues $\tan \delta_\alpha^{(s)}$ and the eigenvectors $U_{i\alpha}$ of the same matrix. Since the pair of eigenvectors $U_{i\alpha}$ form a 2×2 orthogonal matrix

$$U_{i\alpha} = \begin{vmatrix} \cos\theta & \sin\theta \\ -\sin\theta & \cos\theta \end{vmatrix}, \tag{11.2}$$

the end result of the radial integration is represented by the three parameters $\delta_1^{(s)}$, $\delta_2^{(s)}$, and θ. Values of these parameters calculated for different energies [2] and by the alternative procedure of Section 9.3.3 [5] are shown in Fig. 11.6.

The most important result, that the mixing angle θ settles at $\sim\pi/4$ at $\epsilon_1 = E - I(Be^+2s) \gtrsim 0$, as anticipated in Sections 11.2.3 and 8.5.8, is borne out by both sets of data. Both sets also show θ to vanish at $E \ll I$, but the new data rise more smoothly beginning at $E \sim I - 6$ eV, whereas those of Greene [2] rise more abruptly closer to

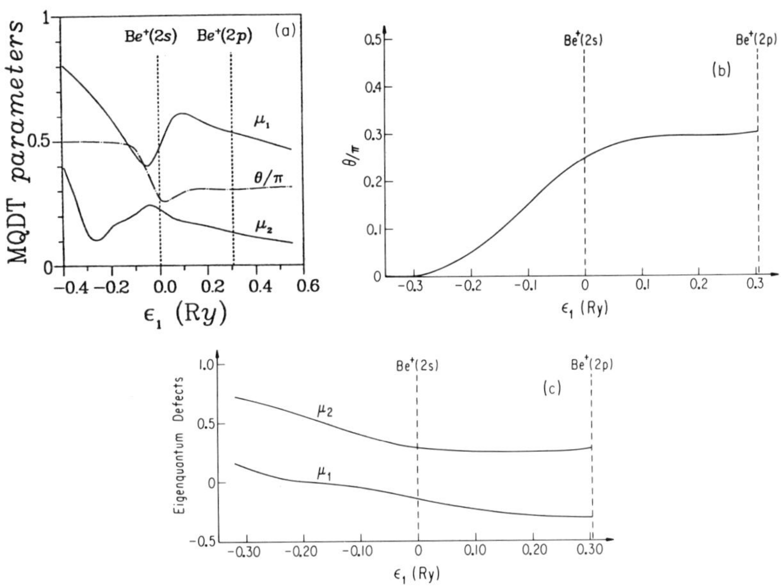

Fig. 11.6. Quantum defect parameters for Be as functions of the energy ϵ_1 relative to the 2s threshold. (a) Mixing angle θ and eigenquantum defects μ_1, μ_2 from hyperspherical calculations of Greene [2], (b) θ, and (c) μ_1, μ_2 from R-matrix direct variational approach. (From O'Mahony and Greene [5].)

the threshold. Similarly, both sets of eigenphases $\delta_\alpha^{(s)}$ decrease generally with increasing energy, but the newer data do so smoothly while those of Greene [2] follow a jagged course. The eigenchannels labeled by $\alpha = 1, 2$ turn out to be represented very approximately by wave functions (11.1) that reduce at small R to $F_1\Phi_1$ and $F_2\Phi_2$, respectively. Probability densities in the eigenchannels, $|\Phi_1|^2$ and $|\Phi_2|^2$, are shown in Fig. 11.7. Results for Mg, analogous to Fig. 11.6 for Be, are shown in Fig. 11.8.

Entering the parameters $\delta_\alpha^{(s)}$ and $U_{i\alpha}$ in the Jost matrix [(8.32)] lays the basis for calculating discrete levels as well as collision and photoabsorption cross sections by the procedures described in Chapters 7 and 8. The energy levels for our two-channel atoms, Be and Mg, are given by the pair of equations analogous to (7.48) and (7.49)

$$\mathrm{Det}|U_{i\alpha}\sin(\pi\nu_i + \delta_\alpha^{(s)})| = 0, \qquad (i, \alpha) \equiv (1, 2). \tag{11.3a}$$

For Be in particular,

$$\nu_2^2 = \nu_1^2/(1 + 0.29\nu_1^2), \tag{11.3b}$$

where $0.29 = [I(\mathrm{Be}^+2\mathrm{p}) - I(\mathrm{Be}^+2\mathrm{s})]/13.6\,\mathrm{eV}$. Values of the roots ν_{1n} of (11.3a), expressed as effective quantum defects of the channel $i = 1$, namely as $\mu_n = -\nu_{1n}$

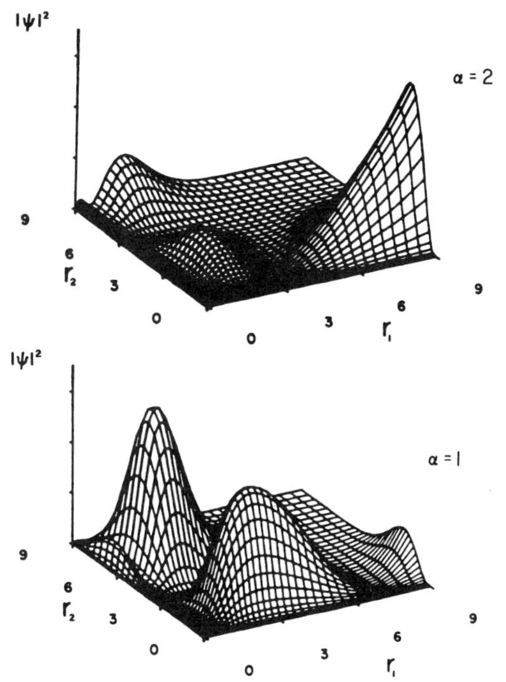

Fig. 11.7. The sp component of the charge density for the eigenchannels α of Be at an energy $\epsilon_1 = 0.02$ Ry. Note the roughly equal amplitude in each valley ($r_1 \sim 0, r_2 \sim 0$) as a result of encountering the potential ridge at $R \sim 4$ a.u. (From O'Mahony and Greene [5].)

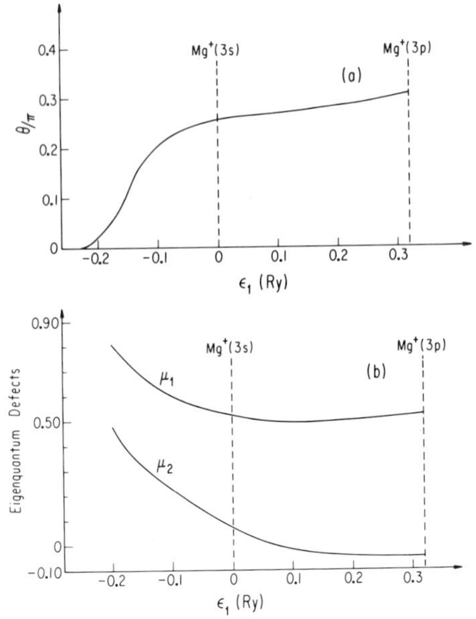

Fig. 11.8. MQDT parameters for Mg. (a) Mixing angle θ and (b) eigenquantum defects μ_1 and μ_2. [From O'Mahony and Greene [5].)

(mod 1), obtained from the data in Fig. 11.6, are shown in Table 11.1 together with experimental values and values from alternative calculations.

Coupling to the third excitation channel $Be^+(2p)\epsilon_2 d$ becomes relevant as the energy E rises to approach the upper threshold $I(Be^+ 2p)$ or the corresponding threshold of Mg; a third eigenphase is then introduced and the matrix $U_{i\alpha}$ regains an additional row and column. Collision cross sections also involve, of course, channels with classifications other than $^1P^o$. The heavier alkaline-earth atoms (Ca, Sr, Ba, Ra) also involve additional $^1P^o$ channels in the valence shell, with $nd\epsilon l'$ fragmentation. No

Table 11.1

$^1P^o$ Excitation Levels of Be Above Its Ground State[a]

Level	Experiment [6]	Theory [5]	Theory [7]
2s2p	0.166	0.085	0.151
2s3p	0.296	0.235	0.283
2s4p	0.332	0.319	0.322
2s threshold		0.389	0.374

[a] In atomic units.

analysis of collision data has yet been conducted for alkalis or for alkaline-earth ions using wave functions with explicit correlation analysis. The interpretation of photoabsorption spectra in terms of eigenfunctions $\Psi_\alpha(\omega, r)$ will be discussed in the next section.

11.3.3 Photoabsorption Spectrum

The major role of channel mixing in the $^1P^o$ states of alkaline earths implies that the intensity of two-electron excitations is comparable to that of single-electron excitations in their photoabsorption spectra. This implication is borne out by the very strong modulation of Be photoabsorption in the spectral range between the $Be^+(2s)$ and $Be^+(2p)$ thresholds (Fig. 11.9), whose major peaks are labeled as 2pns autoionizing levels. This modulation is amplified, in all alkaline-earth spectra, by the accidental occurrence of a Cooper minimum near the threshold of the $ns\epsilon p$ channel (Section 2.7).

Analytically, the dipole matrix elements for transitions to the alternative channels $ns\epsilon p$ and $np\epsilon s$ can be expressed as linear combinations of those for transitions to the eigenchannels of short-range interactions, indicated by

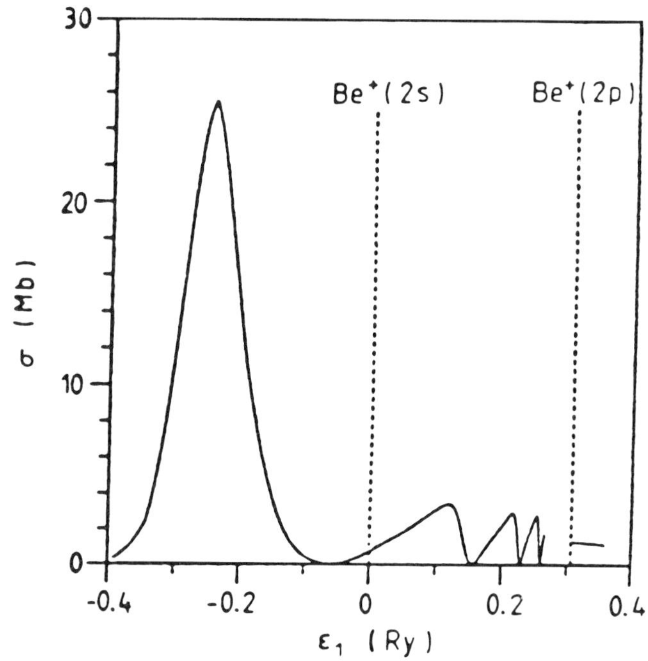

Fig. 11.9. Photoabsorption cross section for Be calculated by the method of hyperspherical coordinates. Relevant parameters are in Fig. 11.6a. (From Greene [2].)

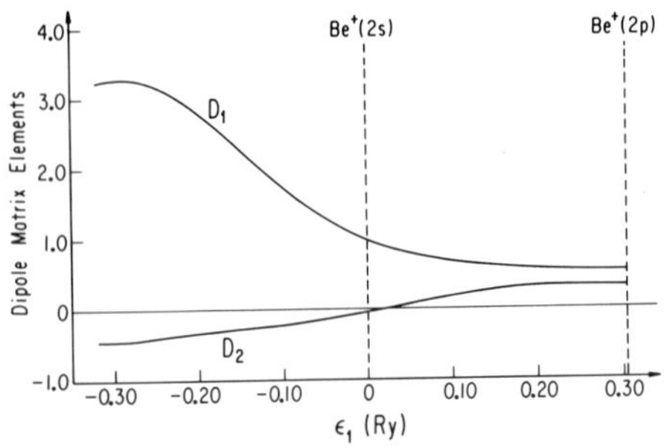

Fig. 11.10. Dipole matrix elements $D_\alpha = (F_\alpha|x|0)$ for Be photoabsorption from R-matrix direct variational approach. (From O'Mahony and Greene [5].)

$(F_\alpha(\omega, R)|x|0)$ in (7.43) and (8.50). In the example of Be we know that the eigenfunctions $F_{\alpha=1,2}$ of the reaction matrix $K^{(s)}$ reduce to the adiabatic functions $F_1\Phi_1$ and $F_2\Phi_2$, respectively, at energies much lower than the ionization threshold. Since the adiabatic potential $U_1(R)$ remains much lower than $U_2(R)$ at low R, where overlap with the ground state occurs, the dipole element $(F_\alpha|x|0)$ for $\alpha = 1$ should dominate over that for $\alpha = 2$ at least at low energies. The data in Fig. 11.10 confirm this surmise.

We might also surmise this relation to hold even near threshold, since the eigenfunction $F_{\alpha=1}(\omega, R)$ coincides with the adiabatic function $F_1\Phi_1$ at low R, $R < R_c$, throughout the energy range considered here. This additional surmise is proved incorrect by the data in Fig. 11.10 where the $\alpha = 1$ dipole drops as the energy approaches the threshold $I(\mathrm{Be}^+2\mathrm{s})$, thus reflecting the occurrence of a Cooper minimum mentioned earlier. This minimum is associated with sign reversal of the dipole moment for transitions to the

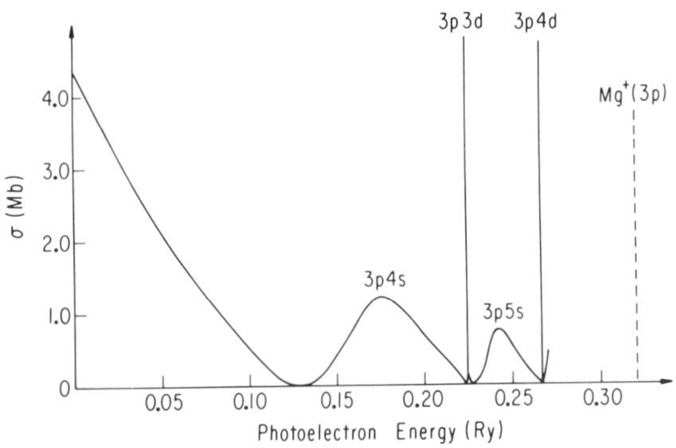

Fig. 11.11. Photoabsorption cross section for Mg from R-matrix direct variational approach. [From O'Mahony and Greene [5].)

fragmentation channel $i = 1$, that is, $2s\epsilon p$, which emerges by combining the two dipoles $(F_\alpha|x|0)$ in Fig. 11.10 linearly with nearly equal coefficients.

The Be calculation has been extended to Mg with analogous results (Fig. 11.11). Extension to the other elements of the IIA group would involve more extensive mixing of channels, since their ions have three valence shell terms (^2S, ^2D, and $^2P^o$), rather than two, with very appreciable fine structure splits. Doubly excited levels with configurations $(n-1)dn'p$ lie in the discrete, causing major perturbations of intensity, shown in Fig. 8.5 for Ba, analogous to the modulation of the Be continuum in Fig. 11.9. Data on Ca have been presented in Fig. 9.19. Spin-orbit coupling also becomes increasingly important in the heavier alkaline earths.

11.4 CORRELATIONS IN THE He$^-$ COMPLEX

The He$^-$ doubly excited complex is generally formed by electron collisions with ground state He that excite one electron of He and retain the incident electron temporarily in an autodetaching state. At least one of these autodetaching states, with the structure $1s2s2p\,^4P^o_{5/2}$, is metastable with a remarkably long lifetime of the order of 1 ms. Autodetaching states with all *three* electrons in the $n = 2$ shell have also been produced by electron collisions in the 55–60-eV range.

These He$^-$ complexes have long been favorite objects of experimental and theoretical study. (The earliest predictions of autodetaching levels date from the 1930s [8]). A calculation by Burke *et al.* of He$^-$ resonances in the spectrum of e + He collisions in the 19–22-eV range, using the close-coupling approach, has reproduced extensive experimental observations as described in Section 9.4.1. The occurrence of metastable states of the complex has even permitted observation of optical transitions between states of this complex [9].

In this section we consider primarily evidence on the electron correlations in doubly excited states of He$^-$ from the point of view developed in Chapter 10. The presence of an unpaired electron in the He$^-$ core has a major influence on the correlations of electrons in a different shell, as anticipated in Section 11.2.2. Basically, the He$^-$ complex may be resolved into either of two alternative structures, indicated by $(He^+ + e) + e$ and $He^+ + (2e)$ respectively, where the parentheses indicate *LS* coupling of two constituents. This alternative both enriches and complicates the electron correlations. The presence of two spins in the neutral He residue at $R \to \infty$ also doubles the number of alternative fragmentation channels by distinguishing singlets from triplets, whose levels are separated by exchange energy contributions much larger than spin-orbit coupling.

Each of the alternative structures predominates in a different region of configuration space. The pair of excited electrons is coupled tightly into a singlet or triplet when their distances from the core are comparable, at $30^\circ \lesssim \alpha \lesssim 60^\circ$, that is, in the region of the potential ridge. One electron is instead coupled to singlet or triplet with the He^+ core in the potential valleys, that is, whenever the other electron lies at much greater distance. The switching of adiabatic states from one structure to the other occurs generally at the transition from the intermediate to the fragmentation region. It implies a change of correlation pattern that may be reflected in an avoided crossing of potential curves. Analogous switches of spin coupling and correlations occur also in triply excited states, studied by the approach of Section 10.7; it had indeed been noticed [10] that each three-electron correlation function $\Phi_\mu(R;\omega)$ represents different coupling structures in different ranges of R.

The abundance of avoided crossings, which results from the changing pattern of spin and orbit coupling, frequently yields major departures from the adiabatic approximation $\Psi \sim F_\mu \Phi_\mu$. These departures require radial integration of two or more coupled equations (10.25), analogous to the treatment of Section 11.3. Collision and spectral data are again obtained by terminating this integration at the separation radius R_s and extracting values of the reaction matrix $K_{ij}^{(s)}$ or the equivalent $\{\delta_\alpha^{(s)}, U_{i\alpha}\}$. A subsequent more elaborate analysis is then required to disentangle the evolution of correlations from the several correlation functions $\Phi_\mu(R;\omega)$ and their average weighted with the radial functions $F_\mu(R)$. Radiative transitions between doubly excited states of the complex also receive contributions from electron motion at large radii ($R \gtrsim 20$ a.u.) in contrast to the narrow range of R involved in transitions from or to the ground state; their study again involves extended analysis of both $\Phi_\mu(R;\omega)$ and $F_\mu(R)$.

Sample results of pilot calculations on He^- are described in the following sections, additional calculations and further details are found in RPP, Section 4.3, and the original literature [11].

11.4.1 Adiabatic Potentials and Radial Propagation

The adiabatic potential functions $U_\mu(R)$ are calculated as eigenvalues of the equation (10.21) at separate values of R. Solution of this problem for a pair of excited electrons in the presence of a He^+ ionic core requires an extension of the representation (10.27) of the eigenfunctions $\Phi_\mu(R;\omega)$ to incorporate the alternative spin-coupling structures $(He^+ + e) + e$ and $He^+ + (2e)$, as anticipated in Section 11.2.2. Alternatively, the procedure indicated in Section 10.7 yields $\Phi_\mu(R;\omega)$ directly for all three electrons of He^-. Both of these procedures have been utilized.

Figure 11.12 shows the sets of potentials $U_\mu(R)$ converging at $R \to \infty$ to levels of He with $N = 1$ or 2, for alternative sets of $\{S, L, \Pi\}$ and calculated for a pair of electrons in the field of He^+. Quartet curves ($S = \frac{3}{2}$) converge at $R \to \infty$ only to triplet

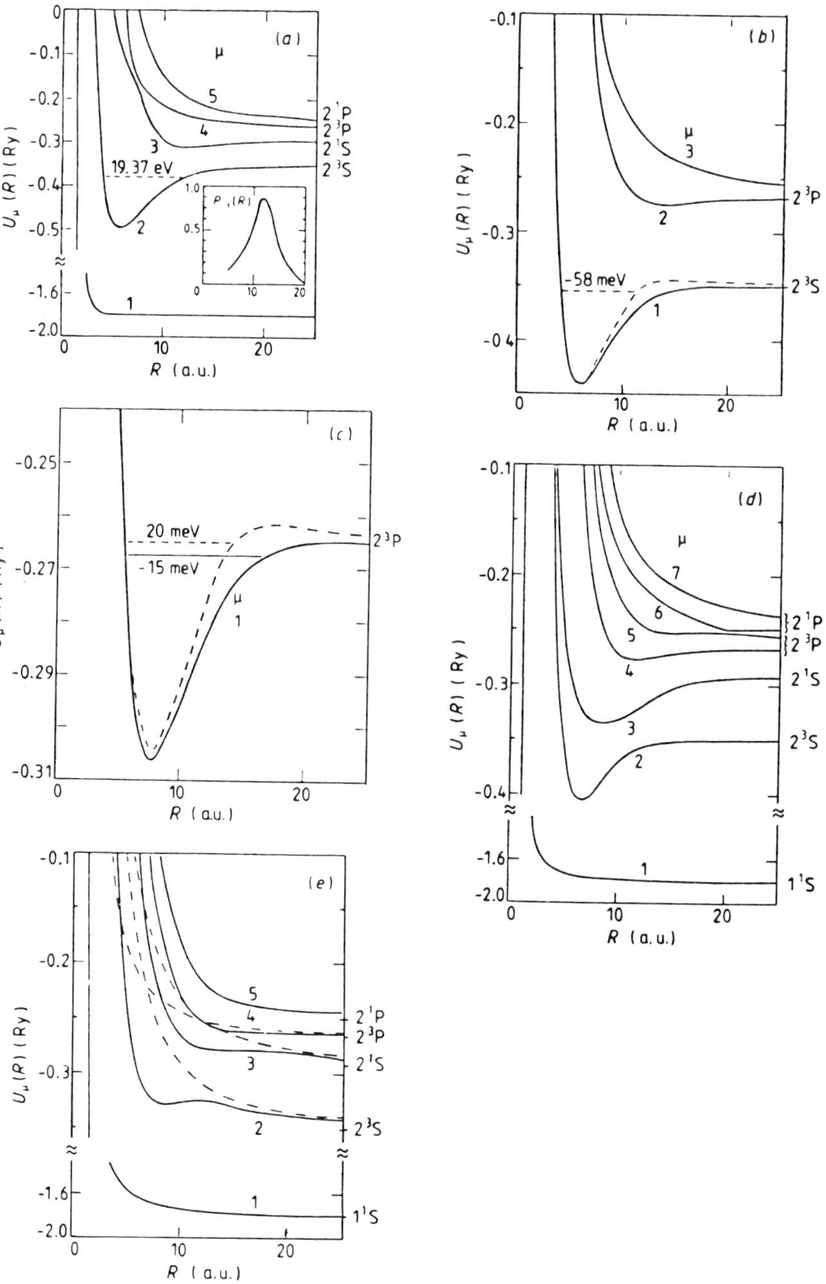

Fig. 11.12. Potential curves $U_\mu(R)$ of He^- converging to He ($N = 2$): (a) $^2S^e$, (b) $^4P^o$, (c) $^4P^e$, (d) $^2P^o$, and (e) $^2D^e$. Here μ values are marked on each curve. Horizontal lines show positions of quasi-bound levels. Broken curves are from an initial approximation (From Watanabe [11].)

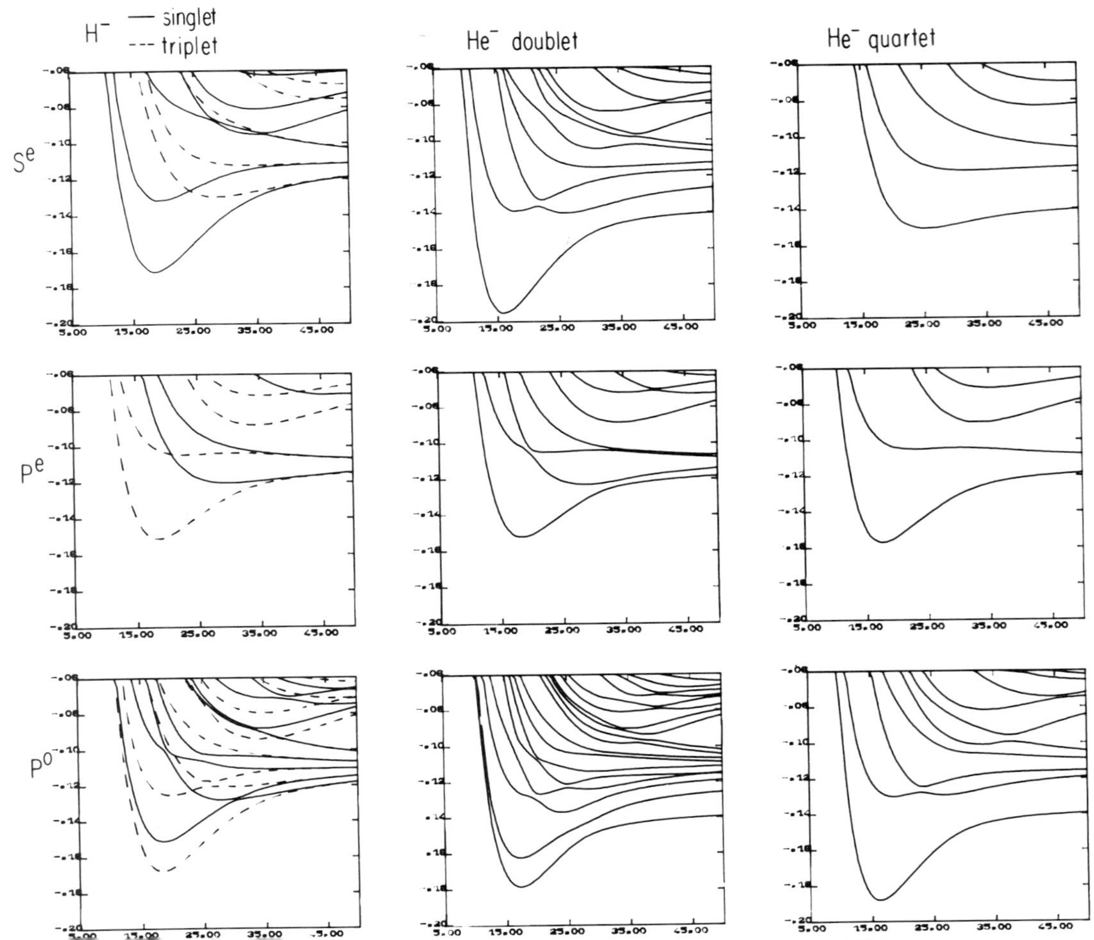

H⁻ — singlet
--- triplet
He⁻ doublet
He⁻ quartet
Se
Pe
Po

levels of He, 2^3S and 2^3P. Experimentally known rather narrow resonance levels of the complex, obtained as eigenvalues of the radial integration, are indicated. Not indicated is the broad $^2P^o$ resonance extending over the 20.0–20.5-eV range, which was discovered long ago [12] and which dominates the cross section for the inelastic collision $He(1s^2) + e \to He(1s2s\,^3S) + e$. Dashed lines represent initial approximations to $U_\mu(R)$ based on a more limited set of $(l_1 l_2)$ channels; comparison of these lines and the full lines traces the origin of avoided crossings of the latter. Note various avoided crossings: (1) the crossing of $\mu = 3$ and 4 in panel *a*, analogous to that of He discussed in Section 10.5.3, (2) the beginning of the crossing arrays in the intermediate region, analogous to those of Section 10.5.4, and (3) the striking avoided crossing in panel d at $R \sim 20$ a.u., which marks a sharp change of spin coupling.

Figure 11.13 shows analogous $U_\mu(R)$ curves converging to He ($N = 3$) levels calculated by a three-electron hyperspherical procedure. Corresponding curves are also shown for H^- singlets and triplets, to illustrate the similarities of He^- quartets and H^- triplets, whereas He^- doublets correspond to both singlets and triplets of H^-. Sharply avoided crossings occur between He^- doublet potentials that correspond to H^- singlets and triplets.

Figure 11.14 shows adiabatic potentials for three-electron excitations. (Analogous potentials are shown in Fig. 10.27.) The dashed lines indicate the energy of the long-lived resonance with the structure $2p^3\,^4S^o$.†

Realistic integration of the set of radial equations (10.25) for a given $\{S, L, \Pi\}$ hardly allows its truncation to exclude any of the adiabatic channels shown in the corresponding panel of Fig. 11.12, as noted earlier. The integration terminates at the separation radius R_s introduced in Section 11.2.1; here the solution $\Sigma_\mu F_\mu(R)\Phi_\mu(R; \omega)$ is to be matched to the fragmentation channel representation (8.53) or (8.56) as outlined in Section 11.3. Two main differences are noted between the application to He^- and the earlier one to Be, owing to the absence of a Coulomb field at $R > R_s$ as an electron escapes from neutral He and the presence in this region of a strong polarization field when the He residue is excited. [The polarizability of He ($N = 2$) is of the order of 1000 a.u., in contrast to its value 4.5 for H ($N = 1$)]. In the absence of a Coulomb field the matching at R_s is appropriately done by using wave functions $\{f^o, g^o\}$ normalized at $R = 0$, as discussed in Chapter 5, thus yielding matching parameters $K_{ij}^{(so)}$ or $\{\delta_\alpha^{(so)}, U_{i\alpha}^{(o)}\}$. The high polarizability requires the base pair of functions $\{f^0, g^0\}$ at $R \gtrsim R_s$ to reflect the influence of the polarization field [(6.68)]; this pair should then consist of Mathieu functions, as indicated in Section 6.9, rather than spherical Bessel functions as envisaged in Chapter 5.

Figure 11.15 shows sample data derived from the fitting of collision parameters for the He^- channels converging to He ($N = 2$). The phase-shift data have been transformed from their initial form $\delta_\alpha^{(so)}$ into $\delta_\alpha^{(s)}$ by (5.23b) and further into the Δ_ρ of Section 8.4 (indicated as δ_ρ in Fig. 11.15) when one or more channels are closed. Note in panel c how the number of open channel Δ_ρ increases as successive channels open up. The data in this panel correspond to those shown in Fig. 9.7 calculated by close coupling. Panel d gives eigenvalues of the time-delay matrix introduced in Section 4.4.

† Calculations of the corresponding curve for the doubly charged complex H^{--} have failed to yield any potential well, in contrast with the reported observation of this complex.

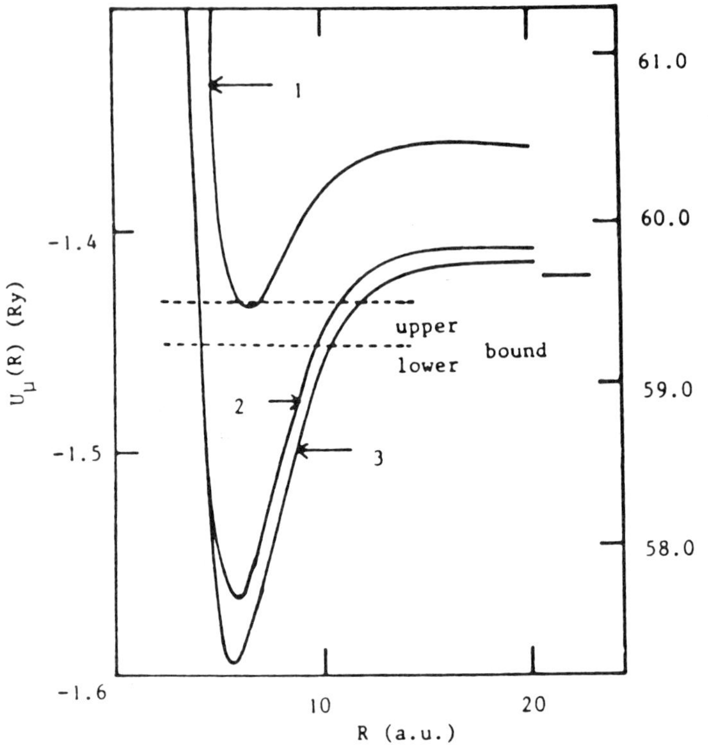

Fig. 11.14. Potential curves of He^- $^4S^o$. Upper and lower bounds on the energy level of the long-lived triply excited state $2p^2$ $^3P^e$ are shown. (From Watanabe *et al.* [13].)

11.4.2 Correlation Analysis

This analysis is complicated for He^- by the extensive mixing of adiabatic channels in the wave functions $\Sigma_\mu F_\mu(R)\Phi_\mu(R;\omega)$. The approach followed by Watanabe [11] involves two successive stages, namely, an initial study of each adiabatic two-electron correlation function Φ_μ through its expansion (10.27) into products $g_{n_{rc}}^{l_1 l_2 \overline{LS}}(R;\alpha)\mathscr{Y}_{l_1 l_2 LM}(\hat{r}_1,\hat{r}_2)$ symmetrized according to the pair's spin $\bar{S}$, followed by consideration of the weight of each of these products.

The correlation represented by each symmetrized product resembles that described in Section 10.5.6 for the corresponding states of doubly excited He at the corresponding radial distances, that is, in the condensation or intermediate region. (The representation does not apply here for $R > R_s$, where the wave function $\Sigma_\mu F_\mu(R)\Phi_\mu(R;\omega)$ is replaced by a fragmentation channel expansion. Correlations between the outer electron and the core are then represented through the coefficients of the channel expansion.)

The weight of each symmetrized product in a given state consists in turn of separate elements. The first of these is the weight $I_\mu(l_1 l_2 \overline{LS})$ of each particular product function (10.27) in a given adiabatic channel Φ_μ. This is obtained by

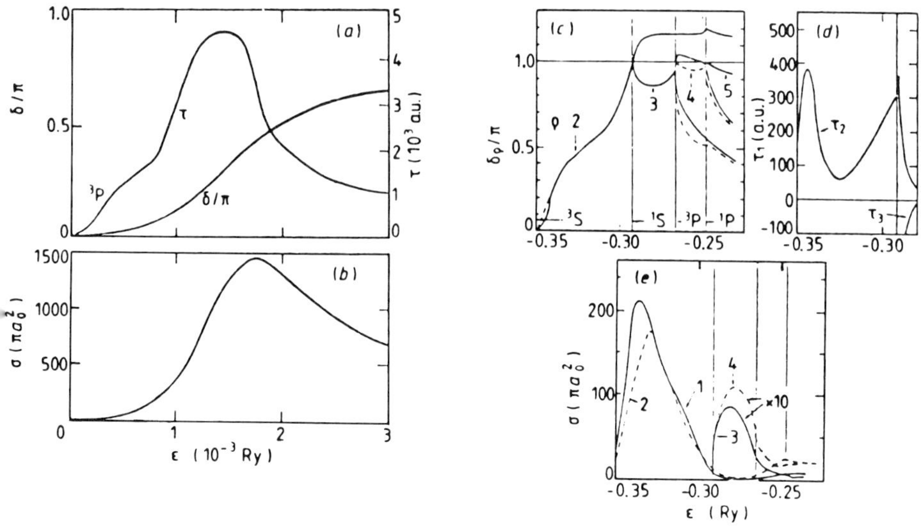

Fig. 11.15. Eigenvalues of phase shifts δ and of time delay τ, and cross sections σ for e + He collisions: (a) and (b) e + He(2^3P)^{4}P^e, (c)–(e) e + He($2^{3,1}$S)^{2}P^o. (From Watanabe [11].)

integrating $|\Phi_\mu|^2$ over all variables except R, utilizing the orthogonality of the product functions. The second step results by weighting the value of $I_\mu(l_1 l_2 \overline{LS})$ by the corresponding radial factors $F_\mu(R)^2$. Figure 11.16 shows a sample of $I_\mu(ll\overline{LS})$ ($\mu = 2, 3$) for ^{2}S states converging to He (2 ^{3}S) and He (2 ^{1}S). In the lower range of $R(\lesssim 10$ a.u.), the state of the excited electron pair in the lower energy channel $\mu = 2$ consists of 80% s^2 ^{1}S and 20% p^2 ^{1}S, which provides a modicum of angular correlation (the quasi-bound state in this channel at 19.37 eV is usually labeled as 2s^2 for simplicity). The corresponding ^{3}S structure of the pair in the $\mu = 3$ channel implies the presence of an additional node in the radial correlation functions $g^{ll}(R;\alpha)$ that raises the kinetic energy contribution to $U_\mu(R)$ increasingly at lower R.

At larger radial distances ($10 \lesssim R \lesssim 13$ a.u.) a change of weights is apparent, which marks the transition from the intermediate to the fragmentation range of R. The dominant contribution in either channel, s^2 ^{1,3}S, decreases slightly from 80 to 75%, while the secondary contribution switches from p^2 ^{1,3}S to the opposite spin coupling, ^{3,1}S and more gradually from p^2 to s^2. The main shift to a 75–25 mix of singlet and triplet in either μ channel is implied by the transition to fragmentation

$$\mathrm{He}^+\ {}^2\mathrm{S} + (2e)^{1,3}\mathrm{S} \rightarrow (\mathrm{He}^+ + e)^{3,1}\mathrm{S} + e, \qquad \mu = 2, 3. \tag{11.3}$$

The replacement of p^2 by s^2 represents a gradual reduction of angular correlation between the excited electrons, which persists to some extent in the fragmentation region ($R \gtrsim 15$ a.u.) as an effect of polarization of He $2^{1,3}$S by the escaping electron.

The particularly transparent example of Fig. 11.16 has been described here as an introduction to the generally more complicated patterns that occur in other channels of He$^-$ and in other multiply excited systems.

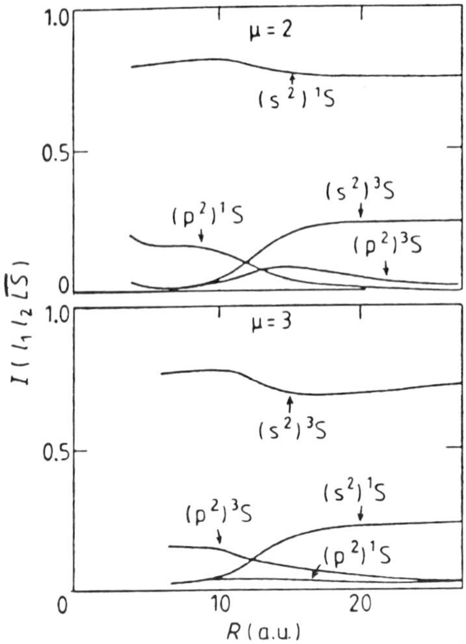

Fig. 11.16. Weights $I_\mu(l_1 l_2 \bar{L}\bar{S})$ of adiabatic channel functions, indicating the evolution of correlations with R. (From Watanabe [11].)

11.4.3 Optical Transitions of Excited Electron Pairs

The observation of photoabsorption by doubly excited He^- has introduced a novel phenomenon to which both alternative structures of the complex $He^+ + (2e)$ and $(He^+ + e) + e$ may contribute. These contributions may be resolved as originating mainly from the intermediate and fragmentation ranges of the radial variable R, respectively. An optical transition in the structure $(He^+ + e) + e$ may arise from the radiation coupling to the outer electron while subject to the polarization field of the He residue or from coupling to the residue with the outer electron as a polarizing spectator.

The contributions may be sorted out by expressing the dipole moment for transitions from an eigenchannel α to an eigenchannel β as a superposition of radial integrals,

$$
\begin{aligned}
(F_\beta(\omega, R)|z|F_\alpha(\omega, R)) &= (F_\beta(\omega, R)|R(\cos\alpha\cos\theta_1 + \sin\alpha\cos\theta_2)|F_\alpha(\omega, R)) \\
&= \sum_{\mu\mu'} (F_\mu^{(\beta)}(R)|d_{\mu\mu'}(R)|F_{\mu'}^{(\alpha)}(R)). \qquad (11.4)
\end{aligned}
$$

Here

$$
d_{\mu\mu'} = R(\Phi_\mu^{(\beta)}(R;\omega)|\cos\alpha\cos\theta_1 + \sin\alpha\sin\theta_2|\Phi_{\mu'}^{(\alpha)}(R;\omega)) \qquad (11.4')
$$

represents the distribution of the transition amplitude over the entire range of R. Some of the components $d_{\mu\mu'}$ displayed in Fig. 11.17 are seen to be nonzero only at $R \lesssim 12$ a.u., others only at $R \gtrsim 12$ a.u., and still others in both of these ranges.

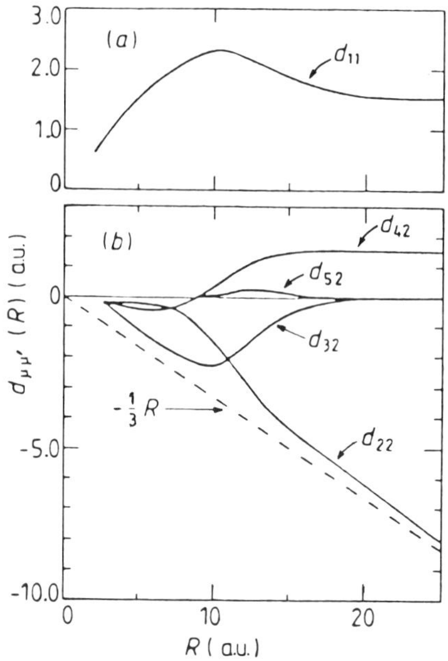

Fig. 11.17. Dipole moments $d_{\mu\mu'}$ for transitions within the $N = 2$ manifold of He^- from Eq. (11.4′): (a) $^4P^o \to {}^4P^e$ and (b) $^2S^e \to {}^2P^o$. (From Watanabe [11].)

11.5 CHANNEL CROSSING IN OPEN VALENCE SHELLS

The spectra of atoms in groups II–VII, as well as in transition and lanthanide groups, display major effects of the electrons' ability to exchange orbital (and/or spin) angular momenta within the radius of the valence shell rather freely, that is, with only modest exchanges of energy. The technique of configuration mixing accounts quantitatively for such of these effects that remain confined to energy levels well below the ionization threshold. Analysis in terms of channel mixing is instead required when the effects extend to or near the continuum. This analysis is essential for collision processes since strong channel mixing implies large inelastic cross sections as noted in Section 11.2.4.

The ^1D excitations of Be afford an initial example that is treated accurately by the multiconfiguration Hartree–Fock procedure (Section 9.3.5). In this case the singly excited series 2s*n*d ^1D is perturbed strongly by admixture of the $2p^2\ {}^1D$ configuration into its lower levels as detailed in Table 11.2. Indeed the lowest level of the series is predominantly $2p^2\ {}^1D$.

The admixture of $2p^2$ involves excitation of a second electron from 2s, the lowest subshell of $n = 2$, to 2p. Analogous perturbers arise in most atoms from excitation of a valence electron from a lower to a higher subshell. The

Table 11.2

Energy and Wave Function Composition of the Three Lowest ^{1}D Levels of Be.[a]

Level	Energy $E_n - I(2s)$ (a.u.)	n^*	$2p^2$ (%) (MQDT)	2s3d (%)	2s4d (%)
1 ^{1}D	−0.08548	2.42	(43) 59.2	25.5	14.5
2 ^{1}D	−0.04905	3.18	(9) 16.0	70.4	12.7
3 ^{1}D	−0.02921	4.16	(3) 4.5	5.8	89.2

[a] From Froese-Fisher [14].

resulting doubly excited configuration generally yields alternative terms with different *LS* quantum numbers and different energies that perturb various channels in different spectral ranges.

The example of Be ^{1}D levels and a few analogs have also been analyzed quantitatively in the alternative framework of two-electron adiabatic channels developed in Chapter 10. Excitation channels $ns\epsilon d$ D and $np\epsilon p$ ^{1}D converge at $R \to \infty$ to the levels ns and np of the ion, respectively, of which ns is energetically lowest when the ionic core generates a non-Coulomb field. At $R \to 0$ the same channels would converge to levels characterized by quantum numbers λ_{sd} and $\lambda_{pp} = \lambda_{sd} - 2$ according to (10.14′), (10.47′), and Section 10.7.5. Accordingly, the two channels would cross in a diabatic treatment; they do not cross in the usual adiabatic treatment, which causes them instead to interchange their quantum numbers as discussed in Section 10.5.3.

Figure 11.18 displays the avoided crossing of the ^{1}D adiabatic potential curves $U_1(R)$ and $U_2(R)$, which converge to the 3s and 3p levels of Mg^+. Avoidance of the crossing is unsharp here, as in the analogous example of Fig. 10.12 for He, extending over a radial range of several atomic units centered at $R \sim 6$ a.u. (The corresponding root *mean* square distance of each electron from the nucleus is $6/\sqrt{2} \sim 4$ a.u., well within the region occupied by the valence shell of Mg.) This location of the avoided crossing—squarely within the valence shell—contrasts with the corresponding location which

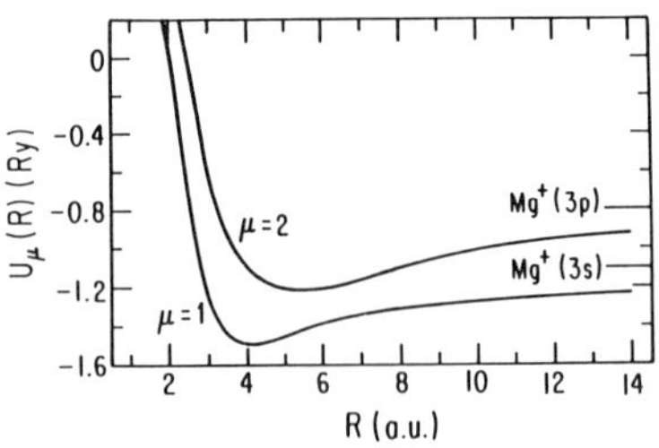

Fig. 11.18. The lowest two $^1D^e$ adiabatic potential curves $U_\mu(R)$ converging to the Mg^+ ($N = 3$) limits. (From O'Mahony [15].)

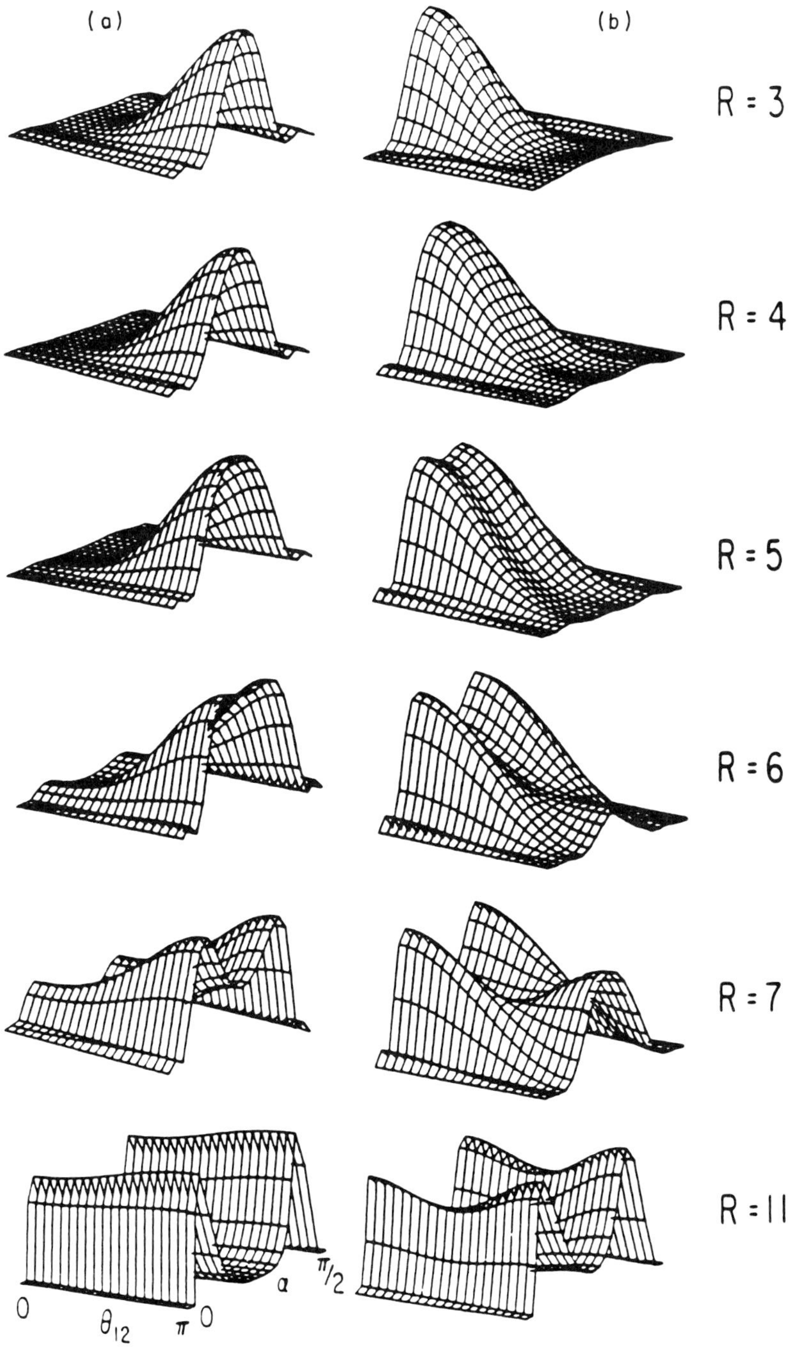

Fig. 11.19. Adiabatic correlation functions $|\Phi_\mu(R;\omega)|^2$ for Mg $^1D^e$ at different values of R: (a) $\mu = 1$ and (b) $\mu = 2$. [From O'Mahony [15].)

was described in Section 10.5.3 as lying in the "condensation range" for an electron pair excited out of the valence shell, that is, at radial distances shorter than normally occupied by the electron pair. The widespread observation of perturbed spectra of the class considered here suggests that avoided crossings occur quite generally in the valence shells of open-shell atoms, but this inference requires verification by an extensive survey.

As noted in Section 11.3, the construction of adiabatic potential curves $U_\mu(R)$ and the corresponding graphical display of eigenfunctions $|\Phi_\mu(R;\omega)|^2$, as in Fig. 11.19, illustrate qualitative features of two-electron states successfully but are less apt to provide accurate values of spectroscopic parameters. Accordingly the MQDT parameters θ [(11.2)], $\delta_1^{(s)}$, and $\delta_2^{(s)}$ have been calculated by the R-matrix method of Section 9.3.2 for the two lowest ^{1}D channels of Be and Mg, as well as for the corresponding ^{2}D channels of Al, namely, $3s^2\epsilon d$ and $3s3p\epsilon p$ [11]. The Al atom includes, of course, an additional valence electron that remains in the 3s subshell somewhat in the role of spectator but nevertheless appreciably perturbed by the "active" ^{1}D pair.

Figures 11.20 and 11.21 show plots of MQDT parameters thus obtained as functions of the total energy of the valence electrons. These illustrations of atomic properties as functions of the energy complement the illustrations provided by adiabatic parameters $U_\mu(R)$ and $|\Phi_\mu(R;\omega)|^2$ plotted as functions of the radial distance R, as in Figs. 11.18 and 11.19, respectively. Note how the occurrence of avoided crossings emerges in the guise of rapid localized variations in the spectra of MQDT parameters. These variations occur at energies barely sufficient for the wave function to spread through the radial range of avoided crossings.

The combined study of excited states by the adiabatic approach of Chapter 10 and the R-matrix method of Section 9.3.2, illustrated by Figs. 11.18–11.21 and by further data in O'Mahony [15], constitutes an important and promising innovation. The R-matrix calculations are remarkably fast and accurate; the results of adiabatic calculations are more pictorial.

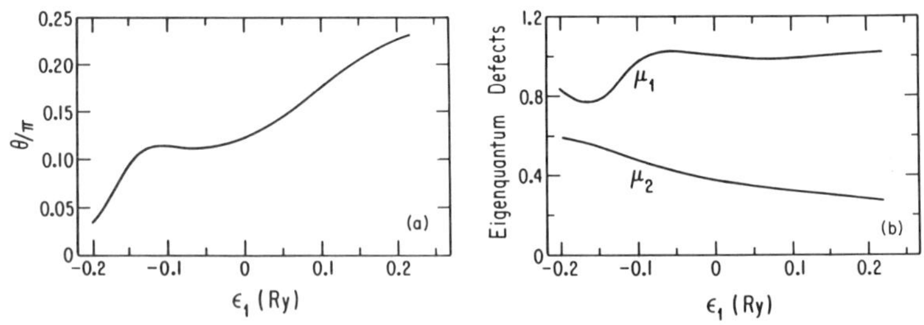

Fig. 11.20. MQDT parameters for Al $^2D^e$ as functions of the energy ϵ relative to the Al$^+$ $^1S^e$ threshold. (a) Mixing angle θ/π and (b) eigenquantum defects μ_1, μ_2. (From O'Mahony [15].)

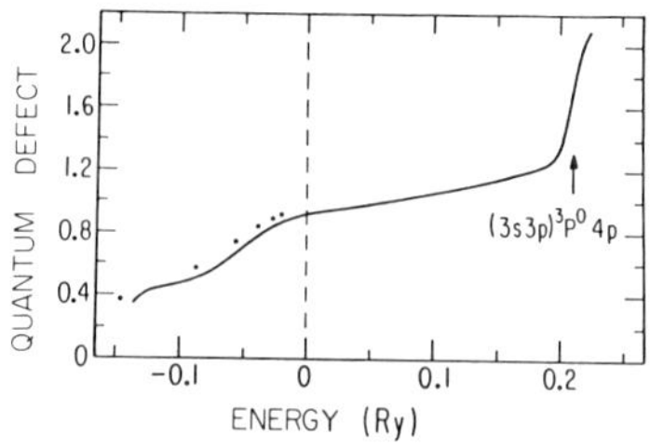

Fig. 11.21. Quantum defect δ/π versus energy ϵ_1 relative to the Al^+ $^1S^e$ threshold. In the discrete region the quantum defect $\mu = \delta/\pi$ is drawn as a continuous function; dots represent experimental values extracted from tables of atomic energy levels. Note the rise in the quantum defect from 0.4 to 1 below the $^1S^e$ threshold due to the $3sp^2$ perturber. The position of the $(3s3p)^3P^o4p$ autoionizing level is shown. (From O'Mahony [15].)

Conceivably, data provided by R-matrix eigenfunctions could be translated into graphical displays of $U_\mu(R)$, $|\Phi_\mu(R;\omega)|^2$, and so on, equivalent to those introduced in Chapter 10. The quantitative study of perturbed series for atoms with more than two or three electrons in their valence shell may require us to develop a procedure that focuses on the pair of electrons that interact strongly, exchanging energy and angular momentum, allowing at the same time for the weaker perturbation exerted by this pair on the other, quasi-spectator, valence electrons.

References

1. G. Wendin, "Photo Electron Spectra, Structure and Bonding," Vol. 45. Springer-Verlag, Berlin and New York, 1981.
2. C. H. Greene, *Phys. Rev A* **23**, 661 (1981); also RPP Sections 4.1 and 4.2.
3. B. Stumpf and A. Gallagher, *Phys. Rev. A* **32**, 3344 (1985); also RPP Section 4.2.2.
4. M. Cavagnero, *Phys. Rev. A* **30**, 1169 (1984).
5. P. F. O'Mahony and C. H. Greene, *Phys. Rev. A* **31**, 250 (1985).
6. C. E. Moore, "Atomic Energy Levels," National Bureau of Standards, Circ. No. 467.
7. D. W. Norcross and M. J. Seaton, *J. Phys. B* **9**, 2983 (1976).
8. T. Y. Wu, *Philos. Mag.* **22**, 837 (1936).
9. R. N. Compton, G. D. Alton, and D. J. Pegg, *J. Phys. B* **13**, L651 (1980).
10. C. W. Clark and C. H. Greene, *Phys. Rev A* **21**, 1786 (1980); C. H. Greene and C. W. Clark, *Phys. Rev. A* **30**, 2161 (1984).
11. S. Watanabe, *Phys. Rev. A* **25**, 2074 (1982).
12. H. Mayer-Leibnitz, *Z. Phys.* **95**, 499 (1936); G. J. Schulz and R. E. Fox, *Phys. Rev.* **106**, 1179 (1957); E. Baranger and E. Gerjuoy, *Phys. Rev.* **106**, 1182 (1957).
13. S. Watanabe, M. LeDourneuf, and L. Pelamourgues, *J. de Physique* **43**, Suppl. 11, C2-223 (1982).
14. C. Froese-Fisher, *Phys. Rev. A* **30**, 2741 (1984).
15. P. F. O'Mahony, *Phys. Rev. A* **32**, 908 (1985); P. F. O'Mahony and S. Watanabe, *J. Phys. B* **18**, L239 (1985).

12 CONCLUSIONS AND PREVISIONS

12.1 DISCUSSION

A review of the themes outlined in the introductory chapter of this book affords us the opportunity to reconsider them with the hindsight provided by their actual development. Our emphasis on the central role of unstable collision complexes holds even though the breakup of a complex into recognizable fragments is generally complete at radial distances less than 1 nm in the examples we have considered. Distances orders of magnitude larger are, in fact, relevant to the balance between the kinetic and potential energies of the fragments, which determines whether the fragmentation will proceed to infinite distance or eventually be reversed with profound influence on the output in other channels.

This contrast between scales of radial distances relevant to fragmentation underlies the separate treatment of interactions prevailing at short and long ranges. It also underlies our joint treatment of collisions and spectra, because the richest structures of atomic and molecular spectra reflect the motion of the fragments at large radial distances in closed channels even though the structures may be observed through fragmentation in open channels. Thus, the treatment centers on channels that encompass discrete and continuum states rather than individual energy levels.

The energy eigenstates of a complex have been normally represented in this book by standing waves rather than by incident and scattered waves, thus emphasizing symmetries (Section 1.4) under interchange of entrance and exit channels (time reversal) and under space rotations (conservation of angular momentum). Standing-wave eigenstates are transformed into scattering states through the simple coefficients of a partial wave expansion (Section 4.3) or their elaborate—but substantially straightforward—generalization described in Section 7.9. This generalization from a single channel to a multichannel setting replaces the familiar phase factors $e^{\pm i\delta_l}$ of the partial wave expansion, together with relevant normalization coefficients, by elements of Jost matrices $J^{\pm}_{i\alpha}$, which play a key role in our treatment. The wave functions for the entire spectral region as well as observable quantities—such as bound-state energy levels, closed-channel resonance positions and profiles, and collision cross sections derived from them—are all constructed as combinations of Jost matrices.

Separating the treatment of long- and short-range interactions articulates the theory into a large number of parameters, independent of one another and individually meaningful. Some of these parameters have analytical expressions; most can be extracted from data banks or alternatively determined by *ab initio* calculation. Their alternative determinations by data analysis and calculation afford an appropriate setting for comparison of theory and experiment. Instead of every independent experimental value (of a cross section for a specific energy and scattering angle, or characteristic of a broad spectral level) it is the finite and often small set of parameters that forms the interface between theory and experiment. A prime example of articulation of theory is seen in the factorization of Jost matrix elements, $J^{\pm}_{i\alpha} = (\pm i)^{l_i} \exp(\pm i\delta^{(\mathrm{l})}_i) U_{i\alpha} \exp(\pm i\delta^{(\mathrm{s})}_\alpha)$ [(8.32a)], which holds when central symmetry prevails in the long-range field. (Failure of this symmetry is discussed and repaired in Section 9.3.7.)

Our treatment of atoms has introduced a single frame transformation matrix $U_{i\alpha}$, the centerpiece in the factorization of the Jost matrix. Multiple couplings of spins, orbitals, vibrational and rotational motions, typically in molecules, will require a sequence of such transformations. Indeed, molecular dissociations proceed generally through several of the alternative couplings involved in molecular spectra that were characterized by Hund in the 1920s. Systematic analysis of this multistage process and standardization of the relevant theory will require further development.

With the long-range interactions standardized and characterized analytically, it is the calculation of short-range interaction parameters (Chapter 9) that presents the most complex task; it also appears to require further development in spite of its long and distinguished history. The R-matrix methods appear most appropriate to this particular task. The contributions to the short-range parameters from the several shells or other elements of atomic and molecular structure might be conveniently sorted out in the future. We have included among these parameters the dipole matrix elements that control the absorption or emission of single long-wave photons. Analogous matrix elements will be appropriate to deal with shorter-wave radiation. Multiphoton processes may similarly be included in the treatment of short-range interactions [1].

Chapters 1–9 have dealt with types of atomic phenomena that have long been familiar, setting them in the framework of theoretical concepts and techniques developed in the last 25 years. These newer methods of analysis are gradually entering into general use as their power becomes better documented and their technology is progressively refined.

On the other hand, the phenomena discussed in Chapter 10 involving the elementary process of decay of a complex into three or more fragments are more novel since their close experimental study dates from the 1960s. Their theoretical analysis remains accordingly still in an early and fragmentary stage of development. Key aspects of these phenomena are nevertheless likely to remain in the lore of atomic physics and, more generally, chemical physics. One such aspect is the potential energy function of three charged particles illustrated in Figs. 10.8 and 10.9. A second one is the mapping of electron correlations in Figs. 10.15–10.17. Still another is the remarkable occurrence of adiabatic propagation of wave functions through their potential field, which preserves their nodal structure over wide ranges. Gross, localized breaches of adiabaticity are equally striking and are, in fact, responsible for the observed changes of structure of atoms and molecules. A more systematic and integrated treatment of these phenomena remains to be developed, but the central relevance of these quantum-mechanical aspects of nonseparable motion to a wide class of atomic phenomena seems clear.

Completion of the study initiated in Chapter 10, concerning the manifold aspects of two-electron motion in a Coulomb field, should lead to a two-electron quantum defect theory. Phenomena involving multiple excitations in atoms or molecules will then be analyzed in terms of long- and short-range influences, analogous to the treatment of single excitations in Part C. (Recall how the extensive analytical development of one-electron wave functions for a Coulomb or other field, in Chapter 5, underlies the multichannel quantum defect theory of Chapters 7 and 8.)

The concepts and techniques introduced in Chapter 10 have, in fact, been applied more recently to the analysis of correlated electron motions in atoms with more than two electrons. These applications, reported in Chapter 11, cast a new view on a class of irregular spectra whose quantitative interpretation had previously been very restricted. They also introduce an effective blend of older and newer theoretical technologies that is likely to find extensive application in coming years to spectroscopy as well as to the less explored theory of slow-electron collisions with open-shell atoms.

Last, but far from least, we recall from the Introduction that the theoretical approach followed in this book aims ultimately at an effective treatment of molecular reactions. Molecular phenomena were nevertheless *not* to be treated explicitly in the book, nor was the book to deal explicitly with atom–atom collisions, which we regard as forming a molecular complex even at very high speeds. A large amount of very competent and imaginative work has, in fact, been devoted over the years to the theoretical study and calculation of molecular collisions and reactions. Yet we feel that work to have been severely handicapped by its stress on the course of reactants from their initial to their final state, solving their combined time-dependent Schrödinger equation semiclassically as well as possible. Consideration of the mechanics of the whole complex formed by the reactants, which seems essential to us, has, of course, no place in such an approach to molecular processes. On the other hand, the few instances of reaction theory that did study the complex eigenstates—as noted in Section 10.7.6—led directly to the key influence of resonances, which had been detected only as incidental features of other calculations.

At this point a future theory of reaction complexes may be visualized within the framework of Chapter 10 as follows [2]. The hyperradial coordinate of the complex is defined, in terms of its moments of inertia and its mass M, by $R^2 = 2(I_x + I_y + I_z)/M$, where (x, y, z) refer to the principal axes of inertia. The structure of the complex is defined by a set of angles and hyperangles, analogous to $\{\theta, \phi, \alpha\}$. At small R the Hamiltonian term Λ^2/R^2 predominates; its eigenvalues and eigenfunctions are known as indicated in Section 10.7. At $R \to \infty$ adiabatic eigenfunctions $\Phi_\mu(R; \omega)$ will represent the structure of alternative fragmentations of the complex. At intermediate R, the complete eigenfunction Ψ will encompass nonadiabatic effects that determine the all-important coupling of alternative adiabatic components $F_\mu(R)\Phi_\mu(R; \omega)$. Initial boundary conditions would be introduced by the procedure of Section 7.9.

The development along this approach presupposes completion of the treatment of Chapter 10 for the much simpler case of an electron pair. Other preliminaries would involve the identification of angular variables that represent effectively the main structural features of a complex in various ranges of R and the most relevant features of potential surfaces, such as their valleys, saddles, and ridges. Judging from the rate of past progress in this area, the goal outlined here appears attainable but very distant.

References

1. P. Zoller, to be published.
2. U. Fano, *Phys. Rev. A* **24**, 2401 (1981); *in* "Electronic and Atomic Collisions," (J. Eichler, I. V. Hertel, and N. Stolterfoht, eds.), p. 629 ff. North-Holland Publ., Amsterdam, 1984.

GLOSSARY OF SYMBOLS

Symbol	Description	Defined on page
β	Angular distribution asymmetry parameter;	176
	Stark field quantum number (Section 8.7 only)	238
$\beta(\epsilon, l)$ or $\beta(\kappa, l)$	Quantum defect theory phase parameter ($\epsilon < 0$)	88, 100
Γ	Resonance width	208, 236
δ_l	Phase shift	63, 94
$\delta_l^{(l)}$	Phase shift, long range part, ($\epsilon > 0$)	81, 87
$\delta_l^{(s)}$	Phase shift, short range contribution	81, 91
$\delta_l^{(so)}$	Phase shift for $\{f^\circ, g^\circ\}$ base pair	89, 91
$\Phi_i(\omega)$	Fragmentation channel wave function in all coordinates except r	152, 162
Φ_μ	Two-electron channel function	317
κ	$(-2\epsilon)^{1/2}$, $\epsilon < 0$ wave number	13, 78
Λ^2	Grand angular momentum for many electrons	313, 355
ξ^2	Mixing parameter	207
μ	Quantum defect	95
ω	Set of angular and spin coordinates	152
σ	Cross section	25
$A(\kappa, l)$	QDT normalization parameter ($\epsilon < 0$)	89
$B(k, l)$	QDT normalization parameter ($\epsilon > 0$)	87
$D(\kappa, l)$	QDT rescaling parameter	88
$C(\alpha, \theta_{12})$	Two-electron potential function in hyperspherical coordinates	311
$f(\epsilon, l, r)$	Regular solution for $v(r)$, normalized per unit energy	82
$f^\circ(\epsilon, l, r)$	Regular solution for $v(r)$ with energy independent normalization	84, 87
$F_\alpha(\omega, r)$	Wave function for eigenchannel α	153
$F_l(r)$	Radial wave function	63, 89
$f^\pm$	Base pair of traveling wave solutions	82
$g(\epsilon, l, r)$	Irregular solution for $v(r)$, normalized per unit energy	90
$g^\circ(\epsilon, l, r)$	Irregular solution for $v(r)$ with energy independent normalization	90

$\mathcal{G}$	QDT parameter relating f,g to f°,g°	90
$J^{\pm}(k, l)$	Jost function	84
$J^{\pm}_{i\alpha}$	Jost matrix	153, 174
j_l	Regular spherical Bessel function	62
K	Two-electron angular quantum number	310, 325
$K_{ij}(E)$	Reaction matrix	69
$K^{(s)}_{ij}$	Reaction matrix for short-range interaction only	118, 256
$K^{(so)}_{ij}$	Reaction matrix for $\{f^\circ, g^\circ\}$ base pair	261
k	$(2\epsilon)^{1/2}$, $\epsilon > 0$ wave number	13
q	Resonance profile parameter	202, 211
Ry	Rydberg energy (13.6 eV)	52
$R_{ij}(E)$	*R*-matrix	124, 260
S_{ji}	Scattering matrix	69, 153
T	Two-electron angular quantum number	310, 325
T_{ji}	Transition matrix	69, 146
$T^{(s)}_{ji}$	Transition matrix for short-range interaction only	119
$U_{i\alpha}$	Frame transformation matrix from eigenchannels α to fragmentation channels i	159
$U_\mu(R)$	Adiabatic potential curve in hyperspherical representation	317
$W(a, b)$	Wronskian, $a(db/dr) - b(da/dr)$	84
y_l	Irregular spherical Bessel function	63

INDEX

S

T

ATOMIC COLLISIONS AND SPECTRA